RAREFIED GAS DYNAMICS

Edited by
J. Leith Potter
ARO, Inc.
Arnold Air Force Station, Tennessee

Volume 51 PART II
PROGRESS IN
ASTRONAUTICS AND AERONAUTICS

Martin Summerfield, Series Editor-in-Chief
Princeton University, Princeton, New Jersey

Technical papers selected from the 10th International Symposium on Rarefied Gas Dynamics, July 1976, subsequently revised for this volume.

Published by the American Institute of Aeronautics and Astronautics.

American Institute of Aeronautics and Astronautics
New York, New York

Library of Congress Cataloging in Publication Data
Main entry under title:

International Symposium on Rarefied Gas Dynamics, 10th, Snowmass-at-Aspen, 1976.
Rarefied gas dynamics.

(Progress in astronautics and aeronautics; v. 51)
Includes bibliographies.
1. Rarefied gas dynamics—Congresses. I. Potter, J. Leith. II. Title. III. Series.
TL507.P75 vol. 51 [QC168]629.1'08s 76-57748 [629.1]
ISBN 0-915928-15-9

Table of Contents

CHAPTER VII—KINETIC THEORY

PROGRESS IN KINETIC THEORY: GENERALIZATIONS OF BOLTZMANN'S EQUATION

Noel Corngold*

California Institute of Technology, Pasadena, Calif.

I. Introduction

Ludwig Boltzmann proposed his remarkable equation in a paper published in 1872, a little over a century ago. Centennial celebrations have been held in Vienna, in Providence, and, doubtless, elsewhere. The equation and the point of view that it represents have been enormously fruitful for physics and engineering. Yet, the equation occupies an unusual place in kinetic theory; it is easy to derive in a heuristic, "hand-waving" manner, but its proper derivation requires long and subtle argument. Thus, its generalization is difficult and its precise range of validity uncertain. Progress in this aspect of kinetic theory generally has been slow and quiet. But, the last decade has been different. New mathematical formalisms have been brought to bear, and we are much closer to an understanding of the kinetic theory of relatively dense systems. I shall try to describe some of the recent progress in this paper. No one interested in rarefied gasdynamics can fail to respond to the challenge of higher densities. The challenge is set not merely by a fascinating problem in mathematical physics but, more to the point, by phenomena and data demanding explanation.

We shall write Boltzmann's equation as

$$(\partial/\partial t)f(\underline{r},\underline{p},t) + (1/m)\underline{p}\cdot(\partial/\partial\underline{r})f = J(f,f) \qquad (1)$$

Presented as Invited Paper A at the 10th International Symposium on Rarefied Gas Dynamics, Aspen, Colorado, July 19 - 23, 1976. This research was sponsored in part by the National Science Foundation. I should like to thank Professor J. V. Sengers for helpful correspondence.

*Professor of Applied Physics.

Its linear version is

$$(\partial/\partial t)h(\underline{r},\underline{p},t)+(1/m)\underline{p}\cdot(\partial/\partial\underline{r})h = \int d^3\underline{p}'K(\underline{p},\underline{p}')h(\underline{r},\underline{p}',t) \qquad (1a)$$

where

$$f(\underline{r},\underline{p},t) = n\,M(\underline{p})[\,1 + h(\underline{r},\underline{p},t)]$$

and $M(\underline{p})$ is the normalized Maxwellian for temperature T. We sometimes shall use the dimensionless momentum $\underline{\xi}$, $\underline{p} = mv_0\underline{\xi}$, with $v_0^2 = k_B\,T/m = 1/m\beta$. Otherwise the notation is standard.

The left-hand side of Eqs. (1) is unobjectionable. It is the right-hand side, Boltzmann's famous collision term, that expresses "the physics." We know, for example, that $M(\underline{p})K(\underline{p},\underline{p}')$ is symmetric, and that K is unaltered by simultaneous rotation of $\underline{p}$ and $\underline{p}'$. Furthermore, K has five eigenfunctions, degenerate with respect to eigenvalue zero. These functions are the collisional invariants $(1,\underline{p},\underline{p}^2)$. Particularly convenient is the orthonormal set

$$\psi_1 = 1 \qquad \psi_2 = \xi_3 \qquad (2a)$$

$$\psi_3 = (1/\sqrt{6})(\xi^2-3) \qquad \int d^3\underline{p}\,M(\underline{p})\,\psi_\alpha\psi_\beta = \delta_{\alpha\beta} \qquad (2b)$$

$$\psi_4 = \xi_1 \qquad \psi_5 = \xi_2 \qquad (2c)$$

(Note that the kinetic energy, rather than the total energy, is conserved.) This brings us to the most important feature of the collision term, so obvious that it usually is ignored in textbooks: the change in the distribution function due to interactions is expressed in terms of "collisions per unit time." An element of "coarse-graining" thereby is introduced into the description of nature. Should a characteristic time t_o be associated with collisions near equilibrium, and this time be compounded of the range of force r_o and a thermal velocity v_o, then $f(\underline{r},\underline{p},t)$, the solution to Eqs. (1) and (2), will have meaning only for $t \gg t_o$. Otherwise, the relation between Boltzmann's $f(\underline{r},\underline{p},t)$ and the one-particle distribution function

$$f_1(\underline{r},\underline{p},t) = \left\langle \sum_i \delta^3[\underline{r}-\underline{q}_i(t)]\,\delta^3[\underline{p}-\underline{p}_i(t)] \right\rangle$$

is not simple. What, for example, is the relation between $f_1(\underline{r},\underline{p},t=0)$ and $f(\underline{r},\underline{p},t=0)$?

A second feature of the collision term is its form, an expression of the "Stosszahlansatz." Correlations in position and velocity of the colliding particles are ignored, and a two-particle distribution function is approximated by the product of one-particle functions. The approximation is reasonable when the gas is dilute. Then another set of scales appears, t_1, the mean time between collisions, and r_1, the mean free path. Boltzmann's equation is accurate in these regimes. As the system becomes denser, $t_1 \to t_o$, $r_1 \to r_o$, and well-known difficulties appear.

Finally, we note that the collision term, giving the "average number of collisions per unit time," causes f to be interpreted as an average density. But where is the equation that determines fluctuations in density? In Boltzmann's formulation, the "noise" that turns out to be so rich in information is absent. Thus, there are three issues that we must deal with: the coarse-graining, the restriction to low density, and the question of fluctuations. (One aspect is, are the approximations independent? Does one imply another?)

The modern theory, which we shall describe ahead, responds to these questions. It is quite general and is in no way limited to certain ranges of time-interval, space-interval, or density. It is also quite abstract. We shall approach it gradually by reviewing attempts, now "classical," to improve upon Boltzmann. Finally, we confront the key question: how valuable are the new insights?

II. Quasiclassical Kinetic Theory

An important goal of kinetic theory is the calculation of transport coefficients. The classical approach, based upon the Chapman-Enskog analysis of the Boltzmann equation, is well known. In this section, I shall present a typical coefficient, the shear viscosity, in the modern manner. The techniques and attitudes introduced here will reappear later in more complicated situations. Furthermore, even in this simple case, they present the functional assumptions of Chapman, Enskog, and Bogoljubov and the notions of time-scaling in a different light.

The formulation of kinetic theory in a Hilbert space of phase functions (dynamical variables) or field functions (distribution functions) and the introduction of operators into the space characterize the "modern" approach. Thus, classical dynamics is made to resemble quantum dynamics. The ingenious use of the simplest operators, the projection operators, further marks the approach.

We ask for the manner of relaxation of a typical Fourier component of $h(\underline{r}, \underline{p}, t)$ when a Boltzmann gas is perturbed slightly from equilibrium. The initial-value problem has the form

$$\left[\frac{\partial}{\partial t} + \frac{i}{m}\underline{k}\cdot\underline{p} - K\right]h(\underline{k}, \underline{p}, t) = \left[\frac{\partial}{\partial t} - H(\underline{k})\right]h(\underline{k}, \underline{p}, t) = 0 \tag{3}$$

$h(\underline{k}, \underline{p}, 0)$ given, and we know that the resolvent operator

$$G(\underline{k}) = [s - H(k)]^{-1} \;;\quad s \text{ complex}$$

is crucial. The operator H has a point spectrum and a continuous spectrum, which, for reasonable potentials and small $\underline{k}$, are disjoint. The "hydrodynamic" points (poles), i. e., those that approach zero as $k \to 0$, are of particular interest. Now consider the Hilbert space of functions of $\underline{p}$ with the norm suggested by Eq. (2): $(\psi_\alpha, \psi_\beta) = \delta_{\alpha\beta}$. Introduce P, the projector onto the space spanned by the ψ_{α}, and $Q = 1 - P$, its complement. The resolvent becomes

$$\frac{1}{s-H} = \frac{1}{s-HQ} + \frac{1}{s-HQ}\, HP\, \frac{1}{s-H} \tag{4}$$

and its matrix elements in the subspace satisfy the equation

$$\langle\alpha|G|\beta\rangle = \frac{1}{s}\,\delta_{\alpha\beta} + \sum_{\alpha'}\left\langle\alpha\left|\frac{1}{s-HQ}\right|\alpha'\right\rangle\langle\alpha'|G|\beta\rangle \tag{5}$$

The projection of $h(\underline{k}, \underline{p}, t)$ into the subspace is important because the components (ψ_α, h) are the hydrodynamical moments, from which one forms the equations of fluid dynamics. Thus, we expect expressions for the transport coefficients to lurk among the $\langle\alpha|G|\beta\rangle$. We shall extract the viscosity. To begin, notice that the behavior of H and the ψ_α upon rotation and reflection causes many of the matrix elements $\langle\alpha|\cdots|\alpha'\rangle$ in Eq. (5) to vanish. For example, if we take $\underline{k}$ to be parallel to the 3-axis, all off-diagonal elements in the fourth and fifth rows of $\langle\alpha|G|\beta\rangle$ vanish. The function $\langle 4|G|4\rangle = \langle\xi_1|G|\xi_1\rangle$ describes the relaxation of transverse momentum. The "collapse" of the sum in Eq. (5) causes it to have the simple form

$$\langle 4|G|4\rangle = \frac{1}{s - D'(\underline{k}, s)}$$

where

$$\begin{aligned} D'(\underline{k}, s) &= \left\langle 4\left|HQ\,\frac{1}{s-QHQ}\,QH\right|\right\rangle \\ &= -k^2 v_o^2 \left\langle\xi_1\xi_3\left|Q\frac{1}{s-QHQ}\,Q\right|\xi_1\xi_3\right\rangle \\ &= -k^2 v_o^2 \left\langle\xi_1\xi_3\left|\frac{1}{s-K+(i/m)Q\underline{k}\cdot\underline{p}Q}\right|\xi_1\xi_3\right\rangle \end{aligned} \tag{6}$$

{We have used the relations $Q[1/(s-HQ)] = Q[1/(s-QHQ)]$, $P\xi_1\xi_3 = 0$, and $QKQ = K$ to reach the final equation.}

We expect to see the linearized Navier-Stokes equation for the transverse momentum in Eq. (6), and indeed it appears as the limit $k \to 0$ is taken in $D'(\underline{k}, s)$. Thus

$$\langle 4|G|4\rangle \to \frac{1}{s-(1/mn)\eta k^2} \tag{7}$$

with

$$\eta = nk_B T \langle \xi_1\xi_3 |(1/K)| \xi_1\xi_3\rangle$$

The expression for η involves the solution of an integral equation. It is precisely the Chapman-Enskog result. The remaining 3×3 matrix yields the longitudinal eigenvalues, two damped sound modes, and a mode describing the diffusion of heat. The Chapman-Enskog expression for λ, the coefficient of heat diffusion, may be read out of the dispersion relation. Thus, the projection-operator treatment is equivalent to the traditional. No hypothesis about "normal solutions," in which the time-dependence enters only through the moments, need be made. Hydrodynamical behavior is associated with the smallest eigenvalues (and eigenvectors) of a scattering operator, which describes dynamics on a finer scale of space and time. The expansion in wave vector, $\underline{k}$, gives a series in the ratio of mean free path to characteristic macroscopic dimension of the system. When this ratio is small, the hydrodynamical eigenvalues are well separated from the higher eigenvalues and the continuous spectrum. Then, there is an epoch $t > t_2 > t_1$ during which the distribution function does indeed relax in the manner of its moments. The modern picture is, overall, clearer.

Equation (7) contains a bonus. We may write it as

$$\eta = nk_B T \lim_{s\to 0}\int_0^\infty dt\, e^{-st}\int d^3\underline{p}\, M(\underline{p})\xi_1\xi_3\, e^{-tK}\xi_1\xi_3 \tag{8}$$

This has the form of a Green-Kubo relation, giving a transport coefficient in terms of an autocorrelation function. Of course, Eq. (8) is somewhat synthetic and does not express truly a "fluctuation-dissipation theorem," for, at the Boltzmann level, we do not discuss exact dynamics. The relevant dynamical variable here is the (1,3) component of the pressure tensor. However, only the kinetic portion $\xi_1\xi_3$ appears here: an expression of the limitations of Boltzmann's equation. We shall have more to say about Eq. (8) later.

III. Significant and Partially Successful Approaches

The success and efficiency of the projection operator technique would suggest that we apply it in a more complicated context. We shall do this in Sec. IV. Section III may be regarded as an interlude, in which we summarize progress in a mixed line of research. It is mixed because on the one hand it deals with the Liouville equation, the full dynamics, but on the other hand it concerns itself principally with corrections to the Boltzmann equation and to transport coefficients. These corrections turn out to be of successively higher order in density. We shall be brief here, because most of the following already has achieved "textbook" status.

Thus, in order to correct and understand Boltzmann's equation, we must begin at the beginning, with exact dynamics. We contemplate the motion of a dynamical variable, $A(t) = A[\underline{q}_1(t), \cdots \underline{q}_N(t); \underline{p}_1(t), \cdots \underline{p}_N(t)] = A(\underline{q}_1^{\,o}, \cdots \underline{p}_N^{\,o}; t)$, and the motion of the phase-space density, $f_N(\underline{r}_1, \cdots \underline{p}_N; t)$. Thus,

$$(d/dt)A = [A, H] = iLA; \ (\partial/\partial t)f_N = -iLf_N; \ f_N(t{=}0) = f_N(0) \qquad (9)$$

where operator iL symbolizes "Poisson-bracketing with the Hamiltonian." For the systems we consider

$$iL\, f = \sum_j \left(\frac{1}{m}\underline{p}_j \cdot \frac{\partial f}{\partial \underline{r}_j} + \underline{F}_j \cdot \frac{\partial f}{\partial \underline{p}_j} \right)$$

for a field function. It has the same form, with $\underline{r}_k \to \underline{q}_k$, when operating upon a dynamical variable. Finally, we note a typical connection

$$f_N(\underline{r}_1 \cdots \underline{p}_N; t) = \left\langle \delta^3[\underline{r}_1 - \underline{q}_1(t)] \cdots \delta^3[\underline{p}_N - \underline{p}_N(t)] \right\rangle_o$$

between distribution function (field function) and the appropriate dynamical variable. The averaging is with respect to the initial ensemble $f_N(0)$.

The dynamical variable

$$\hat{f}_1(\underline{r}, \underline{p}, t) = \sum_{j=1}^{N} \delta^3[\underline{r} - \underline{q}_j(t)]\ \delta[\underline{p} - \underline{p}_j(t)] \qquad (9a)$$

and its Fourier transform

$$\hat{f}_1(\underline{k}, \underline{p}, t) = \sum_{j=1}^{N} e^{-i\underline{k} \cdot \underline{q}_j(t)}\, \delta^3[\underline{p} - \underline{p}_j(t)]$$

are of special interest to us; $\langle \hat{f}_1(\underline{r}, \underline{p}, t)\rangle_o$ is the one-particle distribution function, and the equilibrium correlation function is related to the observed neutron scattering law $S(\underline{k}, \omega)$ through $\delta\hat{f}_1 = \hat{f}_1 - \langle \hat{f}_1 \rangle_{eq}$ and

$$S(\underline{k}, \omega) = \frac{1}{2\pi} \int_{-\infty}^{\infty} dt\, e^{-i\omega t} \int d^3\underline{p}\, d^3\underline{p}' \langle \delta\hat{f}_1^{\,*}(\underline{k}, \underline{p}', 0)\, \delta\hat{f}_1(\underline{k}, \underline{p}, t) \rangle_{eq} \quad (9b)$$

Reduced distribution functions for the N-particle system play an important rôle. In particular, they satisfy a hierarchy of equations (moment equations) of which only the first concerns us here. It is

$$\left(\frac{\partial}{\partial t} + \frac{1}{m}\underline{p}_1 \cdot \frac{\partial}{\partial \underline{r}_1}\right) f_1(\underline{r}_1, \underline{p}_1, t) = -\frac{\partial}{\partial \underline{p}_1} \cdot \int d^3\underline{r}_2\, d^3\underline{p}_2\, F_{12} f_2(\underline{r}_1, \underline{r}_2; \underline{p}_1, \underline{p}_2; t) \quad (10)$$

Although the right-hand side of Eq. (10) describes the change in f_1 due to collisions, the equation is far from Boltzmann's. A conservative view would see it merely as a connection between first and second moments. The search for a true, closed, master equation

$$[(\partial/\partial t) + (1/m)\underline{p}_1 \cdot (\partial/\partial \underline{r}_1)] f_1 = \Phi(\underline{r}_1, \underline{p}_1; f_1; t)$$

has, for many decades, been a search for the philosophers' stone. Like the stone, it cannot exist, but in searching for it one finds much that is valuable.

In his important book Dynamical Problems in Statistical Physics, N. N. Bogoljubov presented an approach to the problem of the master equation which stimulated a decade of research activity in the West. Its essence is that a master equation does exist for the epoch $t > t_o$, and that its form is based upon the Ansatz that

$$f_s(\underline{r}_1 \cdots \underline{p}_s; t) = f_s[\underline{r}_1 \cdots \underline{p}_s; f_1(\underline{r}_1, \underline{p}_1; t)] \;; \quad s \geq 2$$

and upon a rather curious initial condition. Thus, after a collision time, more or less, the higher distribution functions "slave to" f_1 in their temporal behavior. The closed master (kinetic) equation is then

$$[(\partial/\partial t) + (1/m)\underline{p}_1 \cdot (\partial/\partial \underline{r}_1)] f_1 = \Phi_1(\underline{r}_1, \underline{p}_1; f_1) \quad (11)$$

Unfortunately, we do not know the explicit form of $f_2(\;; f_1)$, and we do not know the correct initial condition to be used with Eq. (11). Further approximation is necessary before an explicit theory can be developed. However, it is possible to go directly to the hydrodynamic stage with Eq. (11) by adopting the strategy of Chapman and Enskog. The normal solution, wherein f_1, in turn, "slaves to" the slow, temporal evolution of its moments, is introduced (projection onto hydrodynamical subspace), and expansion is made with respect to smallness of spatial variation of the moments (expansion in powers of $\underline{k}$). The integral equations that result are complicated, but, in principle, a solution exists. Unfortunately, the kernels are not known explicitly.

Explicit evaluation of Eq. (11) has, to date, been carried out only through expansion with respect to density. In fact, a systematic expansion in the density (of a uniform reference state), coupled with an unhappy assumption about the initial state, viz.

$$f_s(\underline{r}_1, \cdots \underline{p}_s;0) = \prod_{i=1}^{s} f_1(\underline{r}_i, \underline{p}_i;0)$$

will close the hierarchy and, in particular, Eq. (10). The final forms are the same, whether the Bogoljubov Ansatz is made or not. The technique is that of cluster expansion, similar to the Ursell-Mayer technique of equilibrium statistical mechanics. With $(\underline{r}_i, \underline{p}_i)$ denoted by x_i, one has

$$f_2(x_1x_2;t) = f_2^{(o)}(x_1x_2;t) + n\, f_2^{(1)}(x_1x_2;t) + \cdots \qquad (12a)$$

$$f_2^{(o)}(x_1x_2;t) = S_t^{(2)}(x_1x_2) f_1(x_1;t)\, f_1(x_2;t) \qquad (12b)$$

$$f_2^{(1)}(x_1x_2;t) = \int dx_3 S_t^{(3)}(x_1x_2x_3) f_1(x_1;t) f_1(x_2;t) f_1(x_3;t) \qquad (12c)$$

The $S_t^{(k)}$ appearing in Eq. (12) are symbolic operators, which refer to the motion of k particles. For example, $S_t^{(3)}(x_1x_2x_3)$ alters the phases that appear in $f_1f_1f_1$ in a well-defined manner, based upon the motion of one-, two-, and three-particle groups (clusters) in the time interval t. Connection with the Boltzmann equation is made by replacing $S_t^{(k)}$ by $S_\infty^{(k)}$ and by neglecting the difference in position of colliding molecules (coarse-graining). Finally, one has

$$[(\partial/\partial t)+(1/m)\underline{p}_1\cdot(\partial/\partial \underline{r}_1)] f_1 = n\,\bar{J}(f_1f_1)+n^2 K(f_1f_1f_1) + \cdots \qquad 13)$$

with or without the Ansatz.

The three-body term and its implications for thermodynamics and hydrodynamics have received considerable attention. The second virial coefficient appears correctly in the equilibrium distribution, and corrections to η and λ have been computed for a gas of hard spheres. However, attempts to continue the series in density have led to unpleasant surprises. In three dimensions, the four-body and all higher-order terms diverge, whereas in two dimensions, the three-body and higher-order terms diverge. The cluster-Bogoljubov approach seems to have come to a dead end.

It is difficult to estimate just how much truth the cluster theory contains. Is it true only in the limit $t \to \infty$, or can the system of Eq. (12) tell us something about the approach to the hydrodynamic stage in a moderately dense gas? Is the natural variable for expansion some combination of time and density? How much has been lost in making the crude approximation of the initial condition? How does one surmount the density divergence? Can the singular terms in the expansion of η and λ (which turn out to be $\sim n^2 \log n$) be seen in experiment? These questions have not been answered thoroughly yet.

If one restricts oneself to transport coefficients, one can arrive at the same point more directly, in a manner that shows that some of our concerns are irrelevant. The calculation is based upon the Green-Kubo expressions, themselves based upon a linear-response theory (small perturbation from equilibrium) that is palatable. One has, for example

$$\eta = \frac{1}{k_B T V} \lim_{s \to \infty} \left\langle T_{13} \middle| \frac{1}{s - iL} \middle| T_{13} \right\rangle \tag{14}$$

T_{ij} being the pressure tensor, and the density-expansion is generated by an expansion of the resolvent called the "binary-collision expansion." As in our Eqs. (7) and (8), the Kubo and Chapman-Enskog methods lead to the same results. The same divergences appear; it is clear that they have nothing to do with choice of initial condition or with coarse-graining. The theory based upon Eq. (14) is "exact," and it yields the same transport coefficients (whenever the results are finite) as do the Boltzmann and generalized Boltzmann theories, which are coarse-grained. How do we avoid paradox? Perhaps we can by asserting that the true smallness-parameter for expansion involves n and t in such a way that it is small only for n small <u>and</u> t large. Another possible clue is that the Laplace transforms of autocorrelation functions associated with transport coefficients have been found to be singular at $s = 0$. A subsequent expansion in density only makes the nonanalyticity clearer.

In any case, Boltzmann's equation appears as the "leading term" in a formal expansion of the exact dynamics with respect to density. There is another expansion, worth investigating, and that is in terms of the strength of forces between the particles. The expansion is limited in its applicability to the physical world because it requires that the potential be bounded or most weakly divergent. Then, the expansion may be thought of as equivalent to an expansion in inverse temperature. When Boltzmann's equation is expanded in this manner, one obtains a kinetic equation, first analyzed by Landau. If one does not expand first in density but expands the exact dynamics in terms of coupling strength, one obtains a system that should illuminate the issues of convergence, initial conditions, etc., raised previously. The leading term is a kinetic equation of generalized Fokker-Planck form. It is of some use in plasma physics, and we shall meet it again.

Finally, we turn to the question of fluctuations. There is no such issue in the exact dynamics of an N-body system. It is only when one contracts or coarsens the description that problems arise. Except for special cases, the contracted system is not in itself closed. Strictly speaking, a "noise" term is always present, for the degrees of freedom, or time scales, "projected out" never can be eliminated completely. The effect of the noise is usually small; hence we have the ability to invoke macroscopic or microscopic physics to describe nature. The study of critical phenomena is a popular counterexample.

To restore the noise term to Boltzmann's equation, in the absence of a systematic theory, is an exercise in informed guesswork. The paradigm is Langevin's equation for the momentum of a macroscopic particle

$$(dp/dt) + (\gamma/m)\,p = F(t) \tag{15a}$$

$$(d/dt)\langle p\rangle + (\gamma/m)\langle p\rangle = 0 \qquad \langle F\rangle = 0 \tag{15b}$$

$$\langle F(t)F(s)\rangle = 2Q\,\delta(t-s); \;\; F \text{ is gaussian} \tag{15c}$$

The equation for the test particle coupled to its N-1 colleagues is replaced by an equation for a single momentum, driven by a "stochastic" force and damped by a macroscopic force. The equation for the averaged momentum is the analog of Boltzmann's equation. The success of the Langevin equation in describing correlated fluctuations, etc., is based upon a crucial point of time scaling: that the correlation time for fluctuations is much shorter than the macroscopic relaxation time. Hence we have the delta function in Eq. (15). The time-scaling argument, although reasonable, is quite difficult to prove. It can describe only a limiting case. As a final point of physics, a "fluctuation-dissipation theorem" connects the amplitude of the force with γ. It reflects the fact that the test particle, whatever its initial state, comes to equilibrium with the "bath" of N-1 other particles.

We shall describe two slightly different approaches to the construction of the "correct" Boltzmann-Langevin equation. Both lead to the same equation and then, via Chapman-Enskog, to equations for fluctuating hydrodynamics which are now classic and have some practical significance. The proper fluctuating quantity is $h(\underline{r}, \underline{p}, t)$ of Eq. (1a). Since $(\underline{r}, \underline{p})$ may be thought of as labels, the kinetic equation resembles a set of first-order equations for quantities $a_i(t)$, $i = (\underline{r}, \underline{p})$, which relax according to

$$(da/dt)_i + \sum_j (A_{ij} + S_{ij}) a_j = 0 \tag{16}$$

where A_{ij} is antisymmetric and S_{ij} symmetric. Here $a_i \leftarrow \rightarrow \sqrt{M(\underline{p})}\, h(\underline{r}, \underline{p}, t)$.

When a stochastic "force" or noise source $F_i(t)$ is added to the right-hand side and the force function is a gaussian process with zero correlation time, $\langle F_i(t) F_j(s) \rangle = 2 Q_{ij}\, \delta(t-s)$, we have a not unusual problem in the theory of gaussian Markov processes. The a_i also are distributed in a gaussian manner, with $W(a_j)$ proportional to $\exp[(-1/2) \sum_{ij} a_i E_{ij} a_j]$. Now the pieces fall into place. $G_{ij} = A_{ij} + S_{ij}$ is known. The stationary nature of the process gives Q_{ij} in terms of G_{ij} and E_{ij}, and the latter is connected with the entropy through the Boltzmann-Einstein relation, $W(a_i)$ proportional to $\exp[(\Delta S/(a_i)/k_B]$. Since the entropy of the distribution may be determined from $f \log f$, the system is closed, and we have a unique prescription for calculating gaussian forces. One finds that the right-hand side of Eq. (1a) should be augmented by $F(\underline{r}, \underline{p}, t)$, where

$$\langle F(\underline{r}, \underline{p}, t) F(\underline{r}', \underline{p}', t) \rangle = 2M^{-1}(\underline{p}') K(\underline{p}, \underline{p}') \delta^3(\underline{r} - \underline{r}') \delta(t-t')$$

This approach has been discussed in detail by Fox and Uhlenbeck.

The second approach, due to Bixon and Zwanzig, is somewhat more intuitive, more "physical." Instead of adding a noise source to the linearized Boltzmann equation and thereby changing its nature, these authors regard the equation as the proper average of another equation, which contains fluctuations. The new equation concerns the dynamical variable $\hat{h}[\underline{r}, \underline{p}; \underline{q}_i(t), \underline{p}_i(t)]$, which is related to $\hat{f}_1(\underline{r}, \underline{p}, t)$ through

$$\hat{f}_1(\underline{r}, \underline{p}, t) = nM(p)[1 + \hat{h}(\underline{r}, \underline{p}, t)]$$

One can write a true equation for $\hat{h}$ via $\hat{f}_1$; it involves $\hat{f}_2$ and resembles the first equation of the hierarchy. An equation for $\hat{h}$ follows at once. Bixon and Zwanzig write this equation in the suggestive form [see Eq. (3)]

$$[(\partial/\partial t)+(1/m)\underline{p}_1 \cdot (\partial/\partial \underline{r}_1) - K]\hat{h} = F$$

where the extremely complicated right-hand side is taken to be the "noise." Its average is assumed to be zero, although it is not, and its correlation time is taken to be zero, although it is not. Yet, to the extent that the linearized Boltzmann equation is a correct description (low densities, low frequencies), these assumptions about the noise are tenable. One has here a theory that is not purely stochastic, that tries to identify the dynamical elements in the low-density limit.

IV. Modern Theories

The treatments of N-body dynamics just sketched have concentrated upon the later stages of evolution. Some procedure of "coarse-graining," of eliminating information corresponding to short time or high frequencies, is a natural part of the formalism. Indeed, John Kirkwood introduced his method of coarse-graining through an argument that it is an essential part of the measuring process and therefore needs to be an essential part of the mathematical formalism.

Certainly, there is a problem of resolution in every experiment. But the issue is of little import in the experiments that have stimulated the development of modern kinetic theory. These are experiments to determine the "scattering law" for neutrons interacting with simple fluids and dense gases, experiments in the scattering of intense beams of light, and, most important, computer experiments in which N-body dynamics is simulated. The interpretation of these requires a theory that is good for arbitrary intervals of time and space. Of course, such a theory is impossibly complicated. But I shall describe a compact formalism that serves as a good platform for leaping to approximations in all directions. As promised, it is based upon the ingenious use of projection operators.

As a first example, consider the kinetic equation for a classical "test particle." In the Boltzmann limit, the system approaches the Lorentz model. One has a linear transport equation, resembling Eq. (1a), for $f(\underline{r}, \underline{p}, t)$. Since only a single conservation law exists, the hydrodynamic pole is nondegenerate. We begin by returning to the Liouville equation and denote the N-body distribution function f_N by $\rho(\cdots;t)$.

The test-particle distribution function is related to f_N or ρ through a contraction or projection. For example

$$Pf_N = \rho_B \int dx_2 \cdots dx_N f_N = \rho_B f(\underline{r}, \underline{p}, t)$$
$$= \rho_o V g(\underline{r}, \underline{p}, t) \qquad (17)$$

where

$$\rho_B = \rho_o V/M(\underline{p})$$

and ρ_o is the canonical, equilibrium distribution. Note that $P^2=P$, as it should, and that the choice of the multiplicative factor in Eq. (17) is not unique.

The key to our analysis is a simple identity, given first by R. W. Zwanzig. If $1 - P$ is denoted by Q, the decomposition $\rho = P\rho + Q\rho$ permits one to integrate the Liouville equation formally to obtain

$$\left(\frac{\partial}{\partial t} + PiL\right) P\rho = \int_o^t dt\, PiLe^{-i\tau QL} QiLP\rho(t-\tau) - PiLe^{-itQL} Q\rho(0) \qquad (18)$$

for <u>any</u> projector P. Since $P\rho$ is proportional to f, Eq. (18) is, except for the last term, an exact, closed equation for the test-particle density. The last term vanishes only if the distribution is initially in the projected subspace, i.e., $Q\rho(0) = 0$. This condition is not a spoiler; it admits initial distributions of the form $\rho_o \varphi(1)$, where $\varphi(1)$ is an arbitrary function of $(\underline{r}_1, \underline{p}_1)$. The initial distribution need not be close to equilibrium. Should the condition not be satisfied, one has the interesting problem of determining how long it takes for memory of the initial state to disappear. (Does the Bogoljubov Ansatz ever hold?) Although abstract, Zwanzig's equation poses the issue of initial conditions quite clearly.

If one exploits the properties of our projector, Eq. (17), one can reduce the kinetic equation to

$$[(\partial/\partial t) + (1/m)\underline{p}\cdot(\partial/\partial \underline{r})] g(\underline{r}, \underline{p}, t) = \int_o^t d\tau\, \Sigma(\tau)\, g(\underline{r}, \underline{p}, t) \qquad (19)$$

+ initial value term

with

$$\Sigma(\tau) g = \int d^3\underline{r}' \int d^3\underline{p}' \left(\frac{\partial}{\partial \underline{p}} - \frac{1}{m}\beta \underline{p}\right) \cdot \underset{\sim}{D}(\underline{r}-\underline{r}', \underline{p}' \rightarrow \underline{p}, \tau) \cdot \frac{\partial}{\partial \underline{p}'} g(\underline{r}', \underline{p}'; t-\tau)$$

Let us choose the initial condition so that we have a closed, exact kinetic equation. Then, the equation is dominated by the kernel, $\underset{\sim}{D}(\underline{r}-\underline{r}', \underline{p} \rightarrow \underline{p}', \tau)$. It is nonlocal in space and time and depends upon a complicated correlation of forces experienced by the test particle. Although the equation is linear in $f(\underline{r}, \underline{p}, t)$, it is extraordinarily complicated and hardly looks as though it will lend itself to a neat analysis on the basis of time scales. The "modified propagator" exp(itQL) is particularly troublesome; one has been able to deal with it only via formal expansion of one sort or another. One knows, through example, that its properties are quite different from exp(itL).

To progress systematically, one would like to expand in terms of a good, small parameter. In the case of the dense gas, we have only the familiar, inadequate pair, density and interaction strength. In lowest order, the density expansion, coupled with a time-scaling argument, reproduces the simple kinetic equation for the Lorentz model. Although density expansions have received considerable attention in the test-particle problem, their effect upon the kinetic equation (19) is, for the most part, unknown. The weak-coupling expansion of $\underset{\sim}{D}(\underline{r}-\underline{r}', \underline{p}' \to \underline{p}, \tau)$ simplifies Eq. (19) dramatically, in lowest order. The equation for a spatial Fourier component of $f(\underline{r}, \underline{p}, t)$ becomes

$$\left(\frac{\partial}{\partial t} + \frac{i}{m}\underline{k}\cdot\underline{p}\right) f(\underline{k}, \underline{p}, t) = \int_0^t dt \frac{\partial}{\partial \underline{p}} \cdot \underset{\sim}{D}(\underline{k}, \underline{p}, \tau) \cdot \left(\frac{\partial}{\partial \underline{p}} + \frac{1}{m}\beta\underline{p}\right) f(\underline{k}, \underline{p}, t-\tau)$$

This has the form of a generalized Fokker-Planck equation, one in which the diffusion tensor is momentum-dependent and has memory. It is simplest to study the case $\underline{k} = 0$, in which a spatially uniform distribution of test particles relaxes to equilibrium in momentum space. Then, we may write

$$\frac{\partial}{\partial t^*} f(\underline{u}, t^*) = \epsilon \int_0^{t^*} d\tau \frac{\partial}{\partial \underline{u}} \cdot \underset{\sim}{D}(\underline{u}, \tau) \cdot \left(\frac{\partial}{\partial \underline{u}} + 2\underline{u}\right) f(\underline{u}, t^* - \tau) \qquad (20)$$

where $t^* = t/t_o = \sqrt{2}(v_0/r_0)t$ is dimensionless, $\epsilon = (1/4)nr_o^3[(\lambda/k_B T)]^2$, $\underline{u} = (\sqrt{2}\, v_0)^{-1}\underline{v}$ is a dimensionless velocity, λ is the strength of the potential, and r_o is its range. Two time scales appear naturally, $t_o = r_0/\sqrt{2}\, v_0$ and $t_1 = (\pi r_o^2 n\sqrt{2}\, v_0)^{-1}$. Their ratio, $\delta = t_o/t_1 = n\pi r_o^3$, has an obvious meaning. The diffusion tensor relaxes with the shorter time scale, and the distribution function with the longer. The Boltzmann limit is produced if one takes the ratio of time scales to be precisely zero. Then one makes the replacement

$$\int_0^t d\tau\, \underset{\sim}{D}(\cdots\tau)\cdots f(t-\tau) \to \left\{\int_0^\infty d\tau\, \underset{\sim}{D}(\cdots\tau)\right\} f(\cdots t)$$

and the kinetic equation is now Markovian and familiar. However, analysis of Eq. (20) shows that, as long as $\delta \neq 0$, the long-time behavior of its solutions differs from that of the

Markovian equation. If this feature is not a consequence of the weak-coupling approximation itself, we have here an example of the intertwinedness of time scales in kinetic theory.

The next example of the generalized theory uses a more abstract setting. In fact, it approaches Koopman's original notion of setting classical mechanics in terms of the mathematics of Hilbert spaces, with benefits accruing as in quantum mechanics. Thus, for dynamical variables A_1, A_2 selected from an appropriate space of dynamical variables, we define the inner product

$$(A_1, A_2) = \int d^3\underline{q}_1 \cdots d^3\underline{p}_N \, \rho_o A_1^* A_2$$

where ρ_o is the canonical distribution. Dynamical variables evolve via the action of the evolution operator $\exp(itL)$, so that we may write an equilibrium autocorrelation function as

$$C_A(t) = \left(A, A(t)\right) = (A, e^{itL}A)$$

Again, projection operators play key roles. As a first example, consider the dynamical variable A, with equilibrium expectation zero and normalized so that $(A, A) = 1$. Let $P_A = |A\rangle\langle A|$ project onto A, and let $Q_A = 1 - P_A$. Then, an ingenious rearrangement of the equation of motion, Eq. (9), leads to the striking form

$$\frac{d}{dt}A(t) - i\Omega A(t) + \int_0^t d\tau \, \varphi(t-\tau)A(\tau) = f(t) \tag{21}$$

where

$$i\Omega = (A, \mathring{A}) = (A, iLA) = (A, PiLPA)$$

$$\varphi(t) = \left(f, f(t)\right) \; ; \; f(t) = e^{itQL} Q\mathring{A} = e^{itQL} QiLA$$

The quantity $i\Omega$ refers to a characteristic frequency; it vanishes when A is real in this simple example. φ is taken to describe damping, or "memory," and f(t) is interpreted as a fluctuating "force," or as "noise." This rôle for f(t) is sensible, for it refers only to motion in the space orthogonal to the projected (contracted) space A. And the equilibrium average of f(t) vanishes.

Equation (21) has two valuable offspring, the equation for the autocorrelation function

$$\frac{d}{dt}C_A(t) - i\Omega \, C_A(t) + \int_0^t dt \, \varphi(t-\tau) \, C_A(t) = 0 \tag{22}$$

and the equation for the evolution of the average of A, $\langle A \rangle$, with respect to an arbitrary initial distribution

$$\frac{d}{dt}\langle A\rangle - i\Omega\langle A\rangle + \int_0^t dt\, \varphi(t-\tau)\langle A\rangle = \langle f\rangle \qquad (23)$$

Equation (21) is a "generalized Langevin equation." To see more clearly, take $A = p_x$, whence $\Omega = 0$, φ is a damping with memory (frequency-dependent damping), which, in some cases, becomes $(\gamma/m)\delta(t-\tau)$, and $f(t)$ has the properties necessary for the fluctuating force. Equation (22), describing the correlation function, is a closed equation, and Eq. (23) is the prototype of a kinetic equation (take $A = \delta\hat{f}_1$). In the latter case, we note that the equation is closed only for certain initial distributions: a familiar point! In fact, when Eq. (23) is the equation for the average test-particle density, the corresponding Fokker-Planck equation for the distribution of density is Eq. (18).

In building a generalized kinetic theory, the dynamical variables that we want to use are collective variables, not those belonging to a unique particle. The one-body-additive functions defining mass, momentum, and kinetic energy densities

$$\begin{pmatrix} m(\underline{r},t) \\ \underline{g}(\underline{r},t) \\ T(\underline{r},t) \end{pmatrix} = \sum_j \begin{pmatrix} m \\ \underline{p}_j \\ (1/2m)p_j^2 \end{pmatrix} \delta^3[\underline{r} - \underline{q}_j(t)] \qquad (24)$$

and their Fourier components

$$\begin{pmatrix} m_k(t) \\ g_k(t) \\ T_k(t) \end{pmatrix} = \sum_j \begin{pmatrix} m \\ \underline{p}_j \\ (1/2m)p_j^2 \end{pmatrix} e^{-i\underline{k}\cdot\underline{q}_j(t)}$$

are particularly important. Denote m_k, g_k, and T_k (adjusted to have zero expectation) as A_1, A_2, A_3. Then, it would appear profitable to project the equation of motion for each onto the three-dimensional subspace spanned by the A's. The system that one obtains is a simple generalization of Eq. (21) and is the basis for "generalized hydrodynamics." The Green-Kubo relations for transport coefficients appear naturally, in this formalism, as the macroscopic limit $\underline{k}\to 0$, $t\to\infty$ is taken.

In our search for generalized Boltzmann equations, the key variable is $\hat{f}_1$, itself one-body additive. The natural decomposition of its motion involves the subspace of one-body additive functions, a space spanned by

$$N(\underline{r}, \underline{p}) = \sum_j \delta^3(\underline{r} - \underline{q}_j)\, \delta^3(\underline{p} - \underline{p}_j) \tag{25a}$$

or

$$N(\underline{k}, \underline{p}) = \sum_j e^{-i\underline{k}\cdot\underline{q}_j}\, \delta^3(\underline{p} - \underline{p}_j) = |\underline{k}, \underline{p}\rangle \tag{25b}$$

where $(\underline{r}, \underline{p})$ and $(\underline{k}, \underline{p})$ are labels. We shall consider this development in some detail, focusing upon the $(\underline{k}, \underline{p})$ basis. Notice first that, although our bases are complete, their elements are not orthogonal. We need the results

$$B(\underline{p}, \underline{p}') \equiv \langle \underline{k}, \underline{p} | \underline{k}, \underline{p}' \rangle = NM(\underline{p}')\,[\,\delta^3(\underline{p}-\underline{p}') + n\,M(\underline{p})\,h(\underline{k})] \tag{26a}$$

$$B^{-1}(\underline{p}, \underline{p}') = [\,1/NM(\underline{p})][\,\delta^3(\underline{p}-\underline{p}') - n\,M(\underline{p})\,c(\underline{k})] \tag{26b}$$

where

$$\int d^3\underline{p}''\, B^{-1}(\underline{p}, \underline{p}'')\,B(\underline{p}'', \underline{p}') = \delta^3(\underline{p}-\underline{p}')$$

and the operator projecting into the subspace is

$$P_1 = \int d^3\underline{p}\, d^3\underline{p}' |\underline{p}\rangle\, B^{-1}(\underline{p}, \underline{p}')\, \langle \underline{p}'| \tag{27}$$

(The index $\underline{k}$ is suppressed here.) The functions $h(\underline{k})$ and $c(\underline{k})$ are related to the static pair distribution function $g(\underline{r})$. $h(\underline{k})$ is the Fourier transform of $g(\underline{r}) - 1$, and $c(\underline{k})$, the direct correlation function, is $h(\underline{k})[\,1+nh(\underline{k})\,]^{-1}$. Now we may proceed to a generalized kinetic equation in either of two ways. We may choose $A = \delta\hat{f}_1(\underline{k}, \underline{p}) = \hat{f}_1(\underline{k}, \underline{p}) - \langle \hat{f}_1 \rangle$ as the dynamical variable, and notice that the corresponding P_A is the P_1 of Eq. (27). Then Eqs. (21-23) apply at once, with the modification that Ω and φ are now operators with respect to momentum labels. Thus

$$\frac{\partial}{\partial t}\delta\hat{f}_1(\underline{p},t) - i\int d^3\underline{p}'\Omega_k(\underline{p}, \underline{p}')\delta\hat{f}_1(\underline{p}',t) + \int_0^t d\tau \int d^3\underline{p}'\varphi_k(\underline{p}, \underline{p}';t-\tau)\delta\hat{f}_1(\underline{p}',\tau) = \cdots \tag{28}$$

and additional equations corresponding to Eqs. (22) and (23) tell us all that we want to know about the one-particle distribution function. The "PiLP" aspect of $i\Omega$ is notable. It is the projected part of iL, that part of the dynamics contained entirely in the projected subspace. We find

$$PiLP\delta\hat{f}(\underline{p}, t) = \frac{1}{m}\underline{k}\cdot\underline{p}\delta\hat{f}(\underline{p}, t) + k_B T n c(\underline{k}) i\underline{k}\cdot\frac{\partial}{\partial \underline{p}} M(\underline{p}) \int d^3\underline{p}'\delta\hat{f}(\underline{p}', t) \tag{29}$$

The first term represents simple streaming, and that is all one obtains in the test-particle problem. The second term reminds one of a corresponding term in Vlasov's kinetic equation for plasmas, a term describing the motion of a particle in the mean field of its neighbors. The corresponding term in the equation for fluids replaces the two-body potential by the direct correlation function, $-V(\underline{k}) \longleftrightarrow k_B T c(\underline{k})$, with advantage.

{The second method, somewhat more abstract, decomposes the evolution operator, $U = e^{itL}$, and notes that the equation of motion for $P_1 U P_1 \equiv U_1$ is

$$\left(\frac{\partial}{\partial t} - iP_1 LP_1\right) U_1(t) - \int_0^t d\tau\, iP_1 LQ_1 e^{i\tau Q_1 LQ_1} Q_1 iLP_1 U_1(t-\tau) = 0 \quad (30)$$

closely related to Eq. (28) and its cousins, Eqs. (21-23). P_1UP_1 describes the projected part of the evolution of any additive, one-body dynamical variable. It appears naturally in the autocorrelation functions for such variables. For example, $(g_k, g_k(t)) = (g_k, P_1 U(t) P_1 g_k)$. Since knowledge of the equilibrium correlation function is equivalent to that of the temporal relaxation of the variable from a particular non-equilibrium state, it is customary to assert that P_1UP_1 describes a kinetic equation for g_k as well. }

The memory function $\varphi_k(\underline{p}, \underline{p}';t)$ expresses dynamical correlations in scattering and is very complicated. It must be approximated, either systematically, through expansion in a small parameter, or through "modeling." In any case, one would like to know as much as possible about φ_k beforehand, to insure that an approximation that appears reasonable has not neglected some important property.

In a recent article, D. Forster has discussed general properties of φ_k, properties that should appear as "requirements" for a sound approximation. These involve symmetries, a positivity property, certain sum rules that stem from short-time behavior, and a structure that would guarantee proper approach to the hydrodynamical limit as $\underline{k} \to 0$ and $t \to \infty$. The analysis uses ideas introduced in Part I and one obtains suitably generalized expressions for transport coefficients. Since the operator $H(\underline{k})$ introduced there is replaced by one dependent upon s as well, the approach is more delicate, and certain analyticity properties are required. Most important is that the limiting kernel $(\underline{k} \to 0,\ s \to 0)$ have the five eigenfunctions of Eq. (2). This may be shown to be so for an arbitrary homogeneous isotropic classical system. It is surprising that kinetic energy appears as an eigenfunction in systems of arbitrary density, but the fact is not inconsistent with conservation of energy.

V. Approximations To The Memory Function

The simplest approximation is surely $\varphi_k = 0$. One gets the generalized Vlasov equation, which, being collisionless, does not give sensible results in the hydrodynamic limit, although it renders short-time, i. e., high-frequency behavior correctly. This approximation may be viewed as one in which motion orthogonal to the projected space is neglected. At once, an improvement suggests itself. Restricting the motion to S_1, the space of one-body additive variables, is too severe. Let P_1 be replaced by $P_{12} = P_1 + P_2$, which projects into S_{12}, the space of one- and two-body additive functions. S_{12} may be constructed by introducing the two-body basis functions for the space S_2

$$N(1,2) = N(\underline{r}_1, \underline{p}_1; \underline{r}_2, \underline{p}_2) = \sum_{i \neq j} \delta^3(\underline{r}_1 - \underline{q}_i)\,\delta^3(\underline{p}_1 - \underline{p}_i)\,\delta^3(\underline{r}_2 - \underline{q}_j)\,\delta^3(\underline{p}_2 - \underline{p}_j)$$

and choosing linear combinations of $N(1,2)$ which are orthogonal to the functions [Eq. (25)] of S_1. Then $S_{12} = S_1 + S_2$. The form of the equation of motion for $\delta\hat{f}_1$ is unaltered; only the meaning of the projectors P and Q is changed. [P_{12} is the obvious generalization of Eq. (27).] The improved approximation consists in setting $Q_{12} = 0$, whence the equation for $\delta\hat{f}_1$ becomes

$$(d/dt)\delta\hat{f}_1 - iL_{12}\delta\hat{f}_1 = 0, \quad L_{12} = P_{12}LP_{12} \tag{31}$$

Equation (31) is Markovian and deceptively simple. If we seek the autocorrelation function for $\delta\hat{f}_1$, $(\delta\hat{f}_1, \delta\hat{f}_1(t)) = (\delta\hat{f}_1, P_1U_{12}P_1\delta\hat{f}_1)$, we are led at once to Eq. (30) with L replaced by L_{12} and U_1 by U_{12}. A little calculation then shows that we may replace L_{12} by L and Q_1 by P_2 throughout. The new equation of motion

$$\left(\frac{\partial}{\partial t} - iP_1LP_1\right)U_{12}(t) - \int_0^t dt\, iP_1LP_2\, e^{i\tau P_2LP_2}P_2 iLP_1U_{12}(t-\tau) = 0 \tag{32}$$

is unusually interesting. It is the second step in a systematic attack upon Liouville's equation. One also may view the approach as an expansion of f_N in terms of one-body, two-body, ••• functions with amplitudes given by the solutions of closed kinetic equations of increasing complexity. This promising scheme was introduced into kinetic theory by E. P. Gross and R. W. Zwanzig. It has been generalized and analyzed by C. D. Boley.

Much remains to be done as far as detailed properties of solutions of Eq. (32) "and beyond" are concerned. For example, not all of the properties described by Forster have been demonstrated for the nth-order memory function. The hydrodynamical limit has not been studied, nor have transport coefficients been calculated and judged, at any stage. However, the theory is intellectually appealing on many counts. At each stage, it is possible to eliminate the potential in favor of static correlation functions, so that the theory is in a sense "renormalized," expressed in terms of quantities that are more accessible. In the second approximation, for example, two particles interact, not in vacuum, but in the presence of a mean field due to the remaining N-2. The scheme is surely superior to direct expansion in density or coupling strength.

Gross' second approximation, although promising, does not describe the sequences of collision which lead to the singular behavior described following Eq. (13). We must refer here to another important series of papers, by G. F. Mazenko, which develop a theory of memory functions and renormalized kinetic theory. Mazenko produces tractable memory functions by truncation and educated physical insight into the meaning of various terms. His methods lie close to those associated with "summation of diagrams" in many-body physics.

Mazenko's first contribution was the correct low-density memory function. This function becomes the linearized Boltzmann-Enskog kernel in the long-time $(s \to 0)$ limit. (Next the order in density was considered by Boley and R. C. Desai. This more complicated memory function, involving three-body collisions, gives the results of the Bogoljubov-Cohen-Green theories when appropriate limits are taken.) The low-density function for hard spheres is particularly simple. Since the collision time vanishes, φ_k is independent of s. The kinetic equation that ensues, the "generalized Enskog equation," has been studied by Mazenko, S. Yip, and associates. Since the equation can describe fluctuations in density of arbitrary wavelength, one can use it to understand neutron scattering data from moderately dense gases (hydrogen, and $n\pi r_0^3 < 1/100$).

A second Mazenko memory function, which is not limited to low densities, augments the generalized Enskog term by contributions from "ring diagrams," mode coupling, and "cross terms." It is designed to describe just about all interesting processes in simple classical fluids. The ring diagrams produce the divergences in the transport coefficients, and the mode-coupling terms are responsible for fluctuations near the critical point. The predictions of the model are being examined at present.

Since we have drifted from formally exact to "model" equations, we should mention the considerable success that very simple, analytical representations of the memory function have had in the interpretation (correlation?) of neutron-scattering and computer-dynamical data. The Zwanzig-Mori form of the equations of motion provides a most appropriate framework for the discussion of N-body dynamics.

To conclude this section, we return to the systematic approach to mention the earliest: the expansion in coupling parameter. The weak-coupling memory function was deduced first by A. Z. Akcasu and J. Duderstadt. The corresponding kinetic equation (accurate, one hopes, for all wavelengths and frequencies, given sufficiently small $\lambda/k_B T$) was analyzed in the original and stimulating work of D. Forster and P. C. Martin. Their discussion of the computation of η, the shear viscosity, which we shall summarize, shows the richness of the generalized method.

Kinetic theory à la Boltzmann-Landau gives

$$\eta/nk_B T = \langle \xi_1 \xi_3 | QK^{-1}Q | \xi_1 \xi_3 \rangle$$

where K is the (weak-coupling) scattering operation. The generalized kinetic equation gives (in the hydrodynamic limit) two terms. The first is

$$\eta'/nk_B T = \langle \xi_1 \xi_3 + T_{13} | QK^{-1}Q | \xi_1 \xi_3 + T_{13} \rangle$$

The second, $\eta''/nk_B T$, involves a matrix element of the memory function in the $k \to 0$, $s \to 0$ limit. The new feature of η' is a contribution to the stress tensor from forces (collision transfer), whereas the η'' represents an additional effect of correlations. Together, the terms modify the kinetic estimate by a factor $[1+B(n/T^2)+C(n^2/T^4)]$.

A final remark about the systematic expansions: one wonders about their uniformity. If one uses the memory function for small $\lambda/k_B T$ or $n\pi r_0^3$, are the solutions of the approximate equations close to the exact solutions for all $\underline{k}$ and s? Or are the approximate solutions useful only in some limit? At this point in the development of the theory, one is not sure, although some unsettling results have appeared recently in the study of a harmonic system.

VI. Summary

The power and utility of generalized kinetic theory is obvious, whether the theory is used systematically or as a framework for modeling. (See Figures 1, 2 and 3). In principle, it describes phenomena occurring in intervals of time and space as large or as small as one desires. In practice, it has been most successful in correlating scattering data, but less so in "first principles" calculation of, say, transport coefficients. (There, description in terms of an equivalent, "classical" gas of hard spheres, augmented by data from computer experiments, still satisfies the practical man.) We are in a time of rapid growth in kinetic theory. Although no formalism can mask the difficulties inherent in 3-, 4-, N-body dynamics, these new tools may help us answer many questions, while minimizing headaches. Although Ludwig Boltzmann is reported to have said, "Elegance ... is for tailors," I think he would be pleased.

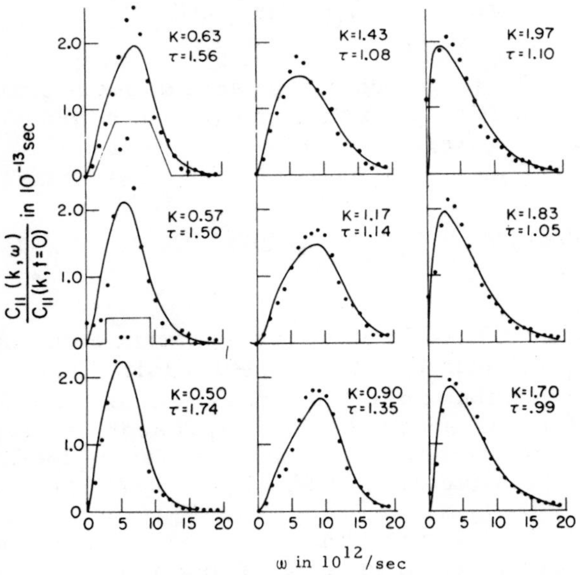

Fig. 1 A simple, modeled memory function is used to compute the frequency spectrum of the time correlation function for a fluctuation in density of wave number K in liquid argon. The dots come from computer dynamics. The solid line is "theory." The τ's are fitted K-dependent relaxation times.

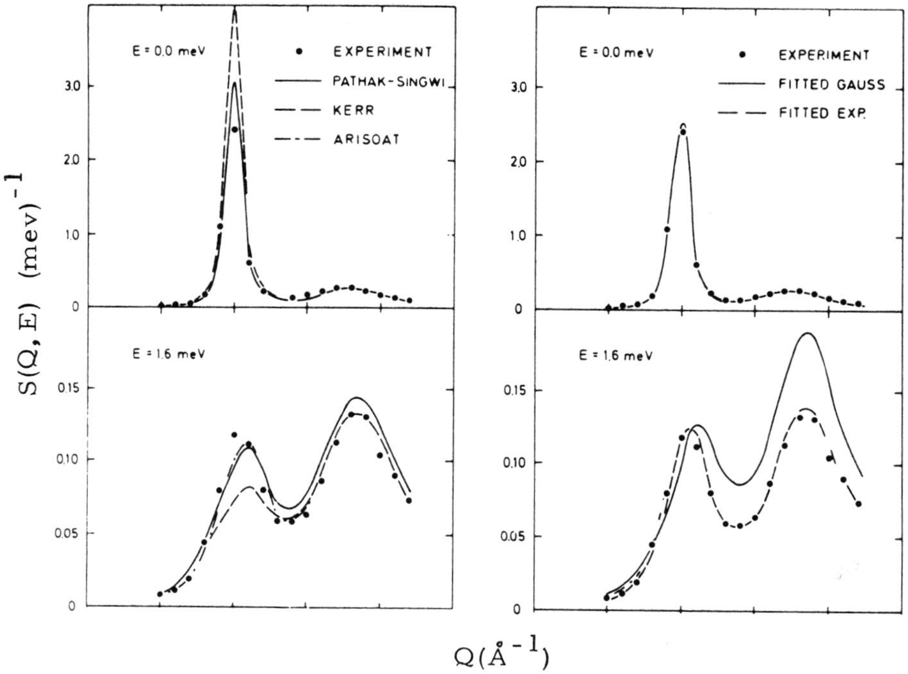

Fig. 2 An example of the efficacy of simple, modeled memory functions in correlating neutron scattering data. The substance is liquid argon (82.5°K). The memory functions are guassian or exponential. E is the energy transfer and $\hbar Q$ the momentum transfer.

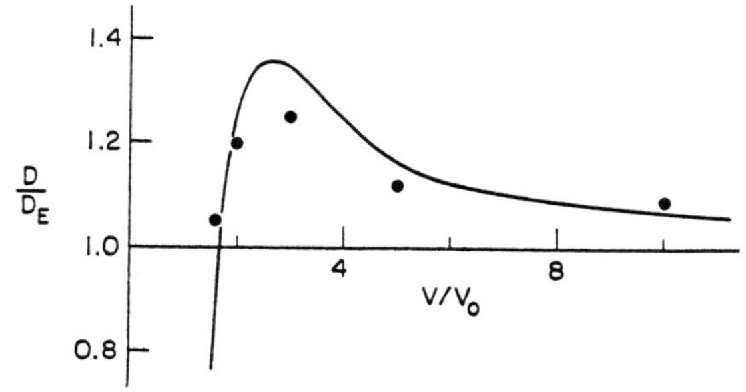

Fig. 3 The coefficient of self-diffusion in a dense gas of hard spheres is displayed, in comparison with the classical Enskog value D_E in two situations. The solid curve is the result of computer dynamics. The points are a semianalytical calculation based upon Mazenko's memory function. The abscissa is inversely proportional to density.

Annotated Bibliography

I. Introduction

An excellent general reference for our purposes is

Ferziger, J. H. and Kaper, H. G., Mathematical Theory of Transport Processes in Gases, North-Holland, Amsterdam, 1972.

II. Quasiclassical Kinetic Theory

The point of view expressed here is derived from

McLennan, J. A., "Correlation Functions for Dilute Systems," The Physics of Fluids, Vol. 9, 1966, p. 1581, and

Forster, D. and Martin, P. C., "Kinetic Theory of a Weakly Coupled Fluid," Physical Review, Vol. A2, 1970, p. 1575.

For a discussion of the Green-Kubo formulas see

Zwanzig, R., "Time Correlation Functions and Transport Coefficients in Statistical Mechanics," Annual Review of Physical Chemistry, edited by H. Eyring, Vol. 16, Annual Reviews Inc., Palo Alto, Calif., 1965.

III. Significant and Partially Successful Approaches

For neutron scattering,

Parks, D. E., Nelkin, M. S., Beyster, J. R., and Wikner, N. F., Slow Neutron Scattering and Thermalization, W. A. Benjamin, New York, 1970, and

Bogoljubov, N. N., "Dynamical Problems in Statistical Mechanics," transl. in Studies in Statistical Mechanics, edited by J. de Boer and G. E. Uhlenbeck, Vol. I, North-Holland, Amsterdam, 1962.

The present state of the dynamical cluster theory is described in articles by E. G. D. Cohen and J. V. Sengers in The Boltzmann Equation, edited by E. G. D. Cohen and W. Thirring, Springer, New York, 1973.

For transport coefficients and the binary collision expansion see

Zwanzig, R., "Density Expansion of Transport Coefficients of Gases," Physical Review, Vol. 129, 1963, p. 486,

Kawasaki, K. and Oppenheim, I., "Correlation Function Method for Transport Coefficients of Dense Gases," Physical Review, Vol. 136, 1964, p. A1519, and

Dorfman, J.R., "Binary Collision Expansion Method in Kinetic Theory," Lectures in Theoretical Physics, edited by W.E. Brittin, Vol. IX C, Gordon and Breach, New York, 1967.

For noise and fluctuations

Fox, R.F. and Uhlenbeck, G.E., "Contributions to Non-Equilibrium Thermodynamics," The Physics of Fluids, Vol. 13, 1970, p. 1893; also Vol. 13, 1970, p. 2881, and

Bixon, M. and Zwanzig, R., "Boltzmann-Langevin Equation and Hydrodynamic Fluctuations," Physical Review, Vol. 187, 1969, p. 267.

IV. Modern Theories

For general theoretical background

Koopman, B.O., "Hamiltonian Systems and Transformations in Hilbert Space," Proceedings of National Academy of Sciences, U.S., Vol. 17, 1931, p. 315,

Zwanzig, R., "Statistical Mechanics of Irreversibility," Lecturers in Theoretical Physics, edited by W.E. Brittin, Vol. III, Interscience, New York, 1961,

Mori, H., "Transport, Collective Motion and Brownian Motion," Progress - Theoretical Physics, Vol. 33, 1965, p. 423, and

Akcasu, A.Z. and Duderstadt, J.J., "Derivation of Kinetic Equations from the Generalized Langevin Equation," Physical Review, Vol. 188, 1969, p. 479.

The kinetic equation for a test particle is noted in

Akcasu, A.Z., Corngold, N., and Duderstadt, J.J., "Self-Diffusion in Classical Fluids," The Physics of Fluids, Vol. 13, 1970, p. 2213,

Kuščer, I. and Corngold, N., "Kinetic Theory of Diffusion in Liquids," Physica, Vol. 82A, 1976, p. 195, and

Corngold, N. and Gomberg, R., unpublished research, California Institute of Technology, 1975.

Forster's paper on properties of the memory function is

Forster, D., "Properties of The Kinetic Memory Function in Classical Fluids," Physical Review, Vol. A9, 1974, p. 943.

V. Approximations

The Zwanzig-Gross-Boley development will be found in

Zwanzig, R., "Approximate Eigenfunctions of the Liouville Operator in Classical Many-Body Systems," Physical Review, Vol. 144, 1966, p. 170,

Gross, E. P., "Approximate Solutions of the Liouville Equation," Annals of Physics, Vol. 69, 1972, p. 42; also Vols. 9, 10, 11 of Journal of Statistical Physics, and

Boley, C. D., "The Approach to Kinetic Theory due to Gross," Annals of Physics, Vol. 86, 1974, p. 91.

Some of Mazenko's papers, in which his theory is developed are

Mazenko, G. F., "A Microscopic Method for Calculating Memory Functions in Transport Theory," Physical Review, Vol. A3, 1971, p. 2121; "Properties of the Low Density Memory Function," Vol. A5, 1972, p. 2545; "Fully Renormalized Kinetic Theory. I. Self-Diffusion," Vol. A7, 1973, p. 209; "Fully Renormalized Kinetic Theory. II. Velocity Autocorrelation," Vol. A7, 1973, p. 222; "Fully Renormalized Kinetic Theory. III. Density Fluctuations," Vol. A9, 1974, p. 360.

Some consequences of the theory are discussed in

Mazenko, G. F., Wei, T. Y. C., and Yip, S., "Thermal Fluctuations in a Hard Sphere Gas," Physical Review, Vol. A6, 1972, p. 1981, and

Furtado, P. M., Mazenko, G. F., and Yip, S., "Hard Sphere Kinetic Theory Analysis of Classical, Simple Liquids," Physical Review, Vol. A12, 1975, p. 1653; also

"Effects of Correlated Collisions on Atomic Diffusion in a Hard Sphere Fluid," (in press 1976).

Higher terms in the density expansion are considered by Boley, C. D. and Desai, R. C., "Kinetic Theory of a Dense Gas: Triple-Collision Memory Function," Physical Review, Vol. A7, 1973, p. 1700; "Kinetic Theory of a Dense Gas: Properties of Collision Kernels of the Bogoljubov Type," Vol. A7, 1973, p. 2192.

The present state of theory vs experiment is described for the case of neutron scattering by

Copley, J. R. D. and Lovesey, S. W., "Dynamic Properties of Monatomic Liquids," Reports of Progress in Physics, Vol. 38, 1975, p. 461.

and, for transport coefficients by

Sengers, J. V., "Thermal Conductivity and Viscosity of Simple Fluids," International Journal of Heat and Mass Transfer, Vol. 8, 1965, p. 1103.

Sengers, J. V., "Transport Properties of Compressed Gases," Recent Advances in Engineering Sciences, edited by A. C. Eringen, Vol. III, Gordon and Breach, New York, 1968.

Sengers, J. V., "Transport Properties of Gases and Binary Liquids near the Critical State," Transport Phenomena - 1973, edited by J. Kestin, American Institute of Physics, New York, 1973.

VI. Figure Captions

Figure 1 is taken from Ailawadi, N. K., Rahman, A., and Zwanzig, R., "Generalized Hydrodynamics and Analysis of Current Correlation Functions," Physical Review, Vol. A4, 1971, p. 1616.

Figure 2 is taken from Sköld, K., Rowe, J. M., Ostrowski, G., and Randolph, P. D., "Coherent and Incoherent-Scattering Laws of Liquid Argon," Physical Review, Vol. A6, 1972, p. 1107.

Figure 3 is taken from Furtado, P. M., Mazenko, G. F., and Yip, S., Physical Review, in press 1976.

VII. More

For the Chapman-Enskog treatment of Bogoljubov's master equation see

García-Colín, L. S., Green, M. S., and Chaos, F., "The Chapman-Enskog Solution of the Generalized Boltzmann Equation," Physica, Vol. 32, 1966, p. 450.

On the issue of the uniformity of a Fokker-Planck equation obtained by expansion see

Chang, E. L., Mazo, R. M., and Hynes, J. T., "The Fokker-Planck Equation for the Linear Chain," Molecular Physics, Vol. 28, 1974, p. 997.

THE SLOWING DOWN OF FAST ATOMS IN A UNIFORM GAS

M. M. R. Williams*

University of London, London, England.

Abstract

The nonlinear Boltzmann equation is modified to describe the slowing down of fast particles, i.e., those with energies very much greater than thermal values. Two situations are discussed: 1) the foreign gas, in which fast alien particles are followed after their injection into the host; and 2) the single gas, in which a fast disturbance in the host is examined as the particles in it share their energies by generating recoils. Using a scattering cross section derived from a modified inverse power law of interaction, we obtain exact solutions by Mellin transforms of the time-energy evolution of the particles. The mean energy dissipation rate, the average number of recoils generated, and the slowing-down times are studied for a variety of scattering models. It is concluded that, in general, the slowing-down time of particles in the single gas is longer than those of the foreign gas, when scaled in units of the mean free time at source energy. We also find that, for a given total cross section energy variation, as the scattering becomes more anisotropic in the centre-of-mass system, the particles of the foreign gas take longer to slow down, whereas those in the single gas take a

---

Presented as paper 17 at the 10th International Symposium on Rarefied Gas Dynamics, July 19-23, 1976, Aspen, Colo. Financial support enabling the author to attend the Symposium was received from the Royal Society, The Nuclear Engineering Department of Queen Mary College and the Symposium Organisors. The author is extremely grateful to these organisations in a time of such economic difficulty.

*Professor, Nuclear Engineering Department, Queen Mary College.

shorter time. An explanation of this anomalous behaviour is advanced.

Introduction

Extensive work has been performed on the relaxation of a uniformly perturbed gas to equilibrium.[1,2] From a study of the Boltzmann equation, we now understand the nature of the associated eigenvalues and eigenfunctions and can follow in some detail the final stages of equilibrium for a variety of intermolecular force laws. The emphasis in this area has, however, been on states close to equilibrium: little or no work has been presented for the case when the perturbation is caused by atoms with initial energies very much greater than kT. Such a "bunch" of fast atoms, although in principle still governed by the same equations, will not be amenable to the eigenfunction expansion treatment. The reason for this is because the perturbed atoms will, for a major part of their relaxation history, be continually slowing down and therefore be virtually unaffected by the thermal motion of the unperturbed atoms. Moreover, the slowing-down mechanism will cause not only the original fast atoms to have their energies reduced in a collision, but will also generate, by virtue of this collision, a cascade of additional, fast-recoil atoms from the stationary collision partners. It is the rate and manner with which these cascades slow down that forms the basis of the present paper.

A general slowing-down equation is used which can be deduced as a limiting form of the multispecies, linearized Boltzmann equation.[3] This equation is then reduced to two special cases: the single-gas and foreign-gas problems. The latter case is, except for a different scattering law, similar to the case of neutrons slowing down in a moderating medium about which much is known.[4] The former case has not been studied in this context before, but it does have important connections with radiation damage theory and sputtering[5] and also with the chemical kinetics of energetic atoms.[6] Its investigation is therefore of more than academic interest.

The basic equations are solved analytically via Mellin transforms using model scattering laws deduced on the basis of an approximation to a general inverse power law of interaction.[5] For certain special cases, e.g., hard spheres, explicit solutions of the velocity-time distribution are obtained. For more general scattering laws, we may calculate the time moments and hence assess the sensitivity of the average slowing-down time to the scattering model. The foreign-gas and single-gas problems are compared, and, as in the case of thermal relaxation,[7] it is found that for hard spheres the atoms of a single gas take longer to slow down to a given velocity. The importance of the slowing-down density is stressed, and its use in calculating slowing-down times is illustrated.

Basic Equation of Transport

The Boltzmann equation describing the particle distribution function $f_i(\underline{v},t)$ for atoms of type i in a mixture can be written as

$$\partial f_i(\underline{v},t)/\partial t = (\partial f_i/\partial t)_{coll} \tag{1}$$

where the collision term on the right-hand side takes the form

$$(\partial f_i/\partial t)_{coll} = \sum_j \int d\underline{v}_1 \int d\underline{v}_1' \int d\underline{v}' \Big[W_{ij}(\underline{v}_1'\to\underline{v}_1;\underline{v}'\to\underline{v}) f_i(\underline{v}',t) f_i(\underline{v}_1',t) - W_{ij}(\underline{v}_1\to\underline{v}_1';\underline{v}\to\underline{v}') f_i(\underline{v},t) f_i(\underline{v}_1,t) \Big] \tag{2}$$

$W_{ij}(\cdots)$ being the transition probability for scattering of particles of type i with those of type j with velocities $(\underline{v}_1',\underline{v}')$ before collision to velocities $(\underline{v}_1,v)$ after collision. Writing

$$f_i(\underline{v},t) = n_i f_{Mi}(\underline{v}) + G_i(\underline{v},t) \tag{3}$$

where n_i is the number of atoms in the equilibrium Maxwell-Boltzmann distribution $f_{Mi}(\underline{v})$, and $G_i(\underline{v},t)$ is the number of atoms disturbed from equilibrium, we may linearize Eq.(2) by neglecting terms involving products G_iG_j. Physically this amounts to assuming that perturbed atoms do not collide with each other but only with the atoms of the unperturbed component. A detailed account of the mathematical procedure involved in this derivation is given in Ref.3. In the limiting case when either the energy of the particles in $G_i(\underline{v},t)$ is very much greater than kT or when $kT\to 0$, we can obtain an equation describing the slowing down of the fast bunch of particles.

Changing to the energy variable and integrating over the angular coordinate, we obtain an equation for the flux of particles ϕ_i defined by

$$\phi_i(E,t) = v\, N_i(E,t) = v^3\, G_i(v,t)$$

The equation can be written as

$$1/v\ \partial/\partial t\ \phi_i(E,t) = \sum_j 4\pi n_j/(1-\alpha_{ij}) \int_E^{E/\alpha_{ij}} dE'/E'\ \phi_i(E',t)$$
$$\sigma_{ij}\left[E',\ \theta_c(E/E')\right] + \sum_j 4\pi n_i/(1-\alpha_{ij}) \int_{E/(1-\alpha_{ij})}^{\infty} dE_1'/E_1' \tag{4}$$
$$\phi_j(E_1',t)\sigma_{ij}\left[E_1',\theta_c(1-E/E_1')\right] - \sum_j 4\pi n_j/(1-\alpha_{ij})\phi_i(E,t)\ 1/E$$
$$\int_{\alpha_{ij}E}^{E} dE'\sigma_{ij}\left[E,\theta_c(E'/E)\right] + S_i(E,t)$$

In this equation, $\sigma_{ij}(E,\theta_c)$ is the scattering cross section in the centre-of-mass system.

$$\alpha_{ij} = \left[(M_i-M_j)/(M_i+M_j)\right]^2$$

where M_i and M_j are the masses of the atoms of species i and j, respectively. Finally, $S_i(E,t)$ is the source of particles i. We note that the second integral on the right-hand side of Eq.(4) is due to recoil atoms, and it is this term which distinguishes the foreign-gas problem from the single-gas one. For example, if we assume that the injected particles are not present in the host gas, then n_i=0, and we have an equation that is identical to that for neutrons slowing down in an infinite medium.[4] In general, however, even foreign particles will cause recoils in the host medium, and the equation for the host atoms is coupled to that for the foreign particles. The situation arises in radiation damage studies when the solid medium can be treated as an amorphous scatterer. In the present work, we shall simplify the analysis by considering only two cases: 1) the foreign gas slowing down in a medium of the same mass number, and 2) a single-species gas.

Scattering Law

It is well known[8] that $\sigma(E,\theta)$ can be obtained once the force law between atoms has been specified. However, for many realistic potentials, it is not possible to obtain explicit expressions for $\sigma(E,\theta)$. For this reason, a number of synthetic modelled kernels have been proposed. A very successful model is that based upon the momentum approximation[5,8] which indicates that, for inverse power law scattering, $\sigma(E,\theta)$ may be approximated by

$$\sigma(E,\theta) = \text{const}/E^{2m}\ 1/\sin^{2m+2}(\theta/2) \tag{5}$$

where $m = 1/g$, g being the power of the potential. Equation (5) is accurate in the range $1 \leq g < 4$. Thus it embraces pure Coulomb scattering but not hard sphere scattering. Our proposal is to write $\sigma(E,\theta)$ in the general form

$$\sigma(E,\theta) = \left[\sigma_t(E)/4\pi \right] f(\theta) \tag{6}$$

where $\sigma_t(E)$ is the transport cross section referred to the centre-of-mass system, i.e.

$$\sigma_t(E) = 2\pi \int_0^\pi d\theta \sin\theta\ (1-\cos\theta)\sigma(E,\theta) \tag{7}$$

and $f(\theta)$ is normalized so that

$$1/2 \int_0^\pi d\theta \sin\theta(1-\cos\theta)f(\theta) = 1 \tag{8}$$

We shall write

$$\sigma_t(E) = \sigma_t(E_0)\ (E_0/E)^{k/2} \tag{9}$$

where E_0 is some reference energy and

$$f(\theta) = \left[(2-S)/2\right] \sin^{-2S}(\theta/2) \tag{10}$$

We choose k and S to suit the physical situation. Clearly k=S=0 corresponds to hard-sphere scattering, although it would be possible to set S=0 and vary k independently should this prove to be useful in practical situations.

Solution of the Equations of Transport

Defining $\Sigma_t(E)=n\sigma_t(E)$ and $\Phi(E,t)=\Sigma_t(E)\phi(E,t)$, we can write Eq.(4), with the scattering model of the previous section as follows

$$\sqrt{E_0/E}\ 1/v_0 \Sigma_t(E)\ \partial/\partial t\ \Phi(E,t) = \int_E^{E_0} dE'/E'\ \Phi(E',t)\ f(E/E')$$
$$+ \lambda \int_E^{E_0} dE'/E' \Phi(E',t) f(1-E/E') - \Phi(E,t)\ 1/E \int_0^E dE' f(E'/E)$$
$$+ \delta(E-E_0)\delta(t) \tag{11}$$

where we have used a single species and have represented the initial conditions by a pulse of monoenergetic particles of energy E_0. The term $f\left[\theta_c(E/E')\right]$ has been abbreviated to $f(E/E')$. Also note the presence of a parameter λ: when $\lambda=1$, we have the single-species problem where recoils are counted; on the other hand, when $\lambda=0$, we have the foreign-gas problem.

An interesting observation is that, when $\lambda=1$, energy is conserved but particle number is not owing to cascade multiplication. When $\lambda=0$, particle number is conserved but energy is not, owing to the neglect of the recoil energy. Such considerations are of crucial importance in radiation damage studies where the number of displacements per incident particle has to be calculated.[5]

Equation (11) can be solved via the Mellin transform. Thus we define

$$\overline{\Phi}(p,t) = \int_0^\infty dE\ E^{p-1}\ \Phi(E,t) \tag{12}$$

and find from Eq.(11) that

$$\tau_0 E_0^{\ x}(\partial/\partial t)\overline{\Phi}(p-x,t) = -\ D(p)\Phi(p,t) + E_0^{\ p-1}\delta(t) \tag{13}$$

where $\tau_0^{\ -1} = v_0\Sigma_t(E_0)$, $x = (1-k)/2$, and

$$D(p) = \int_0^1 dy\ f(y)\ \left[1-y^{p-1} - \lambda(1-y)^{p-1}\right] \tag{14}$$

Equation (13) is a differential-difference equation, which may be solved by a generalization of a method due to Waller.[9] We write

$$\chi(p,t) = E_0^{\ -p}\ \overline{\Phi}(p,t)$$

when (13) takes the form

$$\tau_0(\partial/\partial t)\chi(p,t) = -\ D(p+x)\chi(p+x,t) + (1/E_0)\delta(t) \tag{15}$$

Now define two new quantities $P(p)$ and $\Psi(p,t)$ as follows

$$\chi(p,t) = P(p)\Psi(p,t) \tag{16}$$

where $P(p)$ satisfies the difference relation

$$D(p+x)\ P(p+x) = P(p) \tag{17}$$

Inserting (16) into (15) and using (17) leads to

$$\tau_0(\partial/\partial t)\Psi(p,t) = -\ \Psi(p+x,t) + \left[\delta(t)/E_0P(p)\right] \tag{18}$$

Defining the Mellin transform

$$\Psi(p,t) = \int_0^1 dw\ w^{p-1}\ \Psi_F(w,t) \tag{19}$$

and its inverse

$$\Psi_F(w,t) = (1/\partial\pi i) \int_L dp\ w^{-p}\ \Psi(p,t) \tag{20}$$

we see that Eq.(18) can be written as

$$\tau_o(\partial/\partial t)\Psi_F(w,t) = -w^x\Psi_F(w,t) + (1/2\pi i)\int_L dp\left[w-p/P(p)\right]\left[\delta(t)/E_o\right] \tag{21}$$

Solving (21) as a first-order differential equation, we find

$$\Psi(p,t)=(1/E_o\tau_o)\int_o^1 dw\ w^{p-1}\exp(-w^x t/\tau_o)1/2\pi i \int_{L'} dp' w^{-p'}/P(p') \tag{22}$$

and hence, from (16) and (12)

$$\Phi(E,t) = 1/\tau_o E_o\ 1/2\pi i \int_L dp\ (E_o/E)^p\ P(p)$$
$$\int_o^1 dw\ w^{p-1}\exp(-w^x t/\tau_o)\ 1/2\pi i \int_{L'} dp' w-p'/P(p') \tag{23}$$

Thus we have a complete solution subject to a knowledge of P(p). It is not difficult to show that

$$P(p+nx)/P(p-x) = \prod_{m=0}^{n} 1/D(p+mx) \tag{24}$$

which will be shown to be a valuable relation for obtaining quantities of physical interest.

Slowing-Down Density

The number of atoms slowing down past energy E per unit time per unit volume is called the slowing-down density. A complete derivation of this quantity for an arbitrary mixture has been given by Williams.[11] For Eq.(11), the corresponding slowing-down density is

$$q(E,t) = \int_E^{E_o} dE'\Phi(E',t)\left[\int_o^{E/E'} dxf(x)+\lambda \int_{E/E'}^1 dxf(1-x)\right] \tag{25}$$

which, after Mellin transformation, becomes

$$\overline{q}(p,t) = (1/p)\ D(p+1)\overline{\Phi}(p+1,t) \tag{26}$$

We may therefore write the inverse as

$$q(E,t) = E/\tau_o E_o \; 1/2\pi i \int_L dp (E_o/E)^p \; D(p)/p-1 \; P(p)$$
$$\int_o^1 dw \; w^{p-1} \exp(-w^x t/\tau_o) \; 1/2\pi i \int_{L'} dp' \; w^{-p'}/P(p') \tag{27}$$

We shall make use of this expression to characterize the time moments in a later section.

Exact Solutions

Despite the formidable appearance of Eqs.(23) and (27), Waller[9] has shown that it is possible to cast them into a form convenient for computation. We shall not work through the details here which are described in Waller's paper and for a more general situation in Ref.11. The result is that we may write

$$\Phi(E,t) = e^{-z}/E_o\tau(E) \sum_{n=o}^{\infty} \gamma_n(E) \; L_n(z) \tag{28}$$

$$q(E,t) = E/E_o \; e^{-z}/\tau(E) \sum_{n=o}^{\infty} \beta_n(E) \; L_n(z) \tag{29}$$

where $L_n(z)$ are the Laguerre polynomials[10] and $z = t/\tau(E)$. $\gamma_n(E)$ and $\beta_n(E)$ are defined by

$$\gamma_n(E) = \sum_{v=o}^{n} (1)^v \binom{n}{v} \; 1/2\pi i \int dp \; (E_o/E)^p \prod_{\ell=o}^{v} 1/D(p+\ell x) \tag{30}$$

$$\beta_n(E) = \sum_{v=1}^{n} (-)^v \binom{n}{v} \; 1/2 \; i \int_L dp (E_o/E)^p \; 1/p+vx-1 \prod_{\ell=o}^{v-1} 1/D(p+\ell x) \tag{31}$$

Asymptotic estimates of $\gamma_n(E)$ and $\beta_n(E)$ are readily obtained by examining the simple zeros of $D(p)$. There are two situations for which explicit and simple expressions can be obtained for $\phi(E,t)$, and we shall consider them below.

A. Cross Section Inversely Proportional to Velocity

In this case, x=o, and it is readily seen that the differential-difference equation (13) can be solved directly to give

$$\Phi(E,t) = 1/E_o\tau_o \; 1/2\pi i \int_L dp (E_o/E)^p \exp\{-t/\tau_o \; D(p)\} \tag{32}$$

For a general scattering law, the method of steepest descents can be employed to obtain solutions for $E \ll E_o$. However, for the hard-sphere model, when $f(x)=1$, it readily is shown that

$$D_o(p) = (p-1)/p \qquad D_1(p) = (p-2)/p$$

Noting the asymptotic behaviour of $D(p)$ as $|p| \to \infty$, we can obtain, using $D_o(p)$ and defining $u = \ln(E_o/E)$, $\phi(E,t) = \delta(v-v_o)e^{-t/\tau_o}$

$$+(v/E_o)\ (t/\tau_o u)^{\frac{1}{2}}\ I_1\left[2\ (ut/\tau_o)^{\frac{1}{2}}\right]e^{-t/\tau_o} \qquad (33)$$

Using $D_1(p)$, the result is identical in form, except that τ_o in the square root is replaced by $\tau_o/2$. That is, the slowing down is less rapid for the single-gas problem.

B. Cross Section Independent of Energy

In this special case, $x = (1/2)$. If we further assume hard-sphere scattering, it is easy to show that, in Eq.(28) for $\lambda = o$, $\gamma_n = 0$ for $n>2$. The solution then reduces to

$$\phi(E,t) = \delta(v-v_o)e^{-v_o\Sigma t}+(v\Sigma t/v_o)e^{-v\Sigma t}\{2+v_o\Sigma t\left[1-(v/v_o)\right]\} \qquad (34)$$

This solution is identical with that obtained for neutrons slowing down in a proton moderator.[4]

In the case of $\lambda = 1$, i.e., when recoil is present, the $\gamma_n = 0$ for $n>4$ and the resultant particle flux is

$$\phi(E,t) = \delta(v-v_o)e^{-v_o\Sigma t} + (v\Sigma t/6v_o)e^{-v\Sigma t}$$

$$\{24+36v_o\Sigma t\left[1-v/v_o\right] +12\ v_o^2\Sigma^2t^2\left[1-(v/v_o)\right]^2+v_o^3\Sigma^3t^3\left[1-(v/v_o)\right]^3\} \qquad (35)$$

It should be noted that, in terms of the velocity variable, the particle density $N(v,t)$ is equal to $\phi(E,t)$.

Energy Dissipation Rate and Particle Multiplication

Using the solutions derived in the two previous subsections, we can illustrate two important phenomena: 1) the energy dissipation, $W(t)$ of foreign particles, and 2) the multiplication N(t) of cascade particles. For the foreign gas the energy loss rate is, by definition

$$W(t) = \tfrac{1}{2} \int_0^{Eo} dE\, E \int_0^t dt' \Phi(E,t') \tag{36}$$

Inserting Eq.(33), we find

$$W(t) = Eo\left[1 - \exp(-t/\tau_0)\right] \tag{37}$$

where $\tau_0^{-1} = v_0 \Sigma(v_0)$. In the case of hard spheres, we find

$$W(t) = E_0\{1+12\tau_0^3/t^3 - 6\tau_0^2/t^3-(12\tau_0^3/t^3+6\tau_0^2/t^2)e^{-t/\tau_0}\} \tag{38}$$

where $\tau_0^{-1} = v_0\Sigma$. We note that the loss of energy is much less rapid for constant cross section, a physically acceptable situation in view of the large increase in collision rate for the 1/v scatterer as v decreases. It is readily shown that the total number of particles is conserved during the slowing-down process.

For the single-gas problem where recoil particles are counted, we can calculate the increase in particles per primary as t increases, viz.

$$N(t) = \int_0^{Eo} dE\ 1/v\ \phi(E,t) \tag{39}$$

Using Eq.(33) with τ_0 replaced by $\tau_0/2$, we obtain

$$N(t) = \exp\{t/\tau_0\} \tag{40}$$

i.e., an exponential increase. On the other hand, for constant cross section, using Eq.(35), we find

$$N(t) = 1 + (t/\tau_0) + (1/6)\ (t/\tau_0)^2 \tag{41}$$

a much slower increase consistent with the result of Eq.(38). We note that the value of W(t) for this case remains constant for all times and equal to Eo. This is because all particles are participating and not just the injected ones.

Numerical Results and Time Moments

It is difficult to obtain concise closed-form expressions for $\Phi(E,t)$ for more realistic scattering models than hard sphere. Thus we study an integral parameter, namely, the slowing-down time, which will enable us to assess the influence in a general way of different intermolecular force laws.

Firstly, however, it is necessary to consider how to define the slowing-down time. This is frequently taken to be

$$<t> = \int_0^\infty dt\ t\phi(E,t)/\int_0^\infty dt\ \phi(E,t) \qquad (42)$$

However, further consideration of the physical processes involved indicates that a better definition would be

$$<t> = \int_0^\infty dt\ tq(E,t)/\int_0^\infty dt\ q(E,t) \qquad (43)$$

The slowing-down density is used because it denotes the number of particles crossing a given energy per unit time. On the other hand, the particle density only states how many are present in a given energy interval. The distinction between these two definitions is fairly subtle, but there is little doubt that the slowing-down density average is more convenient mathematically and is consistent with the source condition as we shall show later.

Multiplying Eq.(27) by t^n and integrating over t, i.e.

$$q_n(E) = \int_0^\infty dt\ t^n\ q(E,t)$$

leads to

$$q_n(E) = n!\tau(E)^n/2\pi i \int_L dp(Eo/E)^{p-1}\ 1/(p+nx-1) \prod_{m=o}^{n-1} 1/D(p+mx) \qquad (44)$$

for n>0. It is easily seen that $q_0(E) = 1$, as we would expect from particle balance. The slowing-down time is therefore given by

$$<t> = q_1(E) = \tau(E)/2\pi i \int_L dp\ (E_o/E)^{p-1}\ 1/(p+x-1)\ D(p) \qquad (45)$$

An asymptotic estimate of this integral leads to

$$<t>_1 \sim \tau_0\{1/D_1(1-x) + e^{(1+x)u}/(1+x)D_1'(2)\} \qquad (46)$$

for the single-gas problem [suffix 1 on D(p), λ=1]. For the foreign-gas problem [suffix 0 on D(p), λ=o], we obtain

$$<t>_o \sim \tau_0\{ [1/D_0(1-x)] + [e^{xu}/x\ Do\ (1)]\} \qquad (47)$$

A special case arises in Eq.(47) when x=o; then, taking the limit carefully

$$<t>_o \sim [\tau_0/D_0'(1)]\ \{u-(1/2)\ [D_0''(1)/D_0'\ (1)] \qquad (48)$$

We shall consider some detailed numerical aspects below, but it is interesting to note that, for $x<o$, $<t>_o$ tends to a finite value as $E \to o$, whereas, for $x \geq o$, $<t>_1$ becomes infinite. These results are more a consequence of the mathematics than the physics, since in practice when E falls below several kT the slowing-down kernel is modified by temperature, and the method just described no longer applies.

Two particular cases are worthy of detailed study: hard spheres and Maxwell molecules. Let us consider hard spheres first but generalize to include arbitrary cross section energy dependence. Then the preceding asymptotic results are exact, and we find

$$<t>_1/\tau_o = (1/x)\ (E_o/E) - [(1-x)/x] \tag{49}$$

$$<t>_1/\tau_o = [2/(1+x)]\ (E_o/E)^{1+x} - [(1-x)/1+x)] \tag{50}$$

Note that, when $E=E_o$, both $\langle t\rangle_o$and $\langle t\rangle_o$ are equal to τ_o. This is expected physically, since it is the average time taken for a particle, with source energy, before its first collision.

We consider now a general inverse power scattering law as described by Eq.(10). The corresponding value of f(x) is

$$f(x) = (2-S)/[2(1-x)^S\ ] \tag{51}$$

which, after insertion into Eq.(14), leads to

$$D(p) = (2-S)/2\ [1/(1-S) - \lambda/(p-S) - \Gamma(p)\Gamma(2-S)/(1-S)\Gamma(p+1-S)] \tag{52}$$

where $\Gamma(p)$ is the gamma function. In the special case of S=1

$$D(p) = (1/2)\ \{\psi(p) - \psi(1)\ - [\lambda/(p-1)]\ \} \tag{53}$$

where $\psi(p)$ is the digamma function.

We shall evaluate the slowing-down time <t> as given by Eqs.(46) and (47) for the law given in Eq.(5) and for the parameters used by Lindhart et al. as summarized by Sigmund.[5] Using a Fermi-Thomas model, the power law parameters have been fitted to yield

$$\Sigma_t(E_o) = [\lambda_m/(1-m)](1/R_L)\ (E_L/E_o)^{2m} \tag{54}$$

where λ_m is a tabulated parameter, $R_L = (\pi n\ a^2)^{-1}$, $E_L = (2Z_1Z_2\ell^2/a)$, $\underline{a}$ is the Thomas-Fermi screening length, n the particle number density, and Z_1 and Z_2 the atomic numbers of the incident and struck particles. In our notation, $x = (1/2) - 2m$ and $S = 1+m$. Strictly speaking, $0.25 \leq m < 1$, but we shall consider values of m down to zero. Note that $m = 0.25 (x=o)$ corresponds closely to Maxwell molecules, and, therefore, using Eqs.(46) and (47), we see that $\langle t \rangle_o$ and $\langle t \rangle_1$ are given by

$$\langle t \rangle_o / \tau_o = 1.311 \, (u + 1.415) \tag{55}$$

$$\langle t \rangle_1 / \tau_o = 0.984 \, \left[e^u - (2/3) \right] \tag{56}$$

In Fig.1, we show the hard-sphere results for $\langle t \rangle_o / \tau_o$ $\langle t \rangle_1 / \tau_o$ for a variety of values of x. x=1/2 corresponds to

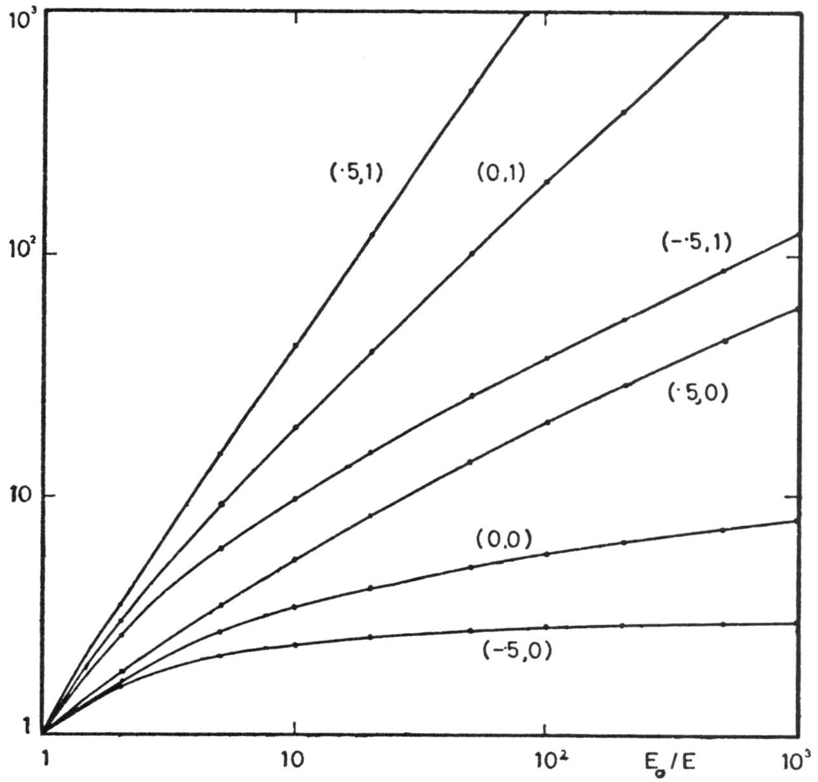

Fig. 1 The slowing-down time $\langle t \rangle / \tau_o$ as a function of energy for hard-sphere scattering. Also shown is the effect of the energy dependence of the total cross section. The designation (α,β) indicates the value of x, denoted by α, and the foreign-gas (β=o) or single-gas (β=1) cases.

constant scattering cross section, x=o to a cross section inversely proportional to v, and x = - 1/2 inversely proportional to v^2. It is clear that in all cases $\langle t\rangle_1 > \langle t\rangle_o$, which is consistent with the mode of relaxation to thermal equilibrium.[7] We also note that, as the cross section becomes a stronger function of v, the slowing-down time reaches its final value more rapidly. This is to be expected in view of the greater collision rate as the particle slows down. In Fig.2, we compare the values of $\langle t\rangle/\tau_o$ for hard spheres (x=1/2) and Maxwell molecules from Eqs.(50) and (51). It is noted that in this case $\langle t\rangle_1 > \langle t\rangle_o$ over most of the slowing-down range, although for E near E_0 there is a reversal in this behaviour which is due to neglect of source transients. However, there are two competing effects influencing the rate of energy loss as particles slow down: 1) increase in scattering cross section as the energy decreases, which tends to increase

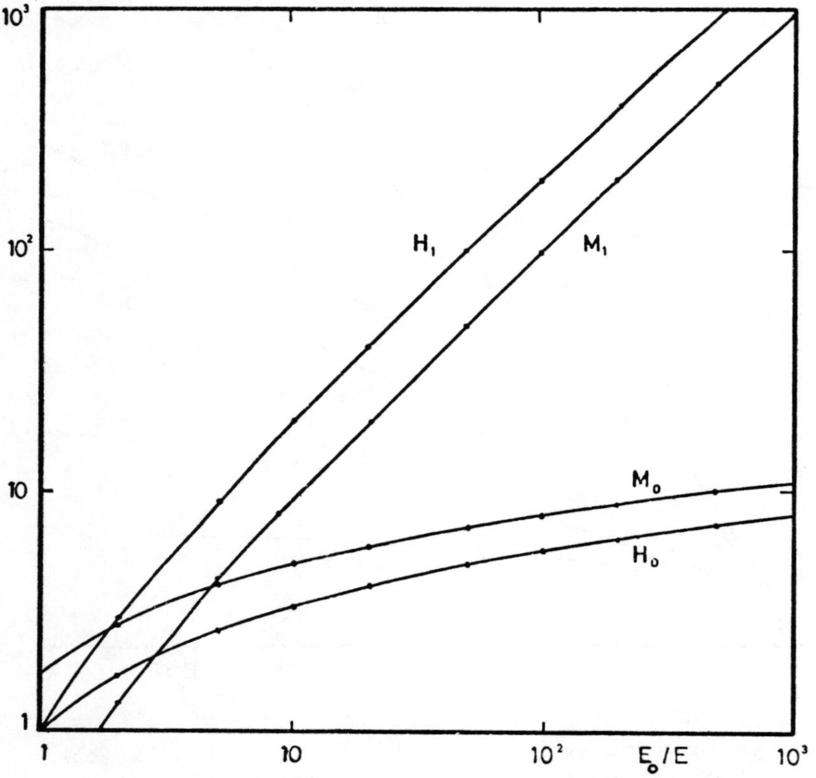

Fig. 2 The slowing-down time $\langle t\rangle/\tau_o$ as a function of energy for hard spheres H_0, H_1(x=0.5, m = -1), compared with Maxwell molecules M_0, M_1 (x=o,m = 0.25).

Table 1 Leading terms in slowing-down time equations for fixed cross section variation x, and increasing anisotropy of scattering m in centre-of-mass system

x	0.5		0	
m	$[x D_0'(1)]^{-1}$	$[(1+x)D_1'(2)]^{-1}$	$D_0'(1)^{-1}$	$D_1'(2)^{-1}$
-1	2	1.333	1	2
-0.5	2.173	1.087	1.087	1.63
-0.25	2.288	0.953	1.144	1.43
0	2.432	0.811	1.216	1.216
0.25	2.624	0.656	1.312	0.984
0.5	2.888	0.481	1.444	0.722

the collision rate and hence the energy loss rate; and 2) increase in forward scattering due to anisotropy in the centre-of-mass system, which leads to smaller energy losses per collision. Clearly, the net result of these effects is a function of energy, and, for certain values of m, their relative importance will change. An interesting comparison is to see how the change in angular distribution affects the slowing-down time for a fixed cross section variation. With this in mind, we take Eqs.(46) and (47) and examine the behaviour of $<t>$ for a fixed x, as the distribution goes from isotropic (m= -1) to highly anisotropic (m = 0.5). In Table 1 we give the leading terms in $<t>_0$ and $<t>_1$ for x = o and x = 0.5 for a range of values of m. We note that, in the case of $<t>_0$, as m increases the leading term increases, which is to be expected physically, since increased anisotropy of scattering implies a smaller energy loss per collision and hence a longer slowing-down time. On the other hand, in the case of $<t>_1$, the leading term decreases as m increases. This somewhat anomolous behaviour can be explained if it is remembered that $<t>_1$ includes recoil particles, and, therefore, as the scattering distribution becomes more biased in the forward direction in the centre-of-mass system, its efficiency for producing recoils of high energy is reduced. Thus there are fewer particles generated at higher energies to slow down.

Finally, we note that all values of $<t>$ are scaled by τ_0, the mean free time for source energy particles. The absolute value of this depends upon m and, for example, can be shown from Eq.(49) to be of order $10^{-11} \sim 10^{-12}$ sec in oxygen at normal temperature and pressure.

References

[1]Cercignani, C., Mathematical Methods in Kinctic Theory,Plenum Press, New York, 1969.

[2]Williams, M. M. R., Mathematical Methods in Particle Transport Theory, Butterworth, London 1971.

[3]Williams, M. M. R., "The Boltzmann Equation for Fast Atoms" Journal of Physics, Part A - Mathematical, Nuclear and General, Vol.9, 1976, pp 771-783.

[4]Williams, M. M. R., The Slowing Down & Thermalization of Neutrons, North-Holland, Amsterdam, 1966.

[5]Sigmund, P., "Collision Theory of Displacement Damage, Ion Ranges and Sputtering," Revue Roumaine de Physique, Vol.17, 1972, pp 823-870, 969-1000, 1079-1106.

[6]Kostin, M. D. and Chapin, D. M., "Collision Density of Hot Atoms," Journal of Chemical Physics,Vol.46, 1967 pp.2506-2510.

[7]Kuscer, I. and Williams, M. M. R., "Relaxation Constants of a Uniform Hard-sphere Gas," Physics of Fluids, Vol.10, 1967, pp. 1922-1927.

[8]Kennard, E. H., The Kinetic Theory of Gases, McGraw Hill, New York 1938.

[9]Waller, I., "An Exact Solution to the Problem of the Time-Energy Distribution of Slowed Down Neutrons," Arkiv fur Fysik Vol.37, 1968, pp 569-580.

[10]Gradshteyn, I. S., and Ryzhik, I. M., Tables of Integrals, Series and Products, Academic Press, New York, 1965.

[11]Williams, M. M. R., "An Exact Solution to the problem of the Time-Energy Distribution of Fast Particles Slowing Down in a Multi-species Medium," Journal of Physics, Part D-Applied Physics Vol.9, 1976, pp 1279-1294.

VISCO-MAGNETIC HEAT FLUX: AN EXPERIMENTAL STUDY IN THE BURNETT REGIME

L.J.F. Hermans*, G.E.J. Eggermont+, P.W. Hermans≠, and J.J.M. Beenakker§

University of Leiden, The Netherlands

Abstract

In a viscous flow of a rarefied gas, the presence of a second-order spatial derivative of the velocity gives rise to a heat flux that is inversely proportional to the pressure. In the case of a polyatomic gas, the application of an external magnetic field gives rise to a transverse component of this heat flux, the "visco-magnetic heat flux". This effect has been investigated experimentally at room temperature for the gases N_2, CO, and CH_4. This heat flux, using a narrow rectangular channel made of a low thermal conductance material, is measured in the stationary situation as a small temperature difference (of the order of 10^{-3} K) between the narrow walls of the channel. Beside the Burnett (bulk) contribution to the effect, which can be calculated from the experimental results of the field effects in the dilute gas regime, boundary-layer phenomena also make an important contribution. A decomposition of the total effect for N_2 into the bulk and boundary-layer contributions is presented.

Introduction

It is well known that experimental verification of the transport theory of rarefied gases is difficult. The main reason is that the typical Burnett features are superimposed as small corrections on the dilute gas behavior. For example,

---

Presented as Paper 72 at the 10th International Symposium on Rarefied Gas Dynamics, Aspen, Colo., July 19-23, 1976.

*Assistant Professor, Huygens Laboratory.
+Ph.D. Candidate, Huygens Laboratory.
≠Graduate Student, Huygens Laboratory.
§Professor, Huygens Laboratory.

the production of a heat flux by second-order spatial derivatives of the velocity in a rarefied gas, as already discussed by Maxwell[1], has been inaccessible to experimental investigation. The reason is that the heat flux occurs in the same direction as the flow velocity of the gas and is overwhelmed by much larger effects such as convection and the expansion of the gas. In a rarefied gas consisting of polyatomic molecules, however, the application of an external magnetic field gives rise to a transverse component of this heat flux perpendicular to the flow direction. This "visco-magnetic heat flux", predicted by Levi et al.[2], is much easier to investigate. It is the purpose of this paper to present experimental results for this effect. These results are closely related to an effect observed by Scott et al.[3], the thermomagnetic torque. In that experiment, a magnetic field was found to cause a torque on a heated cylinder suspended in a rarefied polyatomic gas. This torque can be considered as being caused by the magnetic field influence on the Maxwell stress due to the presence of a second-order spatial derivative of the temperature $\nabla\nabla T$ [4].

In both the visco-magnetic heat flux and the thermo-magnetic torque, boundary-layer effects also play an important role. The boundary-layer contribution to the thermo-magnetic torque could be measured in a separate parallel plate setup, where only a first spatial derivative of the temperature occurs, as the "thermo-magnetic pressure difference"[5].

Theory

If a rarefied polyatomic gas is flowing in the x direction through a rectangular channel, and a magnetic field in the z direction is applied, a heat flux in the y direction will occur. This "visco-magnetic heat flux" can be written phenomenologically as (Fig. 1)

$$q_y = (1/p)\, L_{yxzz}\, (\underline{H})\, (\partial^2 v_x/\partial z^2). \qquad (1)$$

Here q_y is the visco-magnetic heat flux, p is the pressure, v_x is the flow velocity of the gas and L_{yxzz} $(\underline{H})$ is the (field-dependent) coupling coefficient and $\underline{H}$ is the magnetic field taken in the z direction. The bulk contribution to the effect has been treated by Levi et al.[2], using the "third-order" Chapman-Enskog approximation[6]. Like the well-known Burnett coefficients, this contribution is determined completely by quantities known from experiments in the dilute gas regime.

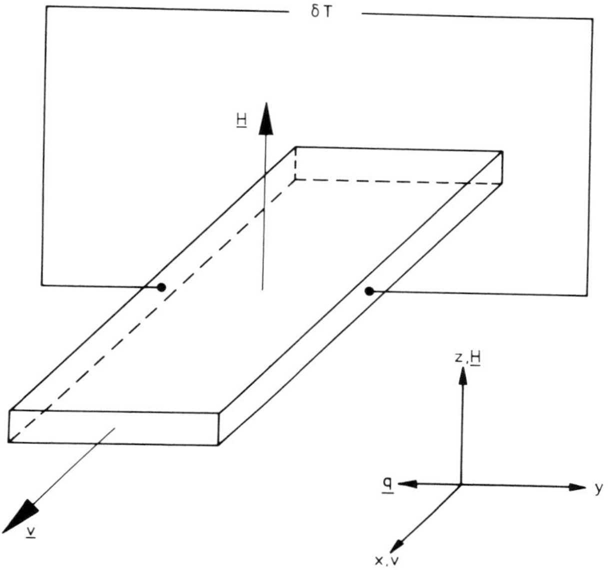

Fig. 1 Schematic diagram of the apparatus and the direction of the visco-magnetic heat flux for N_2 in this configuration.

Beside this bulk effect, boundary-layer contributions play an equally important role in this rarefied gas phenomenon. These have been treated recently by Vestner[7], who distinguished two types of boundary-layer contributions: a) the occurrence of a "surface heat flux", which accounts for the difference between the actual value of the heat flux at the wall and the value of the bulk heat flux extrapolated to the wall (from a thermodynamic point of view this contribution can be linked to the thermal slip) and b) the influence of the boundary layer on the bulk heat flux. This is determined by the behavior at the wall of the polarizations which involve the angular momentum, $\underline{J}$, viz., the tensor polarization of type $\langle \overline{\underline{J}\,\underline{J}} \rangle$ and the Kagan polarization of type $\langle \underline{W}\, \overline{\underline{J}\,\underline{J}} \rangle$. Here, the symbol $\overline{\quad}$ denotes the symmetric traceless part of a tensor, and $\underline{W} = \sqrt{m/2kT}\ \underline{C}$ is the reduced peculiar velocity. The Kagan polarization $\langle \underline{W}\, \overline{\underline{J}\,\underline{J}} \rangle$ can be regarded as the flux of tensor polarization. The influence of the boundary layer on the polarizations is characterized by phenomenological surface parameters in the theory; e.g. the parameters $\tilde{C}_a$ and $\tilde{C}_{am}$ describe the production of tensor polarization at the wall by an incoming flux of tensor polarization (Kagan polarization), and by tangential forces, respectively. Numerical values for these parameters can be found from experiments on the Knudsen effect on the field effect on viscosity[8,9] and from experiments on the thermo-magnetic pressure difference[10,11]. The parameter $\tilde{C}_{ah}$ characterizes the coupling of the surface heat flux with the boundary value of the Kagan polarization. Values for this new parameter can be obtained from the present measurements of the visco-magnetic heat flux.

Experimental Results and Discussion

The experiments were performed using a rectangular channel, schematically drawn in Fig. 1. The length of the channel is 80 mm, the width is 10 mm and the thickness is 1.0 mm. The channel walls were made of low thermal conductance polyester, "mylar" (thickness 75 μm). The heat flux that is generated produces a small temperature difference δT (of the order 10^{-3} K) across the channel. This visco-magnetic heat flux is detected with two thermistors, glued to the narrow walls of the channel. These thermistors are part of a Wheatstone bridge circuit and the field-induced imbalance of the bridge is, in the steady state, proportional to the heat flux. For calibration purposes the heat flux can be simulated using two heaters covering the narrow walls of the channel. They consist of mylar strips (1 mm wide, 80 mm long, and 15 μm thick) covered with a thin layer (5 nm) of aluminum.

Experiments have been performed on the gases N_2, CO, and CH_4, at room temperature. In agreement with theory, the magnitude of the heat flux q_y (subsequently referred to as q) was found to be inversely proportional to the mean pressure p (which is characteristic for a rarefied gas effect). Furthermore, q was found to be proportional to the pressure drop Δp across the channel (the driving force), and odd in the magnetic field direction. The results therefore are presented in terms of the quantity $pq/\Delta p$. This quantity is plotted as a function of H/p, the ratio of magnetic field strength to pressure (Figs. 2-4). As in other odd-in-field effects[3,5], absorptionlike curves are found. The slight pressure dependence of $pq/\Delta p$ at low pressures is caused by Knudsen effects; an extrapolation to zero mean free path length also is given. For this extrapolation the standard technique for Knudsen corrections was used[12,13].

In Fig. 5, the corrected results for N_2 are shown. To demonstrate the considerable boundary-layer contributions to

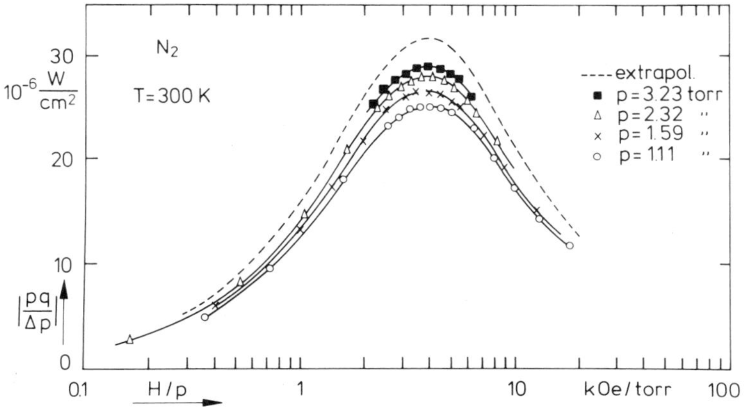

Fig. 2 Results of the visco-magnetic heat flux for N_2 at 300 K. The dotted line is an extrapolation to a mean free path length which is negligibly small compared to the apparatus dimensions.

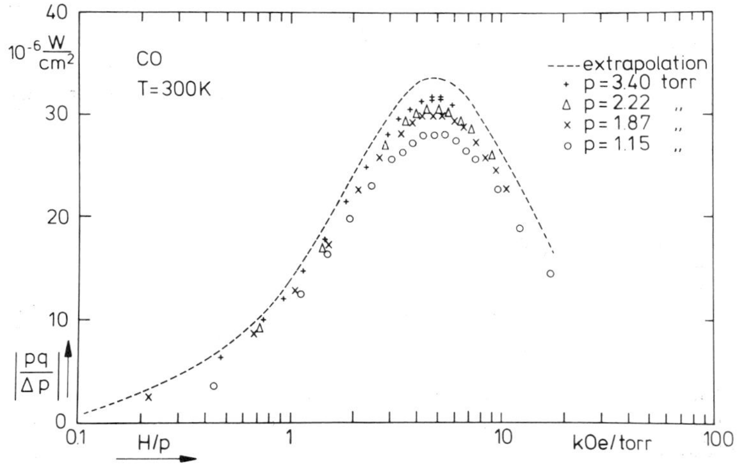

Fig. 3 Results of the visco-magnetic heat flux for CO at 300 K. The dotted line is an extrapolation to a mean free path length which is negligibly small compared to the apparatus dimensions.

the effect, the calculated bulk contributions are shown also. It is now possible to determine the parameter $\tilde{C}_{ah}$ appearing in the description of the boundary-layer contributions[7], using the values for the parameters $\tilde{C}_a$ and $\tilde{C}_{ma}$ found from the experiments on the Knudsen effects on the Senftleben-Beenakker effect on viscosity and from the thermo-magnetic

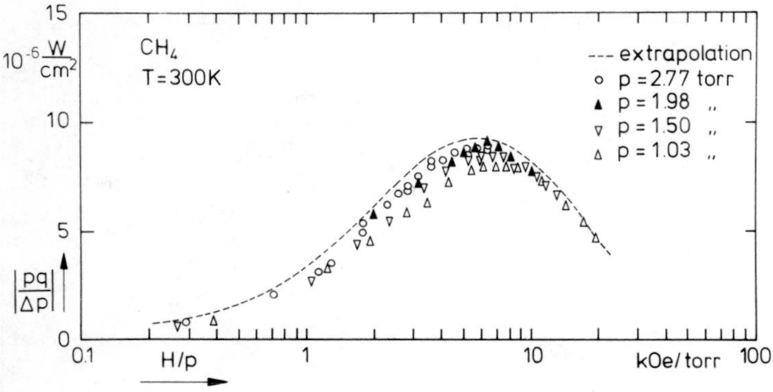

Fig. 4 Results of the visco-magnetic heat flux for CH_4 at 300 K. The dotted line is an extrapolation to a mean free path length which is negligibly small compared to the apparatus dimensions.

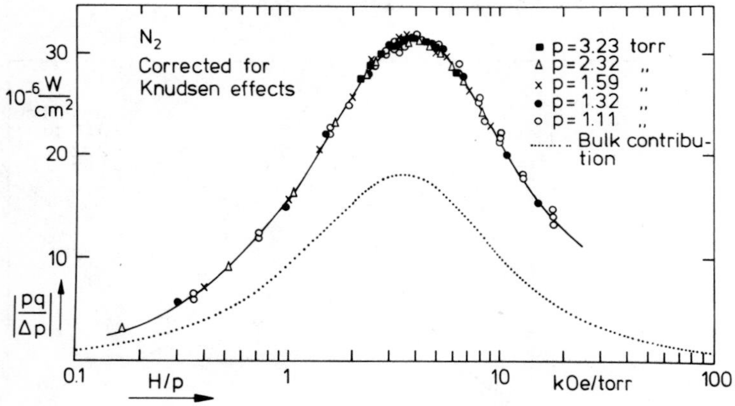

Fig. 5 Knudsen corrected results for the visco-magnetic heat flux for N_2. The dotted line is the calculated Burnett (bulk) contribution to the effect as given in Ref. 2 and Eq. (51) of Ref. 7.

pressure difference. More reliable information on the parameters can be obtained by studying the influence of the orientation of the magnetic field on the visco-magnetic heat flux. Such a study presently is being carried out. More details and a comparison between experimental and theoretical results will be given in due course.

References

[1]Maxwell, J.C., "On Stresses in Rarefied Gases arising from Inequalities of Temperature", Philosophical Transactions of the Royal Society of London, Vol. 170, 1879, pp. 231-256.

[2]Levi, A.C., McCourt, F.R., and Beenakker, J.J.M., "Burnett Coefficients in a Magnetic Field. II. The Linear Effects", Physica, Vol. 42, 1969, pp. 363-387.

[3]Scott, G.G., Sturner, H.W., and Williamson, R.M., "Gas Torque Anomaly in Weak Magnetic Fields", Physical Review, Vol. 158, No. 1, June 1967, pp. 117-121.

[4]Levi, A.C., and Beenakker, J.J.M., "Thermomagnetic Torques in Dilute Gases", Physics Letters, Vol. 25A, No. 5, September 1967, pp. 350-352.

[5]Eggermont, G.E.J., Hermans, L.J.F., Knaap, H.F.P., and Beenakker, J.J.M., "The Thermomagnetic Pressure Difference in Rarefied Polyatomic Gases: a New Type of Thermal Slip", Proceedings of the Ninth International Symposium on Rarefied Gas Dynamics, Vol. 2, 1974, pp. C7.1-C7.9;
Hulsman, H., Bulsing, G.F., Eggermont, G.E.J., Hermans, L.J.F., and Beenakker, J.J.M., "Experiments on the Thermomagnetic Pressure Difference in Polyatomic Gases, a New Type of Thermal Slip", Physica, Vol. 72, 1974, pp. 287-299.

[6]Chapman, S., and Cowling, T.G., The Mathematical Theory of Nonuniform Gases, Cambridge University Press, 1970.

[7]Vestner, H., "Theory of the Viscomagnetic Heat Flux", Zeitschrift für Naturforschung, Vol. 31a, 1976, pp. 540-552.

[8]Vestner, H., "Knudsen Corrections for the Senftleben-Beenakker Effect of Viscosity", Zeitschrift für Naturforschung, Vol. 29a, 1974, pp. 663-677.

[9]Hulsman, H., Van Kuik, F.G., Walstra, K.W., Knaap, H.F.P., and Beenakker, J.J.M., "The Viscosity of Polyatomic Gases in a Magnetic Field", Physica, Vol. 57, 1972, pp. 501-521.

[10]Vestner, H., "On the Behaviour of Rarefied Polyatomic Gases in a Static Magnetic Field", Zeitschrift für Naturforschung, Vol. 28a, 1973, pp. 869-880.

[11]Eggermont, G.E.J., Oudeman, P., Hermans, L.J.F., Vestner, H., and Beenakker, J.J.M., "The Angular Dependence of the Thermomagnetic Pressure Difference", Physica, to be published.

[12]Hermans, L.J.F., Koks, J.M., Hengeveld, A.F., and Knaap, H.F.P., "The Heat Conductivity of Polyatomic Gases in Magnetic Fields", Physica, Vol. 50, 1970, pp. 410-432.

[13]Hulsman, H., and Knaap, H.F.P., "Experimental Arrangements for Measuring the Five Independent Shear-Viscosity Coefficients in a Polyatomic Gas in a Magnetic Field", Physica, Vol. 50, 1970, pp. 565-572.

CLOSURE RELATIONS BASED ON STRAINING AND SHIFTING IN VELOCITY SPACE

J.P. Elliott* and D. Baganoff†
Stanford University, Stanford, Calif.

and

R.D. McGregor‡
University of Victoria, Victoria, B.C.

Abstract

The distribution function is expressed as the sum $f = f_o + \Delta f$, where f_o is the local Maxwellian; and two transformations in velocity space are applied to the conventional polynomial-series expansion, with a Maxwellian weight function, used to represent Δf. The first transformation strains velocity space isotropically by replacing the local temperature in Δf by the hypothetical Rankine-Hugoniot temperature corresponding to local flow conditions. It is shown that, for the upstream singular point in a shock wave (the critical point in the flow), the procedure leads to a solution of Maxwell's moment equations which appears to converge at high order. The second transformation is a shift in velocity space in the amount $u - a^\dagger$, where $a^\dagger$ is the speed of sound corresponding to the same hypothetical temperature. The set of closure relations obtained when both transformations are appli-

---

Presented as Paper 86 at the 10th International Symposium on Rarefied Gas Dynamics, Aspen, Colo., July 19-23, 1976. This work was supported by the National Research Council of Canada under Grant A6006 (JPE) and a National Research Council scholarship (RDM). Support from a University of Victoria faculty research grant (JPE) also is acknowledged.

*Visiting Associate Professor, Department of Aeronautics and Astronautics; on leave from Department of Physics, University of Victoria, Victoria, B.C.

†Professor, Department of Aeronautics and Astronautics.

‡Graduate Student, Department of Physics.

ed leads to a solution of the 13-moment equations which agrees very well with the previous high-order solution. Since both transformations may have a physical basis, it is suggested that a similar operation on the Boltzmann equation may yield new results.

Introduction

Any mathematical representation for the distribution function that is used to develop closure relations for solving Maxwell's moment equations at a given truncation level must embody a reasonable approximation to the outer regions of f (hereinafter loosely designated f_∞). Otherwise, it will yield erroneous predictions for the high-order moments, since they are determined by the behavior of f in these regions; and, consequently, the closure relations will become progressively less accurate with increasing order, leading ultimately to a failure of the solution to converge.

In situations involving a significant degree of translational nonequilibrium, f_∞ may decay much more slowly than the local Maxwellian $f_0 \sim \exp(-C^2/2RT)$, i.e., $f_\infty >> f_0$, so that the conventional polynomial-series expansions having f_0 as a weight factor will fail to reproduce the wings of f; and even the use of an orthonormal expansion may be insufficient to guarantee a good approximation to f_∞ at low order. On the other hand, in situations where $f_\infty \nless f_0$, the conventional expansions are adequate, and no special considerations are required.

To accommodate the outer regions of f in cases where $f_\infty >> f_0$, we shall use the representation $f = f_0 + \Delta f$, since $\Delta f \simeq f_\infty$ for large speeds, and so the behavior of f_∞ can be absorbed into Δf, and since the equilibrium state is recovered conveniently when $\Delta f = 0$. Whereas the mathematical possibilities for representing Δf are virtually innumerable, a particularly easy way to simulate the behavior of Δf in the wings is to expand it in a polynomial series with weight function $\hat{f}_0 \sim \exp(-C^2/2R\hat{T})$, where the temperature parameter $\hat{T}$ is greater than the local temperature T. The introduction of the parameter $\hat{T}$ is far more than a mathematical artifice, for, as we shall see in the next section, a prescription for $\hat{T}$ can be found based primarily on physical concepts.

Introducing the dimensionless thermal velocities $\underline{V} \equiv \underline{C}\,(2RT)^{-\frac{1}{2}}$ and $\hat{\underline{V}} \equiv \underline{C}\,(2R\hat{T})^{-\frac{1}{2}}$, we recognize that the proposed scheme is equivalent to applying the transformation $\hat{\underline{V}} = \alpha\,\underline{V}$ to Grad's representation for Δf, where $\alpha \equiv (T/\hat{T})^{\frac{1}{2}}$; the parameter α thus strains the coordinates of velocity space isotropically. Mathematically, the straining has the effect of changing the scale of the weight function in Grad's expansion such that $\int (\Delta f)^2 \; (\hat{f}_o)^{-1} \; d\hat{\underline{V}}$ exists, and so convergence of the expansion for Δf is guaranteed by Bessel's inequality. Holway was the first to use such a modification to the weight function, in dealing with a time-dependent, spatially homogeneous problem.[1]

Choice of the Straining Parameter α

Bessel's inequality shows that the representation for Δf will be efficient, in the sense of rapid convergence, only if we overestimate $\hat{T}$, and not if we underestimate $\hat{T}$. An upper bound for $\hat{T}$, if it were available, therefore would be the safest choice. Such an upper-bound estimate can be made by considering the maximum temperature that conceivably could be attained by a fluid element having velocity $\underline{u}$ and temperature T. On physical grounds, this is almost certainly the stagnation temperature T_o. However, our experience with the shock-wave problem has shown that $\hat{f}_o$ must reduce to f_o when the local Mach number M is unity, even for a nonequilibrium situation, if the proper closure relations are to be recovered. Since this constraint may not apply in general, we now restrict the discussion to the shock wave, in which case the constraint does not allow the choice $\hat{T} = T_o$. A slightly lower†estimate that does satisfy the constraint is given by $\hat{T} =$ T, where $T^{\dagger}/T = (M^2 + 3)(5M^2 - 1)/16M^2$ is the hypothetical Rankine-Hugoniot temperature ratio corresponding to <u>local</u> flow conditions ($\gamma = 5/3$). We adopt this choice for the moment; later, we shall confirm that it is indeed the better one.

The straining is required only in the supersonic portion of the shock wave, and especially at the upstream singular point, where the difficulty with convergence is most severe. Since there is no convergence problem with Grad's expansion on the subsonic side,[2] the prescription is not needed; but, in order to avoid patching, we shall use the straining there as well. Although the procedure gives $\hat{T}<T$ for $M<1$, convergence of the expansion is not affected in this case, as we shall see.

In order to demonstrate convergence, it is necessary to carry out a solution to rather high order. Since this is practical only at a singular point, we show results for calculations at both singular points in Fig. 1, where the bounded ratio $(M^{-2}\tau/\tau^0)_s$ is displayed vs. M_s to order 13. Here, τ is the axial component of the viscous stress tensor, $\tau^0 \equiv (4/3)\mu(du/dx)$ is the Navier-Stokes prediction for τ, and subscript s takes on the respective values 1,2 at the upstream

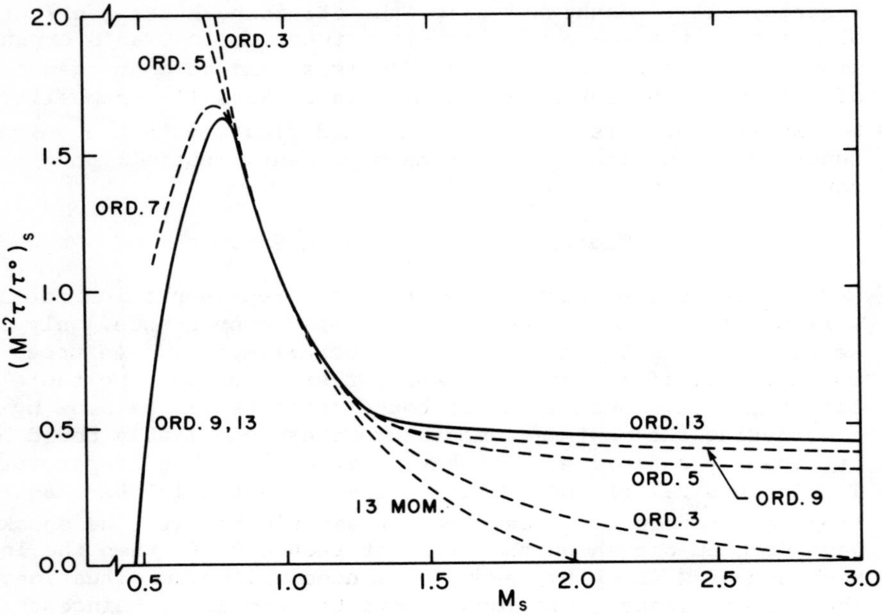

Fig. 1 Bounded stress ratio $(M^{-2}\tau/\tau^{0})_s$ as function of singular-point Mach number M_s to various orders in solution with $\alpha = (T/T^{\dagger})^{\frac{1}{2}}$.

and downstream singular points. In order to avoid cluttering the figure, curves corresponding to several different orders have been omitted.

On the supersonic side ($M_s > 1$), we note the apparent convergence of the solution, in contrast with the case $\hat{T} = T$, for which the upstream solution exhibits a critical Mach number at every order.[2] On the subsonic side ($M_s < 1$), there is no question as to convergence; so the straining, although unnecessary here, is not detrimental. In fact, the downstream thirteenth-order solution shown is identical to that for the case $\hat{T} = T$ and to that obtained by a different procedure considered in our earlier work.[2]

A significant observation from Fig. 1 is that the Stokes relation has approximate validity only for $M_s \simeq 1$ and is grossly in error for strong shock waves, for which $\tau \sim M_1^2\tau^0$ near the upstream singular point, and $\tau \ll \tau^0$ near the downstream singular point.

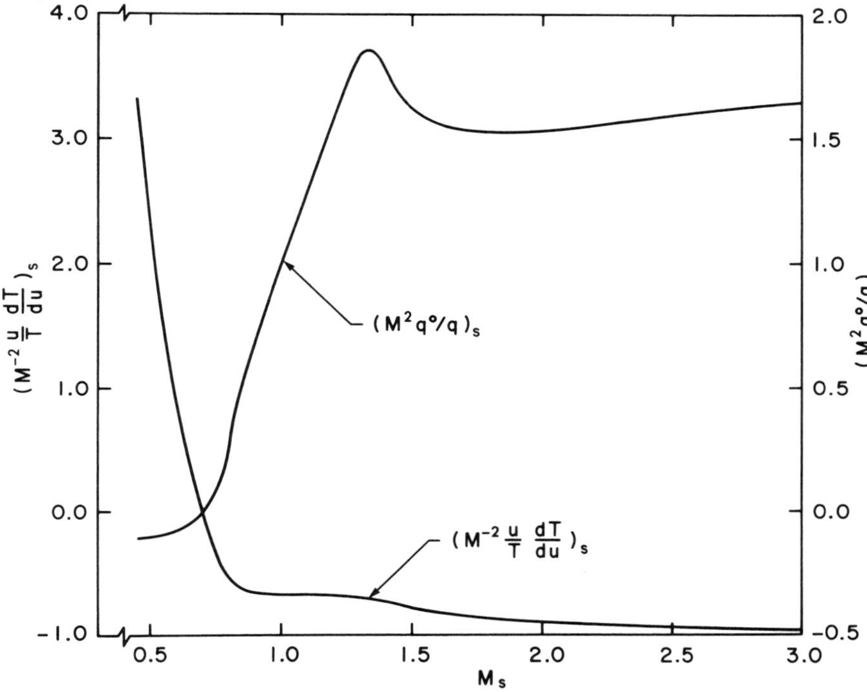

Fig. 2 Bounded heat-flux ratio $(M^2q^0/q)_s$ and temperature derivative $[M^{-2}(u/T)(dT/du)]_s$ as functions of singular-point Mach number M_s in thirteenth-order solution with $\alpha = (T/T^\dagger)^{\frac{1}{2}}$.

Taking the thirteenth-order solution as exact for discussion purposes, we can make two further observations from the quantities plotted in Fig. 2. First, we note that the Fourier heat-flux relation $q = q^0 \equiv -\lambda(dT/dx)$ also is approximately valid only for weak shock waves; and second, that the temperature derivative $(dT/du)_2 > 0$ for $M_2 \lesssim 0.7$, implying a temperature overshoot on the subsonic side for $M_1 \gtrsim 1.55$. Identical conclusions can be drawn from the downstream results with $\hat{T} = T$ and from our earlier work.[2]

In order to compute a shock-wave profile, one is restricted for practical reasons to a low-order (e.g., 13-moment) calculation; but it is clear from Fig. 1 that the straining alone has not produced a sufficiently accurate solution at this level for such a calculation to be justified. One way of improving the solution may be to introduce an additional parameter in the weight function $\hat{f}_o$. To see how this might be accomplished, we point out that an important aspect of the

Mott-Smith distribution is simulated by our representation, $f = f_o + \Delta f$, namely, that near a singular point the "temperature" of Δf corresponds to the temperature near the opposite singular point, whereas f_o has the local temperature. The aspect of the Mott-Smith distribution which is not simulated is that the "mean" of Δf in laboratory space remains at the fluid velocity. The success of the Mott-Smith approach suggests that a second parameter should introduce a shift in the argument of Δf; thus we now propose to use the weight function $\hat{f}_o \sim \exp\{-[(C_x + u - \hat{u})^2 + C_\perp^2] / 2R\hat{T}\}$ in the expansion of Δf, where the amount of shift is given by $(u-\hat{u})$. In terms of the dimensionless velocity components $\hat{V}_x \equiv (C_x + u - \hat{u})(2R\hat{T})^{-\frac{1}{2}}$, $\hat{V}_\perp \equiv C_\perp(2R\hat{T})^{-\frac{1}{2}}$, our scheme is now equivalent to the transformation $V_x = \alpha V_x + \delta$, $\hat{V}_\perp = \alpha V_\perp$, applied to Grad's expansion of Δf. Here, $\alpha \equiv (T/\hat{T})^{\frac{1}{2}}$ as before, and $\delta \equiv \tilde{M}\,\alpha(1 - \hat{u}/u)$, where $\tilde{M} \equiv M(5/6)^{\frac{1}{2}}$.

Choice of the Shifting Parameter δ

If we pick $\hat{u} = u^\dagger$, where $u^\dagger/u = (M^2 + 3)/4M^2$ is the hypothetical Rankine-Hugoniot velocity ratio corresponding to local flow conditions ($\gamma = 5/3$), then a nearly complete correspondence with the Mott-Smith representation is obtained at a singular point, since $u^\dagger = u_2$ for $s = 1$, and $u^\dagger = u_1$ for $s = 2$. In addition, $u^\dagger = u$ for $M = 1$, as required. An alternate, equally justifiable selection meeting this requirement is given by $\hat{u} = a^\dagger \equiv [(5/3)RT^\dagger]^{\frac{1}{2}}$, where $a^\dagger$ is the speed of sound corresponding to $T^\dagger$. Since we seek the most accurate closure relations possible at low order, it is appropriate to be guided by numerical solutions of the Boltzmann equation in deciding between $u^\dagger$ and $a^\dagger$. At the same time, we can re-examine our earlier choice between T_o and $T^\dagger$ for $\hat{T}$.

Leaving the straining and shifting parameters α and δ arbitrary for the moment, we first display the 13-moment closure relations in the form

$$Q \equiv \langle C_x^3\rangle/\langle C^2 C_x\rangle = (3/5) + (9/10)(\delta/\tilde{M}\alpha)(\tau u/q) \tag{1}$$

$$\phi \equiv \langle C^2 C_x^2\rangle/(5R^2T^2) = 1 - (7/5)M^2\left[(M\alpha)^{-2} - (25/21)(\delta/\tilde{M}\alpha)^2\right](\tau/p) - (4/3)M^2(\delta/\tilde{M}\alpha)(q/pu)(1 + Q) \tag{2}$$

where the bracket notation is used for the higher moments. In passing, we note that Grad's 13-moment closure relations are recovered when $\alpha = 1$ and $\delta = 0$, as required.

Written in the preceding form, the closure relations contain the groupings $M\alpha$ and $\delta/\tilde{M}\alpha$. On the basis of the possibilities being considered, these quantities are functions of M alone and are given by

$$M\alpha = \begin{cases} M/(1 + M^2/3)^{\frac{1}{2}} & (\hat{T} = T_o) \\ 4M^2/\left[(M^2 + 3)(5M^2 - 1)\right]^{\frac{1}{2}} & (\hat{T} = T^{\dagger}) \end{cases}$$

$$\delta/\tilde{M}\alpha = \begin{cases} 3(M^2 - 1)/4M^2 & (\hat{u} = u^{\dagger}) \\ 1 - \left[(M^2 + 3)(5M^2 - 1)\right]^{\frac{1}{2}}/4M^2 & (\hat{u} = a^{\dagger}) \end{cases}$$

The two expressions for $M\alpha$ are plotted in Fig. 3, whereas those for $\delta/\tilde{M}\alpha$ appear in Fig. 4; we shall return momentarily to a discussion of these figures.

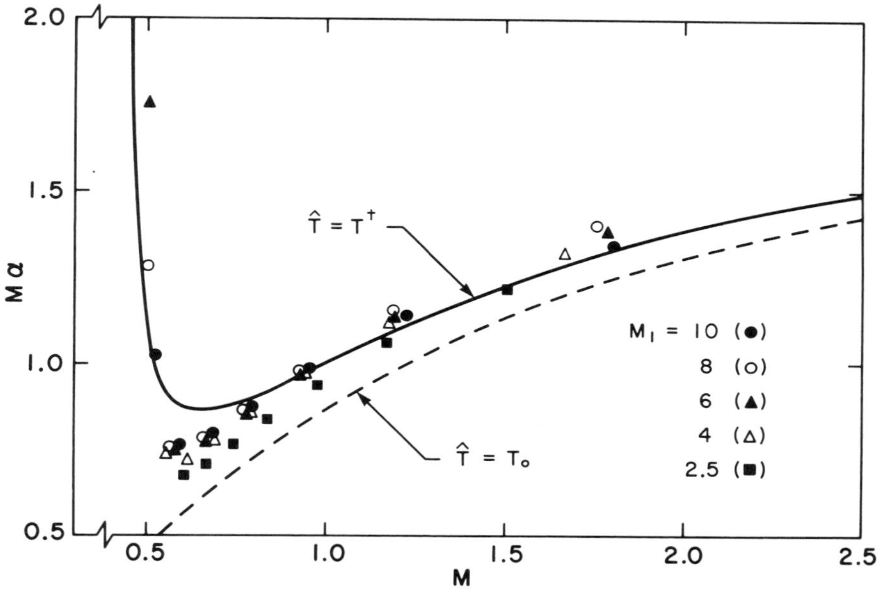

Fig. 3 Comparison of theoretical expressions for straining parameter $M\alpha$ with predictions based on closure relations (1) and (2) and data from numerical solution of the Boltzmann equation.

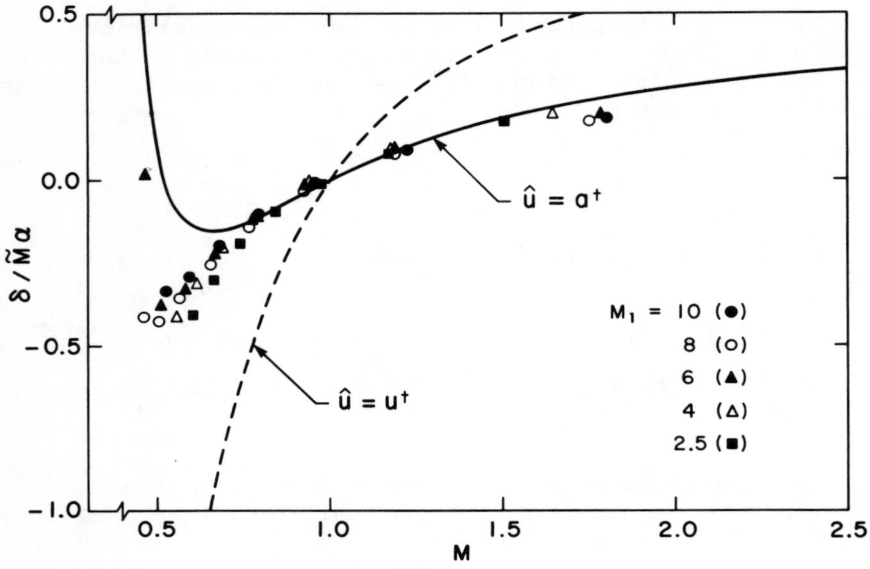

Fig. 4 Comparison of theoretical expressions for shifting parameter $\delta/\tilde{M}\alpha$ with predictions based on closure relations (1) and (2) and data from numerical solution of the Boltzmann equation.

Predictions for $M\alpha$ and $\delta/\tilde{M}\alpha$ from numerical solutions can be made first by solving (1) and (2) for these two quantities, which yields the functional relations

$$M\alpha = \xi(M,\ \tau/p,\ q/pu,\ Q,\ \phi);\ \delta/\tilde{M}\alpha = \eta(\tau u/q),\ Q)$$

The right-hand sides then can be evaluated from numerical data at each station in the flow. (These data, largely unpublished, kindly were made available to the authors by S.M. Yen.) The values for $M\alpha$ and $\delta/\tilde{M}\alpha$ thus obtained are plotted vs. the local Mach number in Figs. 3 and 4, for comparison with the theoretical curves already shown. The data points exhibited are for Maxwell molecules, with shock-wave Mach numbers lying in the range $2.5 \leq M_1 \leq 10$. In both figures, we observe surprisingly little scatter in the numerical data, even though the local Mach number is used as the independent variable merely for comparison with theory.

Of the two theoretical curves in Fig. 3, the one for which $\hat{T} = T^{\dagger}$ shows a better correlation with the numerical results, particularly for $M > 1$. In addition, the data verify that $\alpha \to 1$ for $M \to 1$, as mentioned previously. Thus, the earlier choice for $\hat{T}$ was indeed the correct one. The singularity in the proposed function at $M = (5)^{-1/2}$, although difficult to accept physically, is not ruled out by the data.

Regarding the choice for $\hat{u}$, Fig. 4 shows that $a^\dagger$ clearly is superior to $u^\dagger$, although the agreement with the data is not particularly good for $M < 0.7$; on the basis of this comparison, we make the selection $\hat{u} = a^\dagger$. Since when $\delta > 0$ the shift is in the direction downstream of the fluid velocity, and vice versa, we see that the direction of the shift predicted by the theoretical curve is generally well correlated with the data, except for $M < 0.52$, where a second sign reversal occurs; but, in order to be consistent, we must accept this apparent anomaly in the theory.

Because of the inaccuracy inherent in the closure relations (1) and (2), perfect correlation with the numerical data cannot be expected. Nevertheless, Figs. 3 and 4 indicate that the concepts of straining and shifting are quite reliable, particularly for moderate-strength shock waves. The only uncertainty appears in the region $M < 0.52$; but the potential error here can be assessed by carrying out the 13-moment singular-point calculation on the basis of (1) and (2), and comparing the result with the "exact" thirteenth-order solution shown in Fig. 1.

Figure 5 summarizes the calculations for $(M^{-2}\tau/\tau^0)_s$ in each of four cases. The first (solid line) is the thirteenth-order solution for the case $\hat{T} = T^\dagger$, $\hat{u} = u$ (strain, but no shift), which was shown in Fig. 1; we adopt this result as a standard of comparison. The second (dotted line) is the 13-moment solution with $\hat{T} = T$, $\hat{u} = u$ (no strain, no shift, i.e., Grad's method). Here, we observe a critical upstream Mach number $M_1 \simeq 1.65$ and a wide departure from the exact result downstream. Close agreement with the exact solution occurs only for $M_s \simeq 1$. The third (dashed line) is the 13-moment solution with $\hat{T} = T^\dagger$, $\hat{u} = u$ (strain, but no shift), which also was shown in Fig. 1 for $M_s \geq 1$. The straining has improved the solution by raising the critical Mach number to $M_1 \simeq 2.0$, but agreement with the exact solution is still poor, although better than without straining. The fourth (broken line) is the 13-moment solution with $\hat{T} = T^\dagger$, $\hat{u} = a^\dagger$ (strain, plus shift); and we see that introduction of the shifting has eliminated the critical Mach number altogether. Agreement with the exact result is excellent on the supersonic side. On the subsonic side, the shifting has lowered the solution toward the exact result, but only for $M_s < 0.52$; agreement on this side is therefore only qualitative.

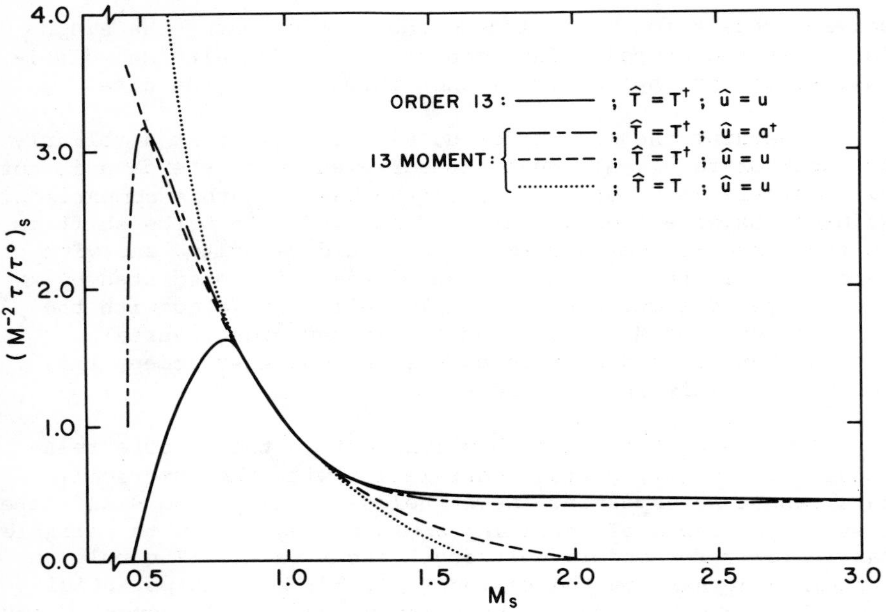

Fig. 5 Bounded stress ratio $(M^{-2}\tau/\tau^{0})_s$ as function of singular-point Mach number M_s in solutions for various orders and indicated choices for straining and shifting parameters.

Conclusions

Our previous work[2] has shown that the singular-point analysis provides an extremely severe test for closure relations. On the basis of the comparison in Fig. 5, the 13-moment theory based on the proposed closure relations (1) and (2) models the upstream flow quite well. The picture downstream remains incomplete although superior to anything previously available. However, since the upstream singular point is the more critical, the present theory removes the difficulties previously encountered in computing shock-wave profiles at the 13-moment level.

The straining operation allows convergence upstream by introducing a reference "temperature" sufficiently high that the series representation of Δf can produce a reasonable model of f_∞. Because the temperature ratio $T/T^{\dagger}$ is a function of local Mach number, the α parameter used in the present formulation automatically introduces M into the closure relations (1) and (2). The necessity for such M-dependence can be inferred by manipulating the moment equations,[3,4] particularly upstream for M>>1; however, until now, it was not quite clear how the introduction of M could be made on a physical basis. The present prescription for the straining parameter α gives $\alpha \sim M^{-1}$ as $M \to \infty$, which is identical to the asymptotic behavior of the similar straining parameters α_+, β_+ used in a recent related study[5]; in that instance, however, the behavior in the limit $M \to \infty$ was the principal guide in making the selection. From a physical point of view, the introduction of $T^\dagger$ does not seem as natural for the subsonic side as for the supersonic side; nevertheless, it changes the solution in the proper direction.

The success of the shifting operation, as seen in Fig. 5, implies that f_∞ can be represented much more efficiently if the expansion for Δf is not centered on the fluid velocity. On the supersonic side, $a^\dagger$ is an excellent choice for $\hat{u}$, as was anticipated from Fig. 4; on the subsonic side, it is a cruder approximation. However, the 13-moment solution has most of the essential ingredients of the exact result; in particular, it predicts temperature overshoot, in approximate agreement with Fig. 2. In order to improve the prescription on the subsonic side, a greater awareness of the underlying physics would have to be developed.

The singular-perturbation character of the Chapman-Enskog procedure, first elucidated by Narasimha[6] for the BGK model, is consistent with the present observation that the Stokes and Fourier relations are not valid in the singular-point limit, even though $|\Delta f|$ becomes vanishingly small near a singular point. Since the Chapman-Enskog theory provides only an inner approximation of f, an analysis of the outer nature of f, valid in the asymptotic limit as $V \to \infty$, would contribute a great deal to our understanding of kinetic theory; such an analysis has yet to be carried out for the Boltzmann equation. The α and δ parameters, although introduced in the spirit of an exact formulation, nonetheless contain some ad hoc elements; yet, because of their evident success, they appear to give a definite direction to the search for an outer solution of the Boltzmann equation.

References

[1]Holway, L.H., "Time-Varying Weight Functions and the Convergence of Polynomial Expansions," The Physics of Fluids, Vol. 10, Jan. 1967, pp. 35-48.

[2]Elliott, J.P. and Baganoff, D., "Solution of the Boltzmann Equation at the Upstream and Downstream Singular Points in a

Shock Wave," Journal of Fluid Mechanics, Vol. 65, Pt. 3, Sep. 1974, pp. 603-624.

[3]Baganoff, D. and Nathenson, M., "Constitutive Relations for Stress and Heat Flux in a Shock Wave," The Physics of Fluids, Vol. 13, March 1970, pp. 596-607.

[4]Nathenson, M., "On Using Maxwell's Moment Equations to Obtain Constitutive Relations for a Gas in a Shock Wave," Ph. D. Thesis, 1970, Stanford University.

[5]Baganoff, D. and Elliott, J.P., "Solution of Maxwell's Moment Equations by the Method of Rational Truncation and Coordinate Straining," The Physics of Fluids, Vol. 18, Dec. 1975, pp. 1660-1665.

[6]Narasimha, R., "Asymptotic Solutions for the Distribution Function in Non-Equilibrium Flows. Part 1. The Weak Shock," Journal of Fluid Mechanics, Vol. 34, Pt. 1, Oct. 1968, pp. 1-24.

INTEGRAL BOLTZMANN EQUATION FOR TEST PARTICLES IN A CONSERVATIVE FIELD: I. THEORY AND EXACT SOLUTIONS TO SOME STATIONARY AND TIME-DEPENDENT PROBLEMS

V. C. Boffi,* J. J. Dorning,+ V. G. Molinari,*
and G. Spiga ǂ

Laboratorio di Ingegneria Nucleare
dell'Università di Bologna,
Bologna, Italy

Abstract

The diffusion of test particles (t. p.) in a random distribution of field particles (f. p.), and in presence of an external conservative force is studied by use of a system of two linear integral equations, which are shown in Sec. I to govern the total density and the total flux vector, respectively, in the case of both isotropic scattering and creation by collision. Applications are then developed with the objective of providing exact solutions for some problems of interest in rarefied gas dynamics. In Sec. II, exact stationary solutions for both total density and flux vector, and for the electrical conductivity, are obtained in the case of a monoenergetic, spatially uniform, isotropic source. In Sec. III, the time-asymptotic behavior of the total density of t. p. is determined for the ca-

Presented as Paper 68a at the 10th International Symposium on Rarefied Gas Dynamics, July 19-23, 1976, Aspen, Colo.

* Professor.

+ C.N.R. Visiting Professor, presently on leave of absence from Nuclear Engineering Laboratory, University of Illinois at Urbana, Ill.

ǂ Assistant.

se of a pulsed, monoenergetic, spatially uniform, isotropic source.

I: Theory

1. Integral Form of the Linearized Boltzmann Equation

The linearized integrodifferential Boltzmann equation governing the density distribution $f(\bar{x}, \bar{v}, t)$ of t. p. having mass m, and diffusing in a random distribution of f. p. (of mass M), is[1,2]

$$[\partial/\partial t + \bar{v}.\bar{\nabla}_{\bar{x}} + (\bar{F}/m).\bar{\nabla}_{\bar{v}}]\, f(\bar{x},\bar{v},t) = (\partial f/\partial t)_c + Q(\bar{x},\bar{v},t) \tag{1}$$

where $\bar{F} \equiv \bar{F}(\bar{x}, t)$ denotes an external conservative force independent of velocity $\bar{v}$, $Q(\bar{x}, \bar{v}, t)$ is an external source of t. p., and the collision term is given as the difference

$$(\partial f/\partial t)_c = \int d\bar{v}' K(\bar{x},\bar{v}' \to \bar{v})\, f(\bar{x},\bar{v}',t) - v\Sigma(\bar{x},v)\, f(\bar{x},\bar{v},t) \tag{2}$$

between the in-scattering + creation integral and the loss term, $v\Sigma f$. In this latter term, we set[3]

$$\Sigma(\bar{x},v) = \Sigma_S(\bar{x},v) + \Sigma_A(\bar{x},v) + \Sigma_I(\bar{x},v) \tag{2a}$$

where Σ_S, Σ_A, and Σ_I — which depend, in general, on both position $\bar{x}$ and speed $v = |\bar{v}|$ — denote the scattering, removal, and creation cross section, respectively. Here the kernel $K(\bar{x},\bar{v}' \to \bar{v})$ is

$$K = K_S + \nu_I\, K_I \tag{2b}$$

where ν_I is the mean number of t. p. produced by creation ($\nu_I =$ $= 2$ for ionization or $\nu_I = 1$ for inelastic scattering).

Equation (1) now can be reformulated as an integral equation

$$f(\bar{x},\bar{v},t) = \int_0^{t-t_0} d\tau \, \exp\left\{-\int_0^{\tau} v(\tau')\Sigma[\bar{x}(\tau'),v(\tau')] \, d\tau'\right\}$$

$$\cdot G[\bar{x}(\tau),\bar{v}(\tau),t-\tau]$$

$$+ f(\bar{x},\bar{v},t_0)\exp\left\{-\int_0^{t-t_0} v(\tau')\Sigma[\bar{x}(\tau'),v(\tau')]\,d\tau'\right\} \tag{3}$$

in which, however, we shall henceforth set the second term on the right-hand side equal to zero, as occurs in most physical cases. In Eq. (3) $G(\bar{x}, \bar{v}, t)$, which combines all of the gain terms, includes all the physical constraints — related to the source and/or the geometry — of the problem, whereas t_0 is a (real) parameter of integration. In the following we shall find it simpler to take for G the expression valid in the case of an infinite medium, and to incorporate the effects of the source and/or the geometry in the upper limit $t - t_0$. The functions $\bar{x}(\tau)$ and $v(\tau)$ can, at last, be obtained by integrating the equation of particle motion.

2. Integral Form in the Case of Isotropic Scattering and Creation by Collision.

If both scattering and creation by collision are taken to be isotropic in the laboratory system, the general kernel K_α of Eq. (2b) (α = S, I) becomes a function of the speeds before and after the collision, namely

$$K_\alpha(\bar{x},\bar{v}' \rightarrow \bar{v}) = (1/4\pi)\ (v'/v^2)\Sigma_{\alpha,0}(\bar{x},v' \rightarrow v) \tag{4}$$

where $\Sigma_{\alpha,0}$ denotes the speed transfer function.

We introduce the new set of independent variables $(\bar{x}, v, \bar{\Omega}, t)$, and denote the corresponding density distribution and source by $n(\bar{x}, v, \bar{\Omega}, t)$ and $S(\bar{x}, v, \bar{\Omega}, t)$ respectively. Then Eq. (3) can be recast as

$$n(\bar{x},v,\bar{\Omega},t) = v^2 \int_0^{t-t_0} \frac{d\tau}{v^2(\tau)} \exp\left\{-\int_0^{\tau} v(\tau')\Sigma[\bar{x}(\tau'),v(\tau')]\,d\tau'\right\}$$

$$\cdot\left\{S[\bar{x}(\tau),v(\tau),\bar{\Omega}(\tau),t-\tau] + \frac{1}{4\pi}\int_0^{\infty} dv'v'\Sigma_0[\bar{x}(\tau),v' \to v(\tau)]\right.$$

$$\left.\cdot\int_{4\pi} d\bar{\Omega}'\,n[\bar{x}(\tau),v',\bar{\Omega}',t-\tau]\right\} \tag{5}$$

where

$$\Sigma_0(\bar{x},v' \to v) = \Sigma_{S,O}(\bar{x},v' \to v) + \nu_I\Sigma_{I,O}(\bar{x},v' \to v) \tag{5a}$$

and the upper limit, $t - t_O$, will be, in general, a function of $\bar{x}$, v, $\bar{\Omega}$, t.

If next we define the moments

$$M_\ell(\bar{x},v,t) = \int_{4\pi} (v\bar{\Omega})^\ell n(\bar{x},v,\bar{\Omega},t)\,d\bar{\Omega} \qquad (\ell = 0,1) \tag{6}$$

of the distribution function [$M_O \Rightarrow N(\bar{x}, v, t)$ and $M_1 \Rightarrow$ $\Rightarrow \bar{J}(\bar{x}, v, t)$ being the total density and the total flux vectors, respectively], it is easily deduced that they satisfy the system of linear integral equations ($\ell = 0, 1$)

$$M_\ell(\bar{x},v,t)$$

$$= v^{2+\ell}\int_{4\pi} d\bar{\Omega}\,\bar{\Omega}^\ell \int_0^{t-t_0} \frac{d\tau}{v^2(\tau)} \exp\left\{-\int_0^{\tau} v(\tau')\Sigma[\bar{x}(\tau'),v(\tau')]\,d\tau'\right\}$$

continued

$$.S[\bar{x}(\tau),v(\tau),\bar{\Omega}(\tau),t-\tau]$$

$$+ \frac{v^{2+\ell}}{4\pi}\int_{4\pi} d\bar{\Omega}\,\bar{\Omega}^{\ell}\int_0^{t-t_0}\frac{d\tau}{v^2(\tau)}\exp\left\{-\int_0^{\tau} v(\tau')\Sigma[\bar{x}(\tau'),v(\tau')]\,d\tau'\right\}$$

$$.\int dv'\; v'\Sigma_0[\bar{x}(\tau),v' \to v(\tau)]\,M_0[\bar{x}(\tau),v',t-\tau] \tag{7}$$

defined for $0 \leqslant v,\ t \leqslant \infty$, and for $\bar{x} \in V$, V being the volume occupied by the f. p. considered. We observe the following: when the external force tends to zero, $v(\tau) \Rightarrow v$. Then all the results quoted since Eq. (3) are easily verified to coincide with those which can be established when the t. p. are taken to be neutrons, and gravitational effects are neglected[3]. Secondly, the dependence of the integrands on the right-hand side of Eq. (7) upon $\bar{\Omega}$ occurs through the functions $v(\tau)$ and $\bar{x}(\tau)$, which, for instance, in the case of a constant external force, are given by

$$v(\tau) = |\bar{v} - (\bar{F}/m)\tau| = \sqrt{v^2 + (F/m)^2\tau^2 - 2\,(\bar{F}/m).\bar{\Omega}v\tau} \tag{8a}$$

$$\bar{x}(\tau) = \bar{x} + (1/2)(\bar{F}/m)\tau^2 - v\tau\bar{\Omega} \tag{8b}$$

3. Scheme of Application of Eqs. (7)

The problems, which we shall treat in the sequel of this report, and which are of basic interest in plasma physics, gaseous electronics, and upper atmospheric investigations, concern essentially the studies of t. p. (electrons) motion in a gas exposed to an electric field. Because of their complexity, these studies usually are carried out on the basis of different simplified hypotheses and models, and are generally solved only approximately or numerically[4].

We shall concentrate on constructing exact analytical solutions to Eqs. (7) with reference to the following two classes of problems: the first is a class of stationary space-dependent problems, which will be considered in Part II of this paper[5], both for the cases of infinite and finite distribution of f. p. The second class of problems to be discussed here is the one in which spatial effects are neglected, both stationary and time--dependent velocity distribution problems are solved exactly, and the solutions are numerically evaluated.

The following additional hypotheses are made: the external conservative force $\overline{F}$ is constant, so that Eqs. (8) hold; both Σ and Σ_0 are independent of position, and, consequently, so are the kernels of Eqs. (7): the external source is isotropic, that is, $S(\overline{x}, v, \overline{\Omega}, t) = (4\pi)^{-1} S_0(\overline{x}, v, t)$.

We prefer to put emphasis on the effects of the external force rather than on a detailed description of the energy exchange between t. p. and f. p. as due to elastic collisions. [Inelastic collisions actually are responsable for the energy exchange between light t. p. (electrons) and heavy f. p. (molecules). In the present approach, an inelastic collision effect is taken into account when we set $\nu_I = 1$.] In this context, the following specializations will be adopted for Σ and $\Sigma_0 = \Sigma_{S,0} + \nu_I \Sigma_{I,0}$: for $\Sigma(v)$, we refer to the model of constant collision frequency[6], namely, we set

$$\Sigma(v) = \Sigma_S(v) + \Sigma_A(v) + \Sigma_I(v) = \frac{C}{v} = \frac{C_S + C_A + C_I}{v} \tag{9}$$

For $\Sigma_{S,0}(v' \to v)$ we refer to the model of t. p. elastically scattered by very heavy f. p., which are in turn free and at rest. In this case $(m/M \to 0)$, we have simply

$$\Sigma_{S,0}(v' \to v) = (C_S/v')\,\delta(v' - v) \tag{10}$$

where δ is the Dirac delta function; for $\Sigma_{I,0}(v' \to v)$ we refer to the model in which t. p. are created by collision (or inelastically scattered) at a fixed speed v_0^I. Thus, we have

$$\Sigma_{I,0}(v' \to v) = (C_I/v')\,\delta(v - v_0^I) \tag{11}$$

II: Exact Stationary Solutions

1. General Remarks

Let us consider the stationary case when a source of the type

$$S_0(v) = Q_0 \ \delta(v_0^Q - v) \tag{12}$$

exists in an infinite random distribution of f. p. In this case $t_0 = -\infty$ in Eq. (5) and $n(\bar{x}, v, \bar{\Omega}, t)$ will be independent of both position and time. The same then holds for the moments M_ℓ defined by Eq. (6). Equations (7) thus become ($\ell = 0, 1$)

$$M_\ell(v) = S_\ell(v) + \int_0^\infty dv' \ V_\ell(v' \rightarrow v) M_0(v') \tag{13}$$

where we set

$$S_\ell(v) = Q_0 \frac{v^{2+\ell}}{4\pi} \int_{4\pi} d\bar{\Omega} \ \bar{\Omega}^\ell \int_0^\infty \frac{d\tau}{v^2(\tau)} \ e^{-C\tau} \delta[v_0^Q - v(\tau)] \tag{13a}$$

and

$$V_\ell(v' \rightarrow v) = \frac{v^{2+\ell}}{4\pi} \int_{4\pi} d\bar{\Omega} \ \bar{\Omega}^\ell \int_0^\infty \frac{d\tau}{v^2(\tau)} \ e^{-C\tau}$$

$$\cdot \Big\{ C_S \delta[v' - v(\tau)] + \nu_I C_I \delta[v_0^I - v(\tau)] \Big\} \tag{13b}$$

With S_0, $V_0 \Rightarrow S_N$, V_N and S_1, $V_1 \Rightarrow \bar{S}_{\bar{J}}$, $\bar{V}_{\bar{J}}$ we find explicitly

$$V_N(v' \rightarrow v) = V_{N,S}(v' \rightarrow v) + \nu_I V_{N,I}(v' \rightarrow v)$$

$$= \frac{1}{2} C_S \frac{m}{F} \frac{v}{v'} \Big\{ E_1\Big[\frac{mC}{F} |v - v'|\Big] - E_1\Big[\frac{mC}{F} (v + v')\Big] \Big\}$$

continued

$$+ \nu_I \frac{C_I}{C_S} \{V_{N,S}(v' \to v)\}_{v'=v_0^I} \tag{14a}$$

$$S_N(v) = (Q_0/C_S)\,\{V_{N,S}(v' \to v)\}_{v'=v_0^Q} \tag{14b}$$

$$\bar{S}_{\bar{J}}(v) = (Q_0/C_S)\,\{\bar{V}_{\bar{J},S}(v' \to v)\}_{v'=v_0^Q} \tag{14c}$$

$$\bar{V}_{\bar{J}}(v' \to v) = \bar{V}_{\bar{J},S}(v' \to v) + \nu_I\, \bar{V}_{\bar{J},I}(v' \to v)$$

$$= \frac{\bar{k}}{4}\frac{v}{v'}\frac{C_S}{C}\Big\{ \mathrm{sgn}(v - v')\, \frac{mC}{F}(v + v')\, E_2\left[\frac{mC}{F}|v - v'|\right]$$

$$- \frac{mC}{F}(v - v')\, E_2\left[\frac{mC}{F}(v + v')\right] + \left[e^{-\frac{mC}{F}|v-v'|} - e^{-\frac{mC}{F}(v+v')}\right]$$

$$+ \nu_I \frac{C_I}{C_S}\{\bar{V}_{\bar{J},S}(v' \to v)\}_{v'=v_0^I} \tag{14d}$$

In these equations $E_n(x)$ is the nth-order exponential integral, and $\bar{k}$ is the unit vector associated with the positive direction of the external force $\bar{F}$.

It can be recalled that the kernel $V_{N,S}$ in Eq. (14a) has been derived previously by Stuart and Gerjuoy in the context of a phenomenological investigation on Townsend's first coefficient in absence of both removal and creation by collision (inelastic scattering)[7]. Supplementary literature on various implications and utilizations of $V_{N,S}$ can be found in Ref. 8.

In the following we shall confine ourselves to the simpler case, when both v_0^Q and v_0^I are zero. Then the source term to be used in Eq. (13) is (ℓ = 0, 1)

$$S_\ell(v) = Q_0 (\bar{k}v)^\ell\, (m/F)\, e^{-(mCv/F)} \tag{14e}$$

2. Exact Solution for the Total Density

Introducing the dimensionless speed $w = (mC/F)v$, Eq. (13) for $\ell = 0$, with $S_0(v)$ supplied by Eq. (14e), can be rewritten as $(-\infty \leqslant w \leqslant \infty)$

$$\psi(w) = \frac{Q}{C}\,\frac{e^{-|w|}}{w} + \frac{1}{2}\lambda \int_{-\infty}^{\infty} E_1(|w - w'|)\,\psi(w')\,dw' \tag{15}$$

where $\psi(w) = w^{-1}N(w)$, the intensity

$$Q = Q_0 + \nu_I\, C_I \int_0^{\infty} N(v')\,dv' \tag{15a}$$

of the source term includes the contributions of both the external source and the creation by collision (inelastic scattering), and the parameter

$$0 \leqslant \lambda = C_S/C \leqslant 1 \tag{15b}$$

accounts for the effects of both removal and creation by collision (inelastic scattering).

Equation (15) can be solved exactly by a technique based on a two-sided Laplace transform. For the details, the reader is referred to Ref. 9. The result-which is plotted in Fig. 1 for different λ - is $(w \geqslant 0)$

$$N(w) = \frac{Q}{C}\,w\,[\psi_p(w) + \psi_{b.p.}(w)] = \frac{Q}{C}\,w$$

$$\cdot\left[\gamma_\lambda(\alpha_0)\,e^{-\alpha_0 w} + \int_1^{\infty} \delta_\lambda(\alpha)\,e^{-\alpha w}\,d\alpha\right]$$

$$= \frac{Q}{C}\,w\left\{\frac{2\alpha_0^2(1-\alpha_0^2)}{\lambda\,[\alpha_0^2-(1-\lambda)]}\,e^{-\alpha_0 w} + \int_1^{\infty}\frac{e^{-\alpha w}\,d\alpha}{\left(1-\frac{\lambda}{\alpha}\tanh^{-1}\frac{1}{\alpha}\right)^2 + \left(\frac{\lambda}{2}\frac{\pi}{\alpha}\right)^2}\right\} \tag{16}$$

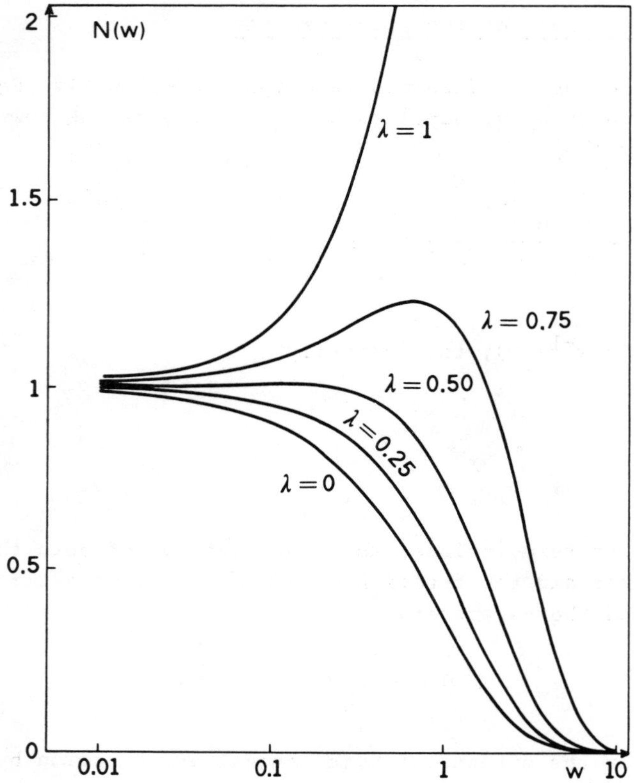

Fig. 1 Total density N(w) vs. ln w according to Eq. (16).

where

$$\alpha_0 = 3(1-\lambda)\ \{1-(4/5)\ (1-\lambda) + (4/175)\ (1-\lambda)^2 \ldots\} \tag{16a}$$

[with $\alpha_0^2 > (1 - \lambda)$] is the root of the transcendental equation $1 - \lambda p^{-1} \tanh^{-1} p = 0$ when $1 - \lambda \ll 1$. [The subscripts p and b. p. refer to the pole and to the branch-point contributions, which occur when inverting the transform $\widetilde{N}(p)$ of N(w).] The effects due to both removal and creation by collision (inelastic scattering) are seen in Fig. 1, where the curves for $0 < \lambda < 1$ are compared with the curve for $\lambda = 1$.

3. Exact Solutions for the Total Flux Vector and for the Electrical Conductivity

With $\ell = 1$, Eq. (13) provides for the magnitude of the total flux vector

$$J(w) = \frac{Q}{mC^2} F\left\{ we^{-w} + \frac{1}{4}\lambda w \int_0^\infty V_{\bar{J},S}(w' \rightarrow w)\left[\psi_p(w') + \psi_{b.p.}(w')\right] dw' \right\}$$

$$= \frac{Q}{mC^2} F\{I_0(w) + I_p(w) + I_{b.p.}(w)\} \qquad (17)$$

where $V_{\bar{J},S}(w' \rightarrow w)$ is the modulus — in terms of w — of the kernel $\bar{V}_{\bar{J},S}$ given by Eq. (14d). The study of J(w) as a function of λ and w is not reported here for the brevity's sake.

From the knowledge of J(w), we pass instead to the calculation of the electrical conductivity σ for electrons (of charge e) in an external constant electrical field $\bar{E}$. By Ohm's law and with F = eE, we derive

$$\sigma = -\frac{1}{E} e\bar{k} \int_0^\infty J(w)\,dw = \sigma_0 + \sigma_p + \sigma_{b.p.}$$

$$= \frac{Qe^2}{mC^2}\left(1 + \lambda \frac{\gamma_\lambda(\alpha_0)}{\alpha_0^2} + \lambda \int_1^\infty \frac{\delta_\lambda(\alpha)}{\alpha^2} d\alpha\right)$$

$$= \frac{Qe^2}{mC^2}(1 + \sigma_p^* + \sigma_{b.p.}^*) \qquad (18)$$

where σ_0, σ_p, and $\sigma_{b.p.}$ are the contribution of the external source, and of the pole, and of the branch-point component of $\psi(w)$, respectively. Thus we recognize the appearance of the new

Table 1 - Values of σ_p^* and $\sigma_{b.p.}^*$

λ	σ_p^*	$\sigma_{b.p.}^*$
0	0	0
0.25	0.0108	0.3225
0.50	0.3992	0.6008
0.75	2.2684	0.7317
1	∞	0.8000

term $\sigma_{b.p.}$ in addition to the usual expression for $\sigma = \sigma_0 + \sigma_p$ as reported in the framework of Langevin's approximation[10]. In Table 1 we list values of σ_p^* and $\sigma_{b.p.}^*$ for some values of λ. We first comment that both σ_p^* and $\sigma_{b.p.}^*$ are monotonically increasing functions of λ. Then we may conclude that, whereas, for small values of $\lambda, \sigma_{b.p.}^* > \sigma_p^*$, the situation is reverted for larger λ until, for λ close to unity, $\sigma_{b.p.}^*$ can be neglected with respect to σ_p^*, and the limiting case corresponding to Langevin's approximation is reproduced. Exact calculations of electrical conductivity have been also carried out by Bakshi and Gross[11]. These authors have considered the source-free case with $\lambda = 1$ and with a suitable separable collision term. Because of the difference between the models used for the collision term, our results for $\lambda = 1$ are not comparable with the ones obtained in Ref. 11, except that — for constant collision frequency — the Ohm's law is rigorous.

III: Time-Dependent Asymptotic Solution

1. General Remarks

We now consider the case when a source of the type

$$S_0(v,t) = Q_0\ \delta(t)\ \delta(v_0^Q - v) \tag{19}$$

acts on an infinite random distribution of f. p. Consequently, in Eqs. (7) $M_\ell(\bar{x}, v, t) \Rightarrow M_\ell(v, t)$ and $t_0 = 0$. Referring from now on to the total density $M_0 \Rightarrow N(v, t)$ in the limit of both

v_0^Q, $v_0^I \to 0$, and accounting for the azimuthal symmetry, from Eqs. (7) we obtain ($\cos\vartheta = \bar{k}\cdot\bar{\Omega} = \mu$)

$$N(v,t) = \frac{Q_0 v^2}{2}\int_{-1}^{1} d\mu \int_0^t \frac{d\tau}{v^2(\tau)}\, e^{-C\tau} \left\{\lim_{v_0^Q\to 0} \delta\left[v_0^Q - v(\tau)\right]\right\} \delta(t-\tau)$$

$$+ \frac{\nu_I C_I}{2} v^2 \int_0^\infty dv' \int_{-1}^{1} d\mu \int_0^t \frac{d\tau}{v^2(\tau)}\, e^{-C\tau} \left\{\lim_{v_0^I\to 0} \delta\left[v_0^I - v(\tau)\right]\right\} N(v',t-\tau)$$

$$+ \frac{C_S v^2}{2} \int_0^\infty dv' \int_{-1}^{1} d\mu \int_0^{\tau_M} \frac{d\tau}{v^2(\tau)}\, e^{-C\tau}\, \delta\left[v'-v(\tau)\right] N(v',t-\tau) \tag{20}$$

where

$$\tau_M \equiv \tau_M(v,t,\mu) = (m/2F)\left[(Ft/m)^2 - v^2\right]\left[(Ft/m) - v\mu\right]^{-1} \tag{21}$$

is the maximum value of τ in the limit of $v_0^Q \to 0$.

If now we exchange the order of integration between μ and τ, and determine correspondingly the limits of v' in both the scattering and the creation integrals $\{0 \leqslant v' \leqslant (1/2)[(F/m)\, t + v]$ and $0 \leqslant v \leqslant (F/m)t$, respectively$\}$, then for $t \gg 5(m/F)v$ we find that

$$N(v,t) = \frac{Q_0}{2}\frac{m}{F} v \left\{\lim_{v_0^Q\to 0} \frac{1}{v_0^Q} \int_{\frac{m}{F}|v-v_0^Q|}^{\frac{m}{F}(v+v_0^Q)} d\tau\, \frac{e^{-C\tau}}{\tau}\, \delta(t-\tau)\right\}$$

continued

$$+\frac{\nu_I C_I}{2}\frac{m}{F}v\int_0^\infty dv'\left\{\lim_{v_0^I\to 0}\frac{1}{v_0^I}\int_{\frac{m}{F}|v-v_0^I|}^{\frac{m}{F}(v+v_0^I)} d\tau\,\frac{e^{-C\tau}}{\tau}N(v',t-\tau)\right\}$$

$$+\frac{C_S}{2}\frac{m}{F}v\int_0^\infty \frac{dv'}{v'}\int_{\frac{m}{F}|v-v'|}^{\frac{m}{F}(v+v')} d\tau\,\frac{e^{-C\tau}}{\tau}N(v',t-\tau) \tag{22}$$

2. Asymptotic Time-Behavior of N(v, t)

We extract now, from Eq. (22), the behavior of N(v, t) for large v and t, and small v/t, by means of a two-fold integral transform technique. In terms of the dimensionless variable w, the Laplace transform (with respect to t) of both sides of Eq. (22) yields first

$$\frac{\tilde{N}(w,s)}{w} = \tilde{\psi}(w,s) = \frac{Q(s)}{C}\frac{e^{-[1+(s/C)]\,|w|}}{w}$$

$$+\frac{1}{2}\lambda\int_{-\infty}^{\infty} E_1\left[\left(1+\frac{s}{C}\right)|w-w'|\right]\tilde{\psi}(w',s)\,dw' \tag{23}$$

where

$$Q(s) = Q_0 + \nu_I\, C_I\, \tilde{N}_0(s) = Q_0 + \nu_I\, C_I\int_0^\infty \tilde{N}(w',s)\,dw' \tag{23a}$$

is the generalization of Eq. (15a) to any $s \neq 0$, and λ is still given by Eq. (15b). By integrating over w, we would get, for the Laplace transform $\tilde{N}_0(s)$ of the total number of t. p., the general relationship

$$\tilde{N}_0(s) = Q_0/[s-(C_S+\nu_I\, C_I - C)]\ (\mathrm{Re}\ s > -C) \tag{24}$$

showing the existence of a pole at $s = -[C_A + (1 - \nu_I)C_I]$, which leads to exponential relaxation at the particle loss frequency.

Now taking a two-sides Laplace transform with respect to w of both sides of Eq. (23) gives

$$\tilde{\psi}_{II}(p,s) = -\frac{2Q(s)}{C}\tanh^{-1}\frac{p}{(1+s/C)}\left[1 - \frac{\lambda}{p}\tanh^{-1}\frac{p}{(1+s/C)}\right]^{-1} \tag{25}$$

which converges for $\mathrm{Re}[1 + (s/C) \pm p] > 0$.

We refer now to the case when $\nu_I \to 0$. By confining the first inversion to the calculation of the contribution due to the pole

$$s_0(p) = -C[1 - p\,\mathrm{ctnh}(p/\lambda)] \tag{26}$$

in the s-plane, we get for $\mathrm{Re}[1 - p\ \mathrm{ctnh}\ (p/\lambda)] \geqslant 0$

$$\psi_{II}(p,t) = -\frac{2Q_0}{\lambda^2}\frac{\sinh^2\left(\frac{p}{\lambda}\right)}{p}\exp\{-C[1 - p\,\mathrm{ctnh}(p/\lambda)]t\} \tag{27}$$

which holds for large t. In inverting with respect to p we utilize the tauberian theorems for small p

$$\psi_{II}(p,t) \simeq -(2Q_0/\lambda^4)\ p\ e^{-C_A t}\ e^{(Ctp^2)/(3\lambda)} \tag{28}$$

and obtain for large w, but small w/t

$$\psi(w,t) \simeq \frac{4Q_0}{\sqrt{\pi}}\left(\frac{C}{C_S}\right)^4 e^{-C_A t}\left(\frac{3C_S}{4C^2 t}\right)^{3/2} w\ e^{-(3C_S w^2)/(4C^2 t)} \tag{29}$$

Returning to the original speed v and to the total density N(v, t), we rewrite Eq. (29) as

$$N(v,t) = 4(\pi)^{-1/2}\ Q_0^*\ e^{-C_A t}\ (4D_0 t)^{-3/2}\ v^2\ e^{-v^2/(4D_0 t)} \tag{30}$$

where $Q_O^* = Q_O(C/C_s)^4$, and $D_O = F^2/(3m^2C_s)$ is a diffusion coefficient in velocity space. Equation (30) represents a Maxwellian with most probable speed given by $v_p = \sqrt{4D_O t} \propto \sqrt{t}$. The intensity of the corresponding maximum is proportional to $e^{-C_A t}\, t^{-1/2}$. The equivalent temperature is $T(t) = (2mD_O t)/k$. It is interesting to note that the function $\psi(v, t) = v^{-1}N(v, t)$, with N(v, t) expressed by Eq. (30), is in turn the exact solution to the partial differential equation of second order

$$\frac{\partial\psi(v,t)}{\partial t} = D_O \frac{\partial^2\psi(v,t)}{\partial v^2} - C_A\,\psi(v,t) + Q_O^* \frac{\partial(v)}{v}\,\partial(t) \qquad (31)$$

References

[1] Kogan, M. N., Rarefied Gas Dynamics, Plenum Press, New York, 1969.

[2] Grad, H., in Handbuch der Physik, edited by S. Flugge, Springer-Verlag, Berlin, 1958, Vol. 12.

[3] Weinberg, A. M., and Wigner, E. P., The Physical Theory of Neutron Chain Reactors, The University of Chicago Press, Chicago, 1958.

[4] Loeb, L. B., Basic Processes of Gaseous Electonics, 2nd ed., University of California Press, Berkeley, 1960.

[5] Boffi, V. C., Molinari, V. G., Pescatore, C., and Pizzio, C., "Integral Boltzmann Equation for Test Particles in a Conservative Field: II. Exact Solutions to Some Space-Dependent Problems," 10th International Symposium on Rarefied Gas Dynamics, July 19--23, 1976, Aspen, Colo.

[6] Mc Daniel, E. W., Collision Phenomena in Ionized Gases, Wiley, New York, 1964.

[7] Stuart, G. W., and Gerjuoy, E., The Physical Review, Vol. 119, No. 3, August, 1960, p. 892.

[8] Paveri Fontana, S. L., Lettere al Nuovo Cimento, Serie I, Vol. 4, Dicembre 1970, p. 1259.

[9] Boffi, V. C., and Molinari, V. G., Il Nuovo Cimento, Serie XI, Vol. 34, Agosto 1976.

[10] Holt, E. H., and Haskell, R. E., Plasma Dynamics, Mc Millan, New York, 1965.

[11] Bakshi, P. M., and Gross, E. P., Annals of Physics, Vol. 49, 1968, p. 513.

INTEGRAL BOLTZMANN EQUATION FOR TEST PARTICLES IN A CONSERVATIVE FIELD: II. EXACT SOLUTIONS TO SOME SPACE-DEPENDENT PROBLEMS

V. C. Boffi,* V. G. Molinari,* C. Pescatore+
and F. Pizzio ǂ

Laboratorio di Ingegneria Nucleare
dell'Università di Bologna,
Bologna, Italy

Abstract

In this paper, space-dependent stationary problems related to the diffusion of test particles (t. p.) in a random distribution of field particles (f. p.), and in the presence of an external conservative force, are discussed along the lines of the general theory and philosophy illustrated in Part I of this study. In Sec. I, the case of a point-wise monoenergetic isotropic source of t. p. in an infinite medium (three-dimensional problem) is considered, and it is complete with a study of the geometrical location of the trajectories of the t. p. generated by the source itself. In Sec. II, the two monodimensional plane problems for the case of a plane infinite monoenergetic iso-

---

Presented as Paper 68b at the 10th International Symposium on Rarefied Gas Dynamics, July 19-23, 1976, Aspen, Colo.

The authors thank F. Santarelli and C. Stramigioli who supervised the numerical data processing.

* Professor.

+ C.N.R. fellow.

ǂ Associate scientist.

tropic source in a slab of both finite and infinite thicknesses are solved exactly when creation by collision is neglected.

I: The Three-Dimensional Problem

1. Linear Integral Equation for the Total Density

For the stationary case and for an infinite random distribution of f. p., $t_0 = -\infty$. Equations (7) of Ref. 1 then become ($\ell = 0, 1$)

$$M_\ell(\bar{x},v) = S_\ell(\bar{x},v)$$

$$+ \frac{v^{2+\ell}}{4\pi} \int_{4\pi} d\bar{\Omega}\, \bar{\Omega}^\ell \int_0^\infty \frac{d\tau}{v^2(\tau)} \exp\left\{- \int_0^\tau v(\tau')\Sigma[\bar{x}(\tau'), v(\tau')]\, d\tau'\right\}$$

$$\cdot \int dv'v'\, \Sigma_0\, [\bar{x}(\tau), v' \to v(\tau)]\, M_0[\bar{x}(\tau), v'] \tag{1}$$

where $S_\ell(\bar{x}, v)$ denotes the source term and $\Sigma_0(\bar{x}, v' \to v)$ is given by Eq. (5a) of Ref. 1.

We shall handle Eq. (1) for the case of $\ell = 0$ and of an external source of the type

$$S_0(\bar{x},v) = Q_0\, \delta(\bar{x})\, \delta(v_0^Q - v) \tag{2}$$

and for the same physical model as outlined in Sec. I.3 of Ref. 1.

Now taking a three-dimensional Fourier transform with respect to $\bar{x}$ of both sides of Eq. (1) yields, for the transform $\widetilde{N}$ of $N(\bar{x}, v)$

$$\widetilde{N}(\bar{B},v) = \widetilde{S}_N(\bar{B},v) + \int_0^\infty \widetilde{V}_N(v' \to v; \bar{B})\, \widetilde{N}\, (\bar{B},v')\, dv' \tag{3}$$

where

$$\tilde{S}_N(\bar{B},v) = \frac{v^2 Q_0}{4\pi} \int_0^\infty d\tau \; e^{-C\tau} \int_{4\pi} d\bar{\Omega} \; e^{i\bar{B}\cdot[\bar{x}(\tau)-\bar{x}]} \; \frac{\delta[v_0^Q - v(\tau)]}{v^2(\tau)} \tag{4a}$$

and

$$\tilde{V}_N(v' \to v;\, \bar{B}) = \tilde{V}_{N,S}(v' \to v;\, \bar{B}) + \nu_I \, \tilde{V}_{N,I}(v' \to v;\, \bar{B})$$

$$= \frac{v^2}{4\pi} \int_0^\infty d\tau \; e^{-C\tau} \int_{4\pi} d\bar{\Omega} \; \frac{e^{i\bar{B}\cdot[\bar{x}(\tau)-\bar{x}]}}{v^2(\tau)}$$

$$\cdot \; \{C_S \; \delta[v' - v(\tau)] + \nu_I \; C_I \; \delta[v_0^I - v(\tau)]\} \tag{4b}$$

with $v(\tau)$ and $\bar{x}(\tau)$ expressed by Eqs. (8) of Ref. 1.

If, in performing the integral over $\bar{\Omega}$, the positive direction of the external force $\bar{F}$ is taken as a polar z axis, for the elastic scattering transformed kernel $\tilde{V}_{N,S}(v' \to v;\, \bar{B})$ there results

$$\tilde{V}_{N,S}(v' \to v;\, B) = \frac{mv}{2Fv'} \; C_S \; e^{-imB_Z(v^2-v'^2)/(2F)}$$

$$\cdot \int_{\frac{m}{F}|v-v'|}^{\frac{m}{F}(v+v')} d\tau \; \frac{e^{-C\tau}}{\tau}$$

$$\cdot \; J_0 \left\{ \frac{m}{2F} \sqrt{(B_x^2 + B_y^2) \left[-\frac{F^4}{m^4}\tau^4 + \frac{2F^2}{m^2}(v^2+v'^2)\tau^2 - (v^2-v'^2)^2 \right]} \right\} \tag{5a}$$

where J_0 is the Bessel function of zero order and of first kind, and $B_\alpha(\alpha = x, y, z)$ is the general Cartesian component of the vector $\bar{B}$ associated with the three-dimensional Fourier transform.

Once $\tilde{V}_{N,S}$ is known, we find

$$\tilde{S}_N(\bar{B},v) = (Q_0/C_S)\,\{\tilde{V}_{N,S}(v' \to v;\ \bar{B})\}_{v'=v_0^Q} \tag{5b}$$

and

$$\tilde{V}_{N,I}(v' \to v;\ \bar{B}) = (C_I/C_S)\,\{\tilde{V}_{N,S}(v' \to v;\ \bar{B})\}_{v'=v_0^I} \tag{5c}$$

respectively.

2. Explicit Calculation of the Source Term

A final step can be made for the source term inasmuch as the residual integral over τ can be performed after returning to the original space. It is found that

$$S_N(x,y,z;\ v) = \frac{Q_0}{4\pi}\,\frac{F}{m}\,\frac{v}{v_0^Q}\,\delta\left[z - \frac{m}{2F}\left(v^2 - v_0^{Q^2}\right)\right] \cdot \frac{G^+(F/m;\ v,v_0^Q;\ x,y) - G^-(F/m;\ v,v_0^Q;\ x,y)}{\sqrt{(v_0^Q v)^2 - (F/m)^2 (x^2 + y^2)}} \tag{6}$$

with

$$G^{\pm}\left(\frac{F}{m};\ v,v_0^Q;\ x,y\right) = \frac{\exp\left\{-(mC/F)\sqrt{v^2 + v_0^{Q^2} \pm 2\sqrt{(v_0^Q v)^2 - (F/m)^2 (x^2+y^2)}}\right\}}{v^2 + v_0^{Q^2} \pm 2\sqrt{(v_0^Q v)^2 - (F/m)^2 (x^2+y^2)}} \tag{6a}$$

In Eq. (6) we observe that $z \gtrless 0$ according to whether $v_0^Q \lessgtr v$: in the case $z < 0$, $v_0^Q > v$, we take into account a t. p. that leaves the source in a direction opposite to the one of the external force.

For the monodimensional plane case, where $\bar{B}$ is parallel to $\bar{F}$, one gets an expression that can be inverted to give

$$S_N(z,v) = \frac{1}{2} Q_0 \frac{m}{F} \frac{v}{v_0^Q} \delta\left[z - \frac{m}{2F}(v^2 - v_0^{Q^2})\right]$$

$$\cdot \left\{ E_1\left[\frac{mC}{F}|v - v_0^Q|\right] - E_1\left[\frac{mC}{F}(v + v_0^Q)\right]\right\} \tag{7}$$

where $E_1(x)$ is the first-order exponential integral. In the limit of $v_0^Q \to 0$, one has instead

$$S_N(z,v) = Q_0(m/F)\, e^{-(mC/F)v}\, \delta[z - mv^2/(2F)] \tag{7a}$$

3. Geometrical Location of the Trajectories

Going back to the general expression, Eq. (6), for the source term $S_N(\bar{x}, v)$, we realize that, for its reality, and consequently in order to guarantee its positivity, it is required that

$$x^2 + y^2 \leqslant [(m/F)\, v_0^Q\, v]^2 \tag{8}$$

Furthermore, we can argue that, at a fixed plane,

$$z = (m/2F)\,(v^2 - v_0^{Q^2}) \tag{9a}$$

the other two coordinates, x and y, of the general field point $\bar{x}$, must lie in the interior or at the boundary of the circle described by the equation

$$x^2 + y^2 = [(m/F)\, v_0^Q\, v]^2 \tag{9b}$$

Eliminating v from Eqs. (9a) and (9b), we obtain the equation

$$z \equiv \sigma(x,y) = [F/(2mv_0^{Q^2})]\,(x^2 + y^2) - m/(2F)\, v_0^{Q^2} \tag{10}$$

which, for any fixed v_0^Q and F/m, represents the surface of a paraboloid, which includes all of the possible trajectories of t. p. generated by the source [Eq. (2)]. Normalizing to $mv_0^{Q^2}/(2F) = 1$. Eq. (10) can be simply restated as

$$z = (1/4)\ (x^2 + y^2) - 1 \tag{10a}$$

II: Monodimensional Plane Problems

1. Linear Integral Equation for the Total Density

Let us consider the case of a slab having a finite thickness 2T along the z axis. A monoenergetic source $S_0(z, v) = Q_0\delta(z)\delta(v_0^Q - v)$ is located at the plane of symmetry, $z = 0$. The linear integral equation for the total density N(z, v) is now $(-T \leqslant z \leqslant T)$, assuming $v_0^I=0$

$$N(z,v) = S_N(z,v)$$

$$+ \frac{1}{2} v^2\ \nu_I C_I \int_{-1}^{1} d\mu \int_0^{\infty} \frac{d\tau}{v^2(\tau)}\ e^{-C\tau} \int_0^{v_M} dv'\ \delta[v_0^I - v(\tau)]\, N[z(\tau),\, v']$$

$$+ \frac{1}{2} v^2\ C_S \int_{-1}^{1} d\mu \int_0^{t-t_0} \frac{d\tau}{v^2(\tau)}\ e^{-C\tau} \int_0^{v_M} dv'\ \delta[v' - v(\tau)]\, N[z(\tau),\, v'] \tag{11}$$

where $S_N(z, v)$ is given by Eq. (7) and $\mu = (\bar{F}/F)\cdot\bar{\Omega}$. As for $S_N(z, v)$ we assume that the abscissa at which $v = 0$, namely, $z_{min} = -\ mv_0^{Q^2}/(2F)$, lies in (- T, 0) so that t. p. cannot escape from the left edge of the slab. In Eq. (11) there remains however to determine the consistent range of v' as well as the upper limit $t - t_0$ of the in-scattering integral.

For the range of v' we find that

$$0 \leqslant v' \leqslant v_M = \sqrt{(2F/m)\, T + v_0^{Q^2}} = \sqrt{(2F/m)\,(T - z) + v^2} \tag{12}$$

as follows by accounting for the delta spatial behavior of $S_N(z, v)$.

For the range of τ in the in-scattering integral we find that

$$0 \leqslant \tau \leqslant t - t_0 \Rightarrow \tau_M =$$

$$= (m/F) \left\{ v\,\mu \pm \sqrt{v^2\,\mu^2 - (v^2 - v'^2)} \right\} \tag{13}$$

in which we must take $\mu^2 > (v^2 - v'^2)/v^2$ when $0 \leqslant v' \leqslant v$.

In force of Eqs. (12) and (13) both creation and in-scattering integrals can be manipulated in such a way that Eq. (11) itself can be recast as

$$N(z,v) = S_N(z,v) + \nu_I (C_I/C_S) \left\{ V_{N,S}(v' \rightarrow v) \right\}_{v'=v_0^I}$$

$$\cdot \int_0^{v_M} N[z - m(v^2 - v_0^{I^2})/(2F),\, v']\, dv'$$

$$+ \int_0^{v_M} V_{N,S}(v' \rightarrow v)\, N\,[z - m(v^2 - v'^2)/(2F),\, v']\, dv' \tag{14}$$

where the elastic scattering kernel $V_{N,S}(v' \rightarrow v)$ is given by Eq. (14a) of Ref. 1.

As it stands, Eq. (14) is not of immediate and easy solution. It becomes simpler when $\nu_I \rightarrow 0$; that is, when creation by collision is ignored. The resulting equation thus is verified to admit a solution of the form

$$N(z,v) = N(v)\ \delta[z - m(v^2 - v_0^{Q^2})/(2F)] \tag{15}$$

We have studied the case also when $v_0^Q \to 0$. The function N(v) in Eq. (15), which represents the velocity distribution integrated over the thickness of the slab, is defined by

$$N(v) = Q_0(m/F)\ e^{-(mC/F)v} + \frac{1}{2}\frac{m}{F} v\, C_S \cdot \int_0^{\sqrt{(2F/m)T}} \frac{dv'}{v'} \left\{ E_1\left[\frac{mC}{F}|v - v'|\right] - E_1\left[\frac{mC}{F}(v + v')\right]\right\} N(v')\ dv' \tag{16}$$

As done in Ref. 1, we recall the dimensionless speed $w = (mC/F)v$, and then rewrite Eq. (16) as $(- w_M \leqslant v \leqslant w_M)$

$$\frac{N(w)}{w} = \psi(w) = \frac{Q_0}{C}\frac{e^{-|w|}}{w} + \frac{1}{2}\lambda \int_{-w_M}^{w_M} E_1(|w - w'|)\psi(w')\ dw' \tag{17}$$

where $\lambda = C_S/(C_S + C_A) \in (0, 1)$ and $w_M = C\left[(2m/F)T\right]^{1/2}$. By introducing the characteristic function $p_{w_M}(w)$ of the interval $(- w_M, w_M)$, the validity of Eq. (17) can be extended to the whole real axis $(- \infty, \infty)$ according to

$$\psi(w) = \frac{Q_0}{C}\frac{e^{-|w|}}{w} p_{w_M}(w) + \frac{1}{2}\lambda \int_{-\infty}^{\infty} E_1(|w - w'|)\psi(w')p_{w_M}(v')\ dv' \tag{18}$$

In this form, Eq. (18) can be treated by a Fourier integral transform technique, and can be given a rigorous constructive solution, which, in the limit of the approximation of order $N \geqslant 0$, reads as $(w \geqslant 0)$

$$N^{(N)}(w) = \frac{Q_0}{C} e^{-w} - \frac{\lambda w}{4\pi\sqrt{w_M}} \sum_{n=0}^{[(N-1)/2]} (-1)^n (4n+3)^{1/2} \beta_{2n+1}^{(N)} \cdot \int_{-w_M}^{w_M} E_1(|w-w'|) P_{2n+1}(w'/w_M)\, dw' \tag{19}$$

where $[(N-1)/2]$ is the maximum integer contained in $[(N-1)/2]$, $P_n(x)$ is the nth Legendre polynomial, and the coefficients $\beta_{2n+1}^{(N)}$ are in turn solutions to the linear system of simultaneous algebraic equations $(m = 0, 1, \ldots [(N-1)/2])$

$$\beta_{2m+1}^{(N)} - \lambda \sum_{n=0}^{[(N-1)/2]} A_{2n+1,2m+1} \beta_{2n+1}^{(N)} = B_{2m+1} \tag{20}$$

Both the matrix elements $A_{2m+1,2n+1}$ and the free terms B_{2m+1} of the system [Eq. (20)] are known explicitly, and are given by

$$A_{2n+1,2m+1} = (-1)^{m-n} (4m+3)^{1/2} (4n+3)^{1/2} w_M \cdot \left\{ \delta_{2n+1,0}\, \delta_{2m+1,0}\, E_1(2w_M) + \sum_{\ell=0}^{2m+2n+3} \gamma_\ell^{2m+1,2n+1} V_\ell(2w_M) \right\} \tag{21a}$$

and

$$B_{2m+1} = 2\sqrt{\frac{\pi}{w_M}} \sum_{k=0}^{m} \frac{(4k+1)!!}{2k+1} C_{2m-2k}^{k+1/2}(0)\, V_{2k}(w_M) \tag{21b}$$

respectively.

In Eqs. (21), C_m^ν indicates a Gegenbauer ultraspherical polynomial, the coefficients $\gamma_\ell^{2m+1,2n+1}$ are given by the recursion formula

$$\gamma_\ell^{2m+1,2n+1} = \begin{cases} (2\delta_{2m+1,2n+1})/(4m+3) & \text{for } \ell = 0 \\ -1 & \text{for } \ell = 1 \\ \dfrac{(-1)^\ell 2}{(\ell+1)(\ell-1)!} \prod_{k=1}^{\ell-1} (2m+2n+3+\ell-2k) & \\ \cdot\, (|2m-2n| + \ell - 2k) & \text{for } \ell > 1 \end{cases} \tag{22a}$$

and the factor $V_\ell(x)$ is given by

$$V_\ell(x) = x^{-(\ell+1)} \left(1 - e^{-x} \sum_{h=0}^{\ell} \frac{x^h}{h!}\right) \tag{22b}$$

In addition, since the integral of the last factor in the sum on the right-hand side of Eq. (19) can be performed explicitly by parts, we may conclude by observing that the solution $N^{(N)}(w)$ of Eq. (19) can be analytically evaluated in an explicit manner.

In Figs. 1a and 1b, we plot the result of Eq. (19) (times w_M^{-1}) as a function of $\zeta = w/w_M = (z/T)^{1/2}$ for different λ and for $w_M = 0.1$ and 1, respectively. It can be seen how the velocity spectra depend strongly on the thickness of the slab considered (compare also Fig. 1 of Ref. 1, which refers to the case of an infinite medium).

In Figs. 1c and 1d we plot instead the total density of t. p. [times $(C^2m/F)^{-1}\, w_M$] — which is obtained by integrating over v both sides of Eq. (15) — as a function of ζ for the same λ and w_M of Figs. 1a and 1b.

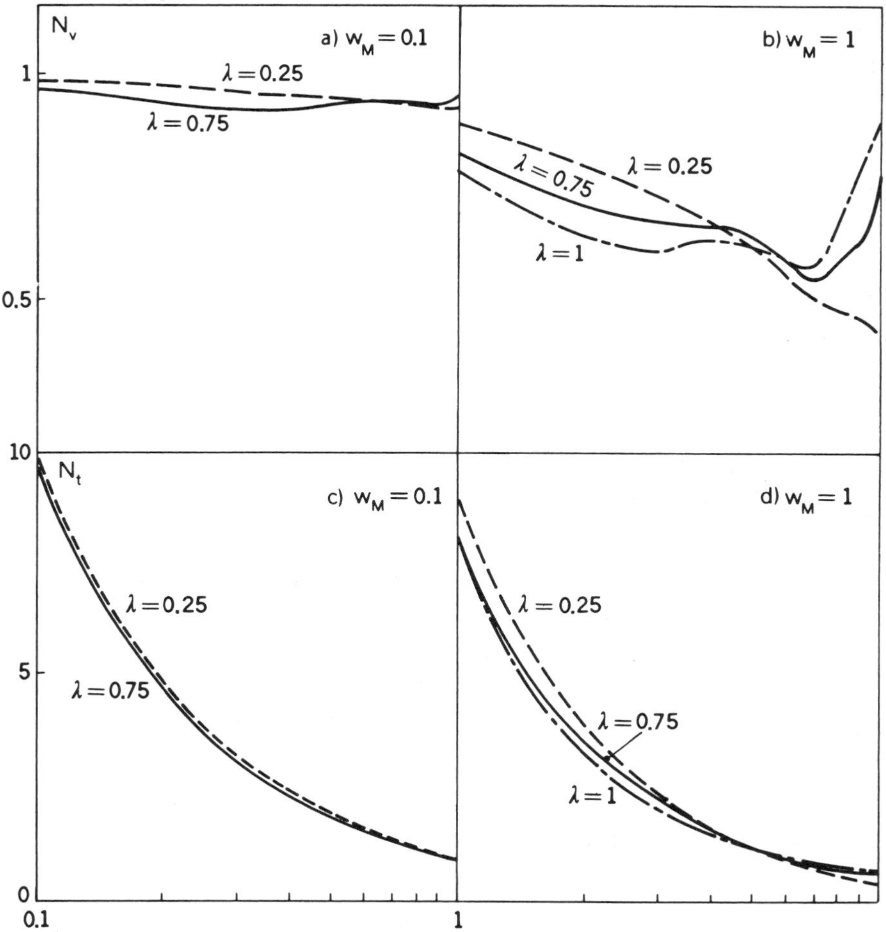

Figs. 1a) and 1b) The velocity distribution (times w_M^{-1}) as a function of $\zeta = w/w_M$ for different λ, and for $w_M = 0.1$ and 1, respectively; c) and d) The total density [times $(C^2m/F)^{-1}w_M$] as a function of $\zeta = (z/T)^{1/2}$ for different λ, and for $w_M = 0.1$ and 1, respectively.

2. Slab of Infinite Thickness

The case of the slab of infinite thickness follows just as the limit of $T \to \infty$ of the preceding treatment for the slab of finite thickness. In particular, Eq. (16) becomes the same equation that was solved in Sec. II of Ref. 1, provided $Q \Rightarrow Q_0$.

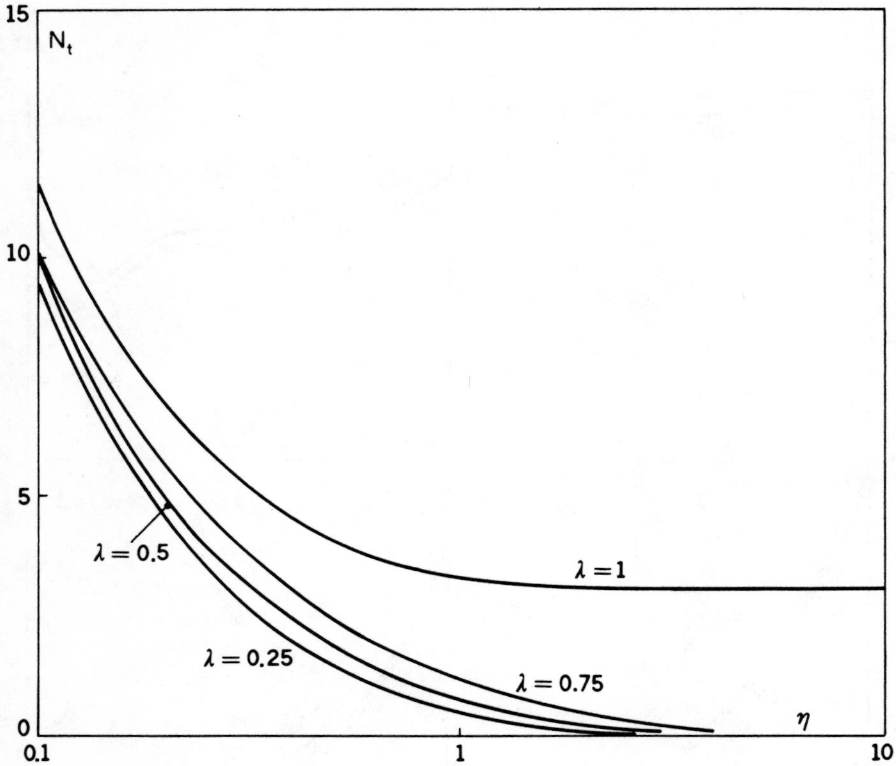

Fig. 2 The total density [times $(C^2m/F)^{-1}$] as a function of $\eta = C(2mz/F)^{1/2}$ for different λ.

On the basis of the results for N(w) plotted in Fig. 1 of Ref. 1 we can thus construct Fig. 2, in which the total density [times $(C^2m/F)^{-1}$] is drawn as a function of $\eta = C(2mz/F)^{1/2}$ for different λ. Comparison between Fig. 2 and Figs. 1c and 1d illustrates the effects of the leakage of t. p. through the edges of a slab of finite thickness.

References

[1] Boffi, V. C., Dorning, J. J., Molinari, V. G., and Spiga, G., "Integral Boltzmann Equation for Test Particles in a Conservative Field: I. Theory, and Exact Solutions to Some Stationary and Time-Dependent Problems "10th International Symposium on Rarefied Gas Dynamics, July 19-23, 1976, Aspen, Colo.

A CONTINUOUS SLOWING-DOWN THEORY FOR TEST PARTICLES

J. Dorning,* C. Pescatore,† and G. Spiga≠

Laboratorio di Ingegneria Nucleare dell'Università di Bologna, Bologna, Italy

Abstract

The slowing down and migration of a rarefied foreign gas of energetic test particles introduced into an equilibrium neutral "host gas" in the presence of a conservative external force field are studied. Approximate solutions for the speed-dependent number density of test particles are generated for the phsyically interesting case in which scattering via hard collisions is the dominant interaction and the change in the number density between collisions due to the force field is small. The multiple integral that appears in the solutions is developed in a series expression for the constant collision frequency case. The development is valid for the widest range of particle speeds when the external field is weak and/or the

---

Presented as Paper 70 at the 10th International Symposium on Rarefied Gas Dynamics, Aspen, Colo., July 19-23, 1976. We very gratefully acknowledge numerous stimulating and informative discussions, in connection with the material presented in this paper, with V. Boffi and V. Molinari. One of us (J.D.) also wishes sincerely to acknowledge the support of the Italian National Research Council (Consiglio Nazionale delle Ricerche) through its visiting professorship program, and the splendid hospitality provided by V. Boffi and the other members of the Laboratorio di Ingegneria Nucleare dell'Universita di Bologna during the course of this work. During the final phases of this work one of the authors (J.D.) was supported by the U.S. National Science Foundation under Grant No. NSF/ENG 74-13029.

*C.N.R. Visiting Professor of Mathematical Physics; Professor of Nuclear Engineering on leave of absence from the University of Illinois at Urbana-Champaign, Urbana, Ill.

†C.N.R. Fellow.

≠Assistant.

average particle collision time is small. For fixed field and collision time, it converges for large test particle speeds.

I. Introduction

The integral form of the linearized Boltzmann equation[1] for isotropic scattering (hard collisions) is used to study the slowing down and migration of a rarefied foreign gas of energetic test particles introduced into an equilibrium neutral "host gas" in the presence of a conservative external force field when elastic scattering is the dominant interaction. The equation is simplified by expanding the inscattering source integral in a Taylor series about the field point variables. This expansion is justified when the number density is a slowly varying function of position, speed, and time between collisions, which is the case when the external field is weak, or the mean collision time is short, or the test particle speed is large.

The lowest-order form of this equation then is applied to the case in which the test particle mass m is vanishingly small compared to the host gas field particle mass M. The solution for this case is immediate and is written in terms of a multiple integral which accounts for the attenuation, due to total collisions (out-scattering, absorption, etc.), of the inscattered particles along their trajectories from the scattering point to the field point. This approximate solution has the general form of the direct contribution due to the transport and attenuation of the external source of particles through the force field plus a correction term that accounts to first order for the particles that are scattered into trajectories in the force field in such a way that they contribute to the number at the field point under consideration. The multiple integral that appears in the solution then is developed in a series for the constant-frequency case. This development, which is consistant with the earlier Taylor series, is valid for the widest range of particle speeds when the external field is weak and/or the mean test particle collision time is small. For fixed field strength and collision time, it converges for large test particle speeds.

Next, the case in which the test particle mass is small, but not vanishingly small, compared to the field particle mass is examined using the same approximate form for the integral equation. Here, since the energy losses due to collisions are nonzero but small, the collision density of particles undergoing inscattering collisions is expanded in a Taylor series about the final collision speed. Zeroth-order truncation leads to the zero-mass-ratio result already obtained. Truncation of this expansion after the first-order term in the average logarithmic energy decrement due to collisions leads to a continuous slowing-down theory for test particles that undergo hard collisions with a statistical system in an external force field. The resulting first-order differential equation in the speed variable is solved for the test particle collision density and number density in terms of the same multiple integral that appeared in the zero-mass-ratio limit. Thus, this solution also is evaluated explicitly for large test particle speeds by using the weak-field/short-collision-time or large-speed development already obtained.

Physical interpretations of the results obtained are proffered, and comments are made on the effects of retaining higher-order terms in the expansions. The special case of test particles that have masses equal to the field particles and that slow down via hard-sphere collisions also is discussed briefly.

II. Integral Boltzmann Equation

The integrodifferential linearized Boltzmann equation for the single-particle distribution function or velocity-dependent number density of test particles (foreign gas) introduced into a statistical system of scattering and absorbing field particles (host gas) in the presence of an external force field is given by

$$[Dn(\underline{r},\underline{v},t)/Dt] + \nu(\underline{r},|\underline{v}|)n(\underline{r},\underline{v},t)$$
$$= \int d^3\underline{v}' \ \nu_s(\underline{r},|\underline{v}'|)\Pi_s(\underline{v}'\to\underline{v})n(\underline{r},\underline{v}',t) + S(\underline{r},\underline{v},t) \qquad (1)$$

where

$$D/Dt \equiv (\partial/\partial t) + \vec{v} \cdot \vec{\nabla}_{\underline{r}} + (1/m)\vec{F}(\underline{r},t)\cdot\vec{\nabla}_{\underline{v}} \qquad (1a)$$

is the substantial derivative, $\nu(\underline{r},|\underline{v}|)$ is the collision frequency, $\Pi_s(\underline{v}'\to\underline{v})$ is the probability distribution function of final velocities of scattered particles, and $S(\underline{r},\underline{v},t)$ is the fixed distributed source of test particles being introduced into the system. This equation can be converted to an integral equation by integrating the substantial derivative[1]

$$n(\underline{r},\underline{v},t) = \int_0^{t-t_1(\underline{r},\underline{v})} d\tau \ \exp[-\ I(\tau,\underline{r},\underline{v})]\{Q[\underline{r}(\tau),\underline{v}(\tau),t-\tau]$$
$$+ S[\underline{r}(\tau),\underline{v}(\tau),t-\tau]\} \qquad (2)$$

Here $Q(\underline{r},\underline{v},t)$ is the inscattering source contribution to the velocity-dependent number density

$$Q(\underline{r},\underline{v},t) \equiv \int d^3\underline{v}' \; \nu_s(\underline{r},|\underline{v}'|)\Pi_s(\underline{v}'\to\underline{v})n(\underline{r},\underline{v}',t) \tag{3}$$

The integral $I(\tau,\underline{r},\underline{v})$, which accounts for the attenuation (due to outscattering and absorption) of the number density as the particles travel along trajectories through the force field to the field point of interest from the point at which they were scattered into or born on this trajectory, is given by

$$I(\tau,\underline{r},\underline{v}) = \int_0^{\tau} \nu[\underline{r}(\tau'),|\underline{v}(\tau')|]d\tau' \tag{4}$$

and τ is the backward elapsed time along these trajectories to all particle inscattering points and to the points of initial introduction of the particles into the system. The quantity $t_1(\underline{r},\underline{v})$ determines when the integration of the backward elapsed time terminates. Since it is affected by the system boundaries and the geometry of the distributed source via $\underline{r}$ and $\underline{v}$ it has been written as a function of these variables. However, this functional dependence is redundent with the dependences of the collision frequencies and the fixed distributed source upon $\underline{r}$ and $\underline{v}$. Thus, t_1 can be treated as a constant if the explicit dependences of the collision frequencies and the fixed distributed source upon geometry are included through their arguments $\underline{r}(\tau)$ and $\underline{v}(\tau)$ in Eq. (2). After defining the angular number density $n(\underline{r},v,\hat{\Omega},t) = v^2 n(\underline{r},\underline{v},t)$ and its integral over all particle directions, the speed-dependent scalar number density $N(\underline{r},v,t)$, integration of Eq. (2) over all direction vectors $\hat{\Omega}$ yields an integral equation for the speed-dependent number density of test particles slowing down and migrating in the host gas when scattering is isotropic

$$N(\underline{r},v,t) = \frac{v^2}{4\pi}\int_{4\pi} d\hat{\Omega}\int_0^{t-t_1} d\tau \; \exp[-\, I(\tau,\underline{r},\underline{v})] \; \frac{1}{v^2(\tau)}$$
$$\cdot \; \{Q[\underline{r}(\tau),v(\tau),t-\tau] + 4\pi \; S[\underline{r}(\tau),v(\tau),\hat{\Omega}(\tau),t-\tau]\} \tag{5}$$

where, for isotropic scattering

$$Q(\underline{r},v,t) = \int_0^{\infty} dv' \; \nu_s(\underline{r},v')\Pi_s(v'\to v)N(\underline{r},v',t) \tag{6}$$

Although the source term is known in general and therefore can be treated explicitly, it is taken as isotropic in the remainder of this work for simplicity. The final form of the integral equation is then

$$N(\underline{r},v,t) = \frac{v^2}{4\pi}\int_{4\pi} d\hat{\Omega}\int_0^{t-t_1} d\tau \frac{\exp[-\ I(\tau,\underline{r},\underline{v})]}{v^2(\tau)}\ Q[\underline{r}(\tau),v(\tau),t-\tau]$$

$$+\ S_N(\underline{r},v,t) \qquad (7)$$

where $S_N(\underline{r},v,t)$ is given explicitly in Eq. (5). The functions $\underline{r}(\tau)$ and $\overline{v}(\tau)$ are determined[1] from the equations of test particle motion in the force field, which is taken to be constant here, as

$$\vec{r}(\tau) = \vec{r} + (1/2)(\vec{F}/m)\tau^2 - \vec{v}\tau \qquad (8)$$

$$v(\tau) = \surd[v^2+(F^2/m^2)\tau^2-2(\vec{F}/m)\cdot\hat{\Omega}\, v\, \tau] \qquad (9)$$

For anisotropic scattering $Q(\underline{r},v,t)$ will involve higher angular moments of $n(\underline{r},v,\hat{\Omega},t)$, and Eq. (7) will be replaced by a set of coupled equations for these angular moments. Here, however, we shall limit our considerations to isotropic scattering.

When a collision frequency is not well-defined, such as in the case of a long-range soft potential (e.g., the coulomb potential), the original integral equation [Eq. (2)] cannot be obtained in the form given here. However, alternative integral equations which are analogous can be obtained for most specific cases of interest by anticipating the approximate representation of the collision term and letting the multiplicative operator which appears in this representation play the role of the collision frequency.

Since elastic scattering is the dominant interaction, when the external field acting upon the test particles is weak [or the test-particle field-particle collision frequency is large (collision time is short), or the test particle speed is large], the number density of test particles will vary very little as the test particles travel along their trajectories between collisions. Thus the source of inscattered particles per unit elapsed time $Q[\underline{r}(\tau),v(\tau),t-\tau]$ will be a slowly varying function, and it can be expanded in a Taylor series about the field point variables $\underline{r}$, v, and t when the fixed source is smoothly varying or located far from the field points of interest. The result is

$$N(\underline{r},v,t) = \frac{v^2}{4\pi} \int_{4\pi} d\hat{\Omega} \int_0^{t-t_1} d\tau \, \frac{\exp[-\, I(\tau,\underline{r},\underline{v})]}{v^2(\tau)} \quad \{Q(\underline{r},v,t)$$

$$+ \left(\frac{1}{2}\frac{\vec{F}}{m}\tau^2 - \vec{v}\tau\right)\cdot \vec{\nabla}_{\underline{r}}\, Q(\underline{r},v,t) + [v(\tau)-v]\frac{\partial}{\partial v} Q(\underline{r},v,t)$$

$$- \tau \frac{\partial}{\partial t} Q(\underline{r},v,t) + \ldots\} + S_N(\underline{r},v,t) \tag{10}$$

Although the first integral on the right-hand side of Eq. (10) is now merely a coefficient, the equation is still an integral equation in the speed variable because of the definition of $Q(\underline{r},v,t)$. The lowest-order truncation of this expansion leads to the approximate integral equation

$$N(\underline{r},v,t) \sim \left\{\frac{v^2}{4\pi}\int_{4\pi} d\hat{\Omega} \int_0^{t-t_1} d\tau \, \frac{\exp[-\, I(\tau,\underline{r},\underline{v})]}{v^2(\tau)}\right\} \int_0^{\infty} dv' \; \nu_s(\underline{r},v')\Pi_s(v'\to v)$$

$$\cdot N(\underline{r},v',t) + S_N(\underline{r},v,t) \tag{11}$$

in which the effect of the field acting upon the particles during their trajectories between collisions only appears in the coefficient in the braces which multiplies the inscatter source. This coefficient simply represents the attenuation of particles due to outscattering and absorption events that remove them from the (generally curved) trajectories along which they otherwise would travel from the point of their inscattering to the field point under consideration. For the case of a smoothly varying source $S[\underline{r}(\tau),v(\tau),t-\tau]$, it might be desirable also to expand it about the field points $\underline{r}$, v, and t so that the two terms can be treated to the same order in a consistent manner. Whether this is done or not, the source term is known explicitly in general, and $S_N(\underline{r},v,t)$ will be treated as a known function in what follows. The actual construction of this function, which corresponds to the solution in a purely absorbing system in the presence of an external field, is not a trivial task especially in finite-system geometries. It has been derived for the case of the time-dependent problem in an infinite medium[1] and for the steady-state problem in slab geometry.[2]

III. Zero-Mass-Ratio Case

When the mass of the test particles is vanishingly small compared to that of the host particles (i.e., $m/M \to 0$), the energy lost by the energetic test particles in elastic collisions with the host particles approaches zero, and the kernel $\Pi_S(v' \to v)$ approaches a delta function $\delta(v'-v)$. In this case, the solution to Eq. (11) can be written immediately

$$N(\underline{r},v,t) = S_N(\underline{r},v,t)$$

$$+ \left\{ \frac{[\nu_s(\underline{r},v)/\nu(\underline{r},v)]f[\nu(\underline{r},v),t-t_1,|\vec{F}|/m;v]}{1 - [\nu_s(\underline{r},v)/\nu(\underline{r},v)]f[\nu(\underline{r},v),t-t_1,|\vec{F}|/m;v]} \right\} S_N(\underline{r},v,t) \qquad (12)$$

Here the second term represents a correction to the direct trajectory contribution [i.e., $S_N(\underline{r},v,t)$] which accounts for particles that are scattered into trajectories that lead them to field point $(\underline{r},v,t)$. The transcendental function that appears here is a function of $\nu(\underline{r},v)$, $|\vec{F}|/m$, $t-t_1$, and v and is given by

$$f[\nu(\underline{r},v),t-t_1,|\vec{F}|/m;\ v]$$

$$= \frac{\nu(\underline{r},v)}{4\pi} \int_{4\pi} d\hat{\Omega} \int_0^{t-t_1} d\tau \ \frac{\exp\left\{-\int_0^{\tau} \nu[\underline{r}(\tau'),v(\tau')]d\tau'\right\}}{1 - 2(\vec{F}/mv)\cdot\hat{\Omega}\tau + (F/mv)^2\tau^2} \qquad (13)$$

It is shown in the Appendix that, when the collision frequency is independent of position (uniform host gas), this function depends only upon three variables: the collision frequency, the maximum fractional change of test particle energy (due to the external field) between collisions

$$\beta^2(v) = \left(\frac{(|\vec{F}|/m)\tau_c(v)}{v} \right)^2 \qquad (14)$$

and the mean number of collisions experienced by particles with speed v from the time the source was initiated

$$\bar{n}_c(v) = (t-t_1)/\tau_c(v) \qquad (15)$$

where $\tau_c(v)$ is the mean time between collisions defined as $\tau_c(v) = 1/\nu(v)$. In this case, the function becomes

$$f[\nu(v),\beta^2(v),\overline{n}_c(v)] = \frac{1}{2}\int_{-1}^{+1} d\mu \int_0^{\overline{n}_c} dy \, \frac{\exp\left\{-\int_0^y \{\nu[v(y')]/\nu[v]\}dy'\right\}}{1-2\beta(v)\mu y+\beta^2(v)y^2} \tag{16}$$

Although it is obvious that the second term on the right-hand side of Eq. (12) is a correction term that includes the effect of inscattered particles, the present form of the function $f[\nu(v),\beta^2(v),\overline{n}_c(v)]$ does not lead easily to the physical interpretation of this correction term. For the case of constant collision frequency (Maxwell molecule interaction), this function can be developed in a series from which its physical interpreation can be deduced more easily. This is done in the Appendix. It first is shown in the Appendix that Eq. (16) can be written as

$$f(\beta^2,\overline{n}_c) = \frac{1}{\beta}\int_0^{\overline{n}_c} dy \, \frac{e^{-y}}{y} \, \text{Arth}(\beta y) = \frac{1}{2\beta}\int_0^{\overline{n}_c} dy \, \frac{e^{-y}}{y} \, \log\left(\frac{1+\beta y}{1-\beta y}\right) \tag{17}$$

and that it is related to the exponential integral function by

$$f(\beta^2,\overline{n}_c) = \frac{1}{2}\left\{\int_{-\infty}^{-\frac{1}{\beta}} + \int_{\frac{1}{\beta}}^{\infty}\right\} \frac{e^{\eta}}{\eta}\left\{E_1(\eta)-E_1(\eta+\overline{n}_c)\right\} d\eta \tag{18}$$

In the final part of the Appendix, this function is developed in a series in powers of β^2. For convergence we must have $\beta^2<1$, which is consistent with the original Taylor expansion of the inscatter source made in Sec. II. However, there is also an additional requirement that $\beta\overline{n}_c<1$, which implies that the maximum total change of speed (or energy) of the test particles due to the force field (which causes the only energy change, since m/M = 0) along their complete trajectories from the time the source was initiated t_1 to the time they reach the field point t is less than the speed (or energy) that they have at the field point. (Since the field is conservative and no energy is lost during collisions when m/M = 0, test particles that arrive at a particular field point and were introduced at the same source point must have the same energy independent of how torturous their trajectory was. Their attenuation probability along different trajectories, of course, would be very different in general.) The series expression is

$$f(\beta^2,\bar{n}_c) = f_0(\bar{n}_c) + f_2(\bar{n}_c)\beta^2 + f_4(\bar{n}_c)\beta^4 + O(\beta^6) \tag{19}$$

where

$$f_0(\bar{n}_c) = 1 - \exp(-\bar{n}_c) \tag{19a}$$

$$f_2(\bar{n}_c) = (1/3)\{f_0(\bar{n}_c)-\bar{n}_c(\bar{n}_c+2)\exp(-\bar{n}_c)\} \tag{19b}$$

$$f_4(\bar{n}_c) = (1/5)\{36\ f_2(\bar{n}_c)-\bar{n}_c^3(\bar{n}_c+4)\exp(-\bar{n}_c)\} \tag{19c}$$

Substituting this expansion into the coefficient of the correction term in Eq. (12) and recalling that the denominator must be cleared so that only terms of order β^4 are retained in the coefficient leads to

$$\begin{aligned} N(\underline{r},v,t) &= S_N(\underline{r},v,t) + \{C_0+(1+C_0)C_2\beta^2 \\ &\quad + (1+C_0)(C_2^2+C_4)\beta^4 + O(\beta^6)\}\ S_N(\underline{r},v,t) \end{aligned} \tag{20}$$

where

$$C_n = C_n\left(\frac{\nu_s}{\nu};\ \bar{n}_c\right) = \frac{(\nu_s/\nu)f_n(\bar{n}_c)}{1-(\nu_s/\nu)f_0(\bar{n}_c)} \tag{20a}$$

The correction term in the braces, which accounts for particles scattered into trajectories that carry them to the field point of interest, to lowest order in β^2, is

$$C_0\left(\frac{\nu_s}{\nu};\ \bar{n}_c\right) = \frac{(\nu_s/\nu)f_0(\bar{n}_c)}{1-(\nu_s/\nu)f_0(\bar{n}_c)} \tag{20b}$$

Here ν_s/ν is the probability that a collision is a scattering collision, $f_0(\bar{n}_c) = 1 - \exp(-\bar{n}_c)$ is the probability that the scattered test particle survives subsequent collisional attenuation along its trajectory to the field point, and the denominator corrects $S_N(\underline{r},v,t)$, which includes attenuation of the source due to all collisions, for the fact that scattering collisions effectively do not contribute to this attenuation. Since ν_s/ν and $f_0(\bar{n}_c)$ are both less than unity, the denominator also can be expanded. This leads to a series representation for this lowest-order correction term in powers of $(\nu_s/\nu)f_0(\bar{n}_c)$ which represents the additional contributions to the number density at the field point due to once-scattered particles, twice-scattered particles, etc. Equation (20) then becomes

$$N(\underline{r},v,t) = \left\{ \sum_{n=0}^{\infty} \left[\frac{\nu_s}{\nu} f_0(\overline{n}_c)\right]^n \right\} S_N(\underline{r},v,t) \tag{21}$$

which corresponds to virgin particles plus once-scattered particles, etc., to zeroth order in β^2. Similar physical interpretations can be made for the higher-order terms in β^2 that appear in Eq. (20) which explicitly includes the effect of the field through β^2. On the other hand, the two terms in Eq. (20) can be combined to form

$$N(\underline{r},v,t) = (1+C_0)\{1+C_2\beta^2+(C_2^2+C_4)\beta^4+0(\beta^6)\}S_N(\underline{r},v,t) \tag{22}$$

where the factor $(1+C_0)$ represents the corrections just discussed, and the higher-rder terms in β^2 represent corrections due to the combination of force field and the collisions to various orders in the force field via the parameter β^2. Equations (21) and (22) suggest the direct formal expansion of Eq. (12) in powers of $[\nu_s(\underline{r},v)/\nu(\underline{r},v)]\ f[\nu(\underline{r},v),t-t_1,|\vec{F}|/m;v]$ to lead to a similar but more general result, corresponding to various order terms in scattering collisions and subsequent transport along trajectories through the field. However, such an expansion cannot be justified a priori, since $[\nu_s(\underline{r},v)/\nu(\underline{r},v)]\ f[\nu(\underline{r},v),t-t_1,|\vec{F}|/m;v]$ is not less than one in general. (In some cases that are consistent with, but more restrictive than, the original Taylor expansion this development can be justified.) Similarly, Eqs. (21) and (22) also suggest a Neumann series solution of Eq. (11) which would result in terms that would correspond to various orders of combined inscatterings followed by attenuation along a transport trajectory. However, again we lack a priori justification, since the operator defined in Eq. (11) is not obviously a contraction operator.

In the next section, we relax the restriction m/M=0 but still require that this ratio be small. Since the same multiple integral as examined previously occurs in that case, much of the preceding discussion will apply there also.

IV. Continuous Collisional Slowing Down

In order to study the case when the mass ratio m/M is small but finite, it is convenient to return to Eq. (7) and introduce the scattering collision frequency, defined by

$F_s(\underline{r},v,t) = \nu_s(\underline{r},v)N(\underline{r},v,t)$, as a new dependent variable and and the lethargy, defined by $u=2\ell og(v_1/v)$, where v_1 is some reference speed, as a new independent variable. [A similar definition also is introduced for $u(\tau)$.] These variables are used widely in the theory of neutron slowing down[3] and are useful in particle slowing down problems in general. This results in

$$F_s(\underline{r},u,t) = \frac{v^2(u)\nu_s(\underline{r},u)}{4\pi} \int_{4\pi} d\hat{\Omega} \int_0^{t-t_1} d\tau \, \frac{\exp[-I(\tau,\underline{r},\underline{v})]}{v^2(\tau)} \frac{dv}{dv(\tau)}$$

$$\cdot \frac{du(\tau)}{du} \int_{-\infty}^{\infty} du' \Pi_s[u' \to u(\tau)] F_s[\underline{r}(\tau),u',t-\tau] + \nu_s(\underline{r},u) S_N(\underline{r},u,t) \quad (23)$$

Since the mass ratio is small, the maximum lethargy gain (energy or speed loss) during a collision is small, and, since scattering is the dominant interaction, the collision density will be a slowly varying function of lethargy over the interval $u'-u(\tau)$ (provided that the source of test particles is either slowly varying or distant). Thus $F_s[\underline{r}(\tau),u',t-\tau]$ is expanded in a Taylor series about $u(\tau)$. Noting that, since $\Pi_s(u' \to u)$ is a probability distribution function, it is normalized to unity and

$$\int_{-\infty}^{+\infty} (u-u')\Pi_s(u'-u)du' = \overline{(\Delta u)_{coll}}(u) = \xi(u) \quad (23a)$$

where $\xi(u)$ is the average gain in lethargy (or average logarithmic energy decrement) per collision, and, in general

$$\int_{-\infty}^{+\infty} (u-u')^n \Pi_s(u' \to u)du' = \overline{(\Delta u)^n_{coll}}(u) \quad (23b)$$

we substitute the expansion, truncate it after two terms, and carry out the integrations in the u' variable to obtain

$$F_s(\underline{r},u,t) \sim \frac{v^2(u)\nu_s(\underline{r},u)}{4\pi} \int_{4\pi} d\hat{\Omega} \int_0^{t-t_1} d\tau \, \frac{\exp[-I(\tau,\underline{r},\underline{v})}{v^2(\tau)} \frac{dv}{dv(\tau)} \frac{du(\tau)}{du}$$

$$\cdot \left\{ F_s[\underline{r}(\tau),u(\tau),t-\tau] - \xi(u(\tau)) \frac{\partial}{\partial u(\tau)} F_s[\underline{r}(\tau),u(\tau),t-\tau] \right\}$$

$$+ \nu_s(\underline{r},u) S_N(\underline{r},u,t) \quad (24)$$

Converting back to the speed variable gives

$$F_s(\underline{r},v,t) \sim \frac{v^2 \nu_s(\underline{r},v)}{4\pi} \int_{4\pi} d\hat{\Omega} \int_0^{t-t_1} d\tau \, \frac{\exp[-\,I(\tau,\underline{r},\underline{v})]}{v^2(\tau)}$$

$$\cdot \left\{ \left[1 + \frac{1}{2}\,\xi(v(\tau))\right] F_s[\underline{r}(\tau),v(\tau),t-\tau] \right.$$

$$\left. + \frac{1}{2}\,v(\tau)\xi[v(\tau)] \,\frac{\partial}{\partial v(\tau)}\, F_s[\underline{r}(\tau),v(\tau),t-\tau] \right\} + \nu_s(\underline{r},v) S_N(\underline{r},v,t) \tag{25}$$

where the factor in braces is now the source of inscattered particles given in Eq. (6) in Sec. II. Repeating the steps that were followed in that section to expand that quantity about the field point variables leads to Eq. (10), with $Q(\underline{r},v,t)$ replaced by the factor in braces in Eq. (25). [Alternatively, the function $Q(\underline{r},v,t)$ in Eq. (10) could have been written in terms of $F_s(\underline{r},u',t)$ and the collision density then expanded about u. That is, the order in which the two Taylor expansions, approximating the collision process and transport process, are carried out is unimportant.] Finally, truncation of the resulting equation at lowest order yields

$$F_s(\underline{r},v,t) \sim [\nu_s(\underline{r},v)/\nu(\underline{r},v)] f[\nu(\underline{r},v),t-t_1,|\vec{F}|/m;v]$$

$$\cdot\{[1+(1/2)\xi(v)]F_s(\underline{r},v,t) + (1/2)\xi(v)v(\partial/\partial v)F_s(\underline{r},v,t)$$

$$+ \nu_s(\underline{r},v) S_N(\underline{r},v,t) \tag{26}$$

where $f(\cdots)$ is the same function that appears in the zero-mass-ratio case discussed in Sec. III. In the limit of zero mass ratio, $\xi(v)$ goes to zero, and Eq. (26) reduces to the case discussed in Sec. III and leads to the solution given by Eq. (12). When $\xi(v)$ is not zero, Eq. (26) still can be solved immediately. The result, obtained using the property that $F_s(\underline{r},v,t) \to 0$ as $v \to \infty$, is

$$F_s(\underline{r},v,t) = \frac{2}{v} \int_v^{\infty} dv' \, \frac{\nu(\underline{r},v')}{\xi(v')f(v')} \exp\left\{ 2\int_v^{v'} \frac{dv''}{v''} \left[\frac{1}{\xi(v'')}\left(1 - \frac{1}{f(v'')}\right) \right.\right.$$

$$\left.\left. - \frac{\nu(\underline{r},v'')-\nu_s(\underline{r},v'')}{\xi(v'')\nu_s(\underline{r},v'')f(v'')} \right] \right\} S_N(\underline{r},v',t) \tag{27}$$

this gives the next higher-order solution than the m/M=0 case and represents a continuous collisional slowing down theory solution, with the effects of the attenuation of the test particles, as they move along trajectories determined by the force field, represented by the presence of the function $f(\cdots)$ which is defined in Eq. (13) and which is developed in powers of β^2 for the constant-collision-frequency case in Eq. (19).

When the test particles slow down according to elastic hard-sphere collisions, the average logarithmic energy decrement becomes a constant[3] ξ. Then, for the constant-collision-frequency case, and when $f(\beta^2,\bar{n}_c)$ is truncated to zeroth order in β^2, this expression for the continuous slowing-down solution becomes

$$F_s(\underline{r},v,t)= \frac{2\nu}{v\xi f_0(\bar{n}_c)} \int_v^\infty dv' \left(\frac{v'}{v}\right)^2 \left\{\frac{1}{\xi}\left[1 - \frac{1}{f_0(\bar{n}_c)}\right] - \frac{\nu-\nu_s}{\xi\nu_s f_0(\bar{n}_c)}\right\} \cdot S_N(\underline{r},v',t) \tag{28}$$

The integral that appears in the exponential in Eq. (27) also can be evaluated explicitly when higher-order terms in β^2 are retained, since $\beta^2 \quad 1/v^2$, and the integrand will be a rational function of v''.

Finally, rewriting Eq. (27) for the test particle number density gives

$$N(\underline{r},v,t)= \frac{2}{v\nu_s(\underline{r},v)} \int_v^\infty dv' \frac{\nu(\underline{r},v')}{\xi(v')f(v')} \exp\left\{2\int_v^{v'} \frac{dv''}{v''} \left[\frac{1}{\xi(v'')} \cdot \left(1 - \frac{1}{f(v'')}\right) - \frac{\nu(\underline{r},v'')-\nu_s(\underline{r},v'')}{\xi(v'')\nu_s(\underline{r},v'')f(v'')}\right]\right\} S_N(\underline{r},v',t) \tag{29}$$

which accounts for the loss of test particles via nonscattering collisions, during the continuous collisional slowing-down process, by the exponential factor in the integral. This factor, however, also includes the effect of the external field upon this loss through the function $f(\cdots)$, which reflects the change of the test particle removal probabilities during transport due to the modification of the particle trajectories by the field.

V. Equal-Mass Case

In the special case that occurs when the test particles and the field particles have equal mass and they interact according to elastic hard-sphere collisions, Eq. (11) can be solved exactly. Rewriting it for this case in terms of the scattering collision density and lethargy variables and defining

$$g(u) = [\nu_s(\underline{r},u)/\nu(\underline{r},u)]f[\nu(\underline{r},u),t-t_1,|\vec{F}|/m;u]e^{-u} \tag{30}$$

yields an integral equation which can be converted to a differential equation and solved immediately.[3] The result is

$$F_s(\underline{r},u,t) = \int_{\infty}^{u} du' \exp\left\{\int_{u'}^{u} du'' \left[\frac{1}{g(u'')}\frac{dg(u'')}{du''} + g(u'')e^{u''}\right]\right\} \cdot \left\{\frac{d}{du'}[\nu_s(\underline{r},u')S_N(\underline{r},u',t)] - \frac{\nu_s(\underline{r},u')}{g(u')} S_N(\underline{r},u',t)\frac{dg(u')}{du'}\right\} \tag{31}$$

where, after simplification, similar physical interpretations can be given for the various factors as those given for the continuous slowing-down solution in Sec. IV. This result, of course, represents an exact hard-sphere elastic slowing-down solution rather than a continuous slowing-down theory approximation. However, the particle transport and field effects still are included only in an approximate way because of the Taylor expansion originally used in obtaining Eq. (11). Moreover, this solution applies only for the very special case that exists when the test and field particles have equal mass and interact according to a hard-sphere interaction potential [in which case $\xi(u) = 1$]. The restriction to isotropic scattering, which is a very poor assumption in the equal mass case, has also been retained here.

VI. Summary

A continuous slowing-down theory has been developed for energetic test particles interacting via hard collisions with a system of host particles in the presence of an external field by utilizing an integral form of the linearized Boltzmann equation for the test particles combined with extensions of methods for treating the collisions which were developed in the theory of neutron slowing down many years ago.[3] Solutions to the two lowest-order approximate equations were immediate. These were given, and their physical interpretation was discussed. In the constant-collision- frequency case, an explicit series was developed for the transcendental function that appears in the solutions.

Well-known steady-state results from the theory of neutron slowing down are recovered when the external field first is made zero ($\beta^2 \to 0$) and then the time between collisions τ_c is let go to zero, or $t_1 \to -\infty$, or $t \to +\infty$ (i.e., $\bar{n}_c \to \infty$) [see Eqs. (A6) and (19)]. For example, in the continuous slowing-down case discussed in Sec. IV. $f(\cdots) \to 1$ in Eq. (27), and the neutron slowing-down result[3,4] is recovered. [The limit $\bar{n}_c \to \infty$ alone transforms the series in β^2 given by Eq. (19) into an asymptotic series.]

In the developments presented here, the effect of the field upon the test particles is included only indirectly via the manner in which it modifies the removal of test particles as they travel along trajectories in the force field between collisions. Hence, the direct effect of the field on the acceleration of the particles and the resulting dispersion in speed are not included. In order to include these effects, higher-order terms in the Taylor expansion made in Sec. II must be retained. The effects of these higher-order terms and the solutions to the resulting equations remain to be investigated.

Appendix

In this Appendix, we examine the transcendental function defined by the multiple integral given in Eq. (13) of the text. After defining the collision time $\tau_c(\underline{r},v) = 1/\nu(\underline{r},v)$, the total number of collisions $\bar{n}_c(\underline{r},v) = (t-t_1)/\tau_c(\underline{r},v)$, the ratio of the maximum change in speed between collisions to the speed of the particle at the field point under consideration $\beta(\underline{r},v) = |\vec{F}|\tau_c(\underline{r},v)/mv$, and the new integration variable $y = \tau/\tau_c(\underline{r},v)$, this function can be written as

$$f[\nu(\underline{r},v),t-t_1,|\vec{F}|/m;v]$$

$$= \frac{1}{4\pi}\int_0^{2\pi} d\phi \int_{-1}^{1} d\mu \int_0^{\bar{n}_c(\underline{r},v)} dy \quad \frac{\exp\left\{-\int_0^y \{\nu[\underline{r}(y'),v(y')]/\nu(\underline{r},v)\}dy'\right\}}{1-2\beta\mu y+\beta^2 y^2} \qquad \text{(A1)}$$

where $\mu = \vec{F}\cdot\hat{\Omega}/|\vec{F}|$. When $\beta(\underline{r},v)$ is replaced by $-\beta(\underline{r},v)$ and the integration variable μ is replaced by $-\mu$, this function remains unchanged except for $\underline{r}(y')$ [see Eq. (8) of text]. Thus, when the collision frequency is independent of position (uniform host gas), this function does not depend upon the sign of $\beta(v)$. Moreover, it does not depend upon v except through $\nu(v), \bar{n}_c(v)$, and $\beta^2(v)$. Thus

$$f[\nu(\underline{r},v),t-t_1,|\vec{F}|/m;v] = f[\nu(v),\beta^2(v),\bar{n}_c(v)] \tag{A2}$$

Furthermore, the orders of integration now can be inverted. Finally, for the constant-collision-frequency case, this expression also becomes independent of the collision frequency except through $\bar{n}_c$ and β^2. Thus, for constant collision frequency

$$f(\beta^2,\bar{n}_c) = \frac{1}{2}\int_0^{\bar{n}_c} dy e^{-y} \int_{-1}^{1} d\mu \frac{1}{1-2\beta y\mu+\beta^2 y^2} \tag{A3}$$

Carrying out the inner integration yields

$$f(\beta^2,\bar{n}_c) = \frac{1}{2\beta}\int_0^{\bar{n}_c} dy \frac{e^{-y}}{y} \ \ell og\left(\frac{1+\beta y}{1-\beta y}\right) = \frac{1}{\beta}\int_0^{\bar{n}_c} dy \frac{e^{-y}}{y} \text{Arth}(\beta y) \tag{A4}$$

Under the condition $\beta\bar{n}_c < 1$, the Arth (βy) can be expanded and the series integrated term by term to obtain an expansion in β^2 which converges for $\beta^2 < 1$ (weak field, or short collision time, or large speed). The result is

$$\begin{aligned} f(\beta^2,\bar{n}_c) &\sim \sum_{n=0}^{\infty} \frac{\beta^{2n}}{(2n+1)} \int_0^{\bar{n}_c} dy \ e^{-y} y^{2n} \\ &\sim (1-e^{-\bar{n}_c}) + \sum_{m=1}^{\infty} \frac{\beta^{2m}}{(2m+1)} \left\{ (2m)! - \left(\bar{n}_c^{2m} \right.\right. \\ &\quad \left.\left. + \sum_{k=1}^{2m} 2m(2m-1)\ldots(2m-k+1)\bar{n}_c^{2m-k} \right) e^{-\bar{n}_c} \right\} \end{aligned} \tag{A5}$$

$$\begin{aligned} f(\beta^2,\bar{n}_c) &\sim \{1-\exp(-\bar{n}_c)\} + (1/3)\{2!-(\bar{n}_c^2+2\bar{n}_c+2)\exp(-\bar{n}_c)\}\beta^2 \\ &+ (1/5)\{4!-(\bar{n}_c^4+4\bar{n}_c^3+12\bar{n}_c^2+24\bar{n}_c+24)\exp(-\bar{n}_c)\}\beta^4 + 0(\beta^6), \quad \beta\bar{n}_c < 1 \end{aligned} \tag{A6}$$

which, of course, requires $\beta^2 < 1$ for convergence. The physical interpretation of the leading term is discussed in Sec. III of the text. The higher-order terms represent further corrections, which directly bring in the effect of the force field ($|\vec{F}|$) during particle motion between collisions (τ_c) upon inscattering, to various orders in $(|\vec{F}|\tau_c/mv)^2$. As v decreases, the higher-order terms in β^2 become larger. This appears to be physically correct because for smaller v more particles are outscattered from direct trajectories [i.e., $S_N(\underline{r},v,t)$ becomes smaller], since for constant collision frequency the mean free path decreases with v. Hence the size of the correction term, to account for the particles that are rescattered (after removal from their direct trajectories) so that they ultimately again enter trajectories that carry them to the field point under consideration, must become larger.

Introducing the variable $\alpha = 1/\beta$ leads to

$$f(\alpha^2,\bar{n}_c) = \frac{\alpha}{2}\int_0^{\bar{n}_c/\alpha} dx\, \frac{e^{-\alpha x}}{x}\, \ell og\left(\frac{1+x}{1-x}\right) = \frac{\alpha}{2}\, I(\alpha,\bar{n}_c) \tag{A7}$$

We examine the integral $I(\alpha,\bar{n}_c)$ by first differentiating it with respect to α and then integrating the resulting expression by parts to obtain

$$\frac{dI(\alpha,\bar{n}_c)}{d\alpha} = -\frac{1}{\alpha}\int_0^{\bar{n}_c/\alpha} \exp(-\alpha x)\left(\frac{1}{1+x} + \frac{1}{1-x}\right) dx \tag{A8}$$

or

$$\frac{dI(\alpha,\bar{n}_c)}{d\alpha} = -\frac{1}{\alpha}\, e^{\alpha}[E_1(\alpha)-E_1(\alpha+\bar{n}_c)] + \frac{1}{\alpha}\, e^{-\alpha}[\bar{E}_1(-\alpha)-\bar{E}_1(-\alpha+\bar{n}_c)] \tag{A9}$$

where

$$\bar{E}_1(-z) = -\bar{E}_i(z) = -(1/2)\lim_{\varepsilon\to 0}\,[E_i(z+i\varepsilon)+E_i(z-i\varepsilon)]$$

$$= (1/2)\lim_{\varepsilon\to 0}\,[E_1(-z-i\varepsilon)+E_1(-z+i\varepsilon)]$$

Using the large argument expansion for the E_1 function

$$E_1(z) = \frac{e^{-z}}{z}\left\{\sum_{m=0}^{M-1}\frac{m!}{(-z)^m} + O(|z|^{-M})\right\}, \quad |z|\to\infty\,,$$

$$-\frac{3}{2}\pi < \arg(z) < \frac{3}{2}\pi, \quad M = 1,2,\ldots$$

combined with Eqs. (A8) and (A7) leads to the result already obtained in Eq. (A6). Integrating Eq. (A9) from α to ∞ (since $I(\alpha,\bar{n}_c) \to 0$ as $\alpha \to \infty$), changing the sign of α in the second term, and substituting into Eq. (A7) yields

$$f(\alpha^2,\bar{n}_c) = \frac{\alpha}{2}\left\{\int_{-\infty}^{-\alpha} + \int_{\alpha}^{\infty}\right\} \{\frac{e^{\eta}}{\eta} [E_1(\eta)-E_1(\eta+\bar{n}_c)]\} \, d\eta \qquad \text{(A10)}$$

which directly gives Eq. (18) of Sec. III.

We close this appendix with a final observation that did not seem of sufficient importance to include in the main text. If $f(\beta^2,\bar{n}_c)$ given in Eq. (19) also is expanded in powers of $\bar{n}_c$, which implies that $t-t_1$ must be very small since τ_c already has been taken to be small, Eq. (12) becomes

$$N(\underline{r},v,t) \sim S_N(\underline{r},v,t) + (\nu_s/\nu)\bar{n}_c[1+(\nu_s/\nu)\bar{n}_c+\ldots]S_N(\underline{r},v,t)$$

which includes a very simple multiple scattering correction to account for particles that have been removed from their direct trajectories to the field point by scattering collisions but subsequently are multiply scattered back into trajectories that carry them to the field point.

References

[1]Boffi, V. C., Dorning, J. J., Molinari, V. G., and Spiga, G., "Integral Boltzmann Equation for Test Particles in a Conservative Field: I. Theory, and Solutions to Some Stationary and Time-Dependent Cases," published elsewhere in this volume.

[2]Boffi, V. C., Molinari, V. G., Pescatore, C., and Pizzio, F., "Integral Boltzmann Equation for Test Particles in a Conservative Field: II. Solutions to Space-Dependent Cases," published elsewhere in this volume.

[3]Weinberg, A. M., and Wigner, E. P., The Physical Theory of Neutron Chain Reactors, The University of Chicago Press, Chicago, Ill., 1958.

[4]Dresner, L., Resonance Absorption in Nuclear Reactors, Pergamon Press, New York, 1960.

THE DISTRIBUTION FUNCTION CONTINUITY METHOD: A LOCAL GREEN'S FUNCTION TECHNIQUE FOR THE NUMERICAL SOLUTION OF THE BOLTZMANN EQUATION

J. Dorning,* C. Pescatore,† and G. Spiga≠

Laboratorio di Ingegneria Nucleare dell'Università di Bologna, Bologna, Italy

Abstract

A new coarse-mesh computational method for the solution of the Boltzmann equation which utilizes a formalism based upon local Green's functions of part of the adjoint Boltzmann operator is developed. The resulting sets of integral equations for each volume element into which the system is decomposed are coupled to those for adjacent elements via distribution function continuity (DFC) conditions. The local integral equations then are reduced to low-order local matrix equations by applying a weighted-residuals solution method. A source iteration scheme, based upon the inscatter source, is developed to treat the matrix equations. The general formalism then is specialized to the Couette flow problem with a BGK scatter-

---

Presented as Paper 106 at the 10th International Symposium on Rarefied Gas Dynamics, Aspen, Colorado, July 19-23, 1976. We very gratefully acknowledge numerous discussions with V. Boffi, R. D. Lawrence, and V. Molinari. One of us (J.D.) also wishes sincerely to acknowledge the support of the Italian National Research Council (Consiglio Nazionale delle Ricerche) through its visiting professorship program, and the splendid hospitality provided by V. Boffi and the other members of the Laboratorio di Ingegneria Nucleare dell 'Università di Bologna during the course of this work. During the final phases of this work one of the authors (J.D.) was supported by the U.S. National Science Foundation under Grant No. NSF/ENG 74-13029.

*C.N.R. Visiting Professor of Mathematical Physics; Professor of Nuclear Engineering on leave of absence from the University of Illinois at Urbana-Champaign, Urbana, Ill.

†C.N.R. Fellow.

≠Assistant.

ing model. Explicit expressions are developed, using local coordinates, for all of the matrix elements that occur when polynomial expansion functions are used, and relationships among these matrix elements, and relationships to tabulated transcendental functions are derived.

Introduction

A new coarse-mesh computational scheme based upon the formal use of local Green's functions of part of the adjoint Boltzmann operator is developed. After decomposing the gaseous system under study into volume elements, the local Green's function (within a particular volume element) for part of the adjoint Boltzmann equation (streaming term plus a death term) is utilized to convert the integrodifferential Boltzmann equation into separate local integral equations within each volume element. Local integral equations then are written also for the outgoing components of the distribution function on the surface of each volume element. These local surface-integral equations are coupled to the local volume-integral equations for a given volume element. This set of simultaneous integral equations for the distribution function in the interior and on the surface of each volume element then is connected to those for the adjacent volume elements by requiring distribution function continuity (DFC) for velocity directions pointing into each volume element. This formalism results in a form of the Boltzmann equation which has the advantages of an integral equation over a differential equation for numerical solution while still retaining an important characteristic of differential equations which is very desirable from the standpoint of numerical solution, viz. local or nearest-neighbor coupling (i.e., no long-range spatial coupling remains).

The local integral equations then are reduced to low-order local matrix equations by expanding the local distribution function in the interior and on the surface of each volume element independently, and locally weighting and integrating the equations. (That is, a separate weighted residuals method is applied locally to each of the volume element interior and surface equations.) This results in simple low-order local matrix equations for the local expansion coefficients. A source iteration technique based on the source due to collisions then is developed to treat the matrix equations (which do not involve matrix inversions) element by element during spatial iterative sweeps that start from the boundary and then move through the system. In multidimensional applications, such as flow in two-dimensional channels these iterative sweeps begin from a "corner" volume element which has sides on two adjacent boundaries. The formalism is developed in detail for the general form of the Boltzmann equation, and the minor modifications necessary for the discretized (multigroup) form[1] are indicated.

The general formalism then is specialized to the Couette flow problem,[2-4] with the collisions described by the BGK model.[2-5] Explicit expressions are developed for all of the matrix elements that occur when polynomial expansion functions are used. These are developed using local spatial coordinates, which, in turn, are related explicitly to the global spatial coordinates. Finally, relationships among the matrix elements and recursion formulas are given, and all of the elements thereby are reduced to tabulated transcendental functions[6] $T_n(z)$.

This computational method is based upon ideas that have been presented recently[7,8] and developed for neutron diffusion problems of interest in nuclear reactor statics and dynamics, where they have been demonstrated to be far more efficient computationally than traditional finite-difference methods and also more efficient and more flexible than finite-element methods.[8] The ideas also have been proposed but not developed for neutron transport problems.[7]

II. Formalism

The formalism begins by decomposing the gaseous system under consideration into K volume elements V_k. The steady-state Boltzmann equation for the distribution function in the kth volume element is then (the time-dependent equation can be treated analogously)

$$\vec{v}\cdot\vec{\nabla}f^k(\underline{r},\underline{v}) + \nu_r^k(\underline{v})f^k(\underline{r},\underline{v}) = S^k\{[f^k(\underline{r},\underline{v})];\ \underline{r},\underline{v}\},\quad \underline{r}\ \varepsilon\ V_k \tag{1}$$

$$S^k\{[f^k(\underline{r},\underline{v})];\ \underline{r},\underline{v}\} = \int d^3v'K^k\{[f^k(\underline{r},\underline{v})];\ \underline{v}'\rightarrow\underline{v}\}f^k(\underline{r},\underline{v}') + S_o^k(\underline{r},\underline{v}),\quad \underline{r}\ \varepsilon\ V_k \tag{2}$$

Here $\nu_r^k(v)$ is some "reference" effective collision frequency that is independent of the distribution function $f^k(r,v)$, and $K^k\{[f^k(\underline{r},\underline{v})];\ \underline{v}'\rightarrow\underline{v}\}$ represents the original collision integral plus this effective collision frequency term. [The definition of the reference effective collision frequency will depend upon the problem under consideration. For example, it could be defined as the multiplicative coefficient in the leading term in a Fokker-Planck expansion of the collision integral in a problem that involves particles which interact according a coulomb potential. For the linearized Boltzmann equation, $\nu_r^k(v)$ would become the ordinary collision frequency $\nu^k(v)$, and $K^k\{[f^k(\underline{r},\underline{v})];\ \underline{v}'\to\underline{v}\}$ would become the "scattering kernel", which is independent of $f^k(\underline{r},\underline{v})$. The "reference" collision frequency might also be taken as a constant in order to simplify the construction of the matrix elements which will be introduced into the formalism below. This, of course, generally will cause the complexity of the generalized source term to be increased.] Thus the left-hand side of this equation is linear in the distribution function $f(\underline{r},\underline{v})$, the nonlinear terms all being included in the generalized source term. The Green's function for the adjoint operator to that on the left-hand side of Eq. (2) now is introduced

$$-\vec{v}\cdot\vec{\nabla}G_k^+(\underline{r},\underline{v}|\underline{r}_o) + \nu_r^k(v)G_k^+(\underline{r},\underline{v}|\underline{r}_o) = \delta(\underline{r}-\underline{r}_o), \quad \underline{r}_o \in V_k \tag{3}$$

so that the distribution within the kth volume element can be written as

$$f^k(\underline{r},\underline{v}) = \int_{V_k} d^3r_o\ G_k^+(\underline{r}_o,\underline{v}|\underline{r})S^k(\underline{r}_o,\underline{v})$$

$$+\int_{S_k} d^2r_o[-\hat{n}_k(\underline{r}_o^s)\cdot\vec{v}]G_k^+(\underline{r}_o^s,\underline{v}|\underline{r})f^k(\underline{r}_o^s,v), \quad \underline{r} \in V_k \tag{4}$$

where Gauss' theorem has been used to write the last term as a surface integral, and $\hat{n}_k(\underline{r}_o^s)$ is the outward normal from the kth volume element at the surface point $\underline{r}_o^s$. In order to couple this equation to the distribution function in the adjacent elements by using the continuity of the distribution function, we choose the boundary condition on the adjoint Green's function to be

$$G_k^+(\underline{r}^s,\underline{v}|\underline{r}_o) = 0, \quad \hat{n}_k(\underline{r}^s)\cdot\vec{v} > 0 \tag{5}$$

Then Eq. (4) can be rewritten as

$$f^k(\underline{r},\underline{v}) = \int_{V_k} d^3r_o\ G_k^+(\underline{r}_o,\underline{v}|\underline{r})S^k(\underline{r}_o,\underline{v})$$

$$+\int_{S_k} d^2r_o[-\hat{n}_k(\underline{r}_o^s)\cdot\vec{v}]G_k^+(\underline{r}_o^s,\underline{v}|\underline{r})f_{out}^{\ell k}(\underline{r}_o^s,\underline{v}), \quad \underline{r} \in V_k \tag{6}$$

which connects it to the outgoing $[\hat{n}_k(\underline{r}_o^s)\cdot\underline{v} < 0$ or $\hat{n}_\ell(\underline{r}_o^s)\cdot\underline{v} > 0]$ components of the distribution function from the ℓth adjacent volume element. Here $S_{\ell k}$ is the surface that is common to the kth and the ℓth volume elements, and $f_{out}^{\ell k}(\underline{r}_o^s,\underline{v})$ is the distribution function of particles passing across the surface from the ℓth volume element into the kth volume element in the direction defined by $\underline{v}$. The equation for the outgoing components of the distribution function across the km surface now can be written down immediately

$$f_{out}^{km}(\underline{r}^s,\underline{v}) = \int_{V_k} d^3r_o\; G_k^+(\underline{r}_o,\underline{v}|\underline{r}^s)S^k(\underline{r}_o,\underline{v})$$

$$+\int_{S_{\ell k}} d^2r_o[-\hat{n}_k(\underline{r}_o^s)\cdot\vec{v}]G_k^+(\underline{r}_o^s,\underline{v}|\underline{r}^s)f_{out}^{\ell k}(\underline{r}_o^s,\underline{v}), \quad \underline{r}^s \in S_{km} \tag{7}$$

The coupled local integral equations [Eqs. (6) and (7)] for the distribution function in the kth volume element and the outgoing distribution function on the kmth surface of that volume element now are solved by applying standard methods of approximation theory locally in space. We make independent local expansions of these distribution functions

$$f^k(\underline{r},\underline{v}) = \sum_{j=1}^{N_k} X_j^k\; P_j^k(\underline{r},\underline{v}), \quad \underline{r} \in V_k \tag{8}$$

and

$$f_{out}^{km}(\underline{r}^s,\underline{v}) = \sum_{n=1}^{M_{km}} Y_{out,n}^{km}\; Q_n^{km}(\underline{r}^s,\underline{v}), \quad \hat{n}_k(\underline{r}^s)\cdot\vec{v} > 0, \quad \underline{r}^s \in S_{km} \tag{9}$$

and use a local form of the weighted-residuals method to generate a set of local matrix equations to determine the local expansion coefficients. Thus, weighting Eq. (6) by $U_i^k(\underline{r},\underline{v})$ and integrating over $\underline{r} \in V_k$ and all $\underline{v}$, and weighting Eq. (7) by $W_s^{km}(\underline{r}^s,\underline{v})$ and integrating over $\underline{r}^s \in S_{km}$ and all $\underline{v}$ such that $\hat{n}_k(\underline{r}^s)\cdot\underline{v} > 0$, after substituting the preceding expansions, yields the matrix equations

$$\underline{\underline{A}}^k\underline{X}^k = \underline{\underline{G}}_{VV}^{k\leftarrow k}\;\underline{S}^k(\underline{X}^k) + \sum_{\ell=1}^{L} \underline{\underline{G}}_{VS}^{k\leftarrow \ell k}\;\underline{Y}_{out}^{\ell k}, \tag{10}$$

$k = 1, \ldots, K$; L = number of elements adjacent to V_k

and

$$\underline{\underline{B}}^{km}\underline{Y}^{km}_{out} = \underline{\underline{G}}^{km\leftarrow k}_{SV}\ \underline{S}^{k}(\underline{X}^{k}) + \sum_{\ell=1}^{L} \underline{\underline{G}}^{km\leftarrow \ell k}_{SS}\ \underline{Y}^{\ell k}_{out}$$

$k = 1, \ldots, K$; L = number of elements adjacent to V_k (11)

The A and B matrices are simple low-order matrices that involve integrals of products of expansion and weight functions. (They are identity matrices if the expansion and weight functions are orthonormal.) Thus they can be inverted immediately and incorporated in the definitions of the G matrices. For the linearized Boltzmann equation, the G matrices are double integrals of the adjoint Green's function over volume or surface expansion and weighting functions, and represent transport (and attenuation) from volume to volume (VV), surface to volume (VS), etc. Their elements have physical meaning and satisfy simple separate equations and can be evaluated using variational methods.[9] For example, the elements of G_{VV} are given by the expansion coefficients of weighted-residuals solutions to the equations

$$\vec{v}\cdot\vec{\nabla}g^{+}_{k,i}(\underline{r},\underline{v}) + \nu^{k}_{r}(v)g^{+}_{k,i}(\underline{r},\underline{v}) = U_i(\underline{r},\underline{v})$$

$$i = 1, \ldots, N_k$$

in which the same expansion functions and weight functions used in the original problem in the foregoing are utilized with their roles reversed. Moreover, because of the boundary conditions used on the adjoint Green's function, it satisfies equivalent-infinite-medium equations, which is particularly important for multidimensional applications. The expressions for the matrix elements which appear in Eqs. (10) and (11) are given in Appendix A for the linear Boltzmann equation. (For the nonlinear Boltzmann equation, the exact form of the collision integral must be stated before the matrix elements can be developed explicitly.

If a discretized multigroup form[1] is used for the Boltzmann equation, Eqs. (10) and (11) change only very slightly. A group index is added to all of the variables that appear, and the matrix elements change slightly, as indicated in Appendix A. The main advantage of this is that the number of expansion functions used to represent the velocity (then only direction vector $\hat{\Omega}$) can be reduced and thereby the dimension of the resulting matrix equations can be further reduced.

Equations (10) and (11) are in a general form that is amenable to solution by source iteration methods. Because of the local nature and nearest-neighbor coupling of these equations, they can be solved element by element starting from a boundary in the context of a source iteration method. To do this, we introduce iteration indices in Eqs. (10) and (11)

$$\underline{X}^{k(p)} = \tilde{\underline{\underline{G}}}_{VV}^{k\leftarrow k}\ \underline{S}^{k(p-1)}(\underline{X}^{k(p-1)}) + \sum_{\ell=1}^{L} \tilde{\underline{\underline{G}}}_{VS}^{k\leftarrow \ell k}\ \underline{Y}_{out}^{\ell k(p-1,Q)} ,$$

$k = 1, \ldots, K$; L = number of elements adjacent to V_k (12)

$$\underline{Y}_{out}^{km(p-1,q)} = \tilde{\underline{\underline{G}}}_{SV}^{km\leftarrow k}\ \underline{S}^{k(p-1)}(\underline{X}^{k(p-1)}) + \sum_{\ell=1}^{L_b} \tilde{\underline{\underline{G}}}_{SS}^{km\leftarrow \ell k}\ \underline{Y}_{out}^{\ell k(p-1,q)} + \sum_{\ell=L_b+1}^{L} \tilde{\underline{\underline{G}}}_{SS}^{km\leftarrow \ell k}\ Y_{out}^{\ell k(p-1,q-1)} \qquad (13)$$

$k = 1, \ldots, K$; L = number of elements adjacent to V_k;

L_b = number of adjacent elements behind kth in "sweep"

and make initial guesses (zeroth iteration values) for the distribution function within, and on the surfaces of, the volume elements (and, therefore, for the expansion coefficients $\underline{X}^k$ and $Y_{out}^{\ell k}$). Here the inverses of the A and B matrices have been included in the definitions of the $\tilde{G}$ matrices. Equation (13) then can be solved for the outgoing particles from the elements on a boundary, since $Y_{out}^{\ell k}$ on the right-hand side is known from the boundary condition or from the initial guess, and $\underline{S}^k$ also is known from the initial guess for $\underline{X}^k$. In multidimensional geometries, it is most convenient to start from a corner volume element, since $\underline{Y}_{out}^{\ell k}$ will be known on two surfaces from the boundary condition (unless the boundary values are functions of the distribution function). Equation (13) then is solved for the outgoing particles moving away from one of the boundaries for the second element from that boundary which lies along the adjacent boundary, etc., until the opposite boundary (or a symmetry plane, etc.) is reached. Then Eq. (13) is solved element by element, moving back toward the original boundary for particles moving across the surfaces in the opposite direction. In one-dimensional plane geometry, one forward and reverse "sweep" is sufficient. In multidimensional geometries, the next row of volume elements then is treated, etc., until the particles crossing all volume element surfaces in both directions are calculated, and then the entire process is repeated iteratively (inner iterations) until some convergence criteria are reached. After this, Eq. (12) is used to determine new values for the volume distribution function expansion coefficients in each element. These values replace the initial guesses, and the next iteration (outer or source interation) is begun. This entire process of inner and outer iterations then is repeated until some convergence criteria are reached. In Eqs. (12) and (13) the iteration index q corresponds to inner iterations for the values of the expansion coefficients of the distribution functions on the volume element surfaces (Q indicating a converged value), and the index p corresponds to the outer (or source) iterations for the expansion coefficients of the distribution functions in the interiors of the volume elements.

III. A Specific Application

As a specific illustration of the method, we develop it in detail for the BGK kinetic model equation[5] description of the Couette flow problem with diffuse boundary conditions.[2-4] In this case, the Boltzmann equation reduces to [2-4]

$$v_x \frac{\partial\psi(x,v_x)}{\partial x} + \nu\psi(x,v_x) = \frac{\nu}{\sqrt{\pi}} \int_{-\infty}^{\infty} dv_x' \; e^{-(v_x')^2} \psi(x,v_x') \tag{14}$$

with boundary conditions

$$\psi(\pm a, v_x) = \pm 1, \quad v_x \lessgtr 0 \tag{15}$$

After rewriting the channel half-thickness and x in units of the average mean free path, and the global position variable in units of this new half-thickness, we decompose the gaseous system into K plane slab volume elements of half-thickness δ_k, and introduce a local position variable via

$$x^{\text{global}} = \delta_k \, x + (1/2)\,(x_k^{\text{global}} + x_{k-1}^{\text{global}}) \tag{16}$$

Then Eq. (15) can be rewritten within the kth volume element as

$$v_x \alpha^k \frac{\partial}{\partial x} \phi^k(x,v_x) + \phi^k(x,v_x) = \frac{1}{\sqrt{\pi}} \int_{-\infty}^{\infty} dv_x' \; e^{-(v_x')^2} \phi^k(x,v_x') \tag{17}$$

with boundary conditions

$$\phi^k(\pm 1, v_x) = \phi^{k\mp 1}(\mp 1, v_x), \quad v_x \lessgtr 0 \tag{18}$$

where

$$\phi^k(x,v_x) = (1/\delta_k)\ \psi^k(x,v_x) = \psi^k(x^{global},v_x) \tag{19}$$

and

$$\alpha^k \equiv (\sqrt{\pi}/2)\ (\bar{\lambda}/a\delta_k) = (\sqrt{\pi}/2)\ [1/(\delta_k/Kn)] \tag{20}$$

which only depends upon the volume element half-thickness in units of the Knudsen number. Changes of variables analogous to these can also be made in the general case, however the average mfp takes on a different value in general. Here it is given by $\bar{\lambda} = 2/\sqrt{\pi}\ \nu$. For the exterior volume element, the boundary condition is $\phi^1(+1,v_x) = 1$, and at the channel midplane it is $\phi^K(-1,v_x) = -\phi^K(-1,-v_x)$. The appropriate Green's function is

$$G_k^+(x,v_x|x_o) = \begin{cases} \pm\dfrac{1}{\alpha^k v_x}\exp\left[\dfrac{1}{\alpha^k v_x}(x-x_o)\right], & v_x \gtrless 0,\ x \lessgtr x_o \\ 0, \quad \text{otherwise}, & x \in (-1,+1) \end{cases} \tag{21}$$

Repeating steps analogous to those in Sec. II and integrating the resulting integral equation for $\phi^k(x,v_x)$ to take advantage of the simple scattering kernel (which cannot be done in more general applications) yields

$$q^k(x) = \frac{1}{\sqrt{\pi}\alpha^k}\int_{-1}^{+1} dx_o\ T_{-1}\left(\frac{1}{\alpha^k}|x_o-x|\right) q^k(x_o) + \int_0^\infty dv_x\ \exp\left[-\frac{1}{\alpha^k v_x}(1+x) - v_x^2\right]\phi_{out}^{k+1}(+1,v_x) + \int_{-\infty}^0 dv_x\ \exp\left[\frac{1}{\alpha^k v_x}(1-x) - v_x^2\right]\phi_{out}^{k-1}(-1,v_x) \tag{22}$$

and

$$\phi_{out}^k(\pm 1,v_x) = \frac{1}{\sqrt{\pi}}\int_{-1}^{+1} dx_o\ \frac{\pm 1}{\alpha^k v_x}\exp\left[\mp\frac{1}{\alpha^k v_x}(1\mp x_o)\right] q^k(x_o) + \exp\left[\mp\frac{2}{\alpha^k v_x}\right]\phi_{out}^{k\pm 1}(\pm 1,v_x), \quad v_x \gtrless 0 \tag{23}$$

where

$$q^k(x) \equiv \int_{-\infty}^{\infty} dv_x\ \exp[-v_x^2]\phi^k(x,v_x) \tag{24}$$

and

$$T_n(z) \equiv \int_0^\infty dy\, y^n \exp\left[-\frac{z}{y} - y^2\right] \tag{25}$$

is a tabulated function.[6] For k=1, $\phi_{out}^{k-1}(-1,v_x)$ is replaced by +1, and for k=K, $\phi_{out}^{k+1}(+1,v_x)$ is replaced by $-\phi_{out}^{K}(-1,v_x)$ in Eqs. (22,23). [If we were to take δ_k=1 here, Eq. (22) would describe the entire channel, and $\phi_{out}^{0}(-1,v_x)$, $v_x < 0$ and $\phi_{out}^{2}(+1,v_x)$, $v_x > 0$ would become +1 and -1, respectively, and the well-known integral equation in one variable which describes the problem[2-4] would be recovered. But this would defeat our purpose which is to illustrate the details of the method by applying it explicitly and in its complete form to a simple test problem.]

Expanding the unknowns

$$q^k(x) = \sum_{j=1}^{N_k} X_j^k\, P_j(x), \quad x \in (-1,+1) \tag{26}$$

$$\phi_{out}^k(\pm 1, v_x) = \sum_{n=1}^{M_k^\pm} {}^{\pm}Y_{out,n}^{k}\, Q_n(v_x), \quad v_x \gtrless 0 \tag{27}$$

substituting into Eqs. (22,23), weighting by $U_i(x) = P_i(x)$, $W_s(v_x)\exp[-v_x^2] = Q_s(v_x)\exp[-v_x^2]$, and integrating over the appropriate intervals yields

$$\underline{\underline{A}}\,\underline{X}^k = (1/\sqrt{\pi}\,\alpha^k)\,\underline{\underline{G}}_{VV}^k\,\underline{X}^k + {}^{+}\underline{\underline{G}}_{VS}^k\,{}^{+}\underline{Y}_{out}^{k+1} + {}^{-}\underline{\underline{G}}_{VS}^k\,{}^{-}\underline{Y}_{out}^{k-1} \tag{28}$$

$${}^{\pm}\underline{\underline{B}}\,{}^{\pm}\underline{Y}_{out}^k = (1/\sqrt{\pi}\,\alpha^k)\,{}^{\pm}\underline{\underline{G}}_{SV}^k\,\underline{X}^k + {}^{\pm}\underline{\underline{G}}_{SS}^k\,{}^{\pm}\underline{Y}_{out}^{k\pm 1} \tag{29}$$

Again the A and B matrices can be inverted immediately. Explicit expressions for all of the matrix elements are given in Appendix B, where the $P_i(x)$ are taken as the first three Legendre polynomials, and the $Q_s(v_x)$ are taken as the first three powers of v_x. Relationships among these matrix elements also are developed there, and recursion formulas that reduce all of them to the T_n functions are given.

The iterative scheme based on the inscatter source developed for the general case in Sec. II can be repeated here. Since this is a one-dimensional plane geometry problem, no inner iterations are necessary. Rather, a single forward and reverse "sweep" starting from the boundary (k=1) is sufficient.

IV. Summary

A new coarse-mesh method has been developed for the numerical solution of the continuous and discretized forms of the general linear Boltzmann equation, and its extension to the nonlinear Boltzmann equation has been indicated. Local Green's functions were used to generate local integral equations associated with each of the volume elements into which the system was decomposed. The equations for the various volume elements then were coupled by utilizing the distribution function continuity (DFC) in the incoming velocities at the surfaces of the elements. The local Green's functions and matrix elements that arise in the formalism can, in general, be parametrized by the ratio of the dimensionless volume element size to the Knudsen number. (An example was given explicity.) Thus, once a value is determined for this ratio (depending upon desired numerical accuracy, etc.), numerous systems of different sizes and densities can be calculated using the same matrix elements simply by changing the number of volume elements employed, provided that the speed dependence of the reference collision frequency is not changed.

A reference collision frequency was used in the initial formulation of the method. Normally, this would be taken as the actual collision frequency for the linear Boltzmann equation. If not, the inscatter operator is replaced by the total collision operator plus the reference collision frequency. It is of interest to note that then the local Green's functions become independent of the collision physics and depend only upon the reference collision frequency used. The physics of the true collision process is, of course, retained in the source that is used in the iteration solution procedure.

When collision invariants are used as local expansion functions in the velocity variable, the calculation of the matrix elements that involve the inscattering operator is simplified greatly, since the operator then can be eliminated in favor of multiplication by the collision frequency. When this is done in conjunction with the use of a reference collision frequency, the details of the collision physics drop out of the formalism altogether, but this is not surprising, since this corresponds to a local hydrodynamics-type approximation. This appears to indicate that, as long as the outscatter term and the inscatter term are consistent, the solutions for the local expansions in the collision invariants are the same, independent of the scattering potential. On the other hand, when higher eigenfunctions of the collision operator are used as local expansion functions for the velocity dependence, the properties of the collision term then do appear in the formalism through the nonzero eigenvalues of the collision operator, even when a reference collision frequency is used. Since these higher eigenfunctions do not correspond to the formulation of a local hydrodynamics-type approximation, the properties of the specific collision interaction potential should be present.

Explicit application of the method was made to the test problem of Couette flow described by the BGK kinetic model equation,[5] including a detailed evaluation of all of the matrix elements. This was done so that ultimately numerical results could be compared with a very detailed numerical solution of the simple integral equation in one variable to which this problem can be reduced[2-4] [Eq. (22) with K = 1]. This test problem, of course, does not emphasize the advantages of the new method, since this method can efficiently determine more detailed information that is necessary in this particular case [viz., it can calculate $\psi(x,v_x)$ directly whereas here it is only necessary to calculate $q(x)$]. However, in other applications using more general kernels, the equations for the expansion coefficients of $f(\underline{r},\underline{v})$ can be made explicit for any particular type of expansion. For example, in one-dimensional plane geometry problems, half-range velocity expansions of $f(\underline{x},\underline{v})$ in the interior of the volume elements should prove very useful, since the contributions from one of the two surfaces drop out of the integral equation for the half-range distribution function in V_k. This may be particularly useful in transport problems, in which the external particle sources are strongly anisotropic and the particle scattering cross sections are small or forward-peaked. Some specific examples of such problems include many neutron and gamma-ray shielding problems and electron, ion, and neutral beam (rarefied foreign gas) injection problems.

The extension of the ideas developed here to the linear Boltzmann equation for particles migrating in the presence of an external force field (for example, test particles slowing down in a host gas) may be possible by introducing the local adjoint Green's function of the substantial derivative, combined with an outscatter or death term, to convert the Boltzmann equation to a local integral equation. The foundations for such an extension have been laid recently,[10,11] but the objectives of these studies were such that the development of local integral equations and computational methods for numerical solutions was not of particular interest.

Appendix A

The matrix elements of the matrices that appear in Eqs. (10,11) for the linear Boltzmann equation are given by

$$(\underline{\underline{G}}_{VS}^{k\leftarrow\ell k})_{in} = \int_{V_k} d^3r \int d^3v\ U_i^k(\underline{r},\underline{v}) \int_{S_{\ell k}} d^2r_o [\hat{n}_\ell(\underline{r}_o^s)\cdot\vec{v}] G_k^+(\underline{r}_o^s,\underline{v}|\underline{r}) Q_n^{\ell k}(\underline{r}_o^s,\underline{v}) \quad (A1)$$

$$(\underline{\underline{G}}_{SS}^{km\leftarrow\ell k)})_{sn} = \int_{Skm} d^2r \int_{\hat{n}_k(\underline{r}^s)\cdot\vec{v}>0} d^3v\ W_{\underline{s}}^{km}(\underline{r}^s,\underline{v}) \int_{S_{\ell k}} d^2r_o [\hat{n}_\ell(\underline{r}_o^s)\cdot\vec{v}] G_k^+(\underline{r}_o^s,\underline{v}|\underline{r}^s) \cdot Q_n^{\ell k}(\underline{r}_o^s,\underline{v}) \quad (A2)$$

For the linearized Boltzmann equation, $\underline{\underline{G}}\ \underline{S}(\underline{X})$ becomes

$$\underline{\underline{G}}\ \underline{S}(\underline{X}) = \underline{S}_o + \underline{\underline{G}}\ \underline{\underline{K}}\ \underline{X} \quad (A3)$$

where

$$(\underline{\underline{G}}_{VV}^{k\leftarrow k}\underline{\underline{K}})_{ij} = \int_{V_k} d^3r \int d^3v\ U_i^k(\underline{r},\underline{v}) \int_{V_k} d^3r_o\ G_k^+(\underline{r}_o,\underline{v}|\underline{r}) \cdot \int d^3v'\ K^k(\underline{r}_o,\underline{v}'\rightarrow\underline{v}) P_j^k(\underline{r}_o,\underline{v}') \quad (A4)$$

$$(\underline{\underline{G}}_{SV}^{km\leftarrow k}\underline{\underline{K}})_{sj} = \int_{S_{km}} d^2r \int_{\hat{n}_k(\underline{r}^s)\cdot\vec{v}>0} d^3v\ W_s^{km}(\underline{r}^s,\underline{v}) \int_{V_k} d^3r_o\ G_k^+(\underline{r}_o,\underline{v}|\underline{r}^s) \cdot \int d^3v'\ K^k(\underline{r}_o,\underline{v}'\rightarrow\underline{v}) P_j(\underline{r}_o,\underline{v}) \quad (A5)$$

$$(\underline{S}_o^k)_i = \int_{V_k} d^3r \int d^3v\ U_i^k(\underline{r},\underline{v}) \int_{V_k} d^3r_o\ G_k^+(\underline{r}_o,\underline{v}|\underline{r}) S_o^k(\underline{r}_o,\underline{v}) \quad (A6)$$

$$(\underline{S}_o^{km})_s = \int_{S_{km}} d^2r \int_{\hat{n}_k(\underline{r}^s)\cdot\vec{v}>0} d^3v\ W_s^{km}(\underline{r}^s,\underline{v}) \int_{V_k} d^3r_o\ G_k^+(\underline{r}_o,\underline{v}|\underline{r}^s) S_o^k(\underline{r}_o,\underline{v}) \quad (A7)$$

and the normalization matrices are given by

$$(\underline{\underline{A}}^k)_{ij} = \int_{V_k} d^3r \int d^3v\ U_i^k(\underline{r},\underline{v}) P_j^k(\underline{r},\underline{v}) \quad (A8)$$

$$(\underline{\underline{B}}^{km})_{sn} = \int_{S_{km}} d^2r \int_{\hat{n}_k(\underline{r}^s)\cdot\vec{v}>0} d^3v\ W_s^{km}(\underline{r}^s,\underline{v}) Q_n^{km}(\underline{r}^s,\underline{v}) \quad (A9)$$

We note that the G_{SV}, G_{VS}, and G_{SS} matrices are, in general, not square matrices. General boundary conditions are included in the matrices for volume elements on the boundaries by incorporating the appropriate symmetry conditions, anti-symmetry conditions, etc. in velocity into the surface expansion functions $Q_n^{km}(\underline{r}^s,\underline{v})$ which define the elements of the $\underline{\underline{G}}_{VS}$ and $\underline{\underline{G}}_{SS}$ matrices. The definitions of the matrix elements which arise for the discretized (multigroup) form of the linear Boltzmann equation are analogous to those just given, with group indices added and the integrations over $\underline{v}$ replaced by integrations over the direction vector $\hat{\Omega}$.

Appendix B

Taking the expansion functions $P_i(x)$ as the first three Legendre polynomials and the expansion functions $Q_s(v_x)$ as the first three pwers of v_x leads to the following explicit definitions of the matrix elements that appear in Eqs. (26,27) for the Couette flow problem

$$(\underline{\underline{G}}^k_{VV})_{ij} = \int_{-1}^{+1} dx\ P_i(x) \int_{-1}^{+1} dx_o\ T_{-1}(|x-x_o|/\alpha^k) P_j(x_o) \tag{B1}$$

$$(^{\pm}\underline{\underline{G}}^k_{VS})_{in} = \int_{-1}^{+1} dx\ P_i(x) \int_0^{\infty} dv_x\ \exp\left[-\ v_x^2 - \frac{(1\pm x)}{\alpha^k v_x}\right] Q_n(\pm v_x) \tag{B2}$$

$$(^{\pm}G^k_{SV})_{sj} = \int_0^{\infty} \frac{dv_x}{v_x} Q_s(\pm v_x) \int_{-1}^{+1} dx\ \exp\left[-v_x^2 - \frac{(1\mp x)}{\alpha^k v_x}\right] P_j(x) \tag{B3}$$

$$(^{\pm}G^k_{SS})_{sn} = \int_0^{\infty} dv_x\ Q_s(\pm v_x) \exp\left[-v_x^2 - \frac{2}{\alpha^k v_x}\right] Q_n(\pm v_x) \tag{B4}$$

except for the special case that corresponds to elements that border on the boundaries: $x^{global} = +a$ $(k=1)$, for which $Q_n(v_x)$ is replaced by 1 in Eqs. (B2) and (B4) for $^{-}\underline{\underline{G}}^1_{VS}$ and $^{-}\underline{\underline{G}}^1_{SS}$; and $x^{global} = 0$ $(k=K)$, for which $Q_n(v_x)$ is replaced by $-Q_n(-v_x)$ in Eqs. (B2) and (B4) for $^{+}\underline{\underline{G}}^K_{VS}$ and $^{+}\underline{\underline{G}}^K_{SS}$. The normalization matrices are given by

$$(\underline{\underline{A}})_{ij} = \int_{-1}^{+1} dx\ P_i(x) P_j(x) \tag{B5}$$

$$(^{\pm}\underline{\underline{B}})_{sn} = \int_0^{\infty} dv_x\ Q_s(\pm v_x) \exp(-v_x^2) Q_n(\pm v_x) \tag{B6}$$

The following relationships among the matrix elements hold when $(-1)^i P_i(-x) = P_i(x)$ and $(-1)^n Q_n(-v_x) = Q_n(v_x)$

$$(\underline{\underline{G}}^k_{VV})_{ij} = 0, \ i + j = \text{odd}; \quad (\underline{\underline{G}}^k_{VV})_{ij} = (\underline{\underline{G}}^k_{VV})_{ji}$$

$$({}^-\underline{\underline{G}}^k_{VS})_{in} = (-1)^{i+n} \, ({}^+\underline{\underline{G}}^k_{VS})_{in}; \quad ({}^-\underline{\underline{G}}^k_{SV})_{sj} = (-1)^{s+j} \, ({}^+\underline{\underline{G}}^k_{SV})_{sj}$$

$$({}^-\underline{\underline{G}}^k_{SS})_{sn} = (-1)^{s+n} \, ({}^+\underline{\underline{G}}^k_{SS})_{sn}; \quad ({}^-\underline{\underline{B}})_{sn} = (-1)^{s+n} \, ({}^+\underline{\underline{B}})_{sn};$$

$$(\underline{\underline{A}})_{ij} = (\underline{\underline{A}})_{ji}$$

except for k = 1 and k = K, where

$$({}^-\underline{\underline{G}}^1_{VS})_{in} = (-1)^i \, ({}^+\underline{\underline{G}}^k_{VS})_{io}; \quad ({}^-\underline{\underline{G}}^1_{SS})_{sn} = (-1)^s \, ({}^+\underline{\underline{G}}^1_{SS})_{so}$$

and, taking $Q_o(v_x)$ as unity,

$$({}^+\underline{\underline{G}}^K_{VS})_{in} = (-1)^{i+1} \, ({}^-\underline{\underline{G}}^K_{VS})_{in}; \quad ({}^+\underline{\underline{G}}^K_{SS})_{sn} = (-1)^{s+1} \, ({}^-\underline{\underline{G}}^K_{SS})_{sn}$$

If the $P_i(x)$ are the first three Legendre polynomials and the $Q_n(v_x)$ are the first three powers of v_x

$$({}^+G^k_{VV})_{ij} = 4\sum_{m=0}^{i+j+1} (m+1)\beta^{ij}_m \left\{ \left(\frac{\alpha^k}{2}\right)^{n+1} T_m(0) - \sum_{r=0}^{n} \frac{1}{(m-r)!} \left(\frac{\alpha^k}{2}\right)^{r+1} T_r\left(\frac{2}{\alpha^k}\right) \right\}$$

where

$$\beta^{ij}_m = \begin{cases} \dfrac{2}{2i+1}\,\delta_{ij}\,, & m = 0 \\ -1, & m = 1 \\ \dfrac{2(-1)^m}{(m+1)(m-1)!} \displaystyle\prod_{s=1}^{m-1} (i+j+1+m-2s)\,(|i-j|+m-2s), & m \geq 2 \end{cases}$$

$$({}^+\underline{\underline{G}}^k_{VS})_{in} = (-1)^i \alpha^k \, T_{n+1}(0) - \alpha^k \, T_{n+1}(2/\alpha^k) + [i(i+1)/2]\alpha^k ({}^+\underline{\underline{G}}^k_{VS})_{i-1,n+1}$$

where

$$({}^{+}\underline{\underline{G}}^{k}_{VS})_{on} = \alpha^k T_{n+1}(0) - \alpha^k T_{n+1}(2/\alpha^k), \quad T_r(0) = \frac{1}{2}\Gamma\left(\frac{r+1}{2}\right)$$

$$({}^{+}\underline{\underline{G}}^{k}_{SV})_{sj} = (-1)^j ({}^{+}\underline{\underline{G}}^{k}_{VS})_{j,s-1}, \quad ({}^{+}\underline{\underline{G}}^{k}_{SS})_{sn} = T_{s+n}(2/\alpha^k)$$

$$({}^{+}\underline{\underline{B}})_{sn} = \frac{1}{2}\Gamma\left(\frac{s+n+1}{2}\right), \quad (\underline{\underline{A}})_{ij} = \frac{2}{2i+1}\delta_{ij}$$

Thus all of the matrix elements are given in terms of the T_n functions,[6] $n = 0, \ldots, 5$. Finally, the solution for the distribution function in the interior of the volume elements can be reconstructed from

$$\phi^k(x,v_x) = \frac{1}{\sqrt{\pi}}\sum_{j=0}^{N_k} X_j^k \sum_{r=0}^{j} (-1)^r (2r-1)!!\,(\alpha^k v_x)^r\, C_{j-r}^{(r+1/2)}(x)$$

$$\cdot\left\{1 - \exp\left[-\frac{(1\pm x)}{\alpha^k|v_x|}\right]\sum_{s=0}^{r}\frac{1}{s!}\left(\frac{1\pm x}{\alpha^k|v_x|}\right)^s\right\}$$

$$+ \exp -\left[\frac{(1\pm x)}{\alpha^k|v_x|}\right]\phi_{out}^{k\pm 1}(\pm 1, v_x), \quad v_x \gtrless 0$$

where the[6] $C_m^{(\alpha)}(z)$ are the Ultraspherical (Gegenbauer) polynomials.

References

[1]Bell, G. I. and Glasstone, S., Nuclear Reactor Theory, Van Nostrand Reinhold, New York, 1970.

[2]Cercignani, C., Mathematical Methods in Kinetic Theory, Plenum Press, New York, 1969.

[3]Kogan, M. N., Rarefied Gas Dynamics, Plenum Press, New York, 1969.

[4]Williams, M. M. R., Mathematical Methods in the Particle Transport Theory, Butterworths, London, 1971.

[5]Bhatnagar, P. L., Gross, E. P., and Krook, M., "A Model for Collision Processes in Gases. I. Small Amplitude Processes

in Charged and Neutral One-Component Systems," Physical Review, Vol. 94, No. 3, 1954, pp. 511-525.

[6]Abramowitz, M. and Stegun, I. A., Handbook of Mathematical Functions, National Bureau of Standards, Washington, D.C., 1964.

[7]Burns, T. J. and Dorning, J. J., "An Integral Balance Technique for Space-, and Energy-Dependent Reactor Calculations," Mathematical Models and Computational Techniques for Analysis of Nuclear Systems, pp. VII - 162-178, CONF - 730414, U.S. Atomic Energy Commission, Washington, D.C., 1973.

[8]Burns, T. J. and Dorning, J. J., "The Partial Current Balance Method: A New Computational Method for the Solution of Multidimensional Neutron Diffusion Problems," Proceedings of the NEACRP and CSNI Specialist's Meeting on New Developments in Three-Dimensional Neutron Kinetics and Review of Kinetics Benchmark Calculations, D-8406, 1975, Laboratorium für Reaktorregelund und Anlagensicherung, Garching, Germany; also Burns, T. J. and Dorning, J. J., "Multidimensional Applications of an Integral Balance Technique for Neutron Diffusion Computations," Computational Methods in Nuclear Engineering, pp. V-57-67, CONF-750413, U.S. Energy Research and Development Agency, Washington, D.C., 1975.

[9]Burns, T. J. and Dorning, J. J., "Approximate Generation of Coupling Parameters for Multidimensional Neutron Diffusion Calculations," Transactions of the American Nuclear Society, Vol. 19, 1974, p. 174.

[10]Dorning, J., Pescatore, C., and Spiga, G., "A Continuous Slowing-Down Theory for Test Particles," published elsewhere in this volume.

[11]Boffi, V. C., Dorning, J. J., Molinari, V. G., and Spiga, G., "Integral Boltzmann Equation for Test Particles in a Conservative Field: I. Theory, and Solutions to Some Stationary and Time-Dependent Cases," published elsewhere in this volume.

[illegible] and [illegible] Phenomena, [illegible] Physical Re- [illegible]

[illegible], M., and Stegun, I. A., [illegible] Handbook of Mathematical Functions, National Bureau of Standards, Washington, D.C. [illegible]

[illegible] Analysis [illegible]

[illegible] and [illegible] Current Balance [illegible] Method for the [illegible] Research and Development [illegible] Washington, D.C. [illegible]

[illegible]

[illegible] Solutions [illegible] published elsewhere in this volume.

A HIERARCHY KINETIC MODEL AND ITS APPLICATIONS

Takashi Abe* and Hakuro Oguchi+

University of Tokyo, Tokyo, Japan

Abstract

In the present paper we derive the model equations for the nonlinear Boltzmann equation together with a formalism slightly different from that by Shakov, and examine the applicability of the model equations. The three types of model equation are considered according to the choice of collision frequency. These model equations are applied to an analysis of the plane shock structure and temporal relaxation problems. The resulting solutions are found to indicate an appreciable difference among the choices of collision frequency. From comparison, it is shown that the model equation yields a reasonable solution closer to the experiment or Boltzmann solution when the collision frequencies are pertinent to various velocity moments, so far as the present examples are concerned.

Introduction

Regardless of the recent development of a large computer, it still is limited to comparatively simple problems to solve directly the full Boltzmann equation. On the other hand, the kinetic model equation has an appreciable advantage in providing a more tractable way to solve comparatively complex rarefied gas problems. The simplest kinetic model equation was proposed first by Krook et al.[1] and by Welander[2] (hereafter shortly termed the Krook equation). The Krook equation bears a resemblance to the Boltzmann equation concerning the few lower moments, but it contains an incorrect expression

---

Presented as Paper 21 at the 10th International Symposium on Rarefied Gas Dynamics, Aspen, Colo., July 19-23, 1976.

*Graduate Student, Institute of Space and Aeronautical Science.

+Professor, Institute of Space and Aeronautical Science.

for either stress tensor or heat flux vector in going over to the continuum flow regime. To remove this defect of the Krook equation, Holway[3] proposed an ellipsoidal model, which yields the correct expression for both stress tensor and heat flux vector in the continuum flow regime. However, any possibility of further systematic refinement of this model toward higher order was not suggested. Preceding Holway's work, for the linearized Boltzmann equation a systematic construction of the higher-order model equations was made successfully by Gross and Jackson.[4] Following their procedure, one can derive a model equation whose collision term could be simulated to that of the Boltzmann equation associated with the moments of any order as high as desired. In a similar way, to some extent, to the linearized case, Shakov[5-7] and Segal and Ferziger[8] proposed a systematic construction of the model equations for the nonlinear Boltzmann equation. According to their proposal, the second-order model equation reduces to the simplest or Krook equation, and the incomplete third-order (13-moment) model equation leads to the correct expressions for both stress tensor and heat flux vector in the continuum flow regime. Actually, a number of rarefied gas flow problems have been analyzed by applying this type of model equation.[9] In any model equations just referred to, the mathematical ambiguity is reduced for the Maxwell molecules, but, as pointed out by Gross and Jackson, there still remains an arbitrariness in determination of the collision frequency involved in the model equation. In the present paper, we derive the model equations for the nonlinear Boltzmann equation, together with a formalism slightly different from the previous, and examine the applicability of the model equations with three different choices of collision frequency by analyzing some of typical rarefied gas problems: the plane shock structure and temporal relaxation problems.

Systematic Construction of the Model Equation

The Boltzmann equation is written as[10]

$$\frac{\partial f}{\partial t} + v_i \frac{\partial f}{\partial x_i} = \int (f_1' \, f' - f_1 f) g d\Omega d\vec{v}_1 \qquad (1)$$

where $d\Omega$ is the differntial cross section, f the distribution function dependent on seven independent variables (time t, space coordinates x_i, and particle velocity v_i), and g the relative velocity, i.e., $g = |\vec{v} - \vec{v}_1|$. For simplicity, throughout the paper we assume the Maxwell molecules for which $gd\Omega$

in Eq. (1) becomes independent of $\vec{v}_1$. Then the Boltzmann collision integral is rewritten as

$$L(f) = \int (f_1' f' - f_1 f) g d\Omega d\vec{v}_1$$
$$= J_1(\vec{v}) - Knf \tag{2}$$

where $J_1(\vec{v})$ is the reverse collision integral and K is a constant independent of $\vec{v}$. Putting $Kn = \nu$, we can write Eq. (1) formally as

$$\partial f/\partial t + v_i(\partial f/\partial x_i) = \nu(f^+ - f) \tag{3}$$

where

$$f^+ = J_1(\vec{v})/\nu$$

As can be seen easily, Eq. (3) reduces to the simplest or Krook equation if we put

$$f^+ = f_0 = n(m/2\pi kT)^{3/2} \exp(-V^2/2)$$
$$= \nu_K = p/\mu$$

where f_0 is a local Maxwellian distribution, n the number density, p the pressure, T the temperature, V_i the peculiar velocity, μ the viscosity coefficient, m the particle mass, and k the Boltzmann constant. In view of this fact, we expand f around f_0 in terms of the reduced Hermite polynomials of V_i, which are the eigenfunctions of the linearized Boltzmann collision operator; that is

$$f^+ = f_0 \sum_{n,r}^{\infty} c_{n,r} B_i^{(n-2r,r)} H_i^{(n-2r,r)}(V_i) \tag{4}$$

with

$$c_{n,r} = (n - r)!(2n - 4r + 1)!/[(2n - 2r + 1)!r!\{(n - 2r)!\}^2]$$

For abbreviation

$$\sum_{n,r}^{\infty} = \sum_{n=0}^{\infty} \sum_{r=1}^{[n/2]}$$

In Eq. (4), $H_i^{(n-2r,r)}(V_i)$ denotes the reduced Hermite polynomial and the subscript i is the abbreviation of $(\alpha_1, \alpha_2, \cdots, \alpha_{n-2r})$. The mathematics associated with the Hermite polynomials is given in the papers by Grad.[11,12]

Since $H_i^{(n-2r,r)}$ is the nth-degree orthogonal polynomial with the weight function f_0, we have from Eq. (4)

$$B_i^{(m,r)} = \frac{1}{n} \int f^+ H_i^{(m,r)}(V_i) d\vec{v}$$

In a similar way, we expand f around f_0

$$f = f_0 \sum_{n,r}^{\infty} c_{n,r} A_i^{(n-2r,r)} H_i^{(n-2r,r)} \tag{5}$$

and thus we have

$$A_i^{(m,r)} = \frac{1}{n}\int f\, H_i^{(m,r)}\, d\vec{v} \tag{6}$$

Substituting Eqs. (4) and (5) into Eq. (3), and integrating over a whole velocity space after multiplication of $H_i^{(m,r)}$, we obtain

$$B_i^{(m,r)} = (1/n\nu)\, I_i^{(m,r)} + A_i^{(m,r)} \tag{7}$$

where

$$I_i^{(m,r)} = \int L(f)\, H_i^{(m,r)}\, d\vec{v}$$

$$I_i^{(m,r)} = \int L(f)\, H_i^{(m,r)}\, d\vec{v}$$

which can be expressed in terms of $A_i^{(m,r)}$. (The concrete expressions for $I_i^{(m,r)}$ are shown in Appendix.) It follows from Eq. (6) that $A_i^{(m,r)}$ is the moment of the order m+2r; for example

$$A^{(0,0)} = 1,\ A_i^{(1,0)} = 0,\ A^{(0,1)} = 0,\ A_{ij}^{(2,0)} = p_{ij}/p$$

$$A_i^{(1,1)} = (2q_i/p)(m/kT)^{1/2},\ A_{ijk}^{(3,0)} = (S_{ijk}/p)(m/kT)^{1/2}$$

Consequently, an approximate expression for f^+ and thus for the collision term of Eq. (3) has been obtained in terms of the moments $A_i^{(m,r)}$ of the distribution f up to any order as high as desired, so far as the convergence of the series (4) and (5) is assured. If the truncation is made above the Nth order, then Eq. (4) becomes

$$f^+ \approx f_N^+ = f_0 \sum_{n=0}^{N} \sum_{r=0}^{[n/2]} c_{n,r} B_i^{(n-2r,r)} H_i^{(n-2r,r)} \tag{8}$$

Thus, the Nth-order model equation is given as

$$\partial f/\partial t + v_i(\partial f/\partial x_i) = \nu(f_N^+ - f) \tag{9}$$

As can be seen from the foregoing deduction, the collision frequency ν involved could not be determined uniquely unless an additional constraint is introduced. It should be noted that the second-order model (N = 2) reduces to the Krook equation if one chooses $\nu = \nu_K = p/\mu$, and with the same choice of ν, i.e., $\nu = \nu_K$, the incomplete third-order (13-moment) model equation yields one proposed by Shakov.[5] In application to rarefied gas problems, even for the Krook equation the solution relies on numerical computation, along with many iteration processes. Therefore, as suggested from the form of the model equation, the treatment of the higher-order model equation does not make so much difference in computation compared with that of the Krook equation whenever an adequate computer is available.

Choice of the Collision Frequency

As mentioned in the previous section, the choice of ν involved in the model equation is not made uniquely. In what follows, we consider three different types of choice of the collision frequency ν and show the resulting model equations. In a systematic construction of the model equations for the linearized Boltzmann equation, Gross and Jackson presented a proposal for determination of ν. Following the proposal, the ν for the Nth-order model equation may be expressed as

$$\nu = \nu_K \lambda_{0N} \tag{10}$$

where λ_{rm} is the eigenvalue of the linearized collision operator ($\lambda_{02} = 1$). This choice is derived based upon the assumption that all of the eigenvalues λ_{rm} of higher order than N + 1 are equal to λ_{0N}. Since, as will be shown later, the ν given by Eq. (10) is regarded as the collision frequency pertinent to the Nth-order moment, this choice is to place much stress on the Nth-order moment. The details of the derivation are given in the original work by Gross and Jackson.[4]

If the nonlinear model equation should reduce to the linearized model equation for the linear case, the Nth-order collision frequency may be given by the form (10). We now designate this choice as type II, whereas, as done by Shakov, the choice $\nu = \nu_K$ is type I. Here we add one more choice of from a statistical consideration, and designate it as type III. Suppose that there is a time-dependent relaxation of the distribution to equilibrium in homogeneous space. In this case, the linearized Boltzmann equation is written as

$$\partial\phi/\partial t = L_0(\phi) = \int f_{00}(v_1)(\phi_1' + \phi' - \phi_1 - \phi) g d\Omega dv_1$$

for ϕ defined by $f = f_{00}(1 + \phi)$, where f_{00} is the absolute Maxwellian and L_0 the linearized collision operator. Using Eq. (5), we have

$$\phi = \sum_{n=2,r}^{\infty} c_{n,r}\, a_i^{(n-2r,r)}\, H_i^{(n-2r,r)} \tag{11}$$

Since $f_{00}H_i^{(m,r)}$ is the eigenfunction of the operator L_0, we have[12]

$$L_0(f_{00}H_i^{(m,r)}) = -\lambda_{rm}\,\nu_K\,f_{00}\,H_i^{(m,r)} \qquad (12)$$

Multiplying Eq. (12) by $H_i^{(m,r)}$ and integrating over the velocity space, we have

$$(\partial/\partial t)\,a_i^{(m,r)} = -\lambda_{rm}\,\nu_K\,a_i^{(m.r)} \qquad (13)$$

which is exactly the same as one derived by Grad.[11, 12] This implies that the relaxation time relevant to the Hermite coefficients $a_i^{(m,r)}$ or the moments of the distribution f is given by

$$\tau_{rm} = 1/(\lambda_{rm}\,\nu_K)$$

For the simplest case, i.e., r = 0, m = 2, we have $\tau_{02} = 1/\nu_K$. which gives the relaxation time of the stress tensor p_{ij}. The eigenvalues λ_{rm} are shown in the Appendix for the convenience of the application. As seen from the aforementioned argument, the relaxation time of a specified moment is subject to the eigenvalue pertinent to that moment. The third type of choice of ν results from the assumption that the collision frequency ν is given by

$$\nu = -\overline{(\partial\phi/\partial t)/\phi}$$

where the bar means the average over the velocity space. If this is the case, then using Eqs. (11) and (13) we have

$$\frac{\partial\phi/\partial t}{\phi} = -\frac{\sum_{n=2,r}^{\infty}\lambda_{r,n-2r}\,\nu_K\,c_{n,r}\,a_i^{(n-2r,r)}\,H_i^{(n-2r,r)}}{\sum_{n=2,r}^{\infty}c_{n,r}\,a_i^{(n-2r,r)}\,H_i^{(n-2r,r)}}$$

and hence

$$\nu = -\overline{\frac{\partial\phi/\partial t}{\phi}} = \frac{\sum_{n=2,r}^{\infty}\lambda_{r,n-2r}\,\nu_K\,D_{n,r}}{\sum_{n=2,r}^{\infty}D_{n,r}} \qquad (14)$$

Here $D_{n,r}$ is the Hilbert distance introduced by Grad; that is

$$D_{n,r} = c_{n,r}\,a_i^{(n-2r,r)}\,a_i^{(n-2r,r)}$$

For the nonlinear case, as the third type of choice we assume the collision frequency given by Eq. (14), in which $D_{n,\,r}$ is defined as

$$D_{n,r} = c_{n,r}\,A_i^{(n-2r,r)}\,A_i^{(n-2r,r)}$$

In any type of choice of the collision frequency, the second-order model equation (N = 2) reduces to the Krook equation. If we take the incomplete third-order (i.e., 13-moment) equation, then we have from Eqs. (8) and (9)

$$\partial f/\partial t + v_i(\partial f/\partial x_i) = \nu_{N=3}(f^+_{N=3} - f)$$

where

$$f^+_{N=3} = f_0[1 + (1 - \frac{\nu_K}{\nu_{N=3}})\frac{p_{ij}}{2p}(V_iV_j - \frac{V^2}{3}) +$$

$$+ (1 - \frac{\nu_K\,\lambda_{03}}{\nu_{N=3}})(\frac{V^2}{5} - 1)\frac{q_iV_i}{p}(\frac{m}{kT})^{1/2}]$$

According to choice of the collision frequency $\nu_{N=3}$, the preceding model equation is classified by the following three types

Type I[5-7] $$\nu_{N=3} = \nu_K$$

Type II[4] $$\nu_{N=3} = \nu_K\,\lambda_{0N}$$

Type III $$\nu_{N=3} = \nu_K[1 + \frac{\lambda_{11} - 1}{(5/4)(p_{ij}\,p_{ij}/q_iq_i)(kT/m) + 1}]$$

Application to Rarefied Gas Problems

In this section, we apply the various model equations to the shock structure and temporal relaxation problems, and the numerical results are compared with each other.

A. Plane Shock Structure

The plane shock structure of monatomic gases is analyzed numerically by use of the model equations. The solutions are obtained for Maxwell molecules with the shock Mach number M = 2 and 3 by an iteration starting from an appropriate initial guess. The normalized number density $\bar{n}$, velocity $\bar{u}$, and temperature $\bar{T}$ are defined by

$$\bar{n} = (n - n_1)/(n_2 - n_1), \quad \bar{u} = (u - u_2)/(u_1 - u_2)$$

$$\bar{T} = (T - T_1)/(T_2 - T_1)$$

where the subscripts 1 and 2 refer to the quantities associated with the pre- and post shock, respectively. Figures 1 and 2 show these quantities for the Mach numbers M = 2 and 3, respectively, in which the x coordinate is taken normal to the shock

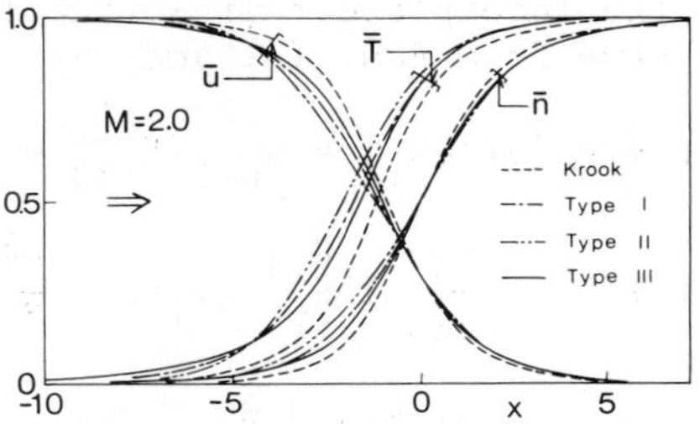

Fig. 1 Shock structre solutions from various types of the incomplete third-order model equation (M = 2.0).

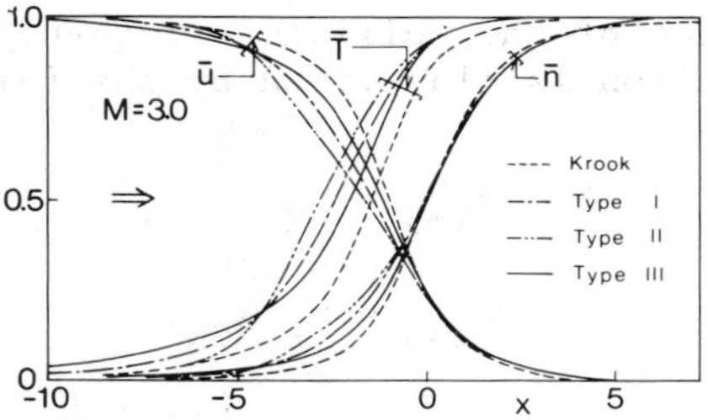

Fig. 2 Shock structure solutions from various types of the incomplete third-order model equations (M = 3.0).

with the origin where $\bar{n} = 0.5$, referred to the upstream mean free path $l_1 = \pi\mu_1/mn_1(2\pi RT_1)^{1/2}$. It can be seen from these figures that there appears to be an appreciable difference among the solutions for various model equations used. The shock thickness Δ, defined by

$$l_1/\Delta = |du/dx|_{max}/(u_1 - u_2)$$

is shown in Table 1.[13] The experiments suggest that the Navier-Stokes solution may provide a proper guess for lower shock Mach numbers, say $M \leq 2.0$. If this is the case, the third type of model equation is likely to lead to a comparatively reasonable solution, so far as the shock structure is concerned.

B. Temporal Relaxation to Equilibrium in Homogeneous Space

As another example, we consider the temporal relaxation of the isotropic distribution function to equilibrium in homogeneous space. Because of the spatial homogeneity and isotropy of the distribution function, all of the odd-order terms in expressions (4) and (5) vanish, and thus the resulting model equations are of the 2Nth order (N = 1, 2,···); the lowest is the Krook equation, and the succeeding ones are the fourth-order, sixth-order, and so on.

Table 1 Inverse shock thickness l_1/Δ

	M = 2.0	M = 3.0
Type I	0.1927	0.2176
Type II	0.1808	0.2085
Type III	0.2088	0.2338
Navier-Stokes	0.2199	0.2562
Krook	0.2457	0.2753

We here apply the second-order or Krook equation and the three types of higher-order model equations to cases of the two kinds of initial distribution. One of the assumed initial distributions is a bimodal one, and this is designated as case I

$$f = \frac{n}{2}\left(\frac{m}{2\pi kT_1}\right)^{1/2} \exp\left(-\frac{m}{2kT_1} v^2\right) + \frac{n}{2}\left(\frac{m}{2\pi kT_2}\right)^{1/2} \exp\left(-\frac{m}{2kT_2} v^2\right)$$

$$T_1 = T/2, \qquad T_2 = 3T/2$$

Another one is of trapezoidal, and this is designated as case II

$$f = \begin{cases} an(m/2kT)^{3/2} & 0 < v < v_1 \\ an(m/2kT)^{3/2}(v - v_2)/(v_1 - v_2) & v_1 < v < v_2 \\ 0 & v_2 < v \end{cases}$$

with the numerals given by

$$a = 0.06389, \quad v_1 = 1.3029(2kT/m)^{1/2}, \quad v_2 = 1.7766(2kT/m)^{1/2}$$

The variation of the distribution f in time is shown in Fig. 3 for case I and in Figs. 4 and 5 for case II, respectively. In these figures f, v, t are dimensionless quantities referred to $n(m/2kT)^{3/2}$, $(2kT/m)^{1/2}$, $(4/5)\mu/p$, respectively. In Fig. 3, there are plotted the results from the Krook equation and the three types of the fourth-order model equation, together with the Boltzmann solution obtained by Rykov.[14] The solutions for three types of the sixth-order model equation

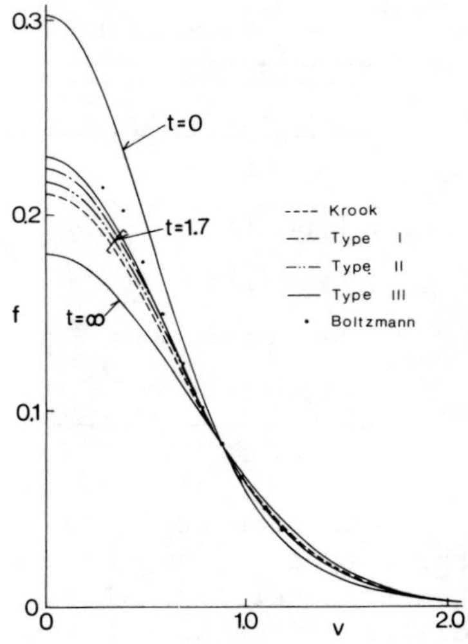

Fig. 3 Temporal relaxation of the distribution function (case I).

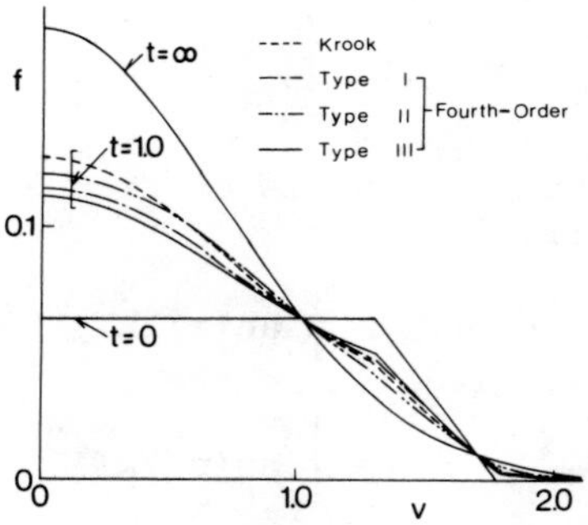

Fig. 4 Temporal relaxation of the distribution function (case II, fourth-order kinetic model).

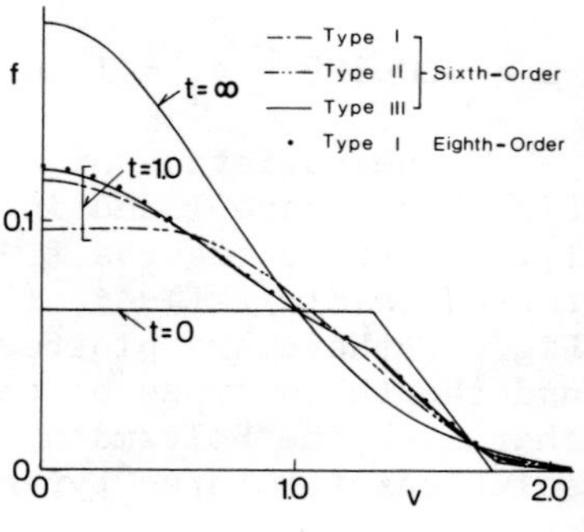

Fig. 5 Temporal relaxation of the distribution function (case II, sixth-order or eighth-order kinetic model).

have been obtained also. (As for the temporal relaxation problem, it can easily be shown that the fourth-order model equation of type I[5-7] is identical with the sixth-order one; that is, the sixth-order terms vanish.) However, it was found that these indicate no appreciable departure from the fourth-order solutions, so that the convergence may be assured. Thus we can say from the results that the type III model equation is likely to provide a more reasonable solution closer to the Boltzmann solution, compared with the other types of model equation.

As regards case II, as can be seen from Figs. 4 and 5, the convergence of the solution is rather poor for any type of the model equations; there appears to be an appreciable departure between the fourth and sixth order for types II and III. This is because in case II, the specified initial distribution has a strong nonequilibrium feature so that the series expression in terms of the Hermite polynomials is insufficient unless a large number of higher-order terms are taken into account.

Concluding Remarks

In the present paper, we derived the higher-order model equations for the nonlinear Boltzmann equation, being classified by three types according to choice of the collision frequency. These model equations were applied to the plane shock structure and temporal relaxation problems and showed that an appreciable difference results in solution, dependent on the type of model equation employed. The type III model equation, in which the collision frequency is chosen as a statistical average, is likely to lead to a more reasonable solution, at least concerning the present examples. Further investigations remain to be made in the future, with progress in the direct solution of the full Boltzmann equation.

Appendix

$$I^{(2,0)}_{ij} = -\nu_K n A^{(2,0)}_{ij}, \quad I^{(3,0)}_{ijk} = -\lambda_{03} \nu_K n A^{(3,0)}_{ijk}$$

$$I^{(1,1)}_{i} = -\lambda_{11} \nu_K n A^{(1,1)}_{i}, \quad I^{(0,2)} = -\lambda_{20} \nu_K n A^{(0,2)}$$

$$I^{(0,3)} = -\lambda_{30} \nu_K n A^{(0,3)}$$

$$I^{(0,4)} = -\lambda_{40} \nu_K n A^{(0,4)} + \lambda' \nu_K (A^{(0,2)})^2$$

$\lambda_{02} = 1.0, \quad \lambda_{03} = 3/2, \quad \lambda_{11} = 2/3, \quad \lambda_{20} = 2/3, \quad \lambda_{04} = 1.2732$

$\lambda_{30} = 1.0, \quad \lambda_{06} = 2.4532, \quad \lambda_{40} = 1.2282, \quad \lambda_{08} = 2.9193$

$\lambda' = 1.3253$

References

[1] Bhatnager, P. L., Gross, E. P., and Krook, M., "A Model Collision Processed in Gases. I.Small Amplitude Processes in Charged and Neutral One-Component Systems," Physical Review, Vol. 94, May 1954, pp. 511-525.

[2] Welander, P., "On the temperature Jump in a Rarefied Gas", Arkiv for Fysik, Vol. 7, No. 44 1954, p. 507.

[3] Holway, L. H., "New Statistical Models for Kinetic Theory: Method of Construction," The Physics of Fluids, Vol. 9, Sept. 1966, pp. 1658-1673.

[4] Gross, E. P. and Jackson, E. A., "Kinetic Models and the Linearized Boltzmann Equation," The Physics of Fluids, Vol. 6, July 1959, pp. 432-441.

[5] Shakov, E. M., "Generalization of the Krook Kinetic Equation," Fluid Dynamics, Vol. 3, Jan. 1968, pp. 95-96.

[6] Shakov, E. M., "Shock Structure in a Monatomic Gas," Fluid Dynamics, Vol. 4, March-April 1969, pp. 44-47.

[7] Shakov, E. M., "Approximate Kinetic Equations in Rarefied Gas Theory," Fluid Dynamics, Vol. 3, Jan.-Feb. 1968, pp. 112-115.

[8] Segal, B. M. and Ferziger, J. H., "Shock-Wave Structure Using Nonlinear Model Boltzmann Equations," The Physics of Fluids, Vol. 15, July 1972, pp. 1233-1247.

[9] Ryzhov, O. S., "Numerical Methods in Rarefied Gas Dynamics Developed at the Computing Center of the Academy of Science of USSR," Rarefied Gas Dynamics, edited by M. Becker and M. Fiebig , Vol. II, DFVLR Press, Porz-Whan, Germany, 1974.

[10] Chapmann, S. and Cowling, T. G., The Mathematical Theory of Non-Uniform Gases, Cambridge University Press, Cambridge, England, 1953.

[11]Grad, H., "Asymptotic Theory of the Boltzmann Equation," The Physics of Fluids, Vol. 6, Feb. 1963, pp. 147-181.

[12]Grad, H., "On the Kinetic Theory of Rarefied Gases," Communiations on Pure and Applied Mathematics," Vol. 2, 1949, pp. 331-407.

[13]Gilbarg, D. and Paolucci, D., "The Structure of Shock Waves in the Continuum Theory of Fluids," Journal of Rational Mechanics and Analysis, Vol. 2, April 1953, pp. 617-642.

[14]Rykov, V. A., "Relaxation of a Gas Described by Boltzmann Kinetic Equation," Journal of Applied Mathematics and Mechanics, Vol. 31, No.4 1967, pp. 770-775.

KNUDSEN LAYERS: SOME PROBLEMS AND A SOLUTION TECHNIQUE

Carlo Cercignani *

Politecnico di Milano, Milano, Italy

Abstract

Some problems are listed which can be investigated by solving half space boundary value problems for the linearized Boltzmann equation, and it is shown how the solution of these problems may, in principle, be obtained in a closed form. The basic contribution is a technique for solving systems of coupled singular integral equations which arise in the process of solving such Knudsen layer problems. After reducing the solution of the system to a problem in analytic function theory, a related matrix problem is shown to be solvable with a diagonalization technique. The particular case of the temperature Jump problem with the BGK model is discussed in some detail.

1. Introduction

Although the Knudsen layers have been studied extensively for more than 20 years[1,2], their study appears to be worthwhile pursuing, expecially after recent experimental contributions[3,4] to the techniques for measuring velocity profiles in the Knudsen layers themselves. Specifically, the results of the experiments performed by Reynolds et al[3] suggest that there is wide space for further research on the theoretical side of the matter, if we want to use the results of experiments to check the accuracy of collision models as well as gas-surface interaction models. As a matter of fact the aforementioned authors found that in the Knudsen layer

---

Presented as Paper 35 at the 10th International Symposium on Rarefied Gas Dynamics, Aspen, Colo., July 19-23, 1976. Performed in the framework of the activity of the Gruppo Nazionale di Fisica Matematica of the Consiglio Nazionale delle Ricerche (Italian Research Council).

*Professor, Istituto di Matematica.

the deviation of the actual velocity profile from a straight line is somewhat different from the BGK model result.[1,2,5,6] A better agreement can be found, according to Loyalka[7], by using a variable collision frequency model.

It has to be remarked, however, that as early as 1966 Cercignani and Tironi[8] pointed out that even if the collision frequency is constant, a correction must be introduced in order to allow a dependence on the Prandtl number. Use of the linearized ellipsoidal statistical (ES) model indicated that the slip coefficient was unaffected by the Prandtl number, whereas the velocity defect in a Knudsen layer was quite sensitive to it. As a matter of fact, it turns out that the layer is modified in size by a factor equal to the Prandtl number, i.e., 2/3 if the value appropriate for a Maxwell gas is adopted. In the latter case, the curve is almost coincident with the one computed by Loyalka[7]. Unfortunalely, a trivial mistake by Cercignani and Tironi[8] led to an indication of 1/Pr as a stretching factor for the Knudsen layer velocity defect; for this reason, probably, this early result was disregarded in the comparisons with experimental data. If we use the argument used by Cercignani and Tironi and correct the trivial mistake, we have two corrections to the BGK results which bring the latter into a reasonably good agreement with the experimental data. If we superpose the two corrections naively (i.e., we modify the lengths by a factor of 2/3 in the direction normal to the plate, in Loyalka's results), we obtain an excellent agreement with experiments. It appears, however, that a more complete investigation may be interesting. A second reason for investigating the Knudsen layers is that they are the seat of some irreversible phenomena, which are not describable in terms of a continuum theory, unless one wants to introduce them as surface phenomena, as indicated in the papers by Waldmann on the non-equilibrium thermodynamics of boundary conditions[9-11]. These phenomena are of importance in order to understand a pair of wellknown phenomena, i.e. thermo-osmosis and mechanocaloric effect, in the nearly-continuum regime. Finally, we mention that the classical problem of sound propagation still is open to a better treatment: the agreement between existing theories and experiment is not as good as some theoretical papers seem to a suggest; for a discussion of this point, the reader is referred to a recent book of the author[1].

All of these problems lead to the solution of some boundary value problems generalizing the simplest of them, i.e. the Kramers' problem for the BGK model. If we try to solve these problems by separating the variables[1,2], we are led to solving a system of singular integral equations. Only in the simplest cases (pure shear flow with the linearized BGK or ES models) is the system made of a single equation and the solution in closed form possible by means of standard techniques[1,2,6,8] In another case (heat transfer with a velocity dependent frequency model) Cassel and Williams were able to show that one can decouple the equations of the system[1,12]. In all of the other cases, including the problem of temperature jump with the BGK model, one must face the problem of solving coupled singular integral equations. The difficulty in finding closed form solutions of the latter led to numerical methods for solving the problem of temperature jump.[13,14] Recently, however, the author has developed a new technique leading to a closed form solution of the problem of the temperature jump with the BGK model.[15] The basic idea of the method was to diagonalize the problem as suggested first by Darrozès in 1967.[16] Darrozès was not able, however, to determine the correct solution.

It might be objected that the technique presented in this paper is very cumbersome and even that a numerical evaluation of the results may turn out to be less convenient than some straigthforward numerical technique. Although the author recognizes the importance of the latter objection, he still believes that analytical techniques are worthwhile investigating, since they may provide interesting information on limiting cases and even suggest improvements for the numerical methods.

2. Review of Standard Results

A general linearized kinetic model constant collision frequency can be written as follows,[1,2] if $(hg)\equiv\int fo\ hg\ d\underline{\xi}$ denotes a scalar product weighted with the basic Maxwellian fo

$$\frac{\partial h}{\partial t} + \xi_1 \frac{\partial h}{\partial x} = \sum_{k,j=1}^{M} \alpha_{jk}(h, \psi_j)\, \psi_k - \nu h \tag{1}$$

where α_{jk} and ν are constants, and the ψ_j the eigenfunctions of the Maxwell collision operator. We briefly recall how one may find the general solution of Eq.[1,2] First one eliminates the transverse components of the molecular velocity by decomposing h according to

$$h = \sum_{j=1}^{n} h_j(x,t,\xi_1)g_j(\xi_2,\xi_3) + h_R(x,t,\underline{\xi}) \tag{2}$$

where g_j is a set of suitable polynomials, and separating the variables according to

$$h_j(x,t,\xi_1) = h_j(\xi_1;\ u,s)\ e^{st-x(s+1)/u} \tag{3}$$

where s,u are constant parameters.

Then h_j can be shown to satisfy

$$(1 - \xi/u)\, \underline{h}\, (\xi;\, u,s) = \underline{\underline{T}}(\,\xi;u,s)\, \underline{A}(u;s) \tag{4}$$

where $\underline{h}$ and $\underline{A}$ are colunn vectors with components h_k and

$$A_k = \Pi^{-\frac{1}{2}} \int_{-\infty}^{\infty} h_k(\,\xi) e^{-\,\xi^2}\, d\xi \;\; / \; D(u,s) \tag{5}$$

and $\underline{\underline{T}}$ is a matrix, whose elements are polynomials in ξ,u, s.D(u,s) is a polynomial in u,s, which reduces to a polynomial in s for Maxwell models. In Eq. (4) we dropped the index in ξ_1 .

In addition to eventual eigenfunctions corresponding to discrete values of u, Eq. (4) has solutions in the form of generalized functions for u ranging from $-\infty$ to$+\infty$ Such generalized eigenfunctions may be written in the following form

$$\underline{h}(\xi;u,s) = \underline{\underline{T}}(\xi;u,s)\, \underline{A}(u,s)\, P\frac{u}{u-s} + \underline{\underline{p}}(u;s)\, A(u,s)\, \delta(u-\xi) \tag{6}$$

where, because of Eq. (5)

$$\underline{\underline{p}}(u,s) = e^{u^2}\, \Pi^{\frac{1}{2}}\, D(u,s)\, \underline{\underline{I}} - P \int_{-\infty}^{\infty} \frac{\underline{\underline{T}}(\xi;u,s)\, u e^{-\,\xi^2}}{u - \xi}\, d\xi \tag{7}$$

In Eq. (6) δdenotes Dirac's delta function and P Cauchy's principal value; $\underline{\underline{I}}$ in Eq. (7) is the identity matrix. Let us introduce the following matrix, which has functions of the complex variable z as elements

$$\underline{\underline{M}}(z,s) = \Pi^{\frac{1}{2}}\, D(u;s)\, \underline{\underline{I}} + \int_{-\infty}^{\infty} \frac{z\, \underline{\underline{T}}(\xi;z,s)\, e^{-\xi^2}}{\xi - z}\, d\xi \tag{8}$$

It is easily seen that the discrete spectrum (if any) is given by the complex values of u solving

$$\text{Det}\, \underline{\underline{M}}\, (u;s) = 0 \tag{9}$$

It is relatively easy to prove the completeness of eigen-solutions in the full range ($-\infty<\xi<\infty$) and construct the coefficients of the eigenfunctions in closed form. It is also an easy guess to anticipate that the elementary solutions having the form shown in Eq. (3) have partial-range completeness properties, i.e. that the following singular integral vector equation has a unique solution

$$\sum_{i=1}^{n} \underline{\underline{g}}_i(\xi;s)\, \underline{A}_i(s) + \int_{-\infty}^{\infty} \underline{\underline{g}}(\xi;u,s)\, \underline{A}(u,s)\, du = \underline{g}(\xi) \tag{10}$$

where

$$\underline{\underline{g}}(\xi;u,s) = \underline{\underline{T}}(\xi,u,s)\, P\, \frac{u}{u-\xi} + \underline{\underline{p}}(u;s)\, \delta(u-\xi) \tag{11}$$

and

$$g_i(\xi;s) = \underline{\underline{T}}(\xi;u_i,s)\, P\, \frac{u_i}{u_i-s} \tag{12}$$

u_i $(i = 1,2, \ldots, m, m+1, \ldots, 2m)$ being the possible solutions of Eq. (9).

However, the straightforward demonstration that holds in the case of a scalar equation is not extended easily to the vector equation, Eq. (10), which is of course equivalent to a system of n singular integral, scalar equations. A valiant attempt toward such a constructive proof was made by Darrozès in 1967[16] when he actually worked with the essentially equivalent method of Wiener and Hopf rather than with the method of elementary solutions and considered the particular case of the temperature jump problem with the BGK model.

The starting point for the procedure suggested by Darrozès as well as for the results presented in this paper is the classical remark[17] that solving Eq. (10) is essentially equivalent to finding a vector valued analytic function $\underline{N}$ (z), which is regular in the complex plane except for a cut along the positive real semiaxis, where the limiting values from above and from below, $\underline{N}^+$ (v) and $\underline{N}^-$ (v) satisfy

$$(\underline{\underline{P}} + \Pi iv\underline{\underline{I}})\, \underline{N}^+ - (\underline{\underline{P}} - \Pi iv\underline{\underline{I}})\, \underline{N}^- = \varphi \tag{13}$$

where

$$\underline{\underline{P}}(v,s) = \underline{\underline{p}}\,(v,s) \left[\underline{\underline{T}}(v,v;s)\right]^{-1} \tag{14}$$

and φ is related to $\underline{f}$. Actually, the foregoing procedure is straightforward only when T does not depend upon u; the case of a polynomial dependence upon u may however reduced to the simplest case by a suitable subtraction procedure.

In addition to satisfying Eq. (13), $\underline{N}$ (z) goes to 0 when $z \to \infty$. We also remark that problems having the form shown in Eq. (13) constitute a subclass of the general problem

$$\underline{\underline{A}}\,\underline{N}^{+} - \underline{\underline{B}}\,\underline{N}^{-} = \varphi \tag{15}$$

considered by Muskhelishvili,[17] because the matrices $\underline{\underline{A}}$ and $\underline{\underline{B}}$ commute.

In order to handle Eq. (1), it is sufficient to be able to master the adjoint homogeneous problem

$$\underline{\underline{X}}^{+}(v)\left[\underline{\underline{P}}(v) + \Pi iv\underline{\underline{I}}\right]^{-1} = \underline{\underline{X}}^{-}(v)\left[\underline{\underline{P}}(v) - \Pi iv\underline{\underline{I}}\right]^{-1} \tag{16}$$

where $\underline{\underline{X}} = \underline{\underline{X}}(z)$ is a matrix-valued unknown with the same properties as $\underline{N}$ (z) except for the behavior at infinity which may be polynomial.

As a matter of fact, if Eq. (16) holds, multiplying Eq. (13) by the left hand side of Eq. (16) gives

$$\underline{\underline{X}}^{+}\,\underline{N}^{+} - \underline{\underline{X}}^{-}\,\underline{N}^{-} = \underline{\underline{X}}^{-}\left[\underline{\underline{P}} - \Pi iv\,\underline{\underline{I}}\right]^{-1}\varphi \tag{17}$$

i.e. a problem that is solved easily by means of the Plemelj formulas.[17] Our task is to find X(z) from Eq. (16); a procedure will be indicated in the next section.

3. The Solution Technique

The idea suggests itself that the matrix problem summarized in Eq. (16) may be diagonalized. In fact, let $\underline{\underline{A}} = \underline{\underline{A}}(z)$ be a matrix diagonalizing $\underline{\underline{P}}(z)$, if any, i.e. such that

$$\underline{\underline{A}}\,\underline{\underline{P}} = \underline{\underline{D}}\,\underline{\underline{A}} \tag{18}$$

where $\underline{\underline{D}}$ is diagonal. If we let

$$\underline{\underline{U}} = \underline{\underline{A}}\,\underline{\underline{X}}\,\underline{\underline{A}}^{-1} \tag{19}$$

then $\underline{\underline{U}}$ satisfies

$$\underline{\underline{U}}^{+}\,(\underline{\underline{D}} + \Pi iv\,\underline{\underline{I}})^{-1} = \underline{\underline{U}}^{-}\,(\underline{\underline{D}} - \Pi iv\,\underline{\underline{I}})^{-1} \tag{20}$$

Now we may look for a solution of this matrix equation in a diagonal form ($\underline{\underline{D}}$ is diagonal). The diagonalization, as previously mentioned, was suggested first by Darrozès[16] in a particular case.

We now must comment upon the difficulties hidden behind the formal transformation just described. First, we have assumed that a matrix $\underline{\underline{A}}$ exists such that Eq. (18) is satisfied. This is not a big requirement in practice; as a matter of fact, if

$\underline{\underline{P}}$ is nondiagonalizable for some values of z, it may be surmised that a careful discussion eventually will show that this only can impose some condition on the right hand side of Eq. (2.13), whereas the case of an identically nondiagonalizable $\underline{\underline{P}}$ may be regarded as a mathematical curiosity in the present context.

There is a more difficult point, however. Both $\underline{\underline{A}}(z)$ and $\underline{\underline{D}}(z)$ will possess new singularities and hence new cuts C_k (k = 1, ..., m) in the complex plane, in addition to those of $\underline{\underline{X}}$. This is evident, e.g., in the case of 2 x 2 matrices. (Such is the case for the problem of temperature jump with the BGK model.) In this case the diagonalization problem requires solving a second degree algebraic equation; apart from trivial cases this solution will introduce branch singularities at the zeroes of the discriminant of the second degree equation. The transformation (19) is chosen in such a way as to have $\underline{\underline{X}}(z)$ free from these new singularities, provided that this is true for $U_1 + U_2$ and $(U_1 - U_2)\sqrt{d}$ (where U_1 and U_2 are the nonzero elements of the diagonal matrix U and d is the aforementioned discriminant). In other words U_1 and U_2 must be two branches of the same analytic function free from singularities (except for the original cut on the real axis) on the two-sheet Riemann surface associated with $\sqrt{d(z)}$.

We restrict ourselves to the particular case of 2 x 2 matrices, which is sufficiently significant both as a practical case and for illustration. In this case, Eq. (20) immediately leads (by taking products, ratios, and logarithms) to the determination of a possible form for $\log(U_1U_2)$ and $\log(U_1/U_2) / \sqrt{d(z)}$ in terms of a standard contour integral. The expressions for U_1 and U_2 then follow.

The problem of the behavior of the solution at infinity now arises: this is the problem that was not mastered by Darrozès,[16] who obtained an incorrect solution. In fact, in general, the solution constructed in the aforementioned fashion has neither zeroes nor poles but presents an essential singularity at infinity if d(z) grows more rapidly than $|z|^2$ for z . If we maintain the condition of neither zeroes nor poles for U_1 and U_2, we can construct new solutions by suitable integrals along the new cuts which might be used to cancel the essential singularity at infinity; in order to have a single valued solution in the cut plane for the original problem, however, the new contour integrals must have integral numbers as coefficients, and this will permit cancellation in very special cases only. It is then necessary to introduce integrals along a portion of the original cut on the real axis, with the lower limit, say different from zero. The numbers z_{ok} are to be chosen in such a way as to cancel the essential singularity at infinity. In the case of a finite number of zeroes of d(z), the latter can be taken to be a polynomial. (The nonzero part can be factored out and does not play any role.) In this case, the determination of the numbers z_{ok} leads to the famous inversion problem of Jacobi.[18]

In conclusion, the following expressions for U_1 and U_2 hold if d(z) is a polynomial

$$U_{1,2} = \exp\left\{\frac{1}{4\Pi i}\int_0^\infty \frac{\log G(t)}{t-z}\,dt \pm \sqrt{d(z)}\ \frac{1}{4\Pi i}\int_0^\infty \frac{\log F(t)}{t-z}\,\frac{dt}{\sqrt{d(t)}} - \sum_{k=1}^{m-1} n_k \int_{C_k} \frac{dt}{\sqrt{d(t)}\,(t-z)} - \sum_{k=1}^{m-1} \int_{z_{ok}}^{\infty} \frac{dt}{\sqrt{d(t)}\,(t-z)}\right\} \qquad (21)$$

where, if $D_1(z)$ and $D_2(z)$ are the nonzero elements of $\underline{\underline{D}}$

$$G(t) = \frac{\left[D_1(t) + \Pi it\right]\left[D_2(t) + \Pi it\right]}{D_1(t) - \Pi it \quad D_1(t) - \Pi it} \qquad (22)$$

$$F(t) = \frac{\left[D_1(t) + \Pi it\right]\left[D_2(t) - \Pi it\right]}{\left[D_1(t) - \Pi it\right]\left[D_2(t) + \Pi it\right]} \qquad (23)$$

the integers n_k and the complex numbers z_{ok} are to be chosen in such a way as to satisfy

$$\sum_{k=1}^{m-1} n_k \int_{C_k} \frac{t^s\,dt}{\sqrt{d(t)}} + \sum_{k=1}^{m-1} \int_{z_{ok}}^{\infty} \frac{t^s\,dt}{\sqrt{d(t)}} = \frac{1}{4\Pi i}\int_0^\infty \frac{t^s \log F(t)}{\sqrt{d(t)}}\,dt \qquad (24)$$

Additional factors in the expressions of U_1 and U_2 must be inserted to cancel singularities of the polar type at $0, 1, z_{ok}$ according to standard rules.

The homogeneous problem thus is solved, and standard methods[17] may be used to solve the inhomogeneous one. It is to be remarked only that the matrix $\underline{\underline{X}}(z)$, in general, will differ from the canonical one in the sense of Muskhelishvili.[17]

4. The Temperature Jump Problem

The technique indicated in the preceding section may be applied to several interesting problems. In this section we shall summarize the results for the particular problem of temperature jump with the BGK model. This problem was investigated by many authors; references and numerical solutions are given in Refs.[14] and [19], whereas Ref.[15] describes in more detail the solution to be summarized presently.

For the treatment of steady problems involving perturbations of density and temperature in the one-dimensional case the BGK model leads to Eq.(1) with M = 2. A suitable transformation[15] leads to solving the following transport equation

$$\xi \frac{\partial Z}{\partial x} + Z = \mathcal{L} Z \qquad (25)$$

where $\mathcal{L}$ is the matrix integral operator defined by

$$\mathcal{L}\underline{Z} = \Pi^{-\frac{1}{2}} \int_{-\infty}^{\infty} e^{-t^2}\, \underline{\underline{T}}(t)\, \underline{Z}(t)\, dt \qquad (26)$$

Here $\underline{\underline{T}}$ is the 2 x 2 matrix

$$\underline{\underline{T}}(t) = \begin{Vmatrix} 1 & \left(\frac{2}{3}\right)^{\frac{1}{2}} \left(t^2 - \frac{1}{2}\right) \\ \left(\frac{2}{3}\right)^{\frac{1}{2}} \left(t^2 - \frac{1}{2}\right) & \frac{2}{3}\left[1 + \left(t^2 - \frac{1}{2}\right)^2\right] \end{Vmatrix} \qquad (27)$$

The temperature perturbation $\tau(x)$ is related to the components (Z_1, Z_2) of $\underline{Z}$ by

$$\tau(x) = \frac{2}{3} \Pi^{-\frac{1}{2}} \int_{-\infty}^{\infty} e^{-\xi^2} \left\{ (\xi^2 - \tfrac{1}{2}) Z_1(x,\xi) + \tfrac{2}{3}\left[(\xi - \tfrac{1}{2})^2 + 1\right] Z_2(x,\xi) \right\} d\xi \qquad (28)$$

The general solution of Eq. (25) is given by

$$\underline{Z} = \underline{B}_o + \underline{b}_1(\xi - x) + \int_{-\infty}^{\infty} e^{-x/v}\, \underline{f}_v(\xi)\, dv \qquad (29)$$

where $\underline{b}_o$ and $\underline{b}_1$ are constant vectors, and $\underline{f}_v$ a vector valued distribution defined by

$$\underline{f}_v(\xi) = P \frac{v}{v-\xi} \underline{\underline{T}}^{-1}(v) + \underline{\underline{p}}(v)\, \delta(v-\xi) \quad \underline{b}(v) \qquad (30)$$

where $\underline{b}(v)$ is a two-component vector arbitrarily dependent upon v, and $\underline{\underline{p}}(v)$ is given by

$$\underline{\underline{p}}(v) = \underline{\underline{T}}^{-1}(v)\, \underline{\underline{Q}}(v)\, \underline{\underline{T}}^{-1}(v) \qquad (31)$$

where

$$\underline{\underline{Q}}(v) = e^{v^2}\, P \int_{-\infty}^{\infty} \frac{t\, e^{-t^2}}{t-v}\, \underline{\underline{T}}(t)\, dt \qquad (32)$$

If we consider the limit $x \to \infty$ of Eq. (29), we obtain $\underline{b}(v)=0$ fo v 0, $\underline{b}_1 = 0$ and the second component of $\underline{b}_o$ turns to be proportional to the temperature jump coefficient.
The boundary condition at x = 0 (perfectly diffusing wall) gives

$$\underline{b}_o + \underline{\underline{p}}(\xi)\,\underline{b}(\xi) + P\int_{-\infty}^{\infty} \frac{v\,\underline{\underline{T}}^{-1}(v)\,\underline{b}(v)}{v-\xi}\,dv = \underline{Z}_o(\xi) \qquad (33)$$

Here $\underline{Z}_o(\xi)$ is given by

$$\underline{Z}_o(\xi) = \frac{3\ell}{\Pi}\,\frac{k}{T_o}\,\xi \begin{vmatrix} -1 \\ \frac{3}{2} \end{vmatrix} \qquad (34)$$

where k is the temperature gradient, T_o the unperturbed temperature, and ℓ the mean free path related to the viscosity coefficient μ, density ρ, and temperature T_o by

$$\ell = (\Pi/2RT_o)^{\frac{1}{2}}\,\rho^{-1} \qquad (35)$$

Equation (33) is a system of two singular integral equations to be solved with the method described in Sects. 2 and 3. As a matter of fact, Eq. (9) leads to solving Eq. (13), where $\underline{\underline{P}}$ is given by $\underline{\underline{P}} = \underline{\underline{p}}\,\underline{\underline{T}}^{-1}$, and

$$\varphi = v\,\underline{Z}_o(v) - \underline{b}_o \qquad (36)$$

The matrices $\underline{\underline{A}}$ and $\underline{\underline{D}}$ appearing in Eq. (18) are now given by

$$\underline{\underline{A}}(v) = \begin{Vmatrix} \frac{1}{2}(\frac{3}{2})^{\frac{1}{2}} & \frac{1}{2}(v^2 - \frac{1}{2}) + d_1(v) \\ \frac{1}{2}(\frac{3}{2})^{\frac{1}{2}} & \frac{1}{2}(v^2 - \frac{1}{2}) + d_2(v) \end{Vmatrix} \qquad (37)$$

$$\underline{\underline{D}}(v) = \begin{Vmatrix} p(v) + \sqrt{\Pi}\,e^{v^2}\,d_1(v) & 0 \\ 0 & p(v) + \sqrt{\Pi}\,e^{v^2}\,d_2(v) \end{Vmatrix} \qquad (38)$$

where

$$p(v) = \sqrt{\Pi}\,(e^{v^2} - 2\,t\int_0^v e^{t^2}\,dt) \qquad (39)$$

and

$$d_{1,2}(v) = \frac{1}{8}(3-2v^2) \pm \frac{1}{8}\,(25-12v^2+4v^4)^{\frac{1}{2}} \qquad (40)$$

The square root appearing here produces two cuts in the complex plane corresponding to the four roots $z_e = \pm 2 \pm i/\sqrt{2}$ of the discriminant

$$d(z) = (25 - 12z^2 + 4z^4) / 64 \tag{41}$$

Since m = 2 in this case, Eq. (24) reduces to a single condition

$$n \int_{z_2}^{z_1} \frac{dt}{\sqrt{d(t)}} + \int_{z_0}^{\infty} \frac{dt}{\sqrt{d(t)}} = \frac{1}{4\pi i} \int_{0}^{\infty} \frac{\log F(t)}{\sqrt{d(t)}} dt \tag{42}$$

z_0 then is determined easily by means of elliptic functions the transformations are straightforward but cumbersome and will not be given here (see, e.g. Ref.[20]).

If one uses Eqs. (19) and (21) to evaluate $\underline{\underline{X}}$, it is convenient to multiply U_1 and U_2 by $(z-z_0)/z$ first. Then it turns out that $\underline{\underline{X}}^-$ has a second order pole at $z = z_0$. We conclude that the matrix constructed in such a way is not the canonical one[17]; we can either proceed with our $\underline{\underline{X}}$ or construct the standard matrix starting from the latter.

If one sticks to the first procedure, thet $\underline{N}(z)$ will have a second order pole at $z = z_0$, unless two conditions are satisfied; these conditions determine $\underline{b}_0$[15] and are given by

$$\underline{A}_1(z_0) \cdot \underline{u} = 0 \tag{43}$$

$$\underline{A}_1(z_0) \cdot \underline{w} + (d\underline{A}_1/dz) \cdot \underline{u} = 0 \tag{44}$$

where $\underline{A}_1(z)$ is the vector formed with the elements of the first row of $\underline{\underline{A}}(z)$ and

$$\underline{u} = P \int_{0}^{\infty} t\underline{\underline{X}}^-(t)\left[\underline{\underline{T}}^{-1}(t)\underline{\underline{Q}}(t) - \pi i t\underline{\underline{I}}\right]\left[\underline{f}(t)-\underline{b}_0\right] \frac{dt}{t-z_0} \tag{45}$$

$$\underline{w} = P \int_{0}^{\infty} \frac{d}{dt} \left\{t\underline{\underline{X}}^{-1}(t) \left[\underline{\underline{T}}^{-1}(t)\underline{\underline{Q}}(t) - \pi i t\underline{\underline{I}}\right]\left[\underline{f}(t)-b_0\right] \frac{dt}{t-z_0}\right. \tag{46}$$

References

[1]Cercignani, C., Theory and Application of the Boltzmann Equation, Scottish Academic Press, Edinburgh, 1965.

[2]Cercignani, C., Mathematical Methods in Kinetic Theory, Plenum Press, New York, 1969.

[3] Reynolds, M.A., Smolderen, J.J., and Wendt, J.F., "Velocity Profile Measurements in the Knudsen Layer for the Kramers Problem," Rarefied Gas Dynamics, Vol. I, edited by M. Becker and M. Fiebig, DFVLR Press, Porz-Wahn, Germany, 1974, p.A-21.

[4] Rixen, W., and Adomeit, G., "Simple Moments of the Molecular Velocity Distribution Function in Plane Poiseuille Flow," Rarefied Gas Dynamics , Vol.I, edited by M. Becker and M.Fiebig DFVLR Press, Porz-Wahn, Germany, 1974, p. B-18.

[5] Willis, D.R., "Comparison of Kinetic Theory Analyses of Linearized Couette Flow," Physics of Fluids, Vol. 5, February 1962, p. 219.

[6] Cercignani, C., "Elementary Solutions of the Linearized Gas-Dynamics Boltzmann Equation and their Application to the Slip Flow Problem," Annals of Physics (New York), Vol. 20, November 1962, p. 219.

[7] Loyalka, S.K., "Velocity Profile in the Knudsen Layer for the Kramers' Problem," Physics of Fluids, Vol. 18, December 1975, p. 1666.

[8] Cercignani, C., and Tironi, G., "Some Applications of a Linearized Kinetic Model with Correct Prandtl Number," Il Nuovo Cimento, Vol. X43, May 1966, p. 64.

[9] Waldmann, L., "Non-Equilibrium Thermodynamics of Boundary Conditions," Zeitschrift für Naturforschung, Vol. 22 a, n. 8, 1967, p. 1269.

[10] Vestner, H., Kubel, M., and Waldmann, L., "Higher Order Hydrodynamics and Boundary Conditions. Application to the Thermal Force," in Rarefied Gas Dynamics, Vol. II, edited by D. Dini, Edizioni Tecniche Scientifiche, Pisa, Italy, 1971, p. 1007.

[11] Waldmann, L., "Kinetic Equations and Boundary Conditions for Polyatomic Gases," Rarefied Gas Dynamics, edited by K. Karamcheti, Academic Press, New York, 1974, p. 431.

[12] Cassel, J.S., and Williams, M.M.R., "An Exact Solution of the Temperature Slip Problem in Rarefied Gas," Transport Theory and Statistical Physics, Vol. 2, n. 1, 1972, p. 81.

[13] Welander, P., "On the Temperature Jump in a Rarefied Gas," Arkiv Fysik, Vol. 7, n. 44, 1954, p. 507.

[14] Bassanini, P., Cercignani, C., and Pagani, C.D., "Comparison of Kinetic Theory Analyses of Linearized Heat Transfer between Parallel Plates," International Journal of Heat and Mass Transfer, Vol. 10, 1967, p. 447.

[15] Cercignani, C., "Analytic Solution of the Temperature Jump Problem for the BGK Model," Transport Theory and Statistical Physics (to be published).

[16] Darrozès, J.S., "Les Glissements d'un Gas Parfait sur les Parois dans un Ecoulement de Couette," La Récherche Aérospatiale, Vol. 119, July-August 1967, p. 13.

[17] Muskhelishvili, N.I., Singular Integral Equations, Noordhoff, Groningen, 1953.

[18] Springer, G., Introduction to Riemann Surfaces, Addison-Wesley, Reading, Mass., 1967.

[19] Kriese, J.T., Chang, T.S., and Siewert, C.E., "Elementary Solutions of Coupled Model Equations in the Kinetic Theory of Gases," International Journal of Enginnering Science, Vol. 12, June 1974, p. 4441.

[20] Abramowitz, M. , and Stegun, I.A., Handbook of Mathematical Functions, Dover, New York, 1965.

CHAPTER VIII—GAS KINETICS

EXPERIMENTAL STUDY OF RELAXING LOW-DENSITY FLOWS

A. K. Rebrov*

USSR Academy of Sciences, Novosibirsk, USSR

Abstract

This paper deals with the experimental investigation of relaxing low-density flows at the Institute of Thermophysics in Novosibirsk. The experimental investigations demonstrate the effects of translational, rotational, and vibrational relaxation and condensation and can be used for the development of calculation models of gasdynamic processes with physical transformations. The presented results concern the physical processes in gases at low temperatures when the molecules are in the ground electron state and chemical reactions are absent. They reflect the current trends of the development of experimental research on physical phenomena in rarefied gases at any degree of nonequilibrium. It is concluded that the improvement of electron-beam and molecular beam measurements could facilitate obtaining more comprehensive information on the energetic balance in gases with internal degrees of freedom and physical and chemical transformations. That could create a base for the experimental determination of relaxation constants and for obtaining precise results needed for numerical methods in the kinetic theory of gases.

I. Features of Relaxation Processes in Low-Density Gas Flows

The estimation of the relaxation situation presents no difficulties if the transition probabilities are known. One

---

Presented as Invited Paper E at the 10th International Symposium on Rarefied Gas Dynamics, Aspen, Colo., July 19-23, 1976.

*Institute of Thermophysics of Siberian Department.

can make the estimation for low-density flows by comparison of the characteristic times of transfer and relaxation processes. The gasdynamic time is defined from the characteristic velocity and either linear size L of the object (body or gas volume) or linear size followed by some parameter gradient [$L = P/(dP/dr)$]. Accordingly, the characteristic gasdynamic time is $\tau_g \approx L/u$ or $\tau_g \approx P/(udP/dr)$. Here u is the average velocity of the flow.

The collision number of the molecules A with the molecules of the sort i for the time τ_g by Maxwellian velocity distribution is

$$Z_{A-i} = \int_0^{\tau_g} n_i \left(\frac{8\ k\ T(t)}{\pi\ m_A} \right)^{1/2} \pi d^2_{A_i} \left(\frac{m_A + m_i}{m_i} \right)^{1/2} dt$$

Let Z be the number of collisions for approaching the equilibrium between A and i steps of freedom. Then, if $Z_{A-i} \gg Z$, the flow is in equilibrium; if $Z_{A-i} \ll Z$, the energy exchange between internal steps of freedom is frozen.

The region of rarefied gas flow is restricted by the Knudsen number values $Kn = \lambda/L > 0.01$. It is easy to show that in the uniform gas the total collision number (including inelastic) in the length L in the flow with Mach number M is $Z_L < 100/M$. It follows that, at supersonic velocities in the gas flow with the effects of violation of the continuum, the molecules undergo not more than 100 collisions.

This means that in rarefied flow the rotational relaxation can be unfrozen except for high rotational quantum levels. All of the channels of the translational-vibrational exchange are practically frozen, with the exception of the case of polar molecules, when, because of the resonance effects and non-adiabatic collision of the particles with the chemical affinity, the transition probabilities are higher than for adiabatic collisions by some orders. The V-V exchange channels can be active also, especially at low resonance defect. In ionized gases, the essential influence of the electron-vibrational and electron-rotational exchange is possible. The ionization and dissociation processes can be considered as frozen; they involve a larger number of collisions than for vibrational and electronic excitation.

The experimental study of relaxation processes is based on measurements that make it possible to get quantitative data, either for a total energy balance or only for some kind of intermolecular motion. If this way fails, one can judge relaxation processes by some integral effects: gas separation, change of current lines, change of temperature, pressure, etc.

For a rarefied gas, the methods of electron beam diagnostics satisfy many requirements; the molecular beam methods are fruitful for study of translational and rotational relaxation. The development of electron beam methods is stimulated essentially by Muntz' work[1]. The molecular beam methods have been developed for a few decades by many authors as a means of the study of collision kinetics. At this time, both methods have a rich history and comprehensive references.

The development of electron beam and molecular beam methods at the Institute of Thermodynamics is described in a collection of papers[2,3]. It is of interest to consider this development. Now we shall restrict our discussion by some principal remarks and references.

II. Interaction of the Expanding Gas with the Background

Most of our investigations of relaxing flows were performed with high-pressure-ratio, low-density freejets. Thus, the gasdynamics of such a flow are worth bearing in mind. When the interaction of the expanding gas with the background cannot be described in the frame of continuum, this problem is a typical one for translational relaxation.

The problem of gas expansion into near vacuum is very well known in industrial, technological, and research practice. But the systematical study of these questions started not long ago, in association with the use of electron beam diagnostics. The history of such work is no longer than 10 years. The main results with an electron beam were interpreted in papers by Ashkenas and Sherman[4], Hamel and Willis[5], and Edwards and Cheng[6]. The radial expansion was studied mostly in theoretical works[5,7,8]; the direct experimental investigation was performed at the Institute of Thermophysics[2,9]. The basic results for jet exapnsion were obtained in experiments[10,11]. The main thrusts of these investigations are the flow structure in the region from continuum to molecular (scattering regime[10]), definition of the region of the collisional interaction of the expanding and background gas, and the behavior of the distribution function in this region.

A. Flow Structure

Muntz et al.[10] have described some important features of the freejets in the transition and scattering regimes. We have made a systematic study of a nitrogen freejet in the region of the flow from continuum laminar flow to a molecular one and have shown the possibilities of the criterion generalization for describing parameter distribution along the jet axis and in the cross sections. This generalization is based, in particular, on the conservatism of the geometry of some characteristic surface of the high-pressure-ratio expanding flows on rarefaction: almost full indepdendence of Mach disk position or shock wave position by radial expansion on rarefaction, and a weak dependence of the barrel shock waves' position on rarefaction.

The studies of low-density freejet structure were presented elsewhere[10-13]. We shall comment here upon the less well-known results. As before[11], the Reynolds number $Re_L = Re_*/\sqrt{P_o/P_H}$ is used as a criterion defining the flow structure in the integral adiabatic conditions. Here Re_* is the Reynolds number by critical parameters, and P_o/P_H is the ratio of stagnation to background pressure. The transition to rarefied flow occurs at $Re_L < 100$ and to scattering regime at $Re_L < 10$.

In the paper by Kuznetsov et al.[14], the monatomic gas (argon) freejet structure was studied over the temperature range up to 5200°K. It is shown that the scaling of parameter distribution by Re_L takes place in the region where there is no heat influence of the background. The influence of the temperature factor T_o/T_H on the mixing zone structure is well illustrated in Fig. 1, where density distribution in jet cross section at Re_L = const and T_o = 290 to 4970°K is shown. The vacuum chamber wall temperature is close to room temperature (at a distance of about 400 mm from the jet). In Fig. 1, the scaling variables $n(\rho/\rho_o)$, $(n = P_o/P_H)$, and $y/d_*\sqrt{n}$ are used for the ordinate axis.

In these experiments, the electron x-ray method of density measurement (by bremstrahlung and characteristic radiation[15,16]) was used. It follows from these data that the scaling of density profile is possible at $T_o = T_H$; at T_o/T_H = const, it needs proof. At great nonisothermality ($T_o >> T_H$), the density variation along the current line is nonmonotonic even at low Re_L in the scattering regime (Fig. 2). The density distribution in the mixing zone is defined by the temperature profile. The density variation is not a direct characteristic of background molecule penetration. For quantitative study of this question, a special experiment was performed by Kislyakov et al.[17]. The free expansion of N_2 into the atmosphere of CO + N_2 was considered. The molecules of these gases have almost identical

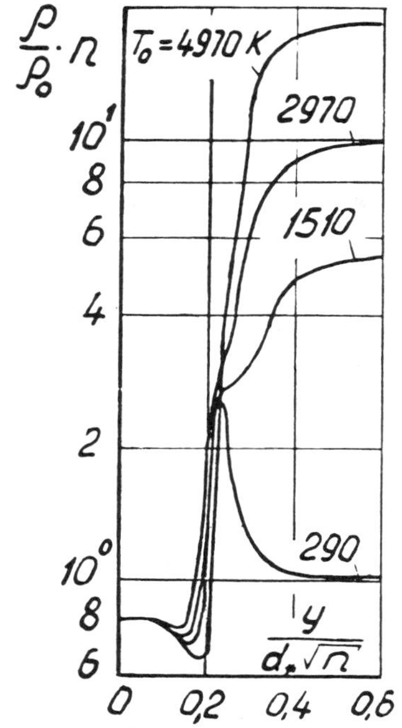

Fig. 1 Shock wave and mixing zone structures in argon jet at strong nonisothermality (Re_L = 102).

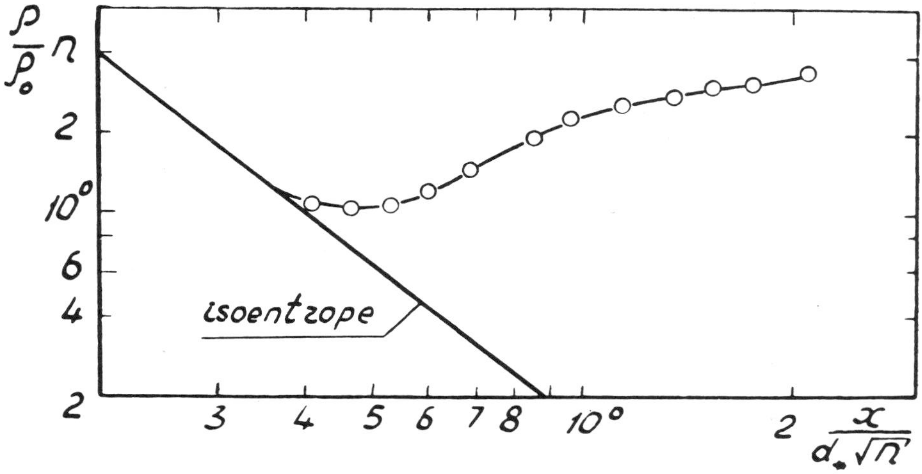

Fig. 2 Density distribution along the jet axis in scattering regime ($Re_L \approx 11$, $T_o = 4710°K$).

collisional characteristics. Figure 3 shows the jet and penetrating gas density distribution at Re_L = 45 and 10 in several cross sections. At Re_L = 45, the background gas does not pene-

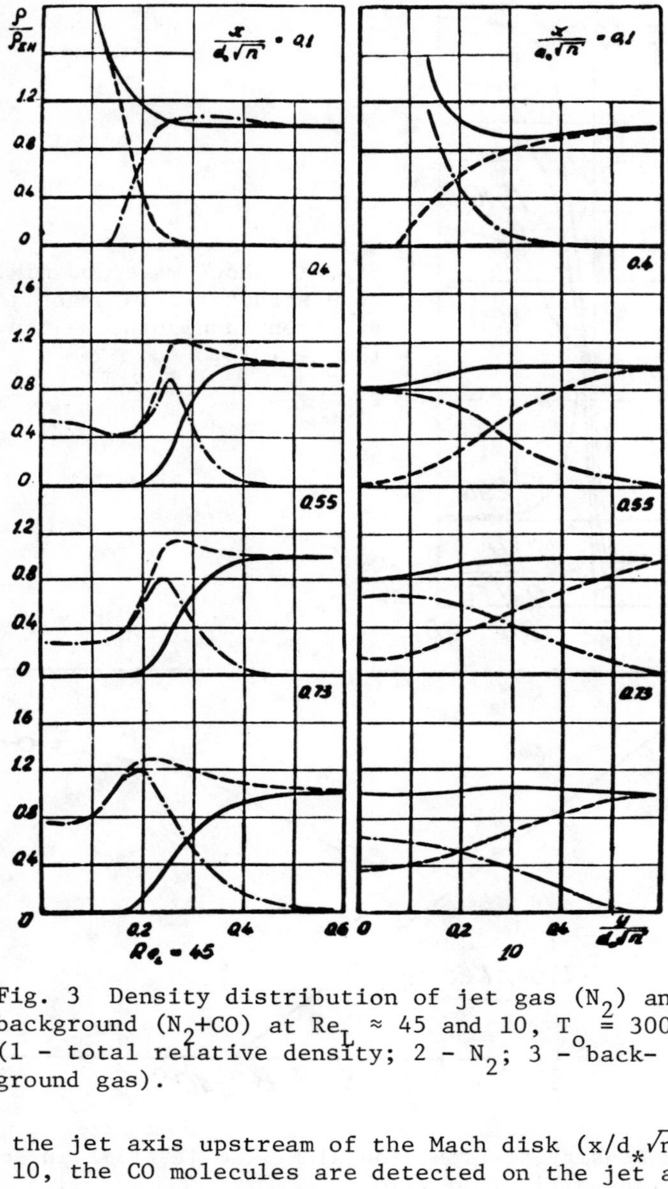

Fig. 3 Density distribution of jet gas (N_2) and background (N_2+CO) at $Re_L \approx 45$ and 10, $T_o = 300°K$ (1 - total relative density; 2 - N_2; 3 - background gas).

trate to the jet axis upstream of the Mach disk ($x/d_*\sqrt{n} < 0.67$). At $Re_L = 10$, the CO molecules are detected on the jet axis at $x/d_*\sqrt{n} \approx 0.4$. In spite of the complicated character of the flow in the mixing zone, the penetrating gas relative density was scaled.

Later on, these investigations proceeded with measurements of rotational temperature and rotational level population and pressure measurements using Pitot tube and free-molecule probe[18]. The Mach number, velocity, density, and static pressure in jet cross section at $Re_L = 45$ are shown in Fig. 4a. The boundary of M = 1 is marked with vertical dashed lines. The Mach number is determined correctly from the data for rotational temperature at M < 1 and in the jet core region. In Fig. 4b, the results of determination of different temperatures and rotational energy by different methods are given.

In the region of M > 1, the total rotational energy and extrapolated value of translational temperature characterize the heat state of the gas. At M < 1, all methods work satisfactorily. In the mixing zone merged with the shock wave front, the essential nonequilibrium with the population of upper levels vs the background temperature takes place. Finally, in the jet core the flow is more an equilibrium one, but the population of the upper rotational levels corresponds to a much higher temperature than the translational one on the jet axis. This effect is discussed in Ref. 19. Nevertheless, a special accent is necessary here. The molecules penetrating into a jet core are excited on a high quantum level and do not have enough collisions to relax. Such molecules, as detected using the electron beam, are a good indicator of diffusing background molecules and are more precise than a concentration of molecules by total signal as in Ref. 17. Essentially it was grounded in Ref. 59. The region of jet core, undisturbed by background and defined by penetration of rotationally excited molecules, is less than one defined by total concentration of a penetrating gas.

B. Scattering of Hypersonic Flow by Background Gas

The matter of principle of translational relaxation for the known jet macrostructure is the behavior of the distribution function in the collisional interaction zone. The development of distribution function anisotropy in radial and jet expansion into vacuum was considered in Refs. 5, 6, and 20. The model of Muntz et al.[10,21] made it possible to obtain the quantitative data on background molecule penetration in the jet. But the mechanism of this process (trifluid or tetrafluid) was chosen intuitively, without detailed study of the distribution function in the zone of molecular-collision interaction of the expanding and background gas. Nevertheless, the results[10] agree well with experimental ones. It is shown by Skovorodko and Chekmarev[22] that, in Navier-Stokes approximation, i.e., in a pure diffusive one, the penetrating component density distribution coincides with the one obtained in Ref. 10.

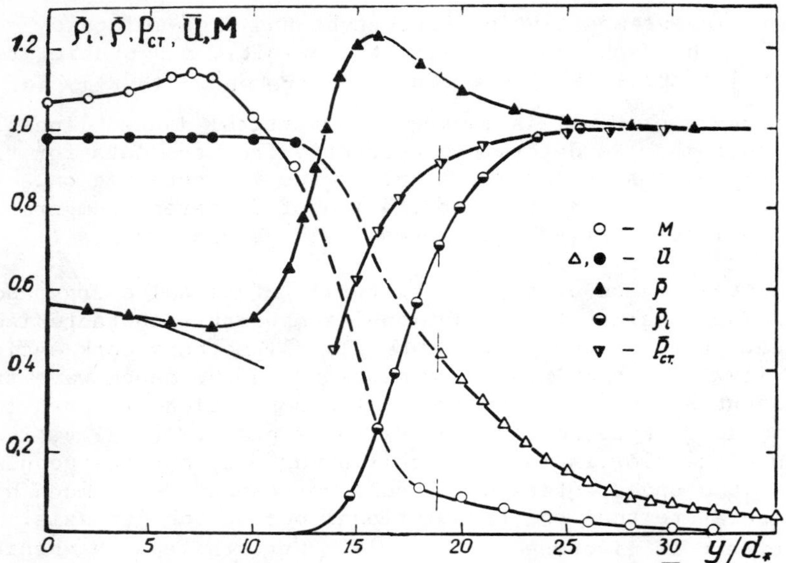

Fig. 4a Mixing zone structure at $Re_L = 45$, $y/d_* \sqrt{n} = 0.4$, $T_o = 285°K$, and $P_o/P_H = 3850$ (M - Mach number, $\bar{u} = u/u_{max}$ - relative velocity, $\bar{\rho} = \rho/\rho_H$ - relative total density, $\rho_i = \rho_{bg}/\rho_H$ - relative background density, P_{st} - relative static pressure).

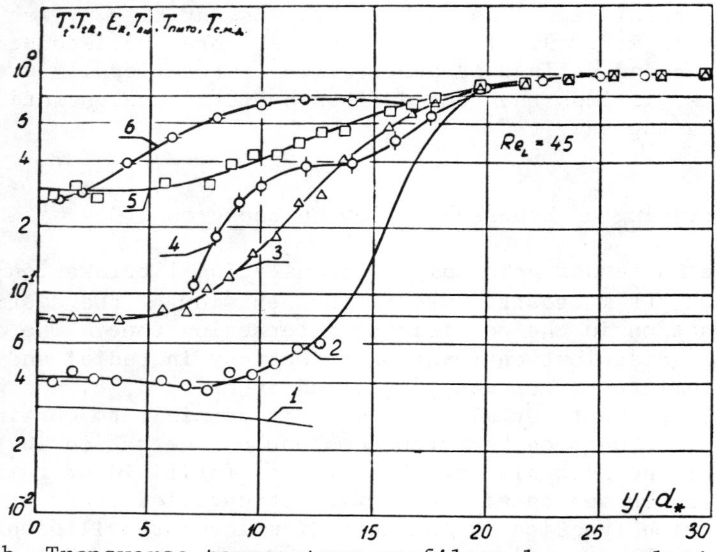

Fig. 4b Transverse temperature profiles: 1 - translational temperature of isentropic flow, 2 - translational temperature extrapolated, 3 - total rotational energy, 4 - translational temperature by Pitot tube, 5 - translational temperature by slope of curve for population, 6 - translational temperature by free-molecular probe.

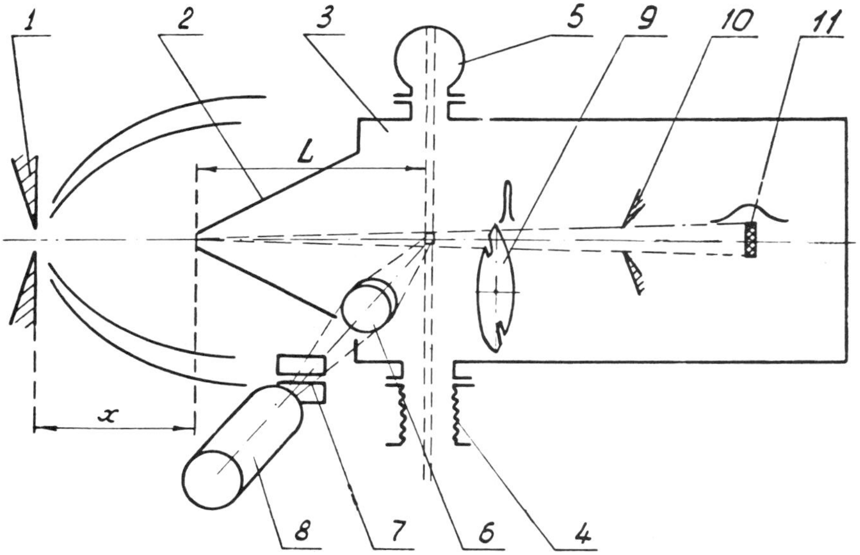

Fig. 5 Sketch of molecular beam parameters measurements: 1 - nozzle, 2 - skimmer, 3 - collimating chamber, 4 - assembly of electron beam supply, 5 - electron collector, 6,7,8 - elements of optical system, 9 - modulator, 10 - collimator, 11 - time-of-flight ionization detector.

As concerns the distribution function, in some of our papers attempts were made to obtain quantitative data on the influence of background gas on the longitudinal and transversal distribution function. The problem of nitrogen molecular beam degradation by background gas in a collimating chamber was solved experimentally. The details of experimental methods are described by Zarvin and Sharaphutdinov[23-25].

The experiments were performed in a low-density wind tunnel equipped with molecular-beam apparatus similar to that used in Bossel's work[26] but essentially improved for measurements inside a collimating chamber by using an electron beam with current up to 10 mA and voltage of 10 kV. The scheme of measurements is shown in Fig. 5. The time-of-flight system was used for measurements of the longitudinal distribution function, and the electron-beam system was used for measurements of the transverse distribution function at the point of skimming. As far as we know, the electron-beam method of measurements of the transverse distribution function by density distribution in a free-molecular beam was used here for the first time. It appeared that both longitudinal and transverse distribution functions did not change in the presence of the background gas in the collimating chamber.

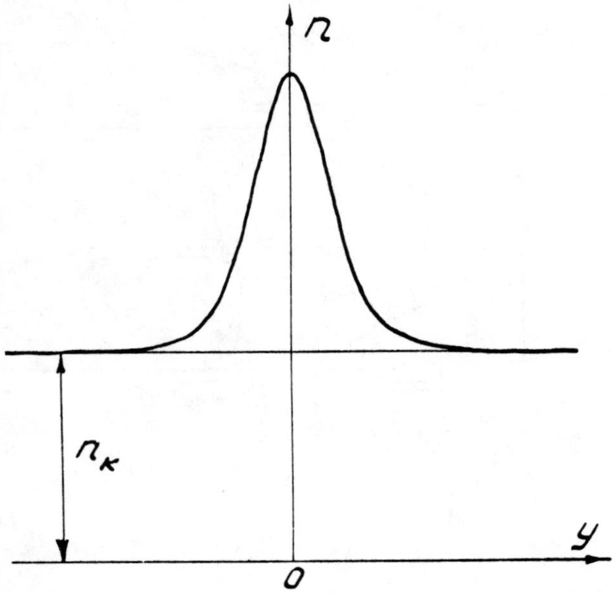

Fig. 6 Transverse density profile in the collimating chamber for conditions d_* - 2 mm, P_o = 150 mm Hg, T_o = 295°K, nozzle-skimmer distance = 140 mm, skimmer-measuring point distance = 225 mm, background pressure P_k = 5 x 10^{-6} mm Hg.

The beam density distribution is separated out easily on the background gas with the help of electron-beam measurements (Fig. 6). If the background molecules' mean free path in the beam is much more than their cross-sectional size, the density distribution in the beam itself is the difference of the total and background density. The beam density decrease of time-of-flight signal due to the scattering of the background molecules is several times more than in electron-beam measurements.

The influence of the background gas in the collimating chamber on the transverse molecular beam profile is illustrated in Fig. 7, where beam molecule density related to the density on the axis is shown to be dependent on the distance from the axis (in skimmer diameters) for the conditions Re_* = 6000, T_o = 295°K, x/d = 70, and P_k = 0.75-4.7 x 10^{-5} mm Hg. The profiles coincide within the limits of experimental error; the data correspond to a translational temperature of about 3.2 ±0.8°K. The fact that this molecule flow is less weakened than that which is going in the time-of-flight detector apparently is indicating that the scattering of the molecules at small angles does not give the essential input in variation of the

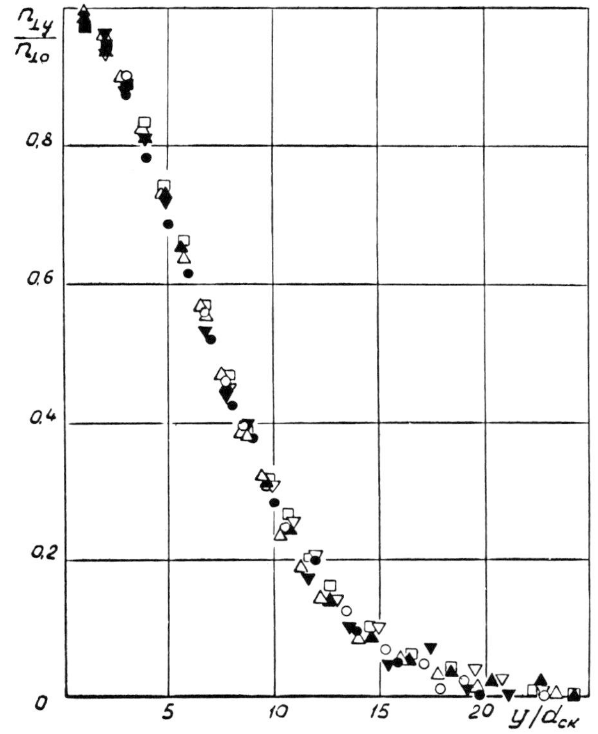

Fig. 7 Molecular beam transverse profile at $Re_* = 6000$; $T_o = 295°K$; nozzle-skimmer distance $x/d_* = 70$; $P_k = 0.75\text{-}4.7 \times 10^{-5}$ mm Hg. Marks correspond to different pressures.

transverse distribution function, but the scattering at big angles transmits the molecules into the background. In a sense, this is supplementary substantiation of the construction of the polyfluid models for shock waves as for the gas flow scattering on a low-density atmosphere.

III. Flows of Relaxing Gas Mixtures

The effect of translational relaxation in a gas mixture is in separation and mixture of components and corresponding variation of total and thermal energy of separate components. As the approximation of continuum, these phenomena can be considered as diffusive because of pressure, temperature, and concentration gradient. These approaches advance in the definite regimes and even in transition from continuum to the molecular ones[27,28]. The highly effective instrument for the study of

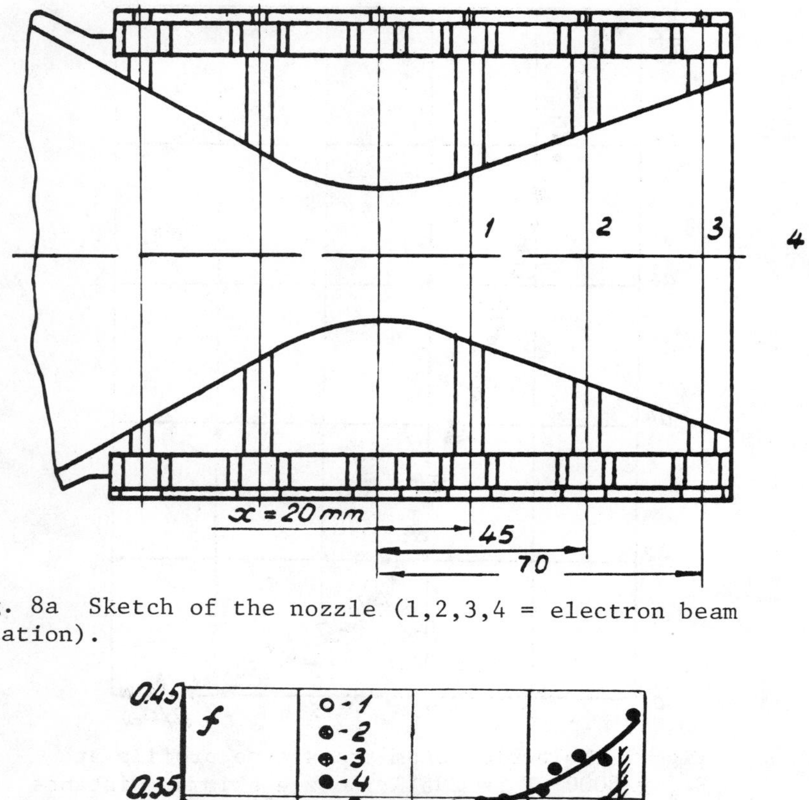

Fig. 8a Sketch of the nozzle (1,2,3,4 = electron beam location).

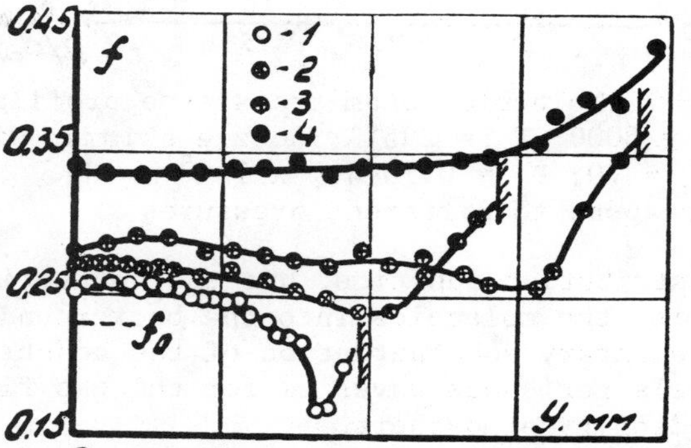

Fig. 8b Transverse argon concentration profiles (P_o = 0.27 mm Hg, f_o = 0.24, P_k = 1.5 x 10^{-3} mm Hg, Re^o_* = 55.6).

diffusive processes is electron-beam apparatus, which enables one to obtain the data on local concentration of separate components practically without mechanical disturbances. One of the first works in this direction was Rothe's investigation[28].

The use of the gas mixture spectrum, excited by an electron beam for concentration measurements, is not always a simple procedure. One should decide on the case of collisional or optical interaction of separate components. For example, in our experiments with the mixture of nitrogen and hydrogen[2], it was found that the hydrogen fluorescence was proportional not to partial concentration of hydrogen molecules but to the total amount of particles. In connection with the growing interest in the gasdynamic method of gas separation at low density[29], the effects of diffusive separation in gradient flow and features of such a flow should be noted.

A. Viscous Flow of Binary Mixtures in Supersonic Nozzles

The nozzle (Fig. 8a) with the toroidal transonic part and geometrical Mach number 4.5 was used in experiments[30] with the mixture Ar + He with the argon mole fraction $f_o = 0.24$, stagnation pressure $P_o = 0.27$ mm Hg, and pressure in the expansion chamber 1.5×10^{-3} mm Hg. The Reynolds number determined by stagnation parameters and critical radius was $Re_* = 55.6$. The nozzle design enabled the electron beam to be brought inside the nozzle through the orifices in several sections of the wall. The observation of the fluorescence was along the jet axis. The transverse argon concentration profile is shown in Fig. 8b. The initial concentration is shown on the ordinate axis. The initial concentration is shown on the ordinate axis. Argon enrichment commonly increases downstream, as observed for all profiles. This is a direct consequence of barodiffusion separation; in other words, this is an effect of the delayed translation relaxation due to the small number of collisions. In the sections 2 and 3, the increase of argon concentration becomes marked, in spite of the concentration gap near the wall, formed in the transonic nozzle part and smoothing downstream. The increase of argon concentration in the region of flow at a distance of 5 mm from the nozzle exit (curve 4 in Fig. 8b) is caused by the freejet expansion outside the nozzle.

B. Gas Separation in the Jet Beyond the Sonic Nozzle

The results of Ref. 30 concerning an argon-helium mixture with the initial concentration $f_o(Ar) = 0.51$, stagnation temperature $T_o = 300°K$, and Reynolds number for the jet $Re_L \approx 60$ differ qualitatively from Rothe's data[28]. The environment of the jet axis, especially the shock-wave front, is enriched by helium. In the shock-wave zone, the helium density increases 4.5-fold (Fig. 9) from a minimal value to the maximal one, but the ratio of maximum density value to the isentropic one at the point of maximum is equal to about 10. Such an enrichment became possible, first, because of the barrel shock wave, whose front is enriched with helium, in spite of its being diffuse,

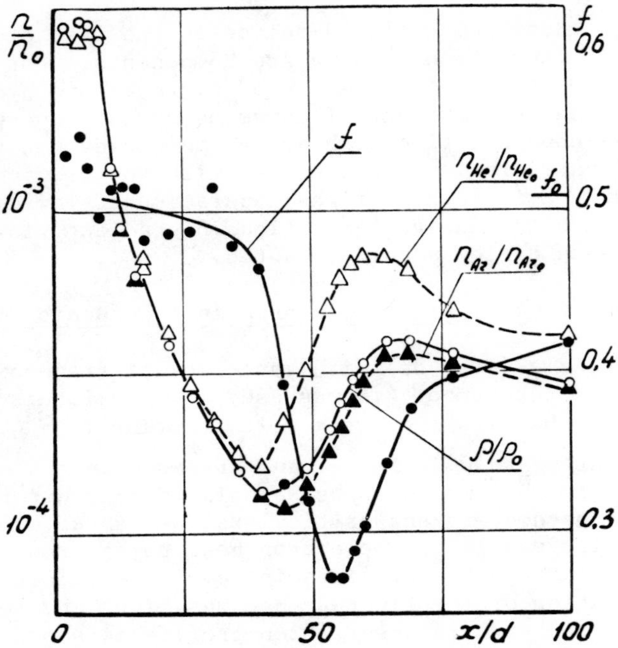

Fig. 9 Density of components and argon concentration along jet axis (P_o = 124.6 mm Hg, T_o = 300°K, f_o = 0.51, $Re_L \approx 60$).

and second, because at $Re_L \approx 60$ the Mach disk front, as well as all of the region downstream, receives the disturbances from the inner part of the barrel shock wave.

C. Radial Expansion in the Medium of Alien Gas

The velocity of diffusion in the rarefied gas can be significant and commensurate with the flow velocity. Therefore, the concentration disturbance diffusion in the gas mixtures upstream and through the shock wave is not surprising. Figure 10 shows measurement data of the component concentrations in the known experiment[9] with the expansion of helium from the punched spherical source in the nitrogen atmosphere. The used sphere was 36 mm diameter and had 133 orifices with diameter 0.85 mm. The pressure inside the spherical shell was 8.5 mm Hg, and in the background it was 0.015 mm Hg. As seen from the figure, in the regime of existence of the diffusive shock wave by density, the nitrogen penetrates through the shock-wave front.

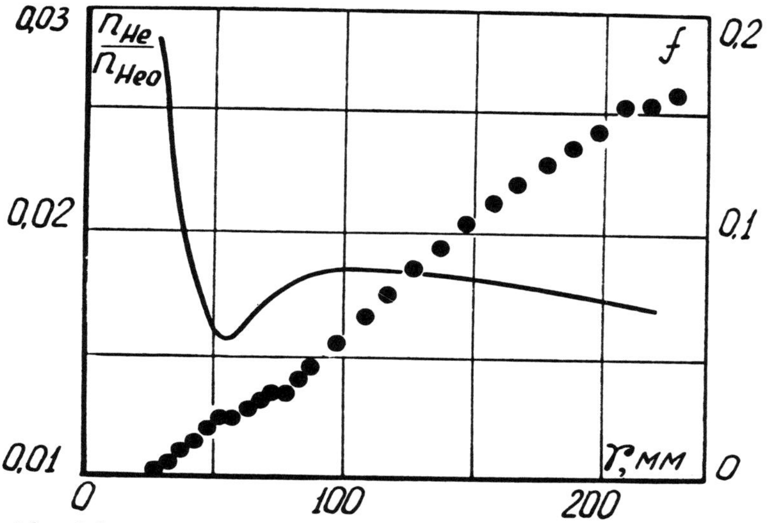

Fig. 10 Spherical expansion of helium into nitrogen atmosphere (P_o = 8.5 mm Hg, P_k = 0.015 mm Hg).

D. Diffusive Separation by Jet Collision

The delayed energy exchange by collision of molecules of different mass with a high parameter gradient is the reason for gas separation which can be used in practice. The great number of schemes for gasdynamic separation are considered in the literature. Here we comment upon the experimental data for two schemes of gas flow collision (of axisymmetrical and plane jets[31]).

Figure 11 shows the circuits of the characteristic surfaces (shock waves and conventional boundaries) of the colliding jets. The experiments were performed with nozzles of 2 mm diameter, with the mixture of argon and helium of different concentration and with various stagnation pressures. The reasonably strong effect of gas separation takes place along the jet axis and in the plane of the fan flow perpendicular to the nozzle axis (Fig. 12) in the regime; when P_o = 135 mm Hg and f_o(Ar) = 0.25, the nozzles are located at a distance of 121 mm. Figure 12 shows the mole argon fraction up to the stagnation point in the center of the flow picture and downstream of the stagnation point perpendicular to the nozzle axis. The argon concentration in the stagnation point is increased 1.45-fold, and then enrichment is increased further, especially in the initial supersonic region at $y/d_* > 20$. The maximum enrichment increase is 1.6-fold. The optimal condition for the gas separation has not been established here.

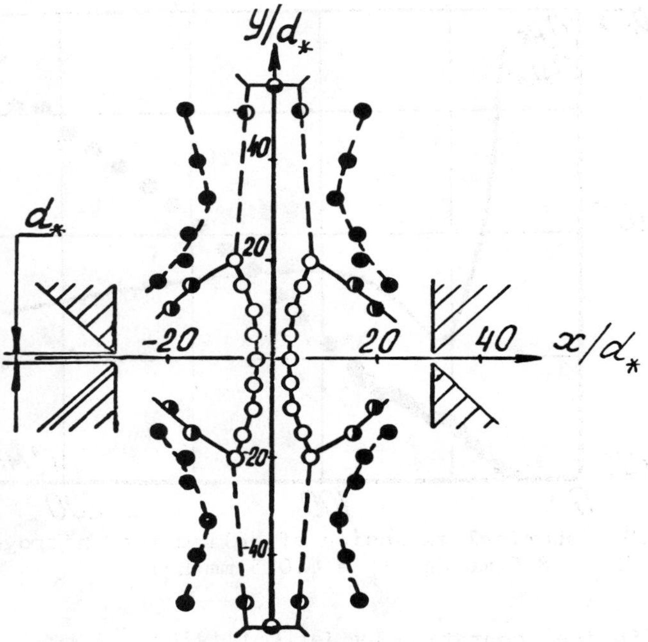

Fig. 11 Flow structure by jet collision.

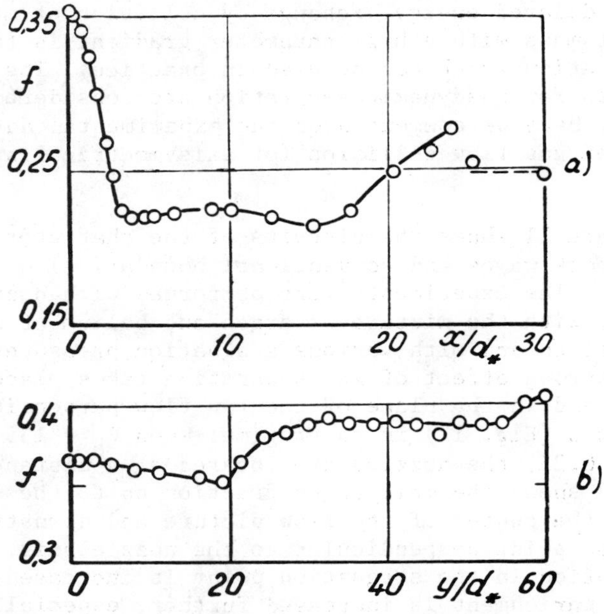

Fig. 12 Argon concentration profile along jet axis (a) and in plane of symmetry (b); d_* = 2 mm, P_o = 135 mm Hg, f = 0.25.

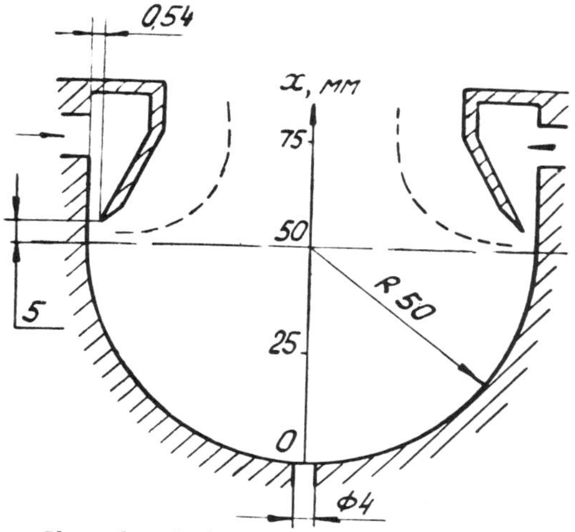

Fig. 13a Sketch of device for studying plane flows.

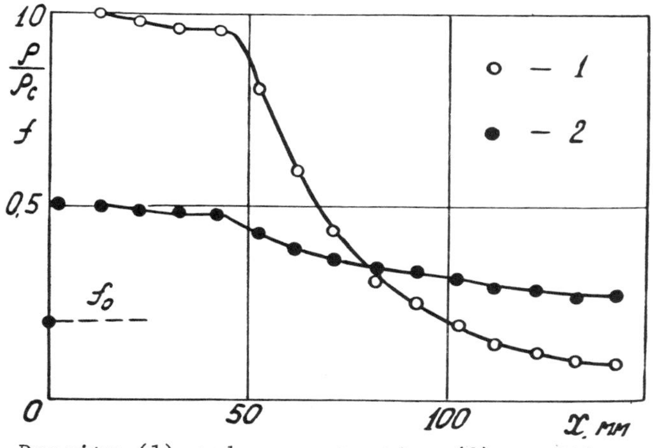

Fig. 13b Density (1) and concentration (2) profiles of argon in the plane of symmetry (P_o = 6.8 mm Hg, f_o = 0.203).

The scheme for studying plane jets collision is shown in Fig. 13a: two jets from the slots are colliding adjacent to the curvilinear surface, where the separation effect should be amplified. The electron beam excites the gas in the plane of symmetry. In Fig. 13b, the concentration profile in the plane of symmetry for regime P_o = 6.8 mm Hg, f_o = 0.203 is shown. At some distance from the wall (12.5 mm), the detected enrichment is about 2.5-fold, this figure not being referred to as maximal.

E. Nonequilibrium Phenomena near the Bluntbody in Supersonic Flow of Gas Mixtures

This investigation is initiated by the paper of Maise and Fenn[32], where the effect of abnormal rising of the temperature recovery factor $r = (T_r - T_\infty)/(T_o - T_\infty)$ in the bluntbody stagnation point in the gas mixture flow with low Reynolds number was described. This curious effect of delayed translational relaxation consists in the stagnation of separating gases, flowing upstream at the same velocity and hence having different Mach numbers. The heavy gas with its higher stagnation temperature is concentrated in the region of the stagnation point, and this creates the atmosphere of the elevated temperature.

These effects have been studied in Ref. 33 for argon-helium and $N_2 + H_2$ mixtures in a freejet (Fig. 14). The conditions of maximum values of recovery temperature were found. It is shown that the effect is more significant at low concentration of heavy gas (about 10%), as in this case the nonequilibrium processes proceed under conditions consistent with light-gas parameters. The maximum value of a recovery factor ($r = 1.8$) is at a Reynolds number behind the shock wave of $10 < Re_L < 20$ [Fig. 15, f_o $(N_2) = 0.1$].

In subsequent works[34,35], a thorough investigation of the flow structure and heat effects in the stagnation point in the flow of a nitrogen-hydrogen mixture was performed. The elec-

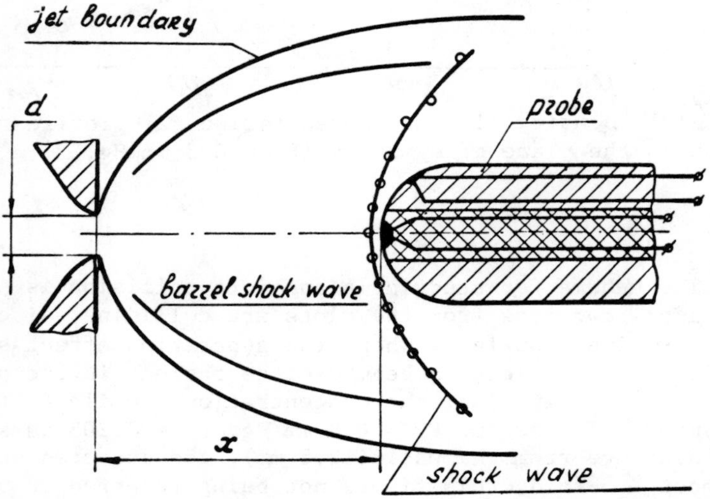

Fig. 14 Sketch of experiment with bluntbody in gas mixture jet.

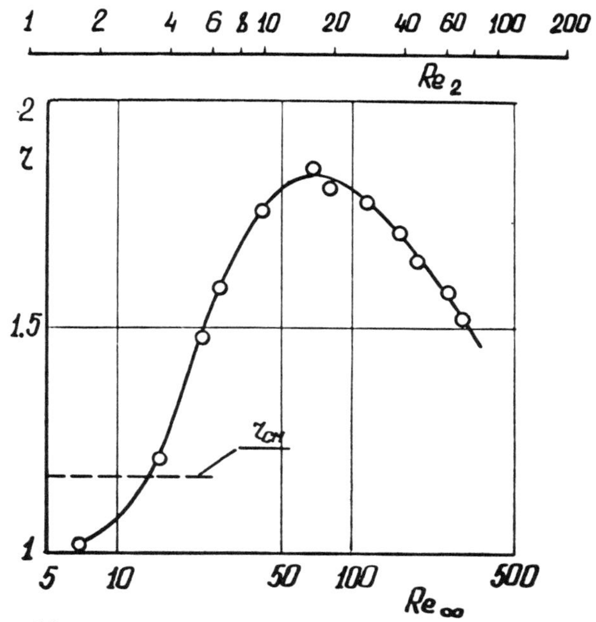

Fig. 15 Temperature recovery factor; nitrogen concentration $f_0 = 0.1$.

tron beam (Fig. 16) was used for measuring the component concentrations and nitrogen temperature, temperature in the stagnation point was measured with a thermocouple, and heat flux measurements were taken with a film resistor by a nonstationary method[36]. Figure 17 shows the appearance of the nitrogen concentration field in the flow, where the initial concentration $f_0 = 0.24$, $M = 3.5$, and $Re_\infty = 140$. The maximum recorded concentration of nitrogen is $\tilde{f} = 0.42$. The nitrogen concentration rise is not localized at a stagnation point; it invades the entire district. The temperature in the stagnation point is 20% higher than in the pure nitrogen, and, at the distance $y/D = 0.75$ (D is the sphere diameter), temperature is the same as in pure nitrogen. In experiments with the electron beam Fig. 17), the copper (nonadiabatic) model was used, which is why the measured temperature was underestimated; nevertheless, it was higher than a limit value for the molecular flow ($r = 1.17$).

The concentration and temperature distribution along the jet axis and upstream of the model at initial concentrations $f_0(N_2) = 0.2$ and 0.1 are shown in Fig. 18. It places emphasis on the fact that recovery temperature is higher in the mixture with low concentration of N_2; in this case, the shock-wave front is steeper.

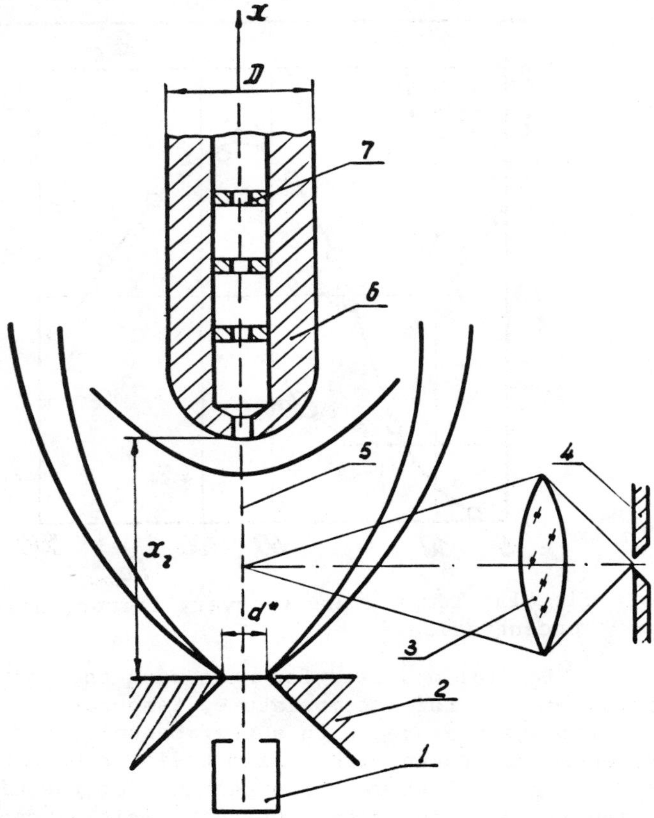

Fig. 16 Flow diagram with electron beam (1 - collector, 2 - nozzle, 3,4 - elements of optical system, 5 - electron beam, 6,7 - elements of model).

The calculations by Sherman's theory give the correct estimation of the level of a nitrogen concentration value. The expansion of nitrogen in the mixture with hydrogen is more in equilibrium in terms of the rotational relaxation. The population of the rotational quantum level in the jet is approaching the Boltzman one, but it is essentially non-Boltzman in the shock wave. (In Fig. 18, the preceding uncertainties in rotational temperature in the shock wave are shown by vertical strokes.) As it approaches the body surface, the nonequilibrium is diminishing, and close to the surface it becomes negligible.

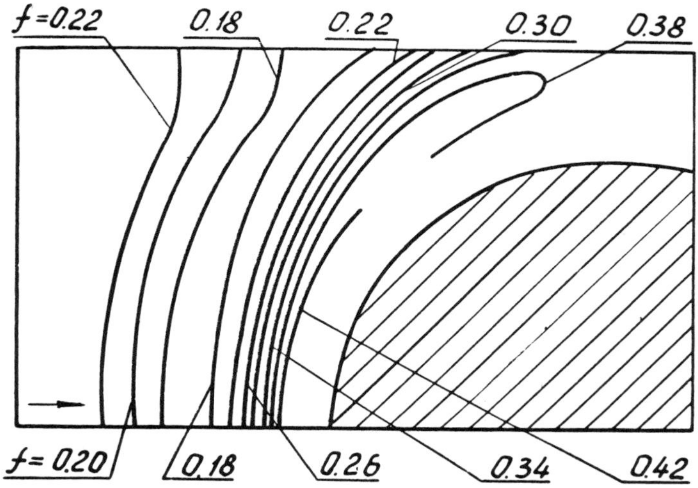

Fig. 17 Nitrogen concentration profile in the mixture flow at f_o = 0.24, M = 3.5, Re_∞ = 140.

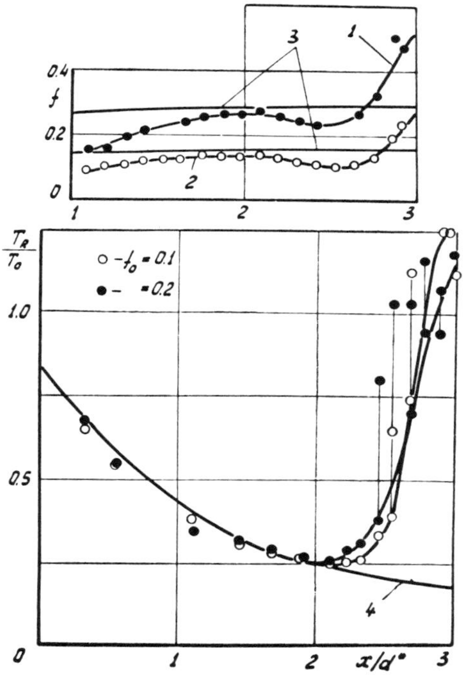

Fig. 18 Nitrogen concentration and rotational temperature upstream of model at f_o = 0.1 (1), $P_o d_*$ = 18.3, Re_o = 337; f_o = 0.2 (2) $P_o d_*$ = 16.5, Re_o = 343 (3 - calculations by Sherman's theory).

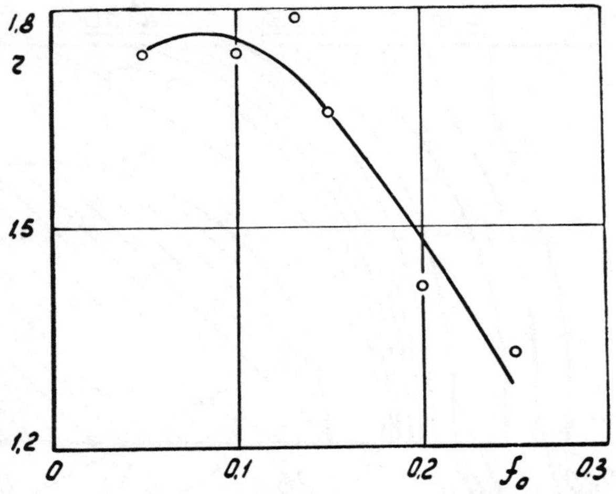

Fig. 19a Dependence of temperature factor on concentration of mixture $N_2 + H_2$ ($Re_o = 372$).

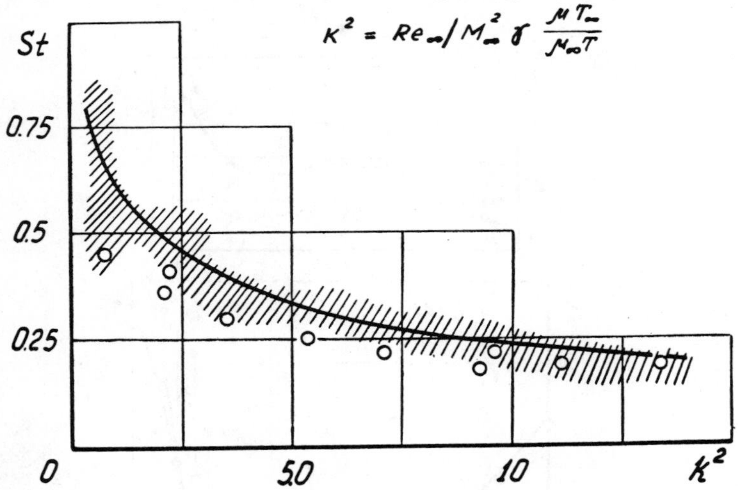

Fig. 19b Heat transfer in low stagnation point (solid line - Cheng theory, shading - Zavarzinas' experiments, points - our experimental data).

The measurements by thin-film heat probe on the glass probe have shown (Fig. 19a) that the maximum temperature recovery factor takes place for nitrogen of a low concentration; at a Reynolds number of parameters beyond the shock wave of $Re_o = 372$, the maximum value corresponds to $f_o \approx 0.07$. The maximum is rather sharp compared to Reynolds number.

The heat effect of nonequilibrium (the rise of heat flux) has its maximum at a definite Reynolds number: for these experiments, at $Re_\infty \approx 75$. Thus, the translational nonequilibrium (mainly barodiffusion effect) has an effect on the heat flux value. However, the processing of heat exchange experimental data, with the recovery temperature taken as reference, shows (Fig. 19b) that the heat transfer is described in scaling parameters by the known law $St = f(K^2)$, where St = Stanton number, $K^2 = Re_\infty/M_\infty^2\gamma\ (\mu T_\infty/\mu_\infty T)$. The shading in Fig. 19b shows the experimental data[36], and the solid line is the thin-shock-layer theory of Cheng[37].

IV. Some Cases of Translational Nonequilibrium in Gasdynamics

A. Gas Mixture Effusion into Vacuum[38]

These simple experiments were performed in checking the mass fraction in a vessel using the electron-beam measurements (for the mixture Ar + He) or by freezing out one of the components (for the mixture He + CO_2). It was established that, in the case of gas effusion from the orifice, the separation effect becomes significant and measurable at $Kn > 1.2 \times 10^{-2}$.

It is shown for the mixture $0.185\ CO_2 + 0.815$ He that the gas separation by discharge from the orifice in the region $0.15 < Kn < 1.33$ is described satisfactorily by the relations for the free molecular flow. This is illustrated by Fig. 20,

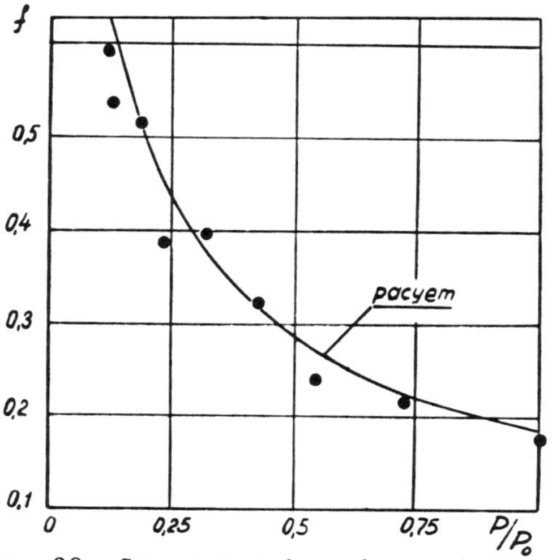

Fig. 20 Concentration change by discharge of mixture from vessel (for CO_2, $f_o = 0.185$, Kn = 0.15 + 1.33).

where the concentration dependence on pressure by the gas discharge from the vessel is shown. The gas separation by effusion from the orifice tends toward maximum in the transition regime at Kn $\lesssim$ 0.15.

B. Shock-Wave Structure in the Mixture Ar + He

The shock wave is the classical object for investigation of translational relaxation. The electron-beam measurements of Center[39] and Harnet and Muntz[40] have given very valuable information on the concentration and temperature distribution to establish the picture of the precise structure of the shock wave. The procedure of measuring temperature and velocity according to Doppler effect was incorporated in one of our installations[41]. Because of this method, Bochkarjov et al.[42] have studied the shock wave structure in the argon-helium mixture at M = 2. There are results of numerical calculation just for this Mach number. Again, the question of the possible consideration of the shock-wave structure at such Mach numbers by methods of continuum gasdynamics was unsolved.

Let us consider immediately the experimental data of Figs. 21a and 21b, without dwelling upon the details of the experiments, the main features of which would relate to the plane shock-wave creation. As seen, at small concentrations of heavy component there is a temperature dispersion ($T_{\parallel} \neq T_{\perp}$). The reason for this is the lack of collisions for establishment of the heat equilibrium in all directions, as in this case the shock-wave structure is close to its structure in the pure gas, and its thickness is small.

The concentration distribution in the shock wave agrees with the Abe and Oguchi calculation by a BGK model[43] for M = 2 and f_o = 0.1 and differs from the experimental one at f_o = 0.5. In the latter case, the experiments agree better with the Bird calculation[44]. It is characteristic that at f_o = 0.5 the temperature anisotropy is as good as unobserved. This is accounted for by the increased thickness of the shock wave and, thus, the number of collisions in the viscous front. The calculations based on the continuum approximation make the structure different from that obtained in the experiments.

V. Rotational Relaxation in the Expanding Nitrogen Flow

The experimental study of rotational relaxation by expansion, based on the pressure measurements in Pitot tube[45] and velocity distribution function[46], makes it possible to deter-

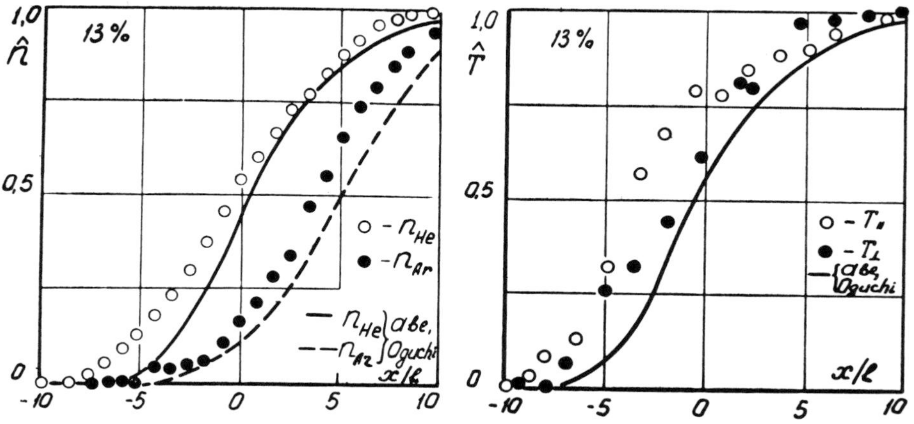

Fig. 21a Component densities and helium temperature in a shock wave [M = 2, $f_o(Ar) = 0.13$].

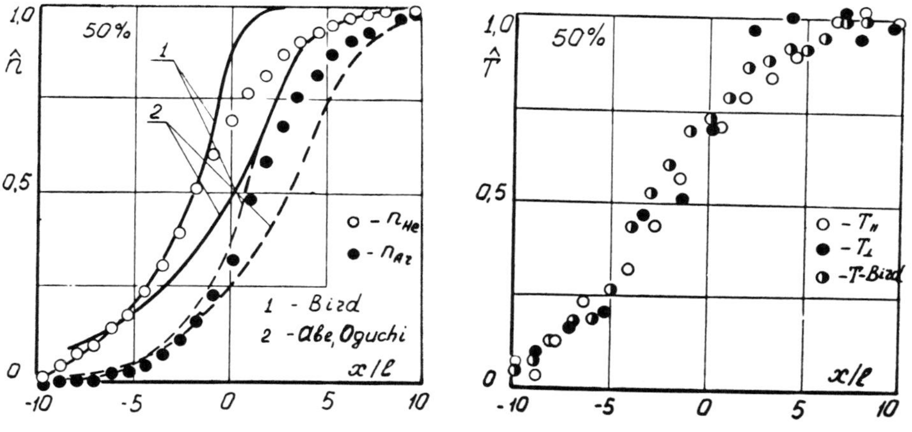

Fig. 21b Component densities and helium temperature in a shock wave [M = 2, fo(Ar) = 0.5].

mine the relaxation time conforming with the ultrasonic measurement data[47]. As the probabilities of monoquantum transitions depend on the quantum level number[48], such a time relaxation is averaged for a number of levels if there is no Boltzman distribution. The rotational nonequilibrium was discovered in the first measurements of rotational level population by Marrone[49]. However, until now authors following the Marrone approximation determined the rotational temperature by the average slope of the dependence

$$\ln \mathfrak{J}/\mathfrak{J}_k \ (k' + k'' + 1) = f \ [k' \ (k' + 1)]$$

(usually by some first quantum level[50]). In this case, the quantitative error is introduced in the rotational energy estimation, and inaccuracy in principle concerning the rotational relaxation treatment does take place.

The study of the rotational level population in the free nitrogen jet at room stagnation temperature in the region $16.5 < P_o d_* < 1340$ was performed using electron-beam measurements[51]. The rotational levels have been recorded up to a level of 21. The following were established: (1) the absence of Boltzman distribution on rotational level in all regimes and all of the flow district studied; (2) the essential exceeding of measured rotational energy over that which corresponds to the rotational temperature according to Marrone measurements; (3) freezing of rotational energy at high quantum levels; (4) abnormally high value of upper-level population at a great value of $P_o d_*$; and (5) the essential effect of the background on the measured population value.

The data on population up to the level 15 are given in Fig. 22 in the form of population temperature T_k calculated from the relation for Boltzman distribution, $N_k = N_o (2k + 1) \exp [-k (k + 1) (\theta/T_k)]$, by substituting measured values of N_k. If the distribution is Boltzman, then for all levels $T_k = T_R$. As seen from the figure, the conventional rotational temperatures of separate levels differ in tens of degrees. The rotational temperature freezing takes place just for the sixth level. The background molecule penetration

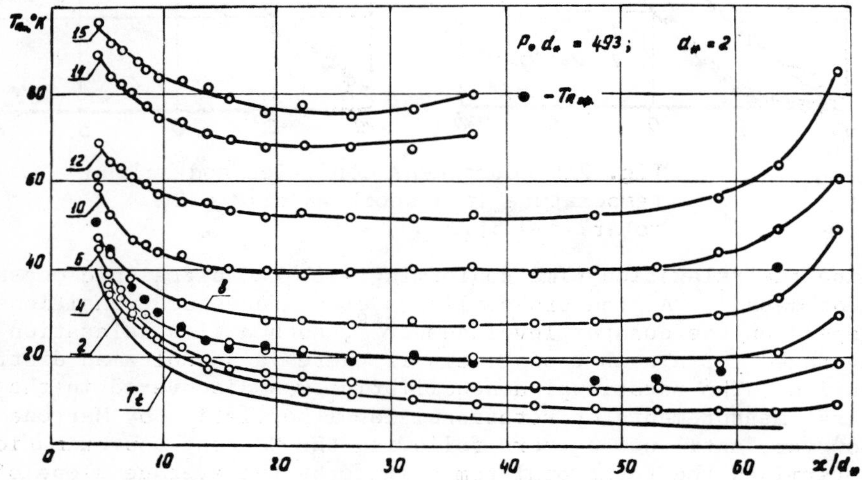

Fig. 22 Temperature profiles of separate rotational levels by expansion. Zero level is the one referenced (black points - rotational energy in degrees Kelvin).

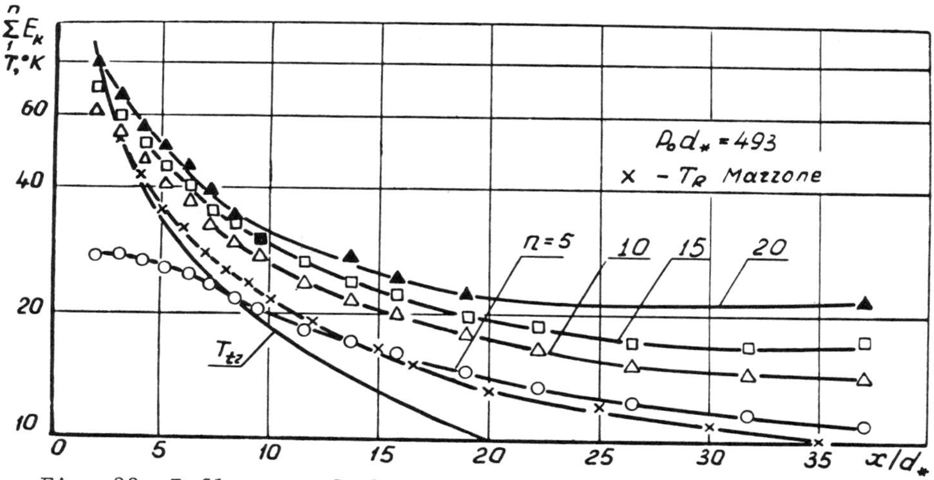

Fig. 23 Influence of the considered quantity of rotational levels on estimation of rotational energy.

effects on the population increase at large distances from the nozzle exit, and this increase occurs earlier, the larger the number of quantum levels.

The estimation of the rotational energy value vs. the number of levels is shown in Fig. 23. For example, if n = 5 at small distances from the nozzle exit, the value of rotational energy

$$E_R = \sum_1^n k\ (k + 1)\ \theta\ N_k / \sum_o^\infty N_k$$

is not worthwhile, as it corresponds to the temperature below the isentropic one. The value of T_k determined by Marrone is shown with crosses. It is lower by 10° than that determined for n = 20. This difference increases downstream in the low-temperature region. The data obtained enable one to determine the rotational transition probabilities as averaged by rotational-rotational and rotational-translational processes.

In conclusion, it should be noted that the rotational relaxation is connected tightly with the translational energy exchange and can be reflected in viscous effects. It depends on the representative number of collisions for rotational-translational exchanges. In the approximation of the Landau-Teller relaxation equation, the only scale criterion for describing rotational relaxation, as shown by Scovorodko[52], for the given viscosity temperature law, is the complex $\alpha \sim Re_*/Zr\ (T_*)$, where Re_* is the Reynolds number by parameters of sonic surface, and Zr is the number of collisions for the translational-rotational equilibrium. In Ref. 52, the relaxation processes have been analyzed in terms of numerical computation in the wide region of α, Zr, and ω (exponent in the temperature dependence for viscosity).

VI. Possibilities for Studying Vibrational Relaxation

The study of vibrational relaxation using electron-beam diagnostics is hampered even with the elaborated methods of population measurement, as these methods are applicable for those conditions where the vibrational-translational exchanges are frozen. However, the use of these measurements is evident even in the case of freezing for estimating the vibrational energy. It was proved repeatedly by different authors and used in one of our papers[53]. In particular, freezing of the relaxation process can be used for studying the mixing processes.

Moreover, by gas expansion into a near vacuum one can obtain information about the frozen vibrational state and use it for studying vibrational relaxation, using the definite calculation model of the process. That is why the development of the electron-beam methods for the measurement of vibrational temperature and vibrational level population is nec
temperature and vibrational level population is a necessary and important problem for the specialist in rarefied gasdynamics. Some results in this direction were obtained by Kosinov[54]. If the intensity ratio of two electron-vibrational bands for the state excited in a direct way and without cascade transition from above at some ground state vibrational temperature T_v is divided by the intensity ratio of the same bands at low temperature (room, for example), and then, the population of the upper level is neglected, the temperature function obtained is

$$F(T_v) = \frac{\mathfrak{J}_{v_2' v_2''}/\mathfrak{J}_{v_{20}' v_{20}''}}{\mathfrak{J}_{v_2' v_2''}/\mathfrak{J}_{v_{20}' v_{20}''}\Big|_{T = 300°K}} = \frac{\sum q_{v_2' 0v_1''} q_{v_2' v_1''} N_{v_1''}}{\sum q_{v_2' 0v_1''} q_{v_{20}' v_1''} N_{v_1''}}$$

The graphic symbols are clear from Fig. 24. This function is determined only by Frank-Condon factors and the ground state population. This is the case provided that the dependence of the electron transition dipole moment on the internuclear distance can be neglected. The preceding relation is valid in a general case, when there is no Boltzman distribution on vibrational levels, and it can be used for determining population if the number of bands recorded with the necessary accuracy is enough for closing the system of equations.

In the particular case of the use of the CO^+ first negative system, the experimental check of the method of vibrational temperature measurement was performed in a low-density facility. The measurements of relative intensity of the bands 01 and 23, 03 and 36 have been made in the heated CO jet at a temperature of 700°-1100°K. The experiments agree well with the calculation, as seen from Fig. 24.

VII. Condensation as a Relaxation Process

It is known that, for the given gas at some stagnation temperature, the complex $P_o d_*$ is the scaling parameter. Many experiments, including ours[55], persuade one that the condensation conditions are scaled by complex $P_o d_*^n$ = const, where $n \neq 1$. The analysis of condensation in terms of the estimations at $P_o d_*^n$ = const show that the same condensation conditions correspond to the situation with the same number of collisions beyond the phase transition curve.

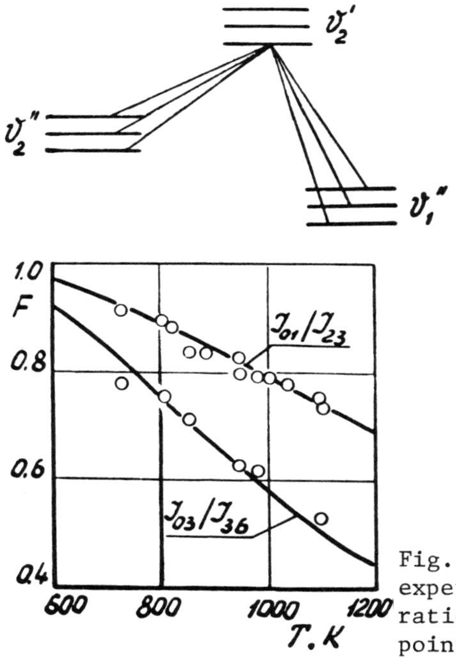

Fig. 24 Comparison of theoretical and experimental values of band intensity ratio (solid lines - calculation, points - experiment).

VIII. Superposition of Relaxation Processes

This problem is of great interest when the relaxation processes connected with energy-containing degrees of freedom are performing simultaneously. Such an example is the expansion of CO_2 into vacuum with nonequilibrium condensation and vibrational relaxation[56]. There are some contradictions in the quantitative presentation of the gas expansion law[57,58] because of the lack of knowledge on the effects of these qualitatively different processes.

In experiments[56], there were used the free CO_2 expansion from the sonic nozzle in the temperature region T_o = 300-2800°K. With the temperature measurements in CO_2 hampered, the investigation was based on the density measurements. They have been taken at the fixed distances x/d_* from the nozzle exit at constant gas consumption. This made it possible to avoid errors in determining the distance from the nozzle exit and almost completely errors caused by the electron-beam scattering. The designated region of temperatures is covered using the resistor and plasma heaters.

The dependence of relative density on stagnation temperature for x/d_* = 20 is shown in Fig. 25. Curves a and c (the

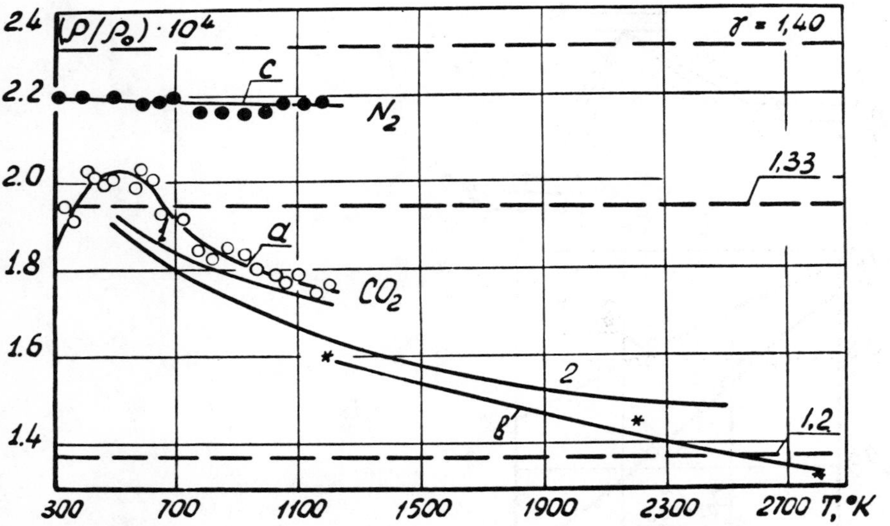

Fig. 25 Influence of condensation and vibrational relaxation on relative density in expanding flow (d_* = 2.85 mm, T_o = 300°K, P_o = 260 mm Hg, 1 - instant freezing at x/d_* = 1, 2 - expansion with finite relaxation time).

latter for instrument calibration) have been obtained from results of x-ray measurements using a resistor heater ($d_* = 2.85$ mm). Curve b was obtained using a plasma torch ($d_* = 2.7$ mm). The horizontal dotted lines show the density value by expansion with $\gamma = C_p/C_v = 1.4, 1.33, 1.2$.

The relative density in a CO_2 jet with variation of temperature essentially changes, and its values are lower than that corresponding to the expansion with $\gamma = 1.4$ at frozen vibrational relaxation and without condensation. The presence of a maximum on the curve (at $T_o = 550°K$) is characteristic. The general view of this dependence remains at various values of x/d_*. It can be explained by the influence of condensation and vibrational relaxation. According to Refs. 29 and 55, at the conditions of these experiments (at $T_o = 300°K$, $P_o = 200$ mm Hg) the initial condensation stage takes place. The estimation by Ref. 29 shows the absence of condensation at $T_o > 490°K$. That is why the behavior of the relative density on the left from the maximum is defined mainly by the change in condensation.

If there is no condensation, the temperature variation can affect the expansion through the vibrational relaxation only in that case where freezing occurs in the supersonic region of the flow. The estimations show that the vibrational relaxation at the conditions of these experiments practically is frozen in the region of critical speed. However, the measurements provide the essential decrease of the relative density with increasing temperature. Consequently, the location of the district of the vibrational relaxation is delayed downstream of the nozzle exit.

These grounds and the character of the dependence of relative density on temperature are confirmed by theoretical calculations: (1) by the model of instant freezing, curve 1 (for freezing at the distance $x/d_* = 1$ and subsequent expansion with $\gamma = 1.4$); and (2) by the model of nonviscous vibrationally relaxing gas with finite relaxation time, curve 2 (for $P_o d_* = 7.6 \times 10^3$ mm Hg). In the region $T_o > 1200°K$, experimental curve b agrees with the observed expansion model, but there is a 10% displacement relative to the dependence obtained using a resistor heater.

It follows from the given data that condensation and vibrational relaxation have a simultaneous influence on the expansion at temperatures less than 500°K. The experiments on division of these effects failed. Their effects are weakened with decreasing stagnation pressure. However, it is impossible to create the condition for expansion of CO_2 as a diatomic gas, since the influence of condensation and vibrational relaxation terminates at low $P_o d_*$ when the freezing of rotational energy becomes essential. In conclusion, it should be remarked that the description of the molecular gas expansion processes with different active steps using some effective heat capacity ratio is rather a rough approximation, as it cannot be connected with any correct physical model of the expansion process.

References

[1]Muntz, E. P., "Static Temperature Measurements in a Flowing Gas," The Physics of Fluids, Vol. 5, No. 1, January 1962, pp. 80-90.

[2]The Experimental Methods in Rarefied Gas Dynamics, Collection of Papers, Institute of Thermophysics, Novosibirsk, 1974 (in Russian).

[3]Some Problems of Hydrodynamic and Heat Transfer, Collection of Papers, Institute of Thermophysics, Novosibirsk, 1976 (in Russian).

[4]Ashkenas, M. and Sherman, F. S., "The Structure and Utilization of Supersonic Free Jets in Low Density Wind Tunnels," Rarefied Gas Dynamics, Proceedings of the Fourth International Symposium on Rarefied Gas Dynamics, Institute of Aerospace Studies, University of Toronto, July 1964, Vol. II, edited by J. H. deLeeuw, Academic Press, New York, 1965, pp. 84-105.

[5]Hamel, B. B. and Willis, D. R., "Kinetic Theory of Source Flow Expansion with Application to the Free Jet," The Physics of Fluids, Vol. 9, No. 5, May 1966, pp. 829-841.

[6]Edwards, R. H. and Cheng, H. K., "Steady Expansion of Gas into Vacuum," AIAA Journal, Vol. 4, March 1966, pp. 558-561.

[7]Gusev, V. N. and Zhbakova, A. V., "The Flow of Viscous Heat-Conducting Compressible Fluids into a Constant Pressure Medium," Rarefied Gas Dynamics, Proceedings of the Sixth International Symposium on Rarefied Gas Dynamics, MIT, Cambridge, Mass., July 1968, Vol. I (Advances in Applied Mechanics, Supplement 5), edited by Leon Trilling and H. Y. Wachman, Academic Press, New York, 1969, pp. 847-862.

[8]Rebrov, A. K. and Chekmarev, S. F., "Spherical Flow of a Viscous Heat-Conducting Gas into an Occupied Space," Journal of Applied Mechanics and Technical Physics, Vol. 12, No. 3, May-June 1971, pp. 455-459. (Translation of Zhurnal Prikladnoi

Mekhaniki i Tekhnicheskoi Fiziki, No. 3, May-June 1971, pp. 122-126.)

[9]Bochkarev, A. A., Rebrov, A. K., and Chekmarev, S. F., "Hypersonic Spherical Expansion of a Gas with Stationary Shock Wave," Journal of Applied Mechanics and Technical Physics, Vol. 10, No. 5, Sept.-Oct. 1969, pp. 742-746. (Translation of Zhurnal Prikladnoi Mekhaniki i Tekhnicheskoi Fiziki, Vol. 10, No. 5, Sept.-Oct. 1969, pp. 62-67.

[10]Muntz, E. P., Hamel, B. B., and Maguire, B. L., "Some Characteristics of Exhaust Plume Rarefaction," AIAA Journal, Vol. 8, September 1970, pp. 1651-1658.

[11]Volchkov, V. V., Ivan, A. V., Kislyakov, N. I., Rebrov, A.K., Sukhnev, V. A., and Sharafutdinov, R. G., "Low Density Jets Beyond a Sonic Nozzle at Large Pressure Drops," Journal of Applied Mechanics and Technical Physics, Vol. 14, No. 2, March-April 1973, pp. 200-207. (Translated from Zhurnal Prikladnoi Mekhaniki i Tekhnicheskoi Fiziki, No. 2, March-April 1973, pp. 64-73.)

[12]Rebrov, A. K., "General Structure of the Low Density High Pressure Ratio Free Jet," unpublished work, 1974.

[13]Rebrov, A. K., "On Gasdynamic Structure of High Pressure Ratio Jets," Problems of Thermophysics and Physical Hydrodynamics, Novosibirsk, 1974 (in Russian).

[14]Kusnetsov, L. I., Rebrov, A. K., and Yarygin, V. N., "High Temperature Jet of Low-Density Argon in the Wake of a Sonic Nozzle," Zhurnal Prikladnoi Matematiki i Tekhniceskoi Fiziki, May-June 1975, pp. 82-87 (in Russian).

[15]Kusnetsov, L. I. and Yarygin, V. N., "The Bremstrahlung of Gas Targets and Its Use for Local Density Measurements," Apparatus and Methods of X-Ray Analysis, Issue XI, Mashinostroenie, Leningrad, 1972 (in Russian).

[16]Kusnetsov, L. I., Rebrov, A. K., and Yarigin, V. N., "Diagnostic of Ionized Gas by Electron Beam in X-Ray Spectrum Range," 11th International Conference on Phenomena in Ionized Gases, 1973, Prague.

[17]Kislyakov, N. I., Rebrov, A. K., and Sharafutdinov, R. G., "Diffusion Processes in the Mixing Zone of a Supersonic Jet of Low Density," Journal of Applied Mechanics and Technical Physics, Vol. 14, No. 1, January-February 1973, pp. 99-104.

(Translated from Zhurnal Prikladnoi Matematiki i Tekhnicheskoi Fiziki, No. 1, January-February 1973, pp. 121-127.)

[18]Karelov, N. V., Kosov, A. V., Rebrov, A. K., and Sharafutdinov, R. G., "The Study of the Mixing Zone of Low Density Jet," Rarefied Gas Dynamics, Institute of Thermophysics, Novosibirsk, 1976 (in Russian).

[19]Borzenko, B. N., Karelov, N. V., Rebrov, A. K., and Sharafutdinov, R. G., "Experimental Investigation of Rotational Level Population in Free Nitrogen Jet," paper presented at Tenth International Symposium on Rarefied Gas Dynamics, 1976, unpublished.

[20]Willis, D. R., Hamel, B. B., and Lin, J. T. "Development of the Distribution Function on the Centerline of a Free Jet Expansion," The Physics of Fluids, Vol. 15, No. 4, April 1972, pp. 573-580.

[21]Brook, J. W., Hamel, B. B., and Muntz, E. P., "Theoretical and Experimental Study of Background Gas Penetration into Underexpanded Free Jets," The Physics of Fluids, Vol. 18, No.5, May 1975, pp. 517-528.

[22]Skovorodko, P. A. and Chekmarev, S. F., "On the Gas Diffusion into Low Density Jet," Rarefied Gas Dynamics, Institute of Thermophysics, Novosibirsk, 1976 (in Russian).

[23]Zarvin, A. J. and Sharafutdinov, R. G., "Molecular Beam Generator for Studying Rarefied Flows," Rarefied Gas Dynamics, Institute of Thermophysics, Novosibirsk, 1976 (in Russian).

[24]Zarvin, A. J. and Sharafutdinov, R. G.," The Determination of Translational Temperature from the Transversal Profile of the Molecular Beam Density," Rarefied Gas Dynamics, Institute of Thermophysics, Novosibirsk, 1976 (in Russian).

[25]Zarvin, A. E. and Sharafutdinov, R. G., "The Measurement of Molecular Beam Parameters in the Presence of Background Gas," Rarefied Gas Dynamics, Institute of Thermophysics, Novosibirsk, 1976 (in Russian).

[26]Bossel, U., Hurlbut, P. S., and Sherman, F. S., Extraction of Molecular Beams from Nearly-Inviscid Hypersonic Free Jets," Rarefied Gas Dynamics, Proceedings of the Sixth International Symposium, MIT, Cambridge, Mass., July 22-26, 1968, Vol. II (Advances in Applied Mechanics, Supplement 5), edited by Leon Trilling and H. Y. Wachman, Academic Press, New York, 1969, pp. 945-964.

[27] Sherman, F. S., "Hydrodynamical Theory of Diffusive Separation of Mixture in a Free Jet," The Physics of Fluids, Vol. 9, No. 5, May 1965, pp. 773-779.

[28] Rothe, D. E., "Electron Beam Studies of the Diffusive Separation of Helium-Argon Mixtures," The Physics of Fluids, Vol. 9, No. 9, September 1966, pp. 1643-1658.

[29] Hagena, O. F. and Obert, W., "Cluster Formation in Expanding Supersonic Jet. Effect of Pressure, Temperature, Nozzle Size and Test Gas," Journal of Chemical Physics, Vol. 56, No. 5, March 1, 1972, pp. 1793-1802.

[30] Bochkarev, A. A., Kosinov, V. A., Prikhodko, V. G., and Rebrov, A. K., "The Structure of Supersonic Jet of Argon-Helium Mixture in Vacuum," Journal of Applied Mechanics and Technical Physics, Vol. 11, No. 5, September-October 1970, pp. 857-861. (Translated from Zhurnal Prikladnoi Mekhaniki i Tekhnicheskoi Fiziki, No. 5, September-October 1970, pp. 158-163.)

[31] Bochkarev, A. A., Kosinov, V. A., Prikhodko, V. G., and Rebrov, A. K., "Effect of Diffusion Separation When Hypersonic Flows of Rarefied Gas Mixture Collide," Journal of Applied Mechanics and Technical Physics, Vol. 12, No. 2, March-April 1971, pp. 149-153

1971, pp. 313-317. (Translated from Zhurnal Prikladnoi Mekhaniki i Tekhnicheskoi Fiziki, No. 2, March-April 1971, pp. 149-153,)

[32] Maise, G. and Fenn, J. B., "Recovery Factor Measurement in Gas Mixtures," The Physics of Fluids, Vol. 7, No. 7, July 1964, pp. 1080-1082.

[33] Kutaledadze, S. S., Bochkarev, A. A., Prikhodko, V. G., and Rebrov, A. K., "Diffusive Effects on Recovery Temperature in Supersonic Flow of Rarefied Gas Mixture," Heat Transfer, 1970, 4th International Heat Transfer Conference, Versailles, France, August 31-September 5, 1970, Vol. 3, edited by Ulrich Grigull and Erich Hahne, Elsevier Publishing Company, Amsterdam, 1970, pp. FC6.4.1-FC6.4.8.

[34] Bochkarev, A. A., Losinov, V. A., Prikhodko, V. G., and Rebrov, A. K., "Flow of a Supersonic Low-Density Jet of Nitrogen-Hydrogen Mixture over a Blunt Body," Journal of Applied Mechanics and Technical Physics, Vol. 13, No. 6, November-December 1972, pp. 804-808. (Translated from Zhurnal Prikladnoi Mekhaniki i Tekhnicheskoi Fiziki, No. 6, November-December 1972, pp. 50-55.)

[35]Bochkarev, A. A., Kosinov, V. A., Prikhodko, V. G., and Rebrov, A. K., "Heat Transfer at the Nose of a Blunt Body in a Rarefied Supersonic Flow of a Nitrogen-Hydrogen Mixture." Journal of Applied Mechanics and Technical Physics, Vol. 14, No. 6, November-December 1973, pp. 809-812. (Translated from Zhurnal Prikladnoi Mekhaniki i Tekhnicheskoi Fiziki, No. 6, November-December 1973, pp. 88-91.)

[36]Zavarzina, I. F., "Experimental Investigations of Local Heat Fluxes on a Sphere and on the Spherical Blunt End of an Axisymmetric Body," Fluid Dynamics, Vol. 5, No. 4, July-August 1970. (Translated from Izvestiya Akademii Nauk SSSR, Mekhanika Zhidkosti i Gaza, No. 4, July-August 1970, pp. 157-161.)

[37]Cheng, H. K., "Hypersonic Shock Layer Theory of the Stagnation Region at Low Reynolds Number," Proceedings of the Heat Transfer and Fluid Mechanics Institute, Stanford University Press, Stanford, California, 1961, pp. 161-175.

[38]Bochkarev, A. A., Kosinov, V. A., Prikhodko, V. G., and Rebrov, A. K., "The Flow of a Rarefied Gaseous Mixture from an Orifice," Journal of Engineering Physics, Vol. 18, No. 4, April 1970, pp. 444-450. (Translated from Inzhenerno-Fizicheskii Zhurnal, Vol. 18, No. 4, April 1970, pp. 653-660.)

[39]Center, R. E., "Measurement of Shock Wave Structure in Helium-Argon Mixtures," The Physics of Fluids, Vol. 10, No. 8, August 1967, pp. 1777-1784.

[40]Harnet, L. N. and Muntz, E. P., "Experimental Investigation of Normal Shock Wave Velocity Distribution Function in Mixtures of Argon and Helium," The Physics of Fluids, Vol. 15, No. 4, April 1972, pp. 565-572.

[41]Timoshenko, N. I., "The Measurements of Translational Temperature," Experimental Methods in Rarefied Gas Dynamics, Novosibirsk, 1974 (in Russian).

[42]Bochkarev, A. A., Rebrov, A. K., and Timoshenko, N. I., "The Shock Wave Structure in the Ar + He Mixture," Izvestiya Sibirskogo Otdeleniya AN SSSR, Seriya Tekhnicheskikh Nauk, Vol. 1, No. 3, 1976 (in Russian).

[43]Abe, J. and Oguchi, H., "Shock Wave Structure in Gas Mixture," Rarefied Gas Dynamics, Proceedings of the Sixth International Symposium on Rarefied Gas Dynamics, MIT, Cambridge, Mass., July 1968, Vol. I, (Advances in Applied Mechanics, Supplement 5) edited by Leon Trilling and H. Y. Wachman, Academic Press, New York, 1969, pp. 425-432.

[44] Bird, G. A., "The Structure of Normal Shock Wave in Binary Gas Mixture," Journal of Fluid Mechanics, Vol. 31, Pt. 4, March 19, 1968, pp. 657-668.

[45] Lefkowitz, B. and Knuth, E. L., "A Study of Rotational Relaxation in a Low-Density Hypersonic Free Jet by Means of Impact-Pressure Measurements," Rarefied Gas Dynamics, Proceedings of the Sixth International Symposium, MIT, Cambridge, Mass., July 22-26, 1968, Vol. 2 (Advances in Applied Mechanics, Supplement 5), edited by Leon Trilling and H. Y. Wachman, Academic Press, New York, 1969, pp. 1421-1438.

[46] Miller, D. R. and Andres, R. P., "Rotational Relaxation of Molecular Nitrogen," Journal of Chemical Physics, Vol. 46, No. 9, 1967, pp. 3418-3423.

[47] Gallagher, R. J. and Fenn, J. B., "A Free Jet Study of Rotational Relaxation of Molecular Nitrogen from 300 to 1000°K," Rarefied Gas Dynamics, Proceedings of the Ninth International Symposium, Göttingen, West Germany, July 15-20, 1974, Vol. I, edited by M. Becker and M. Fiebig, DFVLR Press, Göttingen, 1974, pp. B19-1-B19-10.

[48] Vargin, A. N., Ganina, N. A., Konjukhov, V. K., and Seljakov, V. I., "Calculation of Rotational Transition Probabilities of Diatomic Molecules during Collisions with Heavy Particles," Zhurnal Prikladnoi Mekhaniki i Tekhnicheskoi Friziki, No. 2, March-April 1975, pp. 13-19.

[49] Marrone, P. V., "Temperature and Density Measurements in Free Jets and Shock Waves," The Physics of Fluids, Vol. 10, No. 3, 1967, pp. 521-538.

[50] Lewis, J.W.L., Price, L. L., and Kinslow, M., "Rotational Relaxation of N_2 in Heated Expansion Flow Fields," Rarefied Gas Dynamics, Proceedings of the Ninth International Symposium, Göttingen, West Germany, July 15-20, 1974, Vol. I, edited by M. Becker and M. Fiebig, DFVLR Press, Göttingen, 1974, pp. B17-1-B-17-9.

[51] Borzenko, B. N., Karelov, N. V., Rebrov, A. K., and Sharafutdinov, R. G., "The Experimental Study of Rotational Level Population in Nitrogen," Zhurnal Prikladnoi Mekhaniki i Tekhnicheskoi Fiziki, No. 5, 1976 (in Russian).

[52] Skovorodko, P. A. "The Rotational Relaxation by Gas Expansion into Vacuum," Rarefied Gas Dynamics, Institute of Thermophysics, Novosibirsk, 1976 (in Russian).

[53]Karelov, N. V., Skovorodko, P. A., and Yarygin, V. N., "The Vibrational Relaxation in the Jets Outside the Sonic Nozzles," Rarefied Gas Dynamics, Institute of Thermophysics, Novosibirsk, 1976 (in Russian).

[54]Bochkarev, A. A., Kosinov, V. A., Rebrov, A. K., and Sharafutdinov, R. G., "The Measurements of Gas Flow Parameters with Use of the Electron Beam," Experimental Methods in Rarefied Gas Dynamics, Institute of Thermophysics, Novosibirsk, 1974 (in Russian).

[55]Vostrikov, A. A., Kusner, Y. S., and Semyachkin, B. E., "The Study of CO_2 Condensation by the Method of Molecular Beam," Some Problems of Hydrodynamics and Heat Transfer, Institute of Thermophysics, Novosibirsk, 1976 (in Russian).

[56]Zharkova, N. G., Prokkoev, V. V., Rebrov, A. K., and Yarygin, V. N., "The Nonequilibrium Expansion of CO_2 at the Stagnation Temperature up to 1200°K," Zhurnal Prikladnoi Mekhaniki i Tekhnicheskoi Fiziki, No. 5, 1976 (in Russian).

[57]Rebrov, A. K. and Sharafutdinov, R. G., "On Structure of Freely Expanding Carbon Dioxide Jet in Vacuum," Rarefied Gas Dynamics, Proceedings of the Sixth International Symposium, MIT, Cambridge, Mass., July 22-26, 1968, Vol. 1 (Advances in Applied Mechanics, Supplement 5), edited by Leon Trilling and H. Y. Wachman, Academic Press, 1969, pp. 965-976.

[58]Beylich, A. E., "Experimental Investigation of Carbon Dioxide Jet Plumes," The Physics of Fluids, Vol. 14, No. 5, May 1971, pp. 898-905.

[59]Sharafutdinov, R. G., "Interaction of Background Molecules with a Low Density Free Jet," Rarefied Gas Dynamics, Proceedings of the Seventh International Symposium, University of Pisa, Pisa, Italy, June 29-July 3, 1970, Vol. I, edited by Dino Dini, Editrice Tecnico Scientifica, Pisa, Italy, 1971, pp. 563-574.

VIBRATIONAL RELAXATION TIMES IN CH_4-N_2 AND C_2H_4-N_2 MIXTURES

J.C.F. Wang* and G.S. Springer+

The University of Michigan, Ann Arbor, Mich.

Abstract

Vibrational relaxation times were determined in methane-nitrogen and ethylene-nitrogen mixtures. The experiments were performed by measuring the absorption and dispersion of ultrasonic waves passing through the mixtures. Data are reported for mole fractions of nitrogen ranging up to 50% at temperatures of 300°, 430°, and 630°K for ethylene-nitrogen and at 430° and 630°K for methane-nitrogen mixtures. The results show a strong effect of nitrogen concentration on the relaxation time. The ethylene-nitrogen data also indicate that, as in the case of pure hydrocarbons, the relaxation time in a mixture follows the Landau-Teller expression.

Introduction

Although vibrational relaxation times of hydrocarbon-oxygen mixtures have been studied previously,[1-5] vibrational energy transfer between various hydrocarbons and nitrogen apparently has not been investigated yet. In this paper, vibrational relaxation times are reported for mixtures of methane-nitrogen and ethylene-nitrogen. The experiments were performed for mole fractions of nitrogen varying from 5 to 50%. Measure-

---

Presented as Paper 64 at the 10th International Symposium on Rarefied Gas Dynamics, Aspen, Colo., July 19-23, 1976. The authors wish to thank P. Veenema and H. Chiger for their help in the experiments. This work was supported by the National Science Foundation and by the Environmental Protection Agency.

*Research Associate, Department of Mechanical Engineering, Fluid Dynamics Laboratory; Presently at General Electric Research and Development Center, Schenectady, N.Y.

+Professor, Department of Mechanical Engineering, Fluid Dynamics Laboratory.

ments were made at temperatures of 300°, 430°, and 670°K for ethylene-nitrogen and at 430° and 670°K for methane-nitrogen mixtures. The data were obtained by measuring the absorption of ultrasonic waves passing through the mixture. This technique already has been proved to be useful for studying molecular transfer processes in pure hydrocarbon vapors.[6]

Results and Discussion

The apparatus and procedure were basically the same as those used in our previous investigation[6] of vibrational relaxation of pure CH_4, C_2H_4, C_2H_2, and n-C_6H_6. Consequently, details of the experiment will not be repeated here. Only those aspects of the experiment are given which are needed in disucssing the results. Essentially, in the experiments an ultransonic wave with circular frequency $\omega=2\pi f$ was propagated through the vapor. The absorption coefficient α_M and the speed of sound c were measured at a fixed frequency, but at various pressures P of the mixture (0.1 to 2.5 atm). The absorption coefficient due to the internal degrees of freedom was determined from the expression[6]

$$\alpha_R = \alpha_M - \alpha' - \alpha_c \qquad (1)$$

where α' is a small correction due to the divergence of the sound beam[7] and α_c is the classical absorption coefficient [8-10]

$$\alpha_c = \frac{2\pi^2}{\gamma\lambda}\left(\frac{4}{3}\eta + \frac{\gamma-1}{C_p^\infty}k\right)\left(\frac{f}{P}\right) + \frac{2\pi^2 x_1 x_2}{\lambda} D_{12}\rho\left(\frac{M_2-M_1}{\bar{M}} + \frac{\gamma-1}{\gamma}\frac{k_{th}}{x_1 x_2}\right)\left(\frac{f}{P}\right) \qquad (2)$$

λ is the wavelength of the generated sound beam ($\lambda=c/f$), C_p^∞ and C_v^∞ are the constant pressure and constant volume specific heats of the mixture under "frozen" flow ($\omega=\infty$), γ is the specific heat ratio ($\gamma=C_p^\infty/C_v^\infty$). η, k, D_{12} and k_{th} are the kinematic viscosity, thermal conductivity, diffusivity, and thermal viscosity of the mixture calculated from the Lennard-Jones[6-12] potential using the equation given in Hirschfelder, Curtiss and Bird[11]. ρ is the density of the mixture, x_1 and x_2 are the mole fractions of the components, M_1 and M_2 are the atomic masses of the components, and $\bar{M}$ is the mean atomic mass ($\bar{M}=x_1M_1+x_2M_2$).

The experimentally determined values of $\lambda\alpha_R$ as a function of f/P are represented by circles in Figs. 1 and 2. The f/P ratios that could be established in the apparatus ranged from about 0.2 to 10 MHz/atm. Within this frequency/pressure range a single dominant relaxation maximum was observed for ethylene-nitrogen in the range 300°-630°K and for methane-nitrogen at

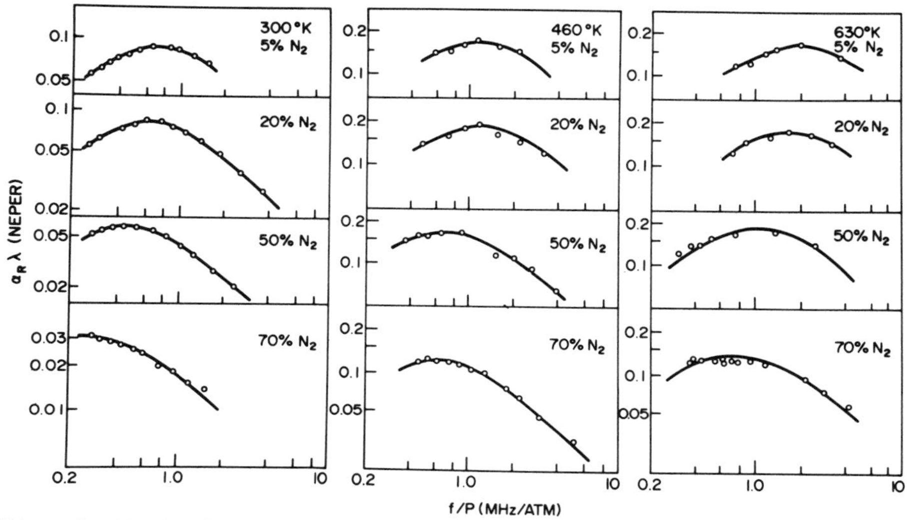

Fig. 1 Variation of the absorption coefficient with the frequency-pressure ratio in ethylene-nitrogen mixtures; O data, —— fit to data [calculated from Eqs. (4) and (5)].

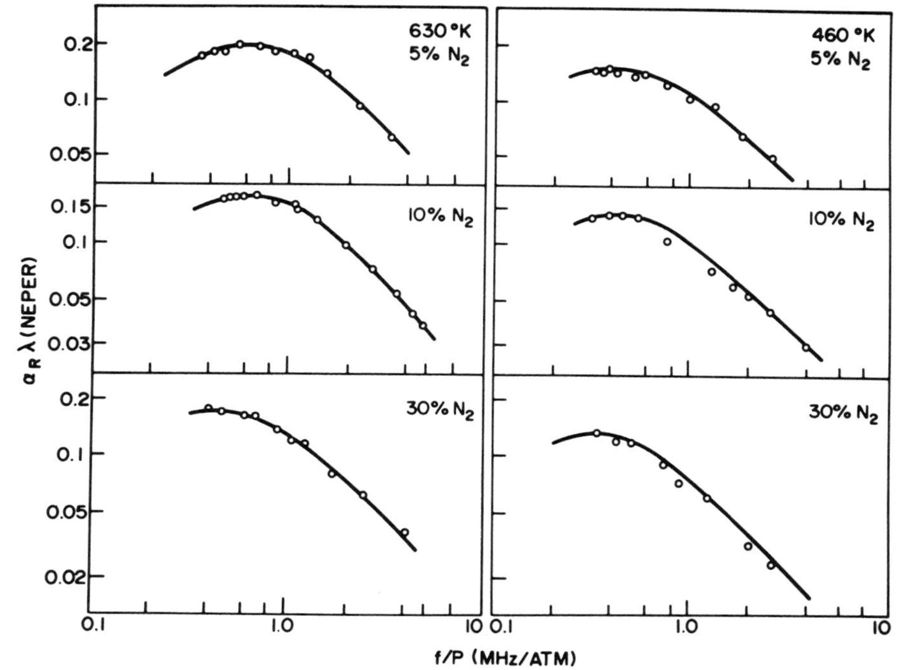

Fig. 2 Variation of the absorption coefficient with the frequency-pressure ratio in methane-nitrogen mixtures; O data —— fit to data [calculated from Eqs. (4) and (5)].

Table 1 Specific heat values used in the calculations (cal/°K-mole)[a]

Mixture	% N_2	Temp, °K	C_p^o	C_v^o	$C_p^o - C_p^\infty$
C_2H_4-N_2	0	300	10.45	8.414	2.506
		460	14.256	12.27	6.312
		630	17.598	15.612	9.654
	5	300	10.276	8.29	2.378
		460	13.851	11.865	5.953
		630	17.032	15.046	9.134
	20	300	9.752	7.766	2.003
		460	12.775	10.789	5.026
		630	16.33	14.344	8.581
	50	300	8.706	6.72	1.255
		460	10.623	8.637	3.172
		630	12.922	10.936	5.471
	70	300	8.008	6.022	9.755
		460	9.189	7.203	1.936
		630	10.65	8.664	3.397
CH_4-N_2	0	300	8.552	6.565	0.604
		460	10.44	8.454	2.492
		630	12.95	10.964	5.002
	5	300	8.472	6.486	0.574
		460	10.27	8.284	2.372
		630	12.665	10.679	4.767
	10	300	8.393	6.407	0.544
		460	10.	8.014	2.151
		630	12.38	10.394	4.531
	30	300	8.075	6.089	0.425
		460	9.42	7.434	1.77
		630	11.238	9.252	3.588

[a]From Ref. 12.

460° and 630°K. For the latter mixture at 300°K the maximum would have occured outside the f/P range of the apparatus.

The relaxation time τ was expressed as

$$\tau = B/P \qquad (3)$$

and the proportionality constant B was determined as follows. The coefficient $\alpha_R\lambda$ was calculated from Eqs. (4-5)[6-9]

$$\alpha_R\lambda = \pi\left(\frac{c}{c_o}\right)^2\left(\frac{C_p^\infty}{C_p^0} - \frac{C_v^\infty}{C_v^0}\right)\frac{(2\pi B)(f/P)}{1 + (C_p^\infty/C_p^0)(2\pi B)^2(f/P)^2} \qquad (4)$$

$$\frac{c}{c_o} = \left\{1 - \frac{(C_p^\infty/C_p^0)(C_p^\infty/C_p^0 - C_v^\infty/C_v^0)(2\pi B)^2(f/P)^2}{1 + (C_p^\infty/C_v^0)^2(2\pi B)^2(f/P)^2}\right\}^{-1/2} \qquad (5)$$

for different values of B. The value of B then was adjusted until the best match was found between the calculated $\alpha_R\lambda$ and the experimental data (solid line in Figs. 1 and 2). The specific heat values used in these calculations are listed in Table 1. The superscript zero indicates equilibrium conditions (ω=0).

Both vibrational and rotational relaxation may contribute to τ. Generally, these two modes of relaxation manifest themselves by separate maximums on the $\alpha_R\lambda$ vs f/P plots.[6] The maximums observed in our experiments are attributed to vibrational relaxation.[9] The maximums due to rotational relaxation would have occured at higher f/P ratios, which were outside the range of the apparatus.

The measured vibrational relaxation times are shown in Figs. 3 and 4 as a function of nitrogen concentration. In order to comment upon these results, we note that, in general, the following five relaxation processes may occur in the mixture[13]

1) $N_2^* + N_2 \rightleftharpoons N_2 + N_2$

2) $N_2^* + HC \rightleftharpoons N_2 + HC$

3) $N_2^* + HC \rightleftharpoons N_2 + HC^*$

4) $HC^* + HC \rightleftharpoons HC + HC$

5) $HC^* + N_2 \rightleftharpoons HC + N_2$

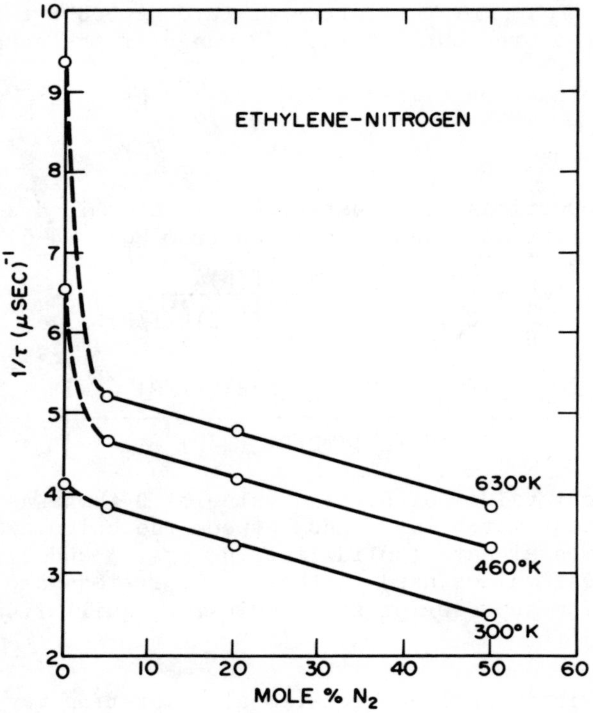

Fig. 3 Relaxation times for mixtures of ethylene and nitrogen.

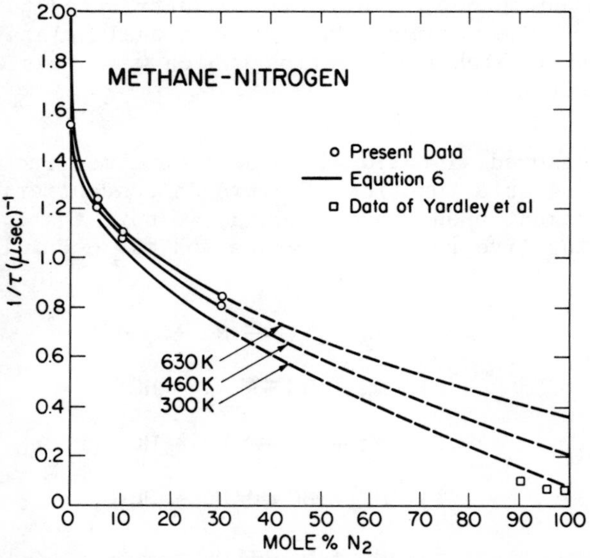

Fig. 4 Relaxation times for mixtures of methane and nitrogen.

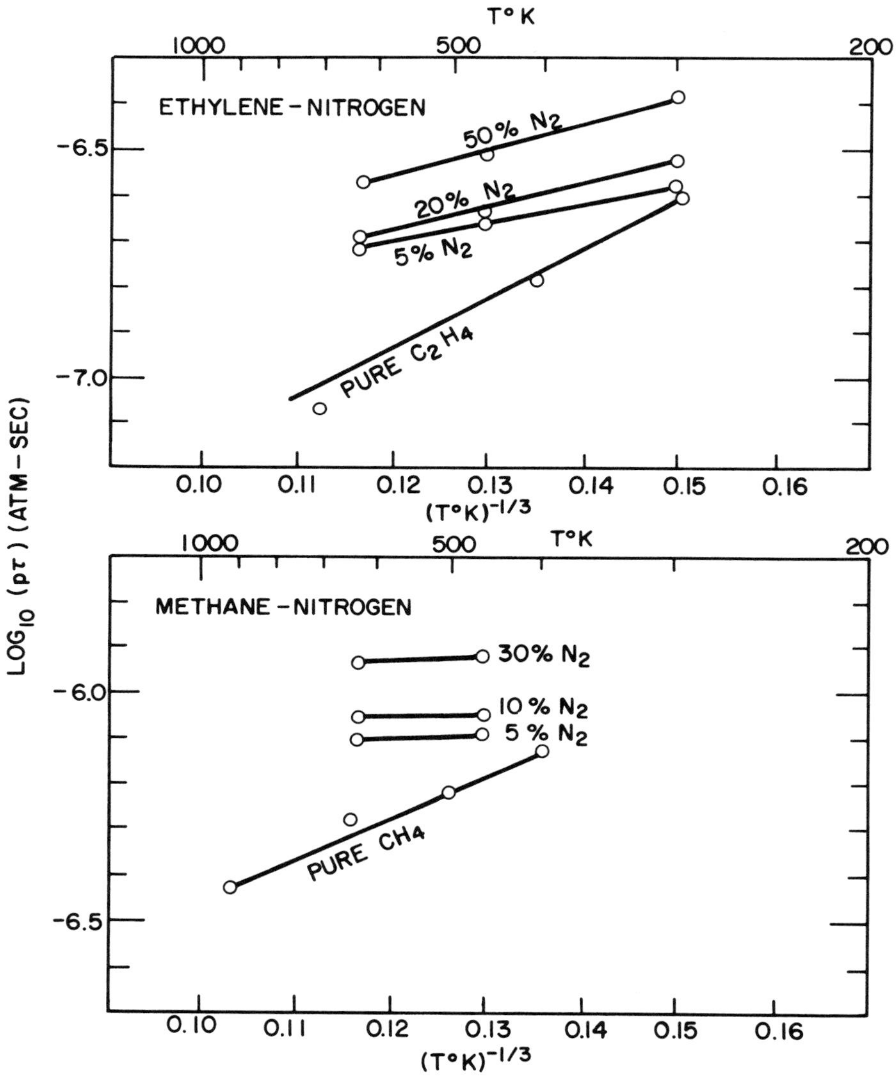

Fig. 5 Dependence of the vibrational relaxation times with temperature; 0 data, —— fit to data [Eq. (6)].

HC refers to the hydrocarbon molecules, and the asterisk denotes a vibrationally excited molecule. When only a small concentration of nitrogen is present in the vapor, the nitrogen may affect the relaxation time of the hydrocarbon through processes 2, 3, and 5. In processes 2 and 5 one would expect the change in relaxation time to be proportional to the nitrogen concentration. The data show, however, that the relaxa-

Table 2 The constants a and b in Eq. (6)

Mixture	% N_2	a	b
C_2H_4-N_2	0[a]	11.33	8.41
	5	3.66	7.134
	20	5.12	7.305
	50	5.75	7.243
	70	10.36	7.79
CH_4-N_2	0[a]	9.15	7.373
	5	0.737	6.18
	10	0.574	6.114
	30	1.01	6.045

[a]Data from Ref. 6.

tion time increases very rapidly with small increases in the nitrogen concentration. This suggests that process 3 is dominant, having the largest effect upon the relaxation time. This is consistent with the analytical results obtained by Stratton for ethylene-oxygen and methane-oxygen mixtures.[13]

The effect of temperature on the relaxation time is demonstrated in Fig. 5. For mixtures of ethylene-nitrogen, the data can be correlated by the expression

$$\log_{10}(\tau_v P) = aT^{-1/3} - b \tag{6}$$

where τ_v is the overall relaxation time (sec), P is the pressure (atm), and T is the temperature (°K). a and b are constants whose values depend on the nitrogen concentration (Table 2). Equation (6) is the Landau-Teller expression for vibrational relaxation.[14] Thus, as in the case of pure hydrocarbons,[6] in mixtures of ethylene and nitrogen $\tau_v P$ varies with temperature, as predicted by the Landau-Teller expression.

Data for methane-nitrogen mixtures were obtained only at two temperatures, and, therefore, for this mixture the validity of the Landau-Teller expression could not be evaluated. Nevertheless, the data also are shown in Fig. 5, and the values of the constants a and b are included in Table 2, indicating the trend in the results. It is interesting to note that at 300°K and 1 atm the relaxation times calculated from Eq. (6) with the values of a and b taken from Table 2 agree well with the data of Yardley et al[15] (Fig. 4).

References

[1]Schnaus, U.E., "Thermal Relaxation in Oxygen with CH_4 and CD_4 Admixtures," The Journal of the Acoustical Society of America, Vol. 37, Jan. 1965, pp. 1-4.

[2]White, D.R., "Vibrational Relaxation of Oxygen by Methane, Acetylene, and Ethylene," The Journal of Chemical Physics, Vol. 42, March 1965, pp. 2028-2032.

[3]Jones, D.G., Lambert, J.D., and Stretton, J.L., "Vibrational Relaxation in Mixtures Containing Oxygen," Proceedings of the Physical Society (London), Vol. 86, Oct. 1965, pp. 857-860.

[4]Cottrell, T.L. and Day, M.A., "Effect of Oxygen on Vibrational Relaxation in Methane," The Journal of Chemical Physics, Vol. 43, Aug. 1965, pp. 1433-1434.

[5]Parker, J.G. and Swope, R.H., "Vibrational Relaxation in Methane-Oxygen Mixtures," The Journal of Chemical Physics, Vol. 43, Dec. 1965, pp. 4427-4434.

[6]Wang, J.C.F. and Springer, G.S., "Vibrational Relaxation Times in Some Hydrocarbons in the Range 300-900°K," The Journal of Chemical Physics, Vol. 59, Dec. 1975, pp. 6556-6562.

[7]Nozdrev, V.F., The Use of Ultrasonics in Molecular Physics, translated by J.A. Cade, revised by E.R. Dobbs, MacMillan Co., New York, 1965.

[8]Herzfeld, K.F. and Litovitz, T.A., Absorption and Dispersion of Ultrasonic Waves, Academic Press, New York, 1959.

[9]Boyer, R.T. and Letcher, S.V., Physical Ultrasonics, Academic Press, New York, 1959, Chap. 4.

[10]Carey, C., Carnevale, E.H., and Marshall, T., "Experimental Determination of the Properties of Gases, Part II. Heat Transfer and Ultrasonic Measurement," AFML-TR-65-141, 1965, Wright-Patterson Air Force Base, Ohio.

[11]Hirschfelder, J.O., Curtiss, C.F., and Bird, R.B., Molecular Theory of Gases and Liquids, John Wiley, New York, 1954.

[12]Rossini, F.D., et al., Selective Values of Physical and Thermodynamic Properties of Hydrocarbons, Carnegie Press, Carnegie Institute of Technology, Pittsburgh, Pa., 1953.

[13]Stretton, J.L., "Vibrational Energy Exchange at Molecular Collisions," Transfer and Storage of Energy by Molecules, Vol. 2, Wiley-Interscience, New York, 1969, pp. 58-178.

[14]Landau, L. and Teller, E., "Theory of Sound Dispersion," Physikalische Zeitschrift der Sowjetunion, Vol. 10, May 1936, pp. 34-43.

[15]Yardley, J.T., Fertig, M.N., and Moore, C.B., "Vibrational Deactivation in Methane Mixtures," The Journal of Chemical Physics, Vol. 52, Feb. 1970, pp. 1450-1453.

VIBRATIONAL RELAXATIONS IN FREEJET FLOWS OF C_3H_8, n-C_4H_{10}, AND BINARY MIXTURES OF CO_2

P. K. Sharma,* E. L. Knuth,† and W. S. Young‡
University of California at Los Angeles, Los Angeles, Calif.

Abstract

Vibrational relaxations of pure species (C_3H_8 and n-C_4H_{10}) and of CO_2 in binary mixtures were studied by comparing measured and predicted terminal vibrational temperatures of these species in freejets. The measurements were made using a supersonic molecular beam and analyzed taking advantage of the temperature dependence of the fragmentation pattern realized when molecules are ionized by electron impact. The predictions are based upon a sudden-freeze model that is applicable to arbitrary relaxation processes in freejet flows. For C_3H_8 and n-C_4H_{10}, vibrational relaxation times were determined by matching the measured and predicted terminal temperatures in the freejet. For the binary mixtures of CO_2, the applicability of a simplified expression for the composition dependence of the relaxation time is tested. An improved correlation for the collision numbers for polyatomic molecules containing H atoms is given.

Introduction

Freejet flows provide a convenient means for studying kinetic processes in gases, e.g., collisional energy transfers between vibrational, rotational, and translational degrees of freedom. Several studies have been made of rotational and translational relaxations in freejet flows. However, studies dealing specifically with vibrational energy transfer have been relatively few.[1,2]

Presented as Paper 26 at the 10th International Symposium on Rarefied Gas Dynamics, Aspen, Colo., July 19-23, 1976.

*Ph.D. Candidate, Energy and Kinetics Department, School of Engineering and Applied Science.

†Professor, Energy and Kinetics Department, School of Engineering and Applied Science.

‡Assistant Professor, Energy and Kinetics Department, School of Engineering and Applied Science.

Using mass-spectroscopic fragmentation measurements from supersonic and effusive beams, Milne et al.[1] studied vibrational cooling in n-butane freejets. Their measurements show significant vibrational cooling in the freejet, indicating a small relaxation time for n-butane. The measurements also demonstrate that the terminal vibrational temperature in the freejet may be significantly lower than the source temperature. Sharma et al.[2] developed a convenient procedure for predicting the terminal vibrational temperature in freejet flows using a sudden-freeze model and demonstrated the applicability of the model using the experimental technique of Milne et al.

In the present work, vibrational energy transfer in freejets of propane, n-butane, and some selected binary mixtures of CO_2 was studied. For propane and n-butane, previous data on relaxation times (or collision numbers) are available only at room temperature. In the present study, temperature dependence of the relaxation time was obtained for these two species by matching the measured and predicted terminal vibrational temperatures. For the binary mixtures of CO_2, the measured terminal temperatures are compared with predicted values based upon a simplified expression for the composition dependence of the relaxation time.

Sudden-Freeze Criterion

For a freejet emanating from an orifice of effective diameter d*, freezing point along the jet centerline is determined by [Ref. 2, Eq. (9)]

$$M(T/T_o)^{\frac{1}{2}} \frac{d(T/T_o)}{d(x/d^*)} = -C \frac{1}{(E_m/kT_o)(T_o/T) + (c_v/k)} \times \frac{T}{T_o} \frac{d^*}{(a\tau_{h,p})_o} \frac{(\tau_{h,p})_o}{\tau_{h,p}} \tag{1}$$

where M is the Mach number, a is the speed of sound, T is the static temperature, x is axial distance from the orifice, E_m is an energy parameter, c_v is the heat capacity at constant volume, τ is the relaxation time, k is the Boltzmann constant, C is a constant equal to 0.5, subscripts h and p refer to enthalpy and pressure, and subscript o refers to stagnation conditions. The Mach number is related to the static temperature ratio T/T_o by

$$T_o/T = 1 + [(\gamma-1)/2]M^2 \tag{2}$$

The dimensionless quantity E_m/kT, called the kinetic parameter, may be approximated by

$$E_m/kT = (1/3)\ln Z_{vib} \tag{3}$$

where Z_{vib} is average vibrational collision number. In order to apply Eq. (1), the temperature dependence of the collision number (alternatively, the relaxation time) is required. For CO_2 and its dilute binary mixtures in H_2 and He, this dependence may be deduced from existing data on relaxation times and collisional deactivation probabilities (Ref. 3, Tables A.3 and A.4). Similar data are scarce for hydrocarbons. Hence, an approximate method for obtaining this temperature dependence of the collision number for hydrocarbons is discussed next.

Temperature Dependence of the Collision Number for Polyatomic Molecules

It has been established from ultrasonic experiments that vibrational energy transfer in most polyatomic molecules can be described with the aid of a single relaxation time. Some important exceptions are SO_2, CH_2Cl_2, and C_2H_6; these molecules exhibit two relaxation times.

The single relaxation time observed in the majority of polyatomic molecules is attributed to a relatively slow relaxation of the mode of lowest frequency ν_1 (with heat-capacity contribution c_1), followed by the much faster collisional participation of other vibrational modes in a series of complex transfer processes. The collision number Z^1_{10} for deactivation in the lowest frequency mode is given by [Ref. 3, Eq. (5.16)]

$$Z^1_{10} = (\tau_{T,p}/\tau_c)(c_1/c_s)\{1-\exp(-h\nu_1/kT)\} \tag{4}$$

where τ_c is mean collision time, h is Planck's constant, and c_s is total vibrational heat capacity.

Values of Z^1_{10} have been obtained for several molecules from measured values of $\tau_{T,p}$ using Eq. (4). Lambert and Salter[4] correlated a large number of experimental data at temperatures close to 300°K by plotting $\log_{10}Z^1_{10}$ against the lowest vibrational frequency ν_1. The correlation is shown in their Fig. 2 and in Fig. 5.15 of Ref. 3. It is seen that the molecules fall into one of two classes, depending on whether or not they contain hydrogen atoms, but that in each case $\log_{10}Z^1_{10}$ is nearly a linear function of ν_1. It is difficult to justify the simplicity of the observed correlation in terms of simple vibration-translation transfer theory. However, it is of interest to compare it with the results of a typical theory, e.g., Eq. (2.24) of Ref. 3. Note that in this equation $p_{1,0} = p_{0,1} = 1/Z^1_{10}$, and the term $\pm h\nu/2kT$ is small compared to the other term under the exponent. Then, at constant temperature, one obtains

$$\ln Z^1_{10} \simeq K\nu^{2/3} \tag{5}$$

where K is a constant. The preceding equation predicts a linear variation of $\log_{10}Z^1_{10}$ with $\nu^{2/3}$, in contrast with ν as used by Lambert and Salter. When $\log_{10}Z^1_{10}$ is plotted against $\nu^{2/3}$, one obtains a better linear fit than the one presented by Lambert and Salter, particularly for the class of molecules containing the hydrogen atoms. This correlation is shown in Fig. 1 at T = 300°K for the more important class of molecules containing hydrogen atoms. Note that all hydrocarbons belong to this class.

The main value of the correlation in Fig. 1 is in predicting Z^1_{10} for a molecule (belonging to the specified class) for which no data on relaxation times $\tau_{T,p}$ are available. At temperatures significantly higher than 300°K, a linear relationship is not expected. However, for temperatures less than 500°K, a linear correlation provides acceptable accuracy for many applications.

For propane and n-butane, values of relaxation time $\tau_{T,p}$ and collision number Z^1_{10} are available only at room temperature. The Z^1_{10} values fit the correlation in Fig. 1 at T = 300°K quite closely. For the relaxation studies in freejet flows, the temperature dependence of the relaxation time is required. This can be obtained for propane and n-butane if Z^1_{10} is known for at least one more temperature. As a first approximation, assume that a linear correlation between $\log_{10}Z^1_{10}$ and $\nu^{2/3}$ exists at 450°K. This correlation may be plotted using the molecules for which Z^1_{10} is either known at 450°K or can be obtained by interpolation from the existing data, e.g., CH_3Br, CH_3Cl, C_2H_4, CH_4, etc. The correlation thus obtained at 450°K also is included in Fig. 1. It is seen that the scatter from the mean straight line is greater in this case. However, approximate values of Z^1_{10} for propane and n-butane at T = 450°K may be obtained from this correlation.

The values for ν_1, $\tau_{T,p}$, c_1/c_s, and Z^1_{10} for propane and n-butane at room temperature are given in Table 1. The values of Z^1_{10} in this table are determined by use of Eq. (4). It is

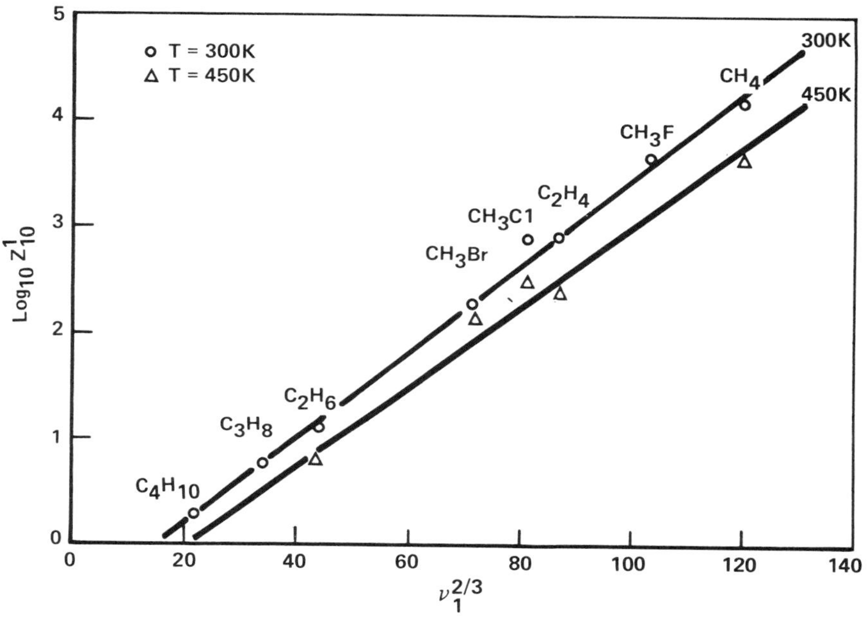

Fig. 1 $\text{Log}_{10}Z'_{10}$ as function of $\nu_1^{2/3}$ for polyatomic molecules (ν_1 = lowest vibrational frequency, cm^{-1}).

seen that these values are in good agreement with the correlation in Fig. 1.

First approximations for the collision number Z_{10}^1 for C_3H_8 and n-C_4H_{10} at 450°K are given by the correlation for 450°K in Fig. 1. These lead to the values given in Table 2. The ratio c_1/c_s is a weak function of temperature and hence may be assumed to be a constant in the narrow temperature range 300°-450°K. The relaxation time $\tau_{T,p}$ now may be obtained at 450°K from Eq. (4). The resulting values also are

Table 1. Selected data on C_3H_8 and n-C_4H_{10}

Molecule	ν_1, cm^{-1}	T, K	$\tau_{T,p}$, sec	c_1/c_s	Z_{10}^1	Ref.
C_3H_8	202	298	$4.1\text{x}10^{-9}$	0.22	5.83	4
n-C_4H_{10}	102	298	$1.5\text{x}10^{-9}$	0.20	2.00	4

Table 2. Values obtained as a first approximation based upon the correlation in Fig. 1

Molecule	Z^1_{10} (450°K)	$\tau_{T,p}$ (450°K), sec	a	b
C_3H_8	2.82	3.16×10^{-9}	13.7	-21.4
n-C_4H_{10}	1.20	1.40×10^{-9}	3.60	-20.9

included in Table 2. The values of $\tau_{T,p}$ at 298° and 450°K may be fitted by the equation

$$\ln(p\tau_{T,p}) = aT^{-1/3} + b \qquad (6)$$

The resulting values of the constants a and b also are included in Table 2.

Application of the Sudden-Freeze Model

Propane and n-Butane

Application of Eq. (1) to freejets requires 1) heat-capacity values, and 2) a description of the variation of static properties of the freejet with axial distance from the orifice. The relaxation time $\tau_{h,p}$ is related to the more commonly reported relaxation time $\tau_{T,p}$ by the relation

$$\tau_{T,p}/\tau_{h,p} = c_p^o/c_p^\infty$$

where c_p^o is the value of the constant-pressure heat capacity with vibrational degrees of freedom active, and c_p^∞ is its value with vibrational degrees of freedom frozen.

Heat capacities for propane and n-butane are given in Ref. 5. Average values (300°-800°K) are given in Table 3. Variation of the static temperature ratio T/T_o with dimensionless distance from orifice x/d* may be obtained by the procedure described in Ref. 2. The resulting variations for propane and n-butane are shown in Fig. 2. These data and relations facilitate the use of Eq. (1) in predicting terminal vibrational temperature for given source conditions.

Binary Mixtures: CO_2-H_2 and CO_2-He

A rigorous treatment of the case of nondilute binary mixtures is difficult because an analytical expression for the relaxation time cannot be formulated easily. However, for low

Table 3. Heat capacity values for C_3H_8 and n-C_4H_{10}

Molecule	γ	c_v/k	c_p^0/c_p^∞
C_3H_8	1.09	23/2	25/8
n-C_4H_{10}	1.06	31/2	33/8

vibrational excitation of the species of interest (M) and with no participating internal modes of the diluent gas X, the relaxation time of species M in the binary mixture $\tau'_{T,p}$ may be described by [Ref. 3, Eq. (3.23)]

$$1/\tau'_{T,p} = [x_M/(\tau_{T,p})_M] + [x_X/(\tau_{T,p})_X] \tag{7}$$

where x_M and x_X are mole fractions of species M and X, respectively, $(\tau_{T,p})_M$ is relaxation time for pure species M, and $(\tau_{T,p})_X$ is relaxation time for M when present as a trace in X. Values of $(\tau_{T,p})_X$ for CO_2-He and CO_2-H_2 binary mixtures may be determined using Eq. (22) of Ref. 2. The heat capacity ratio γ for the binary mixture may be computed using

$$\gamma = (x_M c_{p_M} + x_X c_{p_X})/(x_M c_{v_M} + x_X c_{v_X}) \tag{8}$$

where c_{p_M} and c_{p_X} are constant-pressure heat capacities for species M and X, respectively, and c_{v_M} and c_{v_X} are corresponding constant-volume heat capacities. The application of Eq.

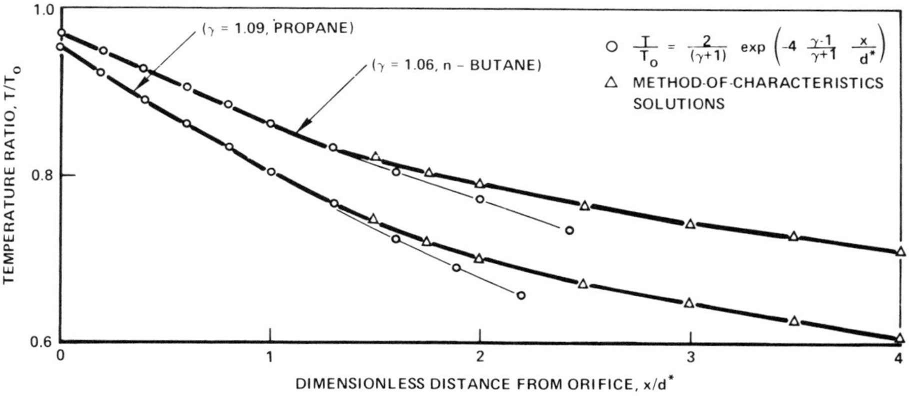

Fig. 2 Static temperature in freejet as function of distance from orifice (γ = 1.06 and 1.09).

(1) then proceeds in a manner similar to the procedure for a pure species.

Experimental Equipment and Procedures

The terminal vibrational temperatures in freejets of propane, n-butane, and binary mixtures of CO_2 were determined experimentally using the supersonic molecular-beam technique suggested by Milne et al.[1] This technique takes advantage of the temperature dependence of the fragmentation pattern realized when molecules are ionized by electron impact. This temperature dependence can be established quantitatively using an effusive molecular beam with a variable-temperature source.

The experimental system used is described in Ref. 2. Mass-spectrometric fragmentations were measured at various temperatures in both supersonic and effusive beams for propane (99.99% pure), n-butane (99.5% pure), and the binary mixtures of CO_2. For the propane and n-butane, the source pressure was maintained at 100 Torr, whereas the source temperature was varied from room temperature to 800°K. The electron energy at the ionizer was maintained constant at 50 eV. In order to minimize the effect of any possible drift in the mass spectrometer, the measurements on the supersonic and effusive beams were made at nearly equal temperatures in rapid succession. For the binary mixtures CO_2-He and CO_2-H_2, the source pressure was maintained at 190 Torr, whereas the source temperature was varied from room temperature to 1300°K. The gas mixtures 10% CO_2-90% He, 10% CO_2-90% H_2, and 50% CO_2-50% H_2 were prepared in the laboratory. Each of the individual gases used in preparing the mixture was of 99.99% minimum purity. The electron energy at the ionizer again was kept constant at 50 eV. For the effusive beam, measurements on CO_2 alone sufficed.

The measurements on propane and n-butane are reported in Figs. 3 and 4, respectively. Typical measurements on one of the binary mixtures (10% CO_2-90% He) are shown in Fig. 5. For a given source temperature T_o, the terminal vibrational temperature T_f in the freejet may be obtained by the procedure illustrated in Ref. 2, i.e., by drawing a vertical line at T_o to meet the fragment curve for the supersonic beam, then drawing a horizontal line to meet the corresponding curve for the effusive beam, and finally reading the effusive-beam temperature at this point of intersection. A check of the consistency of the measurements may be made by carrying out this procedure for each fragment.

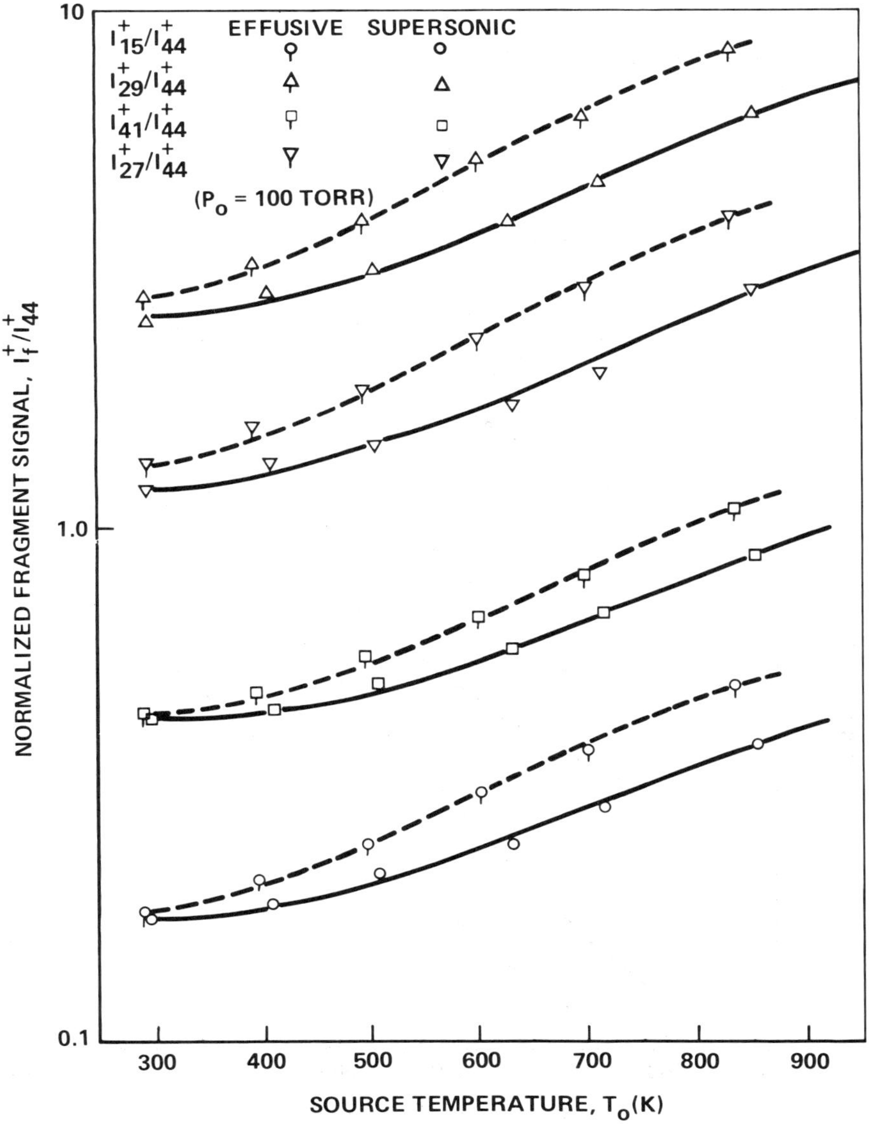

Fig. 3 Fragmentation in ionizer of propane from supersonic freejet and from effusive beam.

Results and Discussion

Propane and n-Butane

Expressions for the temperature dependence of the collision numbers for propane and n-butane were obtained by an

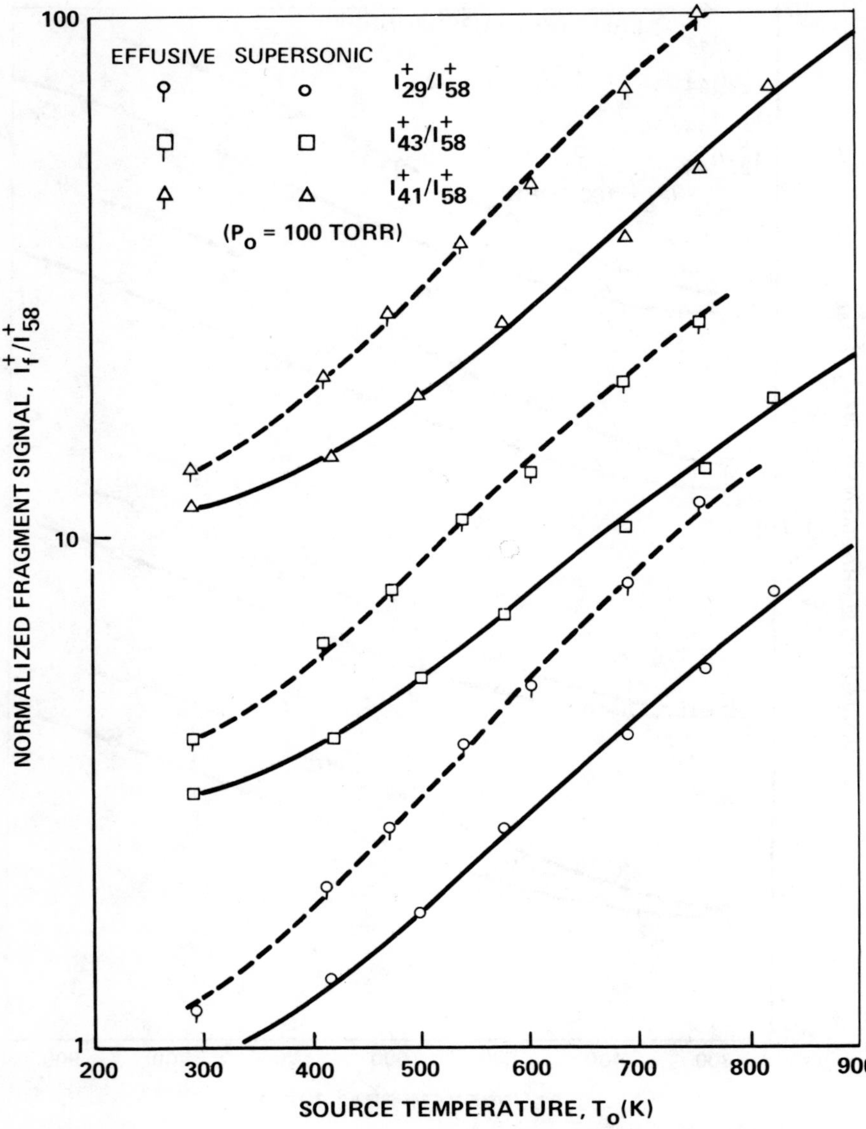

Fig. 4 Fragmentation in ionizer of n-butane from supersonic freejet and from effusive beam.

iterative procedure. In the first step, terminal vibrational temperatures were predicted using the approximate values of a and b given in Table 2 and solving Eq. (1) on the computer. [It has been shown previously[2] that the dimensionless terminal

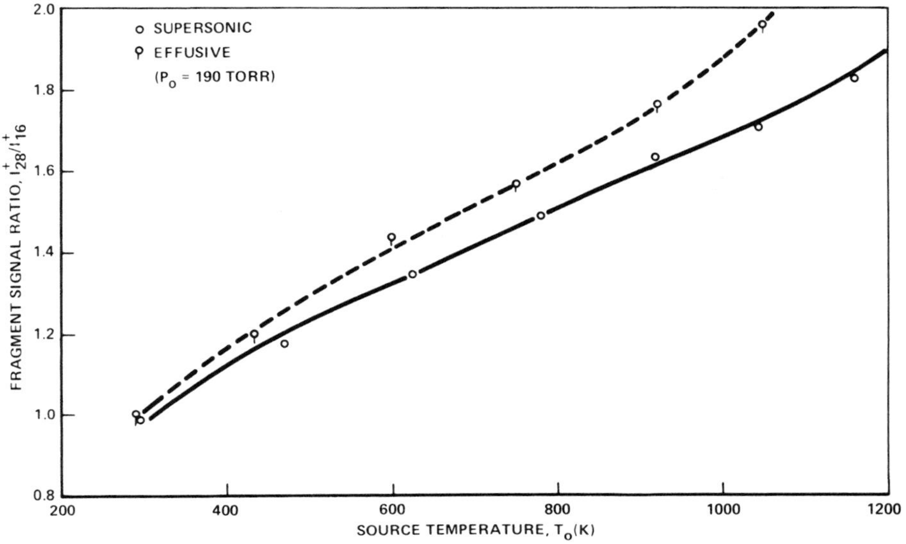

Fig. 5 Fragmentation in ionizer of CO_2 from supersonic 10% CO_2-90% He freejet and from effusive CO_2 beam.

vibrational temperature $(T/T_o)_f$ can be correlated as a function of the dimensionless scaling parameter $(d^*/a\tau_{h,p})_o$.] If the measured values of $(T/T_o)_f$ did not agree with this prediction, then better values of $\tau_{h,p}$ were obtained, for each of several source temperatures, by determining the values of $\tau_{h,p}$ required to shift the measured values of $(T/T_o)_f$ horizontally into agreement with the predicted curve. From these values of $\tau_{h,p}$, new values of a and b were determined. This process was repeated until values of a and b were obtained which did not change during an iteration. For propane, one obtains an a value of 21.62 and a b value of -22.54. For n-butane, one obtains an a value of 6.15 and a b value of -21.24.

Milne et al.[1] discussed possible effects of species decomposition in the source and of condensations and rotational relaxations in the freejet. These effects, however, are not expected to introduce significant errors in the present results in the temperature range selected for the measurements.

CO_2-He and CO_2-H_2 Gas Mixtures

The predicted correlation for CO_2 (either pure or in infinitely dilute binary mixtures) shown in Fig. 6 was obtained by application of Eq. (1). (For details, see Ref. 2.) Note particularly that a common correlation is obtained for pure

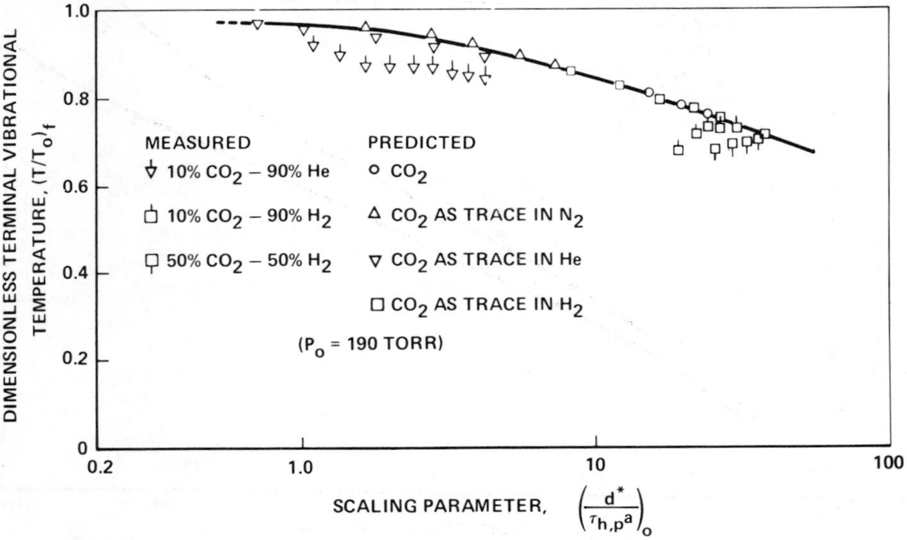

Fig. 6 Comparison of measured and predicted values of terminal vibrational temperatures of CO_2 in freejet flows of several binary mixtures.

CO_2 (0% diluent gas) and its infinitely dilute binary mixtures (∿100% diluent gas). Hence it is hypothesized that the correlation also holds for a binary mixture of any arbitrary fraction of diluent gas.

Measurements of terminal vibrational temperatures for 10% CO_2-90% He, 10% CO_2-90% H_2, and 50% CO_2-50% H_2 obtained in the present study are included in Fig. 6. The values of the scaling parameter $(d^*/a\tau_{h,p})_o$ used here are based upon $\tau_{T,p}$ obtained from Eq. (7).

For 10% CO_2-90% He, the measurements lie about 6-8% below the predicted correlation. For 10% CO_2-90% H_2 and 50% CO_2-50% H_2 mixtures, the agreement between predicted and measured values is very good for source temperatures greater than about 800°K but rather poor for source temperatures close to room temperature. The discrepancy at the lower source temperatures is not explained readily. One possible reason could be CO_2 condensation due to the large cooling effect of H_2 in the CO_2-H_2 binary jet.

The measurements presented in Fig. 6 support the use of 1) the predicted correlation for nondilute as well as dilute binary mixtures of CO_2 and 2) Eq. (7) for predicting the relaxation time of CO_2 in binary mixtures with vibrationally inert diluent gases.

References

[1]Milne, T. A., Beachey, J. E., and Greene, F. T., "Study of Relaxation in Free Jets using Temperature Dependence of n-Butane Mass Spectra," Journal of Chemical Physics, Vol. 56, March 15, 1972, pp. 3007-3013.

[2]Sharma, P. K., Young, W. S., Rodgers, W. E., and Knuth, E. L., "Freezing of Vibrational Degrees of Freedom in Free-Jet Flows with Application to Jets Containing CO_2," Journal of Chemical Physics, Vol. 62, Jan. 15, 1975, pp. 341-349.

[3]Stevens, B., Collisional Activation in Gases, Pergamon Press, New York, 1967.

[4]Lambert, J. D. and Salter, R., "Vibrational Relaxation in Gases," Proceedings of the Royal Society, Vol. A253, Nov. 24, 1959, pp. 277-288.

[5]Tribus, M., Thermostatics and Thermodynamics, D. Van Nostrand, Princeton, N.J., 1961, p. 184.

TRANSPORT COEFFICIENTS AND RELAXATION PHENOMENA IN VIBRATIONAL NONEQUILIBRIUM DIATOMIC GASES

Bernard Zappoli*

Société Européenne de Propulsion, Vernon, France

Abstract

In many physical situations (supersonic expansions of hot gases, shock waves, the molecular collisions involving T-T, T-R, and resonant V-V energy exanges are the most probable. In these cases, the zeroth-order distribution function is shown to be vibrational nonequilibrium. In the present paper, the Chapman-Enskog theory is generalized to a vibrational nonequilirbium gas, in order to obtain the perturbed distribution function. The transport coefficients give correct Prandtl and Lewis numbers. The translational, rotational, and vibrational relaxation equations are found, showing the influence of the characteristic times of the different processes of molecular energy exchanges.

Nomenclature

$\vec{C}_i$ = peculiar velocity of the molecule in the quantum state i
C_t = translational heat capacity
C_r = rotational heat capacity
C_v = vibrational heat capacity
E_i = energy of the ith internal energy level
f_i = distribution function
G_{ij} = relative velocity of two molecules before an encounter
G_o = velocity of the center of mass
i_r = rotational quantum number
i_v = vibrational quantum number
k = Boltzmann constant
m = mass of a molecule
n = number of molecules per unit volume

Presented as Paper 127 at the 10th International Symposium on Rarefied Gas Dynamics, Aspen, Colo., July 19-23, 1976.

*Ingenier, Department of Aeronautical Engineering.

T = translational-rotational temperature
T_v = vibrational temperature
$\vec{\omega}$ = flow velocity
σ_{ij}^{kl} = differential cross section of type I collisions
Γ_{ij}^{kl} = differential cross section of type II collisions
Ω = solid angle of scattering
$\delta_{i\gamma}$ = Kronecker symbol
$\Delta\mathcal{E}$ = total variation of internal energy during an encounter

Introduction

Two types of collisions are taken into account: type I, which are T-T, T-R, and resonant V-V energy exchanges; and type II, which are those encountering all others. In strongly perturbating phenomena, like supersonic expansion of hot gases or shock waves, for which the specific time scale is very short, type I collisions relative to this time scale are the most probable[1].

In such a physical situation, the iterative system obtained by expanding the distribution function as $f_i = f_i^{(0)} + f_i^{(0)}\varphi_i$ and the collision operator as $J = J_I + J_{II}$ can be written as

$$J_I^{(0)} = 0 \tag{1}$$

$$df_i^{(0)}/dt = J_I^{(1)} + J_{II}^{(0)} \tag{2}$$

where

$$J_I^{(0)} = \sum_{jkl}\int \left(f_k^{(0)} f_l^{(0)} - f_i^{(0)} f_j^{(0)}\right) G_{ij}\, \sigma_{ij}^{kl}\, d\Omega\, dc_j$$

$$J_I^{(1)} = \sum_{jkl}\int f_i^{(0)} f_j^{(0)} \left(\varphi_k + \varphi_l - \varphi_i - \varphi_j\right) G_{ij}\, \sigma_{ij}^{kl}\, d\Omega\, dc_j$$

$$J_{II}^{(0)} = \sum_{jkl}\int \left(f_k^{(0)} f_l^{(0)} - f_i^{(0)} f_j^{(0)}\right) G_{ij}\, \Gamma_{ij}^{kl}\, d\Omega\, dc_j$$

Equation (1) gives the zeroth-order vibrational nonequilibrium distribution function

$$f_i^{(0)} = n\left(\frac{m}{2\pi kT}\right)^{3/2} \exp\left(-\frac{m c_i^2}{2kT}\right) \frac{\exp\left(-\frac{E_{ir}}{kT}\right)}{\sum_{i_r}\exp\left(-\frac{E_{ir}}{kT}\right)}\, \frac{\exp\left(-\frac{E_{iv}}{kT_v}\right)}{\sum_{i_v}\exp\left(-\frac{E_{iv}}{kT_v}\right)} \tag{3}$$

The corresponding conservation equations are the Euler equations with the classical relaxation equation

$$n\,\frac{dE_v}{dt} = \sum_i \int E_{iv}\, J_{II}^{(0)}\, dc_i$$

Equation (2) gives the first-order perturbed distribution function. Equation (2) has been solved previously by expanding Ψ_i in terms of the eigenfunction linearized Boltzmann collision operator[2]; the infinite expansion was truncated, and a model equation was found. When the truncation was done for $N=2$, the solution gave Prandtl and Lewis numbers equal to 1; for $N=3$, Prandtl and Lewis numbers had correct values, but the corresponding vibrational relaxation equation was quite untractable. In the present work, Chapman-Enskog theory is generalized to solve Eqs. (2). Transport coefficients and first-order relaxation equations are obtained.

First-order Solution: Generalized Chapman-Enskog Procedure

Equation (2) also can be written as

$$\exp\left(-c_i^2-\varepsilon_{ir}-\varepsilon_{iv}\right)\left[\left\{\left(c_i^2-\frac{5}{2}\right)+\left(\varepsilon_{ir}-\bar{\varepsilon}_r\right)\right\}\frac{\overrightarrow{\partial \ln T}}{\partial r}\cdot\vec{c}_i+\right.$$

$$2\overset{\circ}{c_i c_i}:\nabla u-\frac{C_r}{C_{tr}}\left(\frac{2}{3}\left(\frac{3}{2}-c_i^2\right)+\left(\varepsilon_{ir}-\bar{\varepsilon}_r\right)\right)\cdot\nabla u+$$

$$\left(\varepsilon_{iv}-\bar{\varepsilon}_v\right)\frac{\overrightarrow{\partial \ln T_v}}{\partial r}\cdot\vec{c}_i+\frac{n^2}{n_{iv}\pi^{3/2}Q_r Q_v}\sum_{ir}\int J_{II}^{(0)}\left(f_{i'}^{(0)}f_i^{(0)}\right)dc_i$$

$$\left.+\frac{nkT_v}{\pi^{3/2}Q_r^2Q_v^2C_{tr}T}\left\{\left(c_i^2-\frac{3}{2}\right)+\left(\varepsilon_{ir}-\bar{\varepsilon}_r\right)\right\}I_{\varepsilon iv}^{(0)}\right]=$$

$$\frac{n}{\pi^{3/2}Q_rQ_v}\exp\left(-c_i^2-\varepsilon_{ir}-\varepsilon_{iv}\right)\mathcal{L}(\Psi_i)+\frac{n}{\pi^{3/2}Q_rQ_v}J_{II}^{(0)}\left(f_i^{(0)},f_i^{(0)}\right) \tag{5}$$

where

$$\vec{c}_i=\sqrt{\frac{m}{2kT}}\,\vec{C}_i\ ;\ \varepsilon_{ir}=\frac{E_{ir}}{kT}\ ;\ \vec{g}_{ij}=\sqrt{\frac{m}{2kT}}\,\vec{G}_{ij}\ ;\ \vec{u}=\sqrt{\frac{m}{2kT}}\,\vec{U}$$

$$C_r=\frac{\partial E_r}{\partial T}\ ;\ C_{tr}=\frac{3}{2}k+C_r\ ;\ I_{\varepsilon iv}^{(0)}=\sum_i\int\varepsilon_{iv}J_{II}^{(0)}\,dc_i\ ;\ Q_i=\sum_i\exp(-\varepsilon_i)$$

$$\mathcal{L}(\Psi_i)=\sum_{jkl}\int\exp\left(-c_j^2-\varepsilon_{jr}-\varepsilon_{jv}\right)\left(\Psi_k+\Psi_l-\Psi_i-\Psi_j\right)g_{ij}\sigma_{ij}^{kl}\,d\Omega\,dg \tag{5 a}$$

$$J_{II}^{(0)}(f_i^{(0)} f_i^{(0)}) = \sum_{jkl} \int \Big[\exp(-C_k^2 - C_l^2 - \varepsilon_{kr} - \varepsilon_{lr} - \varepsilon_{kv} - \varepsilon_{lv}) - \exp(-C_i^2 - C_j^2 - \varepsilon_{ir} - \varepsilon_{jr} - \varepsilon_{iv} - \varepsilon_{jv}) \Big] \times g_{ij} \Gamma_{ij}^{kl} \, d\Omega \, dc_j \quad (5\text{ b})$$

The form of Eq. (4) allows us to seek a solution in the following form

$$\mathcal{L}_i = A_i \left(\overrightarrow{\partial \ln T / \partial r}\right) \cdot \vec{C}_i + B_i \, \overset{\circ}{C_i C_i} : \nabla u + D_i \nabla u + F_i \left(\overrightarrow{\partial \ln T_v / \partial r}\right) \vec{C}_i + G_i \quad (6)$$

where A_i, B_i, D_i, F_i, and G_i depend on velocity, internal state, space, position, and time. Substituting this expression in Eq. (4), the following equations for the preceding unknown coefficients are determined

$$\frac{n}{\pi^{3/2} Q_r Q_v} \mathcal{L}(A_i \vec{C}_i) = \vec{C}_i \left(C_i^2 - \frac{5}{2}\right) + \vec{C}_i \left(\varepsilon_{ir} - \bar{\varepsilon}_r\right) \quad (7\text{ a})$$

$$\frac{n}{\pi^{3/2} Q_r Q_v} \mathcal{L}(B_i \overset{\circ}{C_i C_i}) = 2 \overset{\circ}{C_i C_i} \quad (7\text{ b})$$

$$\frac{n}{\pi^{3/2} Q_r Q_v} \mathcal{L}(F_i \vec{C}_i) = \left(\varepsilon_{iv} - \bar{\varepsilon}_v\right) \vec{C}_i \quad (7\text{ c})$$

$$\frac{n}{\pi^{3/2} Q_r Q_v} \mathcal{L}(D_i) = \frac{C_r}{C_G} \left[\frac{2}{3}\left(C_i^2 - \frac{3}{2}\right) - \left(\varepsilon_{ir} - \bar{\varepsilon}_r\right) \right] \quad (7\text{ d})$$

$$\frac{n}{\pi^{3/2} Q_r Q_v} \mathcal{L}(G_i) = \frac{n^2}{\pi^3 Q_r^2 Q_v^2 n_{iv}} I^{(0)}(i_v) - \frac{n}{\pi^{3/2} Q_r Q_v \exp(-C_i^2 - \varepsilon_{ir} - \varepsilon_{iv})} J_{II}^{(0)}$$

$$- \frac{n k T_v}{\pi^3 Q_r^2 Q_v^2 C_G T} \left[\left(C_i^2 - \frac{3}{2}\right) + \left(\varepsilon_{ir} - \bar{\varepsilon}_r\right) \right] I^{(0)}_{\varepsilon_{iv}} \quad (7\text{ e})$$

Then, according to the normal solution procedure, the macroscopic quantities are defined at the zeroth order; that is, the solution is subjected to the following conditions

$$\sum_i \int \begin{bmatrix} 1 \\ C_i \\ C_i^2 + \varepsilon_{ir} \\ \varepsilon_{iv} \end{bmatrix} f_i^{(0)} \mathcal{L}_i \, dc_i = 0 \quad (8)$$

Substituting for from (6), Eq. (8) can be written as

$$\sum_i \int \begin{bmatrix} D_i \\ \\ G_i \end{bmatrix} \begin{bmatrix} 1 \\ \mathcal{E}_{iv} \\ c_i^2 + \mathcal{E}_{ir} \end{bmatrix} f_i^{(0)} \Psi_i = 0 \; ; \; \sum_i \int \begin{bmatrix} A_i \\ \\ F_i \end{bmatrix} f_i^{(0)} \Psi_i \, dc_i = 0 \qquad (9)$$

To solve integral equations (7), A_i, B_i, D_i, F_i, and G_i are written as follows

$$A_i \vec{c}_i = \vec{c}_i \sum_{mnq} \psi^{3/2}_{mnq} a_{mnq} \; ; \; B_i \mathring{c}_i c_i = \mathring{c}_i c_i \sum_{mnq} b_{mnq} \psi^{5/2}_{mnq}$$

$$D_i = \sum_{mnq} d_{mnq} \psi^{1/2}_{mnq} \; ; \; F_i = \sum_{mnq} f_{mnq} \psi^{3/2}_{mnq} \; ; \; G_i = \sum_{mnq} g_{mnq} \psi^{1/2}_{mnq}$$

where

$$\psi^{r}_{mnq} = S^{m}_{r+1/2}(c_i^2) \cdot P^n(\mathcal{E}_{ir}) \cdot P^q(\mathcal{E}_{iv})$$

$S^{m}_{r+1/2}$ are the unnormalized Sonine-Laguerre polynomials, and $P^n(\mathcal{E}_i)$ are those introduced by Wang-Chang and Uhlembeck[3]. A truncation of zeroth order is carried out in the development of B_i, whereas a first-order one is carried out for the others. Then, taking into account the spherical symmetry of σ_{ij}^{kl} and the orthogonality of the ψ^{r}_{mnq} polynomials, a finite system of integral equations is obtained for the $a_{mnq}, b_{mnq}, \ldots,$ coefficients. The parameters $\beta^{rmnq}_{rm'n'q'}$ that appear in the solutions are defined by (see Appendix)

$$\beta^{\,r mnq}_{r m'n'q'} = -\frac{1}{4\pi^{3/2} Q_r Q_v} \sum_{ijkl} \int \delta_{ki}\delta_{lj} \exp\left(-c_i^2 - c_j^2 - \mathcal{E}_{ir} - \mathcal{E}_{jr} - \mathcal{E}_{iv} - \mathcal{E}_{jv}\right) \left[\psi^{r}_{mnq} Y_r\right]$$
$$\times \left[\psi^{r'}_{m'n'q'} Y_{r'}\right] g_{ij} \sigma_{ij}^{kl} \, d\Omega \, dc_i \, dc_j \qquad (10\text{ a})$$

for the elastic part of type I collisions, and by

$$\beta^{\,o r mnq}_{r m'n'q'} = -\frac{1}{4\pi^{3/2} Q_r Q_v} \int \left(1 - \delta_{ki}\delta_{lj}\right) \exp\left(-c_i^2 - c_j^2 - \mathcal{E}_{ir} - \mathcal{E}_{jr} - \mathcal{E}_{iv} - \mathcal{E}_{jv}\right)$$
$$\times \left[\psi^{r'}_{m'n'q'} Y_r\right]\left[\psi^{r}_{mnq} Y_r\right] g_{ij} \sigma_{ij}^{kl} \, d\Omega \, dc_i \, dc_j \qquad (10\text{ b})$$

for the inelastic part of type I collisions. In the last expression, Y_r are the irreductible tensors in peculiar velocity introduced by Waldmann[4].

The solutions are given by

$$a_{100} = \frac{5k}{2}\left[\beta^{1010}_{1010}\beta^{1001}_{1001} - \left(\beta^{1001}_{1001}\right)^2 - C_r\beta^{1010}_{1100}\beta^{1001}_{1001}\right] \Big/ \mathcal{D}$$

$$a_{010} = \frac{\beta^{1001}_{1001}\left[C_r\beta^{1100}_{1100} - (5k/2)\beta^{1010}_{1100}\right]}{\mathcal{D}}$$

$$a_{001} = \frac{\beta^{1001}_{1001}\left[(5k/2)\beta^{1010}_{1100} - C_r\beta^{1100}_{1100}\right]}{\mathcal{D}}$$

With

$$\mathcal{D} = \left(2nk/3\pi^{3/2}Q_rQ_v\right)\left\{\beta^{1001}_{1001}\left[\beta^{1100}_{1100}\beta^{1010}_{1100} - \left(\beta^{1010}_{1100}\right)^2\right] - \beta^{1100}_{1100}\left(\beta^{1001}_{1010}\right)\right\}$$

$$d_{100} = (C_r/C_{tr})\left(\pi^{3/2}Q_rQ_v/n\,\beta'^{0010}_{0010}\right)$$

$$d_{010} = (3k/2C_r)\,d_{100}\ ;\ f_{100} = C_v\beta^{1001}_{1001}\beta^{1010}_{1100}/\mathcal{D}$$

$$f_{010} = -C_v\beta^{1001}_{1010}\beta^{1100}_{1100}/\mathcal{D}$$

$$f_{001} = C_v\beta^{1100}_{1100}\beta^{1010}_{1010}/\mathcal{D}$$

$$b_{000} = \left(5\pi^{3/2}Q_rQ_v/n\,\beta^{2000}_{2000}\right)$$

$$g_{100} = \frac{1}{\pi^{3/2}Q_rQ_v\beta'^{0010}_{0010}}\left(\frac{3}{2}\frac{kC_rT_v}{C_{tr}^2T}I^{(0)}_{\varepsilon_{iv}} + \frac{C_r}{C_{tr}}I^{(0)}_{C_i^2}\right.$$

$$g_{010} = (3k/2C_r)\,g_{100}$$

Thus Ψ_i can be written as

$$\Psi_i = \left(a_{100}\psi_{100}^{3/2} + a_{010}\psi_{010}^{3/2} + a_{001}\psi_{001}^{3/2}\right)\left(\overrightarrow{\partial \ln T/\partial r}\right)\cdot\vec{a} + b_{000}\,\overset{\circ}{a}a + \left(d_{100}\psi_{100}^{1/2} + d_{010}\psi_{010}^{1/2}\right)\nabla u + g_{100}\psi_{100}^{1/2} + g_{010}\psi_{010}^{1/2} + \left(f_{100}\psi_{100}^{3/2} + f_{010}\psi_{010}^{3/2} + f_{001}\psi_{001}^{3/2}\right)\left(\overrightarrow{\partial \ln T_v/\partial r}\right)\cdot\vec{a} \quad (11)$$

Transport Coefficients

The transport coefficients are obtained classically by the knowledge of Ψ_i

$$\mu = -5\sqrt{\frac{m}{kT}}\;\frac{\pi^{3/2} Q_r Q_v kT}{\beta_{2000}^{2000}} \qquad \text{for the viscosity}$$

$$\nu = -\left(\frac{C_r}{C_{ti}}\right)^2\sqrt{\frac{m}{2kT}}\;\frac{\pi^{3/2} Q_r Q_v kT}{\beta_{2000}^{2000}} \qquad \text{for the bulk viscosity}$$

$$\vec{q}_t = \frac{5kn}{4}\sqrt{\frac{2kT}{m}}\left[\left(a_{100}\frac{\partial T}{\partial r} + f_{100}\frac{\partial T_v}{\partial r}\right)\right] \qquad \text{for heat flux due to translational motion}$$

$$\vec{q}_r = -\frac{nC_r}{2}\sqrt{\frac{2kT}{m}}\left(a_{010}\frac{\partial T}{\partial r} + f_{010}\frac{\partial T_v}{\partial r}\right) \qquad \text{for rotational heat flux}$$

$$\vec{q}_v = \frac{nC_v}{2}\sqrt{\frac{2kT}{m}}\left(a_{001}\frac{\partial T}{\partial r} + f_{001}\frac{\partial T_v}{\partial r}\right) \qquad \text{for vibrational heat flux}$$

Relaxation Phenomena

1. Translational-Rotational Relaxation

It is not a relaxation equation that gives the translation temperature T_t, but a new distribution of the energy between translation and rotational energy. We find

$$(3k/2)(T - T_t) = (E_r - \bar{E}_r) = \nu\,\nabla W - nkT g_{100}$$

2. Vibrational Relaxation

In the Boltzmann equation, Ψ_i is expressed in the form $f_i = f_i^{(0)} + f_i^{(0)}\Psi_i$. Multiplying by $\mathcal{E}_{iv}$, integrating on the velocity and summing on the internal states, the following equation is obtained

$$\frac{\partial E_v}{\partial t} + u\frac{\partial E_v}{\partial x} = -\frac{1}{n}\frac{\partial}{\partial x}\sum_i \int E_{i_v} c_i f_i^{(0)} dc_i + \frac{1}{n}\sum \int E_{i_v} J_{II}^{(0)} dc_i$$
$$+ \frac{1}{n}\sum \int E_{i_v} J_{II}^{(1)} dc_i \tag{12}$$

where

$$\sum_i \int E_{i_v} J_{II}^{(0)} dc_i = \sqrt{\frac{2kT}{m}} \frac{n^2 kT_v}{\pi^{3/2} Q_r Q_v} I_{\varepsilon_{iv}}^{(0)}$$

$$\sum_i \int E_{i_v} J_{II}^{(1)} dc_i = \sqrt{\frac{2kT}{m}} \frac{n^2 kT_v}{\pi^{3/2} Q_r Q_v} \sum_i \int \exp(-c_i^2 - \varepsilon_{ir} - \varepsilon_{iv}) \mathcal{L}_1(\Psi_i)\, \varepsilon_{iv}\, dc_i$$

On the right-hand side of the preceding equation, $\mathcal{L}_1(\Psi_i)$ is

$$\mathcal{L}_1(\Psi_i) = \sum_i \int \exp\left(-c_j^2 - \varepsilon_{jr} - \varepsilon_{jv}\right) [\Psi_i]\, g_{ij}\, \Gamma_{ij}^{kl}\, d\Omega\, dc_j \tag{13}$$

The last term on the left-hand side of Eq. (12) is the vibrational heat flux divergence. So, by putting the expression Ψ_i of (11) in Eq. (13), the evolution of the vibrational energy is defined completely. To integrate Eq. (13), the following assumption is made: the differential cross section of type II collisions is assumed to be[5]

$$g^2 \Gamma_{ij}^{kl} = g F(\chi) T_{ij}^{kl} H(-\Delta\varepsilon) + g' F(\chi) T_{kl}^{ij} H(+\Delta\varepsilon)$$

For monatomic gases, it is a common practice to make the assumption of a Maxwell force law to simplify the cross section; for polyatomic gases, no force law yields corresponding simplifications. However, the preceding form satisfies the symmetry requirements of the principle of detailed balancing, while possessing the simple velocity dependence analogous to what Hulembeck calls a quasi-Maxwell model[6]. The translational and internal motion have been separated into a deflection angle function $F(\chi)$ and a transition probability T_{ij}^{kl}, which depends only on the energy levels. $H(\Delta\varepsilon)$ is a heavyside step function. This simplification allows us to perform the integration using the spherical symmetry properties of $\Gamma_{ij}^{k\ell}$, and hence the relaxation equation can be written as

$$\frac{dE_v}{dt} = -\frac{1}{n}\frac{\partial \vec{q}_v}{\partial r} + \sqrt{\frac{2kT}{m}}\,\frac{nkT_v}{\pi^{3/2}Q_rQ_v}\, I^{(0)}_{\varepsilon_{iv}} + nkT_v\left(d_{100}\nabla U + g_{100}\right)\Lambda^{0100}_{0001}$$

$$+\frac{3k}{2C_r}\, nkT_v\left(d_{100}\nabla U + g_{100}\right)\Lambda^{0010}_{0001}$$

where

$$\Lambda^{rmnq}_{rm'n'q'} = -\frac{1}{4}\,\frac{2kT}{m}\,\frac{1}{\pi^{3/2}Q_rQ_v}\sum\int \exp\left(-c_i^2 - c_j^2 - \varepsilon_{ir} - \varepsilon_{jr} - \varepsilon_{iv} - \varepsilon_{jv}\right)$$

$$\times\left[\psi_{rmnq}Y_r\right]\left[\psi_{rm'n'q'}Y_r\right] g_{ij}\,\Gamma_{ij}^{k\ell}\, d\Omega\, dc_i\, dc_j \qquad (14)$$

As the total energy is conserved during a type II collision, the integral Λ^{0100}_{0001} can be written as

$$\Lambda^{0100}_{0010} = \Lambda^{0010}_{0001} + (T_v/T)\,\Lambda^{0001}_{0001}$$

As the energy exchanges $\Delta\varepsilon_r$ and $\Delta\varepsilon_v$ can be either positive or negative, it is possible to neglect their variation compared with that of $(\Delta\varepsilon_r)^2$ and $(\Delta\varepsilon_v)^2$. Thus, the preceding coefficients are approximated by[7]

$$\Lambda^{0100}_{0010} \simeq (T_v/T)\,\Lambda^{0001}_{0001}$$

$$\Lambda^{0010}_{0001} \simeq 0$$

The relaxation equation now is written as

$$\frac{dE_v}{dt} = -\frac{1}{n}\frac{\partial \vec{q}_v}{\partial r} + \sqrt{\frac{2kT}{m}}\,\frac{nkT_v}{\pi^{3/2}Q_rQ_v}\, I^{(0)}_{\varepsilon_{iv}} + nk\frac{T_v^2}{T}\left(\frac{m}{2kT}d_{100}\right.$$

$$\left. + g_{100}\right)\Lambda^{0001}_{0001}$$

where

$$I^{(0)}_{\mathcal{E}_{iv}} = \sum_i \int \mathcal{E}_{iv} \Big[\exp\left(-C_k^2 - C_\ell^2 - \mathcal{E}_{kr} - \mathcal{E}_{\ell r} - \mathcal{E}_{kv} - \mathcal{E}_{\ell v}\right) - \exp\left(-C_i^2 - C_j^2 - \mathcal{E}_{jr} - \mathcal{E}_{ir} - \mathcal{E}_{jv} - \mathcal{E}_{iv}\right) \Big] g_{ij} \Gamma_{ij}^{k\ell} \, d\Omega \, dc_i \, dc_j$$

The term Λ^{0001}_{0001} corresponds to vibrational characteristic time of type II collisions. The relaxation equation for the vibrational energy has been obtained, showing its dependence on the ratio $\Lambda^{0001}_{0001}/\beta'^{0010}_{0010}$ (β'^{0010}_{0010} being contained in d_{100}). The term β'^{0010}_{0010} can be related to a rotational relaxation time; it is possible, using a simple harmonic oscillator model, to calculate the integral Λ^{0001}_{0001}.

When such a model is considered the term $I^{0}_{\mathcal{E}_{iv}}$ reduce to the classical landau-Teller formula. When the gas is at the total equilibrium at the zeroth-order dE_v/dt is undetermined (there is no relaxation equation) and the difference between the vibrational energy (E_v) and the equilibrium vibrational energy ($\bar{E}_v$) is given by a term proportional to ∇u. (as it is written in the present situation for translation and rotation).

Conclusion

By extending the Chapman-Enskog theory to a zeroth-order vibrational nonequilibrium diatomic gas, transport coefficients as well as a simple first-order vibrational relaxation equation can be obtained. When the classical model equation is applied to this particular case[1], it gives correct Prandtl and Lewis numbers but leads to quite an untractable first-order relaxation equation. So, it has been shown that Chapman-Enskog theory and the current model equation give different results for a vibrational nonequilibrium gas. Consequently, a new kinetic model has been constructed to give a better understanding of relaxation phenomena and will be published elsewhere.

Appendix: $\beta^{rmnq}_{rm'n'q'}$ and $\Lambda^{rmnq}_{rm'n'q'}$ Coefficients

As these coefficients depends on the collision cross section, we write them in the notations of Mason and Monchick[7].

To perform the integration, some relations of the inelastic encounter, as written below, are used

$$\vec{c}_i = \vec{G}_0/4 + \vec{g}/2$$

$$\vec{c}_j = \vec{G}_0/4 - \vec{g}/2 \qquad c_i^2 - c_j^2 = 1/2\, \vec{G}_0 \cdot \vec{g}$$

$$\vec{c}_k = \vec{G}_0/4 + \vec{g}'/2 \qquad c_k^2 - c_l^2 = 1/2\, \vec{G}_0 \cdot \vec{g}'$$

$$\vec{c}_l = \vec{G}_0/4 - \vec{g}'/2 \qquad d\vec{c}_i\, d\vec{g} = [(4\pi)^2/64]\, G_0^2 g^2\, dG_0\, dg$$

$$\vec{g}\cdot\vec{g}' = \|\vec{g}\|\cdot\|\vec{g}'\| \cos\chi \qquad g^2 = g'^2 + 2\Delta\varepsilon$$

$$\Delta\varepsilon_v = \varepsilon_{kv} + \varepsilon_{lv} - \varepsilon_{iv} - \varepsilon_{jv}$$

$$\Delta\varepsilon_r = \varepsilon_{kr} + \varepsilon_{lr} - \varepsilon_{ir} - \varepsilon_{jr}$$

$$\vec{\gamma} = \vec{g}/\sqrt{2}$$

Five-of-eight integration can be performed. We have, for example

$$\beta^{2000}_{2000} = -\frac{4\pi\sqrt{2}}{Q_r Q_v} \sum_{ijkl} \int \exp(-\varepsilon_{ir} - \varepsilon_{iv} - \varepsilon_{jr} - \varepsilon_{jv} - \gamma^2)$$
$$\times\left(\gamma^4 \sin^2\chi + \Delta\varepsilon_r^2/3 - \gamma^2 \sin^2\chi\right) \gamma^3 \sigma_{ij}^{kl}\, d\Omega\, d\gamma$$

$$\beta'^{0010}_{0010} = -\frac{\pi\sqrt{2}}{Q_r Q_v} \sum_{ijkl} \int \exp(-\varepsilon_{ir} - \varepsilon_{iv} - \varepsilon_{jr} - \varepsilon_{jv} - \gamma^2)(\Delta\varepsilon_r)^2$$
$$\times\, \gamma^3 \sigma_{ij}^{kl}\, d\Omega\, d\gamma$$

$$\Lambda^{0001}_{0001} = -\frac{2\pi}{Q_r Q_v} \sum_{ijkl} \int \exp(-\varepsilon_{ir} - \varepsilon_{jr} - \varepsilon_{iv} - \varepsilon_{jv})(\Delta\varepsilon_v)^2$$
$$\times\, \gamma^3 \Gamma_{ij}^{kl}\, d\Omega\, d\gamma$$

References

[1]Zappoli, B. and Brun, R., The Physics of Fluids (to be published).

[2]Zappoli, B. and Brun, R., "Model equation and transport coefficients for vibrational nonequilibrium gas", Ninth International Symposium on Rarefied Gas Dynamics, Edited by M. Becker and M. Fiebig, Göttingen, 1974, pp. A 11-1.

[3]Wang-Chang, C.S., and Uhlenbeck, G.E., University of Michigan Engineering Research Report, CM-681, 1951.

[4]Waldmann, L., "Ein mit dem linearen Stossoperator verknüpftes Eigenwertproblem", Hanbuck der Physic, Edited by S. Flügge, Springer Verlag, Berlin, 1958, pp. 368-370.

[5]Hanson, F.B., and Morse, T.F., "Kinetic Model for gas with Internal Structure", The Physics of Fluids, Vol. 10, Oct.1967 p. 345.

[6]Ford, C.W. and Uhlenbeck, G.E., Lectures in Statistical Mechanics, American Mathematical Society, Providance, R.I., 1963.

[7]Mason, E.A. and Monchick, L., "Heat Conductivity of Polyatomic and Polar gases", The Journal of Chemical Physics, Vol. 1 March 1962, p. 1622.

EFFECT OF VIBRATIONAL WALL ACCOMODATION ON SMALL SIGNAL GAIN IN CO_2-N_2-H_2O GASDYNAMIC LASER (GDL)

Nimai K. Mitra*
DFVLR, Cologne, West Germany
and
Martin Fiebig†
Gesamthochschule, Duisburg, West Germany

Abstract

The effects of viscosity, heat conduction, and wall condition involving accomodation of molecular vibrational energy on the flowfield and small signal gain in CO_2-N_2-H_2O gasdynamic lasers are numerically investigated. Slender channel equations are used to describe the flow, and the "two vibrational temperature relaxation model" proposed by Anderson for the gas mixture is employed. The relaxation rate equations are modified with proper diffusion terms. Conventional boundary conditions are used for velocities and the static temperature. Phenomenological wall boundary conditions, incorporating the wall accomodation coefficient, are proposed for vibrational temperatures. Flowfield and small signal gains have been calculated in 2-d, cold-wall nozzles with different accomodation coefficients. Results show that the small signal gain profiles near the wall in the supersonic region depend strongly on the vibrational accomodation coefficients. For GDL performance, the wall layer and vibrational accomodation will become crucial when the chracteristic Reynolds number is such that the wall layer fills on the order of 30% of the cross section, i.e., Re < 8000.

Presented as Paper 53 at the 10th International Symposium on Rarefied Gas Dynamics, Aspen, Colo., July 19-23, 1976. This work was supported by the Deutsche Forschungsgemeinschaft.

*Scientist, Institut fuer Angewandte Gasdynamik.
†Professor, Fachbereich Maschinenbau und Schiffstechnik.

Nomenclature

a_1, a_2, a_3	= mass fraction of CO_2, N_2, and H_2O in the mixture
e	= vibrational energy
G_o	= small signal gain
h	= static enthalpy
p	= static pressure
Pr	= Prandtl number
r_*	= throat height
r, z	= normal and axial coordinates, respectively
Re	= characteristic Reynolds number
T, T_s	= static (translational rotational) temperature
u, v	= axial and normal velocities, respectively
x_{H_2O}, x_{N_2}	= mole fraction of H_2O and N_2 in the mixture
α	= wall accomodation coefficient
μ	= viscosity
ρ	= density
ν_1, ν_2, ν_3	= characteristic vibrational frequencies for symmetric stretching, degenerate bending, and asymmetric stretching of CO_2 molecules
ν_4	= characteristic vibrational frequency for N_2 molecules
τ_v	= vibrational relaxation time
ϑ	= local wall inclination of the wedge nozzle
α_v, K_v	= wall accomodation and wall catalyticty for vibrational energy
τ	= τ_v/τ_o

Subscripts

o	= stagnation chamber condition
r	= rotational modes of motion
tr	= translational rotational modes of motion
1, vu	= lower laser level
2, vo	= upper laser level
w	= wall condition
*	= equilibrium values

Introduction

Gasdynamic lasers require a fast expansion in a large area ratio nozzle (hypersonic Mach number) of a high temperature gas mixture, and appropriate relaxation times of the vibrational modes to reach an inversion. Quasi-one-dimensional inviscid calculations[1] show that the small signal gain G_o is close to optimum when the scaling parameter ($p_o\ r_*$) is on the order of 10^3 N/m (1 atm - cm), and T_o and the expansion ratio are as high as possible. Consequently, the characteristic flow Reynolds number Re, defined as Re = $\sqrt{2h_o}\,\rho_o\, r_*/\mu_o$, becomes on the order of 10^3 to 10^5. For such moderate Reynolds number, viscosity and heat conduction may have significant effects on the flowfield and laser performance. The important physical mechanism associated with fast expansion is rapid cooling, which introduces nonequilibrium expansion necessary for population inversion; hence, one may expect that intense wall cooling of a GDL nozzle will accentuate the nonequilibrium phenomenon. Monsler and Greenberg[2] presented an approximate study and noticed that the inversion was destroyed in the wall layer for hot nozzle walls, whereas wall cooling reduced that negative effect although cooling means a loss of thermal energy. The essential wall effects for a gasdynamic laser are the destruction of the inversion of energy modes due to viscous heating and vibrational energy transfer to the wall. The latter will depend on the gas surface interaction with regard to the different vibrational energy modes.

Mitra and Fiebig[3] presented low Reynolds number vibrational nonequilibirum nozzle flow calculations with nitrogen as flow medium, and showed that the vibrational wall accomodation coefficient has a profound influence on the vibrational temperature profiles. The present state of knowledge about gas surface interaction for polyatomic molecules is far from complete. Oman[4] pointed out that the accomodation coefficient for vibrational temperature is expected to be very small. Unfortunately, Oman did not mention any numerical values. In Ref.5, calculations were made of nozzle wall heat-transfer coefficients for vibrationally relaxing nitrogen flow. The comparison with experimental results showed that, for cold nozzle walls, vibrational accomodation coefficients are indeed expected to be very small.

The purpose of the present work is to use an approximate form of the full Navier-Stokes and energy equations namely, the slender channel equations a consistent set of vibrational relaxation equations, and phenomenological wall boundary conditions incorporating accomodation coefficients for vibrational temperatures to calculate flowfields and small signal gains in a typical GDL nozzle and also to investigate the influences of the vibrational wall accomodation coefficients on the gain profiles.

Basic Equations

The conventional method for nozzle flow computation is to divide the flowfield into an inviscid core and a viscous boundary layer. This procedure is messy and may need a number of simplifying assumptions: 1) The flow is inviscid and in equilibrium up to the geometric throat, which coincides with the saddle point, and the wall effect in the subsonic part can be neglected: 2) Flat plate boundary layer theory can be used along with local similarity, and the wall curvature effect may be neglected.

One can avoid these simplifications by attempting to solve the full Navier-Stokes, energy, and a consistent set of relaxation equations. Such a procedure will be extremely difficult and has never been reported to our knowledge. An order-of-magnitude analysis shows, however, that slender channel equations may be adequate to describe the flowfield in a nozzle for which the ratio of nozzle length and characteristic lateral dimension (throat height for example) is large, and the characteristic flow Reynolds number is at least on the order of the square of this ratio.[6,7] This restriction can be met even in a typical GDL nozzle, i.e., a contoured minimum length 2-d nozzle. In such a nozzle, the angle of expansion is comparatively large in a small region of the supersonic portion. Therefore, the slender channel assumption for a high Mach number nozzle may be violated locally, and still be valid on a global basis.

To analyze the vibrational relaxation of the CO_2-N_2-H_2O gas mixture, Anderson's model[1] was employed. The use of this model was not imperative in the present work, and the more detailed models of Basov[8] or Munjee[9] could be used without difficulty. Anderson's model was chosen for its simplicity and computational economy. This vibrational relaxation model is used in conjunction with Lunkin's[10] formulation of vibrational relaxation equations for viscous flow of a gas mixture. Lunkin[10] treated each vibrational mode in the mixture as an individual component without any internal degree of freedom. The resulting energy and rate equations are modified with proper diffusion terms for vibrational energies. We derived three vibrational rate equations for the CO_2 molecules (ν_1, ν_2, ν_3 modes) and one for the N_2 molecules (ν_4 mode), assuming that the molecules are harmonic oscillators, Boltzmann's distribution exists in each mode, pressure and thermo-diffusion can be neglected, the gas mixture is homogeneous everywhere, and that it behaves as a thermally perfect gas. Then following Anderson,[1] we further assumed that the ν_1 and ν_2 modes can be represented by one vibrational temperature T_{vu} and the ν_3 and ν_4 modes correspondingly by T_{vo}. We added up the rate equations for the ν_1 and ν_2 modes and the ν_3 and ν_4 modes, and arrived at two rate equations; one for the ν_{12} mode, defined by the vibrational energy or temperature e_1 and T_{vu}, and the other for the ν_{34} mode, defined by e_2 or T_{vo}. The forcing functions on the right-hand side of these rate equations could have been described by the detailed scheme of Basov,[8] but for simplicity we replaced

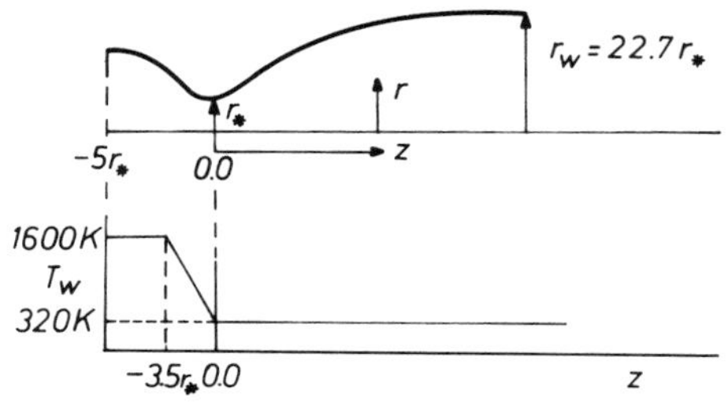

Fig. 1 Schematic of nozzle geometry, coordinate system and assumed nozzle wall temperature distribution.

them with those given by Anderson.[1] The vibrational temperatures of the different modes of the water molecules are assumed to be in equilibrium with the static temperature, i.e., the translational rotational temperature of the mixture.

The slender channel equations are formally identical to the boundary layer equations, except that the axial pressure gradient is not known a priori, and must be calculated at every step. For a GDL flow in a 2-d nozzle (Fig.1), these equations read as follows

Continuity

$$\frac{\partial(\rho u)}{\partial z} + \frac{\partial(\rho v)}{\partial r} = 0 \tag{1}$$

Momentum

$$\rho u \frac{\partial u}{\partial z} + \rho v \frac{\partial u}{\partial r} = - \frac{dp}{dz} + \frac{\partial}{\partial r} (\mu \frac{\partial u}{\partial r}) \tag{2}$$

Energy

$$\rho u \frac{\partial h}{\partial z} + \rho v \frac{\partial h}{\partial r} = u \frac{dp}{dz} + \frac{\partial}{\partial r}\left[\frac{\mu}{Pr_{tr}} \frac{\partial h_{tr}}{\partial r} + \frac{\mu}{Pr_1} \frac{\partial e_1}{\partial r} + \frac{\mu}{Pr_2} \frac{\partial e_2}{\partial r}\right] + \mu \left(\frac{\partial u}{\partial r}\right)^2 \tag{3}$$

Rates

$$\rho u \frac{\partial e_1}{\partial z} + \rho v \frac{\partial e_1}{\partial r} = \rho \frac{e_{*1} - e_1}{\tau_{v_1}} + \frac{\partial}{\partial r} \left(\frac{\mu}{Pr_1} \frac{\partial e_1}{\partial r} \right) \tag{4}$$

$$\rho u \frac{\partial e_2}{\partial z} + \rho v \frac{\partial e_2}{\partial r} = \rho \frac{e_{*2} - e_2}{\tau_{v_2}} + \frac{\partial}{\partial r} \left(\frac{\mu}{Pr_2} \frac{\partial e_2}{\partial r} \right) \tag{5}$$

The viscosity is assumed to be a function of the static temperature, and is calculated from mixture rules.[11] The derivations of the rate equations define the vibrational Prandtl numbers Pr_1 and Pr_2 of ν_{12} and ν_{34} modes as

$$Pr_1 = \frac{c_{v_1} + 2c_{v_2}}{\lambda_{v_1} + 2\lambda_{v_2}} \mu \qquad Pr_2 = \frac{c_{v_3}\, a_3 + c_{v_4}\, a_4}{\lambda_{v_3}\, a_3 \quad \lambda_{v_4}\, a_4} \mu$$

where a_1, a_2, and a_3 are the mass fractions of carbon dioxide, nitrogen, and water vapor in the mixture, and c_{v_n} and λ_{v_n} are vibrational specific heat and heat conductivity of the nth mode.

The vibrational heat conductivity of a component in the mixture can be calculated from complex formulas[10] involving coefficient of self and binary diffusions, which may be calculated from formulas given in Ref.12; however for simplicity, we assumed Pr_t, Pr_1, and Pr_2 to be constant and equal to 0.72.

Boundary Conditions

On the nozzle axis, i.e. at $r = 0$, $v = 0$, and symmetry conditions for other dependent variables were used. On the wall, i.e., at $r = r_w$, the usual boundary conditions for velocities and static temperature for no-slip flow were used. They are

$$u = 0 \qquad v = 0 \qquad T = T_w$$

Following the practice of wall recombination boundary conditions for dissociated flow, a phenomenological boundary condition incorparating the effect of wall catalyticity was proposed in Ref.5 for the vibrational temperature of nitrogen on the wall. This boundary condition reads

$$e_w - e_{*w} = \left[\frac{\mu}{K_v \rho Pr_v} \cos\vartheta \frac{\partial e}{\partial r} \right]_w \qquad (6)$$

Here K_v, the wall catalyticity, has the dimension of velocity, and is zero for a noncatalytic wall and ∞ for a fully catalytic wall. Indirect comparison of calculated and experimental nozzle wall heat-transfer coefficients show that for nitrogen flow in a cold wall nozzle K_v is expected to be very small.[5] For low Reynolds number flow, first-order velocity slip and temperature jump boundary conditions can be used. For vibrational temperature, a jump condition similar for static temperature was derived from the elementary principle of slip flow in Ref.3. This reads as

$$e_w - e_{*w} = \frac{2-\alpha_v}{2\alpha_v}\left[\sqrt{\frac{2\pi}{RT}}\ \frac{\mu}{\rho\ Pr_v}\cos\vartheta\ \frac{\partial e}{\partial r}\right]_w \qquad (7)$$

where R is the specific gas constant.

A comparison of Eqs.(6) and (7) shows that the wall catalyticity K_v should depend on molecular speed $\sqrt{RT}$ and accomodation coefficient α_v. The following qualitative features concerning vibrational relaxation and α_v may be pointed out here:

1) The number of collisions required for vibrational equilibriation is roughly three orders of magnitude larger than that for rotational relaxation, and the local vibrational relaxation time can be used as a measure of this number.

2) Although the physics of vibrational energy exchange in a gas surface interaction may be different from the physics of rotational energy exchange, as a first rough estimate one may assume that α_v is on the order of the rotational accomodation coefficient α_r times the ratio of rotational to vibrational relaxation time. Therefore we may write

$$e_w - e_{*w} = \frac{2-\alpha_r(\tau_r/\tau_v)}{2\alpha_r(\tau_r/\tau_v)}\left[\sqrt{\frac{2\pi}{RT}}\ \frac{\mu}{\rho Pr_v}\ \frac{\partial e}{\partial r}\cos\vartheta\right]_w \qquad (8)$$

where τ_r is the rotational relaxation time. In GDL flow, CO_2 molecules have three vibrational modes, and they may have different wall accomodations. Any theoretical calculation of accomodation coefficients of individual modes will need modelling of the interaction of molecular modes with the wall lattice structure. Such models, in the end, may become highly approximate with limited scope. No experimental data have been reported to our knowledge with regard to vibrational energy accomodation, not to speak of the accomodation of individual vibrational modes. Since we grouped ν_1 and ν_2 modes together into ν_{12}, we would characterize the wall interaction with one accomodation coefficient α_{vu}, and similarly for ν_{34} we would use α_{vo}. In order to calculate T_{vo} and T_{vu} on the wall, we may use Eq.(8) for each mode. To be exact, one should calculate α_r and τ_r, and then compute τ_v. Ref.13 presents some experimental values of τ_r for N_2 and CO_2 which are 1.2 (10^{-9}) sec and 2.3 (10^{-9}) sec, respectively, at room temperature and atmospheric pressure; α_r probably is close to one. If we stick to the no-slip boundary condition, we use 2 for α_r to extract "no jump" of the rotational temperature boundary condition. Noting that $\tau_r \ll \tau_v$ downstream of the throat, we can use $\tau_v/\tau_r \approx \tau$, where τ is the local vibrational relaxation time nondimensionalized by the corresponding vibrational relaxation time at the stagnation chamber condition. In this way we can employ Eq.(7) for a fully catalytic wall by putting $\alpha_v = 2$, for a noncatalytic wall by using $\partial e/\partial r]_w = 0$, and any intermediate catalyticity by using $\alpha_v = 2/\tau$ for each mode of vibration of the GDL mixture, and no-slip boundary conditions for other dependent variables.

Method of Solution

Rae[14] proposed a scheme for the numerical solution of slender channel equations for frozen flow in an axisymmetric nozzle with "wall-slip" boundary conditions. This scheme was modified for frozen and vibrational nonequilibrium hypersonic nozzle flow[5,7] and for no-slip boundary conditions for both 2-d and axisymmetric nozzle.[15] This scheme has been developed further to calculate GDL flowfields in the present work. In the present version of this scheme, an implicit finite-difference technique (Crank Nicholson) with staggered step sizes in both normal and axial directions is a used for computational accuracy and economy. Computation was started $5r_*$ upstream of the throat, and the flow at the initial point was assumed to be in equilibrium. Initial profiles were obtained from asymptotic solutions of the flow equations for slow flow in a wedge.[16] The pressure gradient along the axis was calculated from the global continuity or streamtube relation. First, for a given nozzle with known initial and boundary conditions, a mass flow eigenvalue was calculated. For this mass flow a saddle point, through which the global change from sub- to supersonic condition took place, existed in the flowfield. Then a final run was made for this mass flow. The important features of this calculation are 1) the computation of nonequilibrium relaxation equations can be started anywhere upstream of the throat; 2) the wall effects upstream of the throat are taken into full account; 3) no assumptions regarding the mass flow rate or the location of the saddle point are necessary, and the computed values of these quantities will fully reflect the effect of overall initial and boundary conditions; 4) for experimental cases of GDL nozzle flow, where the stagnation chamber temperature cannot be measured directly, this method can be used to determine a realistic stagnation chamber temperature from experimental mass flow rate and stagnation chamber pressure.[5] The further details of this calculational procedure will be presented elsewhere.

Table 1 Initial conditions, characteristic Re, calculated inviscid and viscous mass flow rates $\dot{m}_{inv}$ and $\dot{m}_v$ respectively, for different runs

r_* mm	P_0 atm	T_0 K	X_{CO_2}	X_{H_2O}	Re	$\dot{m}_{inv}$ g/s	$\dot{m}_v$ g/s
1.1	10	1600	0.07	0.01	82350	1.07110	1.07440
1.1	1	1600	0.07	0.05	8086	0.10684	0.10682
0.11	5	1600	0.07	0.05	4043	...	0.05345

Results and Discussion

Results have been obtained for flow in 2-d contoured, nearly minimum length nozzles, with a nominal exit Mach number of five. The contour was obtained by the method of characteristics and boundary-layer-displacement thickness correction for frozen flow.

Figure 1 shows a schematic of the nozzle geometry and the assumed wall temperature distribution. Equilibrium computations have been started at $z = -5\ r_*$, and solutions of the relaxation rate equations have been started at $z = -4\ r_*$. The wall cooling has been started at $z = -3.5\ r_*$, so that the effect of cold wall on vibrational temperature can be seperated from that on static temperature. The onset of nonequilibrium has been noticed approximately 2 r_* upstream of the throat in the core, but near the wall at 3.5 r_* upstream when less than full catalyticity has been used. Many qualitative features of these GDL flows are similar to the nonequilibrium notrogen flow, as reported in Ref.3. Computations have been made with full, partial, and zero wall catalyticities, with various initial conditions shown in Table 1. The wall catalyticities have no significant influence on the mass flow rate $\dot{m}_v$. The inviscid mass flow rates $\dot{m}_{inv}$ have been obtained from quasi-one-dimensional inviscid calculations.[1] The small signal gain on the axis for viscous flow may become 1 to 2% larger than the gain in inviscid flow. This also was noticed in the approximate calculations of Komimoto et al.[17] Further discussion of this phenomenon will be reported elsewhere.

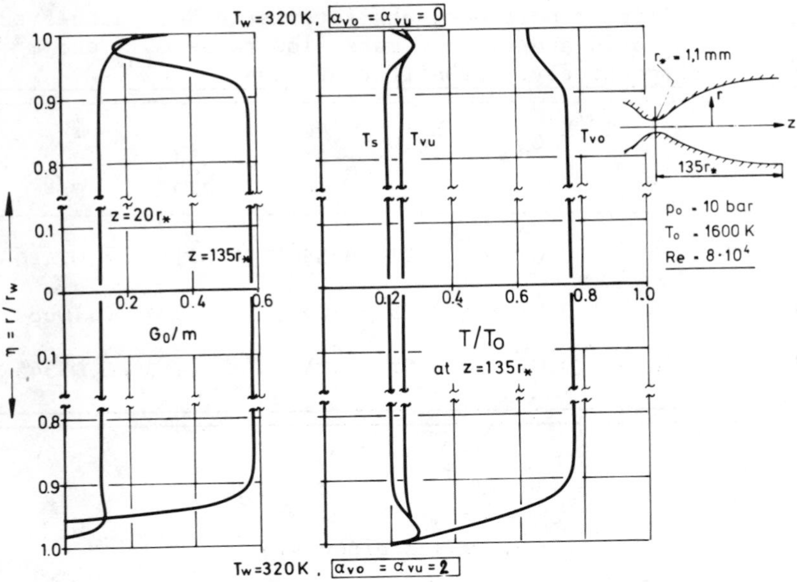

Fig. 2 Gain and temperature profiles at exit.

Figure 2 shows, for the Re = 82350 case, G_o profiles at $z = 20\ r_*$ and at the nozzle exit, and temperature profiles at the exit for a noncatalytic wall [($\alpha_{vu} = \alpha_{vo} = 0$) upper part] and for a fully catalytic wall [($\alpha_{vo} = \alpha_{vu} = 2$) lower part]. Note that the upper and lower parts represent two different runs. For a catalytic wall, calculations show no inversion or gain in the wall layer, whereas for noncatalytic wall the inversion and gain in the wall layer can be larger than that in the core in the early supersonic part ($z = 20\ r_*$). At the exit, the gain profile shows a dip near the wall. This dip corresponds to the bump in the lower laser level vibrational temperature T_{vu} caused by dissipative heating. It may be noted that, at the nozzle exit, the higher laser level temperature represents a relaxation profile between the frozen core and noncatalytic wall temperatures. The layer is a diffusive relaxing zone. Such behavior was noted with nitrogen flow in Ref.3. The lower laser level temperature profile shows that the zone connecting the wall layer and the core is near equilibrium. The reason lies in the smaller departure from equilibrium and the dissipative static temperature bump.

Figure 3 presents results for flow with Re = 8086, for a fully catalytic wall (lower part) and for fractional catalyt-

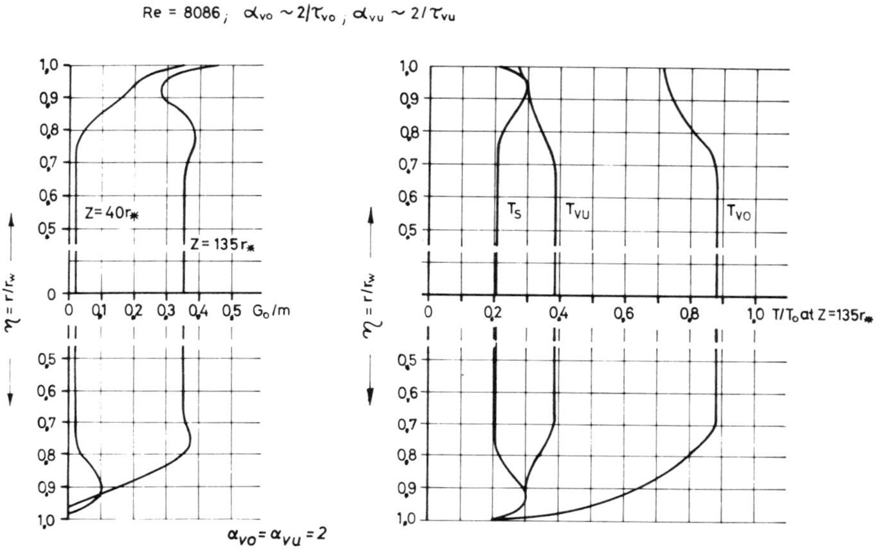

Fig. 3 Gain and temperature profiles at exit.

icity (upper part). The results are qualitatively similar to the Re = 82350 case, except that the wall layer is thicker and its influence is more pronounced.

Figure 4 shows results for Re = 4043 for the fractional wall catalyticity case. Note that, even though p_o is larger than in the Re = 8086 case, Damkoehler numbers defined as $(r_*/\sqrt{2h_o}\ \tau_o)$ for both ν_{12} and ν_{34} modes are smaller than corresponding values of the higher Re case. Hence, both T_{vu} and T_{vo} freeze at the axis at comparatively larger values, and their

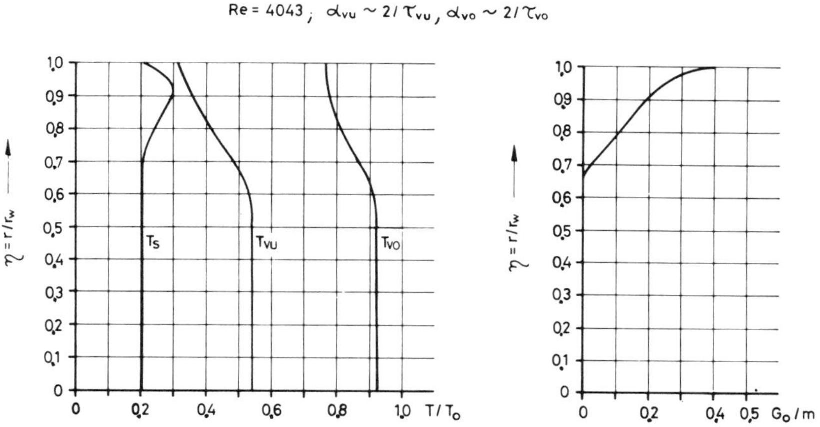

Fig. 4 Gain and temperature profiles at exit.

profiles show frozen layers at the core and wall connected by broad diffusive relaxing zones. At the exit, the inversion and gain appear only in the thick wall layer, and not in the core. Because of the large departure from equilibrium of T_{vu}, the dissipative heating has not produced any bump on the T_{vu} profile, and so a dip in the G_o profile is absent. An inviscid 1-d analysis does not, incidentally, show any gain in this case.

In conclusion, we note that the viscous nonequilibrium flow in GDL presents some interesting phenomena and important questions. The profiles of inverted populations and small signal gains become nonuniform. The gain profiles near the wall depend strongly on the gas surface interaction for the vibrational modes. Little wall accomodation and low wall temperature will counteract the negative effect of dissipative heating.

The calculation procedure of wall accomodation employed in this work to determine the partial wall catalyticity is a crude and better theoretical study, and experimental data are needed for these gas surface interactions. For GDL performance, the wall layers and vibrational accomodations will be of crucial importance only when the Re is such that the wall layer fills on the order of 30% of the cross section, i.e., Re < 8000. For such a flow, the results calculated with little wall accomodation show a gain in the thick wall layer, even when there is no gain in the core.

References

[1]Anderson, J.D., "High Power Gas Lasers," Lecture Series 65, Vol.1, von Karman Institute for Fluid Mechanics, Belgium, March 1974.

[2]Monsler, M.J. and Greenberg, R.A., "The Effects of Boundary Layers on the Gain of Gasdynamic Laser," AIAA Paper 71-24, New York, Jan. 1971.

[3]Mitra, N.K. and Fiebig, M., "Low Reynolds Number Vibrational Nonequilibrium Flow in Laval Nozzle," Proceedings of the 9th International Symposium on Rarefied Gas Dynamics, edited by M. Becker and Fiebig, M., 1974, pp. B.21-1 - B.21.-12.

[4]Oman, R.A., "Calculations of the Interactions of Diatomic Molecules with Solid Surfaces," Proceedings of the 5th International Symposum on Rarefied Gas Dynamics, edited by C.L. Brundin, 1967, pp.83-100.

[5]Mitra, N.K. and Fiebig, M., "Determination of Stagnation Chamber Temperature in High Enthalpy Nozzle Flows," AIAA Journal, Vol.14, March 1976, pp.406-408.

[6]Williams III, J.C., "Viscous Compressible and Incompressible Flow in Slender Channels," AIAA Journal, Vol.1, Jan. 1963, pp.186-195.

[7]Mitra, N.K. and Fiebig, M., "Low Reynolds Number Hypersonic Nozzle Flows," Zeitschrift für Flugwissenschaften, Vol.2, Feb. 1975, pp.39-45.

[8]Basov, N.G., Mikhaylov, V.G., Oraevskii, A.N., and Shcheglov, V.A., "Obtaining Population Inversion of Molecules in a Supersonic Stream of Binary Gas in a Laval Nozzle," Zhurnal Technicheskoi Fiziki, Vol.38, Dec. 1968, pp.2031-2041.

[9]Munjee, S.A., "Numerical Analysis of Gasdynamic Laser Mixture," Physics of Fluids, Vol.15, March 1972, pp.506-509.

[10]Lunkin, Yu.P., "Vibrational Dissociation Relaxation in a Multicomponent Mixture of Viscous Heat-Conducting Gases," Irreversible Aspects of Continuum Mechanics and Transfer Physical Characteristics in Moving Fluids, edited by H. Parkus and L.I. Sedov, Springer, New York, 1966, pp.229-236.

[11]Bird, R.B., Stewart, W.E., and Lightfort, E.N., Transport Phenomena, Wiley, New York, 1964.

[12]Hirschfelder, J.O., Curtiss, C.F., and Bird, R.B., Molecular Theory of Gases and Liquids, Wiley, New York, 1954.

[13]Zel'dovich, Ya.B. and Raizer, Yu.P., Physics of Shock Waves and High-Temperature Hydrodynamic Phenomena, Academic Press, New York, 1966.

[14]Rae, W.J., "Final Report on a Study of Low-Density Nozzle Flows, with Application to Micro- rust Rockets," Cornell Aeronautical Laboratory, Inc., CAL No. A I-2590-A-1, Dec. 1969.

[15]Mitra, N.K. and Fiebig, M., "Flow Field Computation through Sonic Singularity in Viscous Frozen and Nonequilibrium Nozzle Flow," Proceedings of the GAMM Conference on Numerical Methods in Fluid Mechanics, edited by Hirschel, E.H. and Geller, W., Cologne 1975, pp.98-106.

[16]Mitra, N.K. and Fiebig, M., "Comparison of Slender Channel Computations with Experiments for Low Reynolds Number Hyper-

sonic Nozzle Flows," DLR-FB 75-64, Wissenschaftliches Berichtswesen, DFVLR, Cologne 90, 1975.

[17]Kamimoto, G., Matsui, H., and Len, K.T., "A Study of CO_2 Gasdynamic Lasers Part 1, Theoretical and Numerical Analysis," Dept. of Aeronautical Engeneering, Kyoto University, C.P. 38, Sept. 1973, Kyoto, Japan.

THE ENERGY BALANCE AND FREE-JET EXPANSIONS OF POLYATOMICS

P. Poulsen* and D. R. Miller+

University of California at San Diego, La Jolla, Calif.

Abstract

This report concerns experiments in which the rotational temperature of N_2 is measured by the electron beam induced emission technique, and the mean velocity and translational temperature are measured by time-of-flight in the same experimental setup. The objective was to test the total energy balance equation directly, especially the often invoked assumption that the translational temperature is isotropic ($T_{11} = T_{\perp}$). Spectroscopic measurements are made downstream of the beam skimmer on the same molecular beam analyzed by time-of-flight. The data show that the energy balance is satisfied best with $T_{\perp} \rightarrow o$, as is expected from monatomic gas studies. Measurements on N_2/Ar and N_2/He mixtures also are made. An analysis of relaxation in a binary mixture gives the following inelastic cross-sections for rotational relaxation: $\pi\sigma^2(N_2\text{-}N_2) = 14$ Å^2, $\pi\sigma^2(N_2$ - Ar$) = 22$ Å^2, and $\pi\sigma^2(N_2$ - He$) = 3$ Å^2. In addition, we present a simplified model of translational relaxation in polyatomic flows which satisfactorily predicts the terminal translational temperature (T_{11}).

Presented as Paper 54b at the 10th International Symposium on Rarefied Gas Dynamics, July 19-23, 1976, Aspen, Colo. Research supported by NSF Grant No. ENG71-02434A05.

*Ph.D. Candidate, Department of Applied Mechanics and Engineering Sciences.

+Associate Professor, Department of Applied Mechanics and Engineering Sciences.

Introduction

Although freejet expansions of pure monatomic gases appear well-understood both theoretically and experimentally, expansions of polyatomics and mixtures are not yet quantitatively predictable. As Willis and Hamel[1] have shown, multiple relaxation scales complicate the theoretical analysis. Unambiguous experimental data also have been difficult to obtain because of the necessity to measure both the translational distribution of each species and the internal state distributions, simultaneously and independently. It appears that the best experimental results ultimately will be derived from a combination of spectroscopic and velocity selector measurements, and the best theoretical results will be found from Monte Carlo solutions to the kinetic equations. Unfortunately, the latter do not provide analytical results that are useful to the experimentalist for predicting flow (molecular beam) properties.

This paper reports on experiments that use the well-known method of electron-induced fluorescence spectroscopy to measure the rotational temperature of nitrogen in freejet expansions (including mixtures) coupled with an independent time-of-flight measurement of the translational temperature and mean velocity for the same expansions. It also reports on a simplified analysis that permits us to predict the translational temperature in the polyatomic expansions.

Energy Balance

One of our immediate objectives was to test the energy balance technique often used to obtain an estimate of the rotational temperature, T_r, from which the rotational collision number Z_r may be derived.[2] The ellipsoidal velocity distribution function $f(C) = n(\beta_{11}/\pi)^{\frac{1}{2}}(\beta_\perp/\pi) \exp\{-\beta_{11}(C_{11} - V)^2 - \beta_\perp(C_\perp)^2\}$, where V is the mean velocity, C the molecular velocity, $\beta = m/2kT$, and subscript 11 and $\perp$ refer to the parallel and perpendicular components with respect to the radial streamline direction, respectively, has proved useful in freejet calculations[3,4]; this is despite the fact that it does not correctly fit the tails of the velocity distribution,[5] and especially the perpendicular component.[6] For this model of the translational energy distribution, the total energy balance for a diatomic gas, including rotational energy and excluding vibrational energy, is

$$(7/2)kT_o = (1/2)mV^2 + (3/2)kT_{11} + kT_\perp + kT_r \qquad (1)$$

It should be noted that this energy balance is rigorous in general provided that the radial part of the distribution function is even in thermal velocity or that the hypersonic approximation is satisfied.[1]

In using freejet expansions to study rotational relaxation, the velocity V and temperature T_{11} are measured by the time-of-flight technique. It has been customary to assume that $T_{\perp} = T_{11}$. Thus, knowing the stagnation temperature T_o, the rotational temperature T_r may be calculated from the energy balance, Eq. (1). Since this technique is convenient and useful, we felt that it would be important to test not only the energy balance as written in Eq. (1) but also the validity of equating $T_{\perp}$ with T_{11}. Therefore, we set up the experiment shown in Fig. 1 to measure not only V and T_{11} by time-of-flight, but also T_r by the electron beam fluorescence technique for the same expansion. The value of $T_{\perp}$ then may be obtained directly from Eq. (1).

A freejet source with a 0.028-cm-diam orifice was placed 1.9 cm from a conical skimmer, which had a 0.50-cm-diam opening. The stagnation temperature was monitored with a chromel-alumel thermocouple, which was calibrated by time-of-flight measurements on pure argon expansions at high source pressure. The nozzle-skimmer arrangement was left undisturbed throughout the experiment, with the only physical change in the experimental apparatus being the type of chopper disk

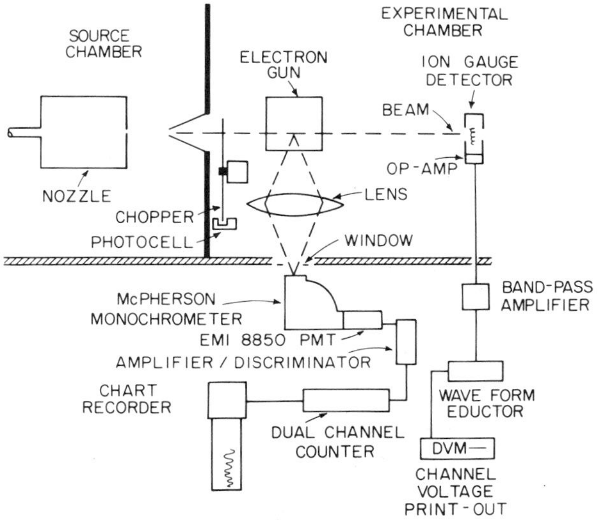

Fig. 1 Schematic of the experimental apparatus.

employed. The spectroscopic measurements were modulated by an equal time on-off chopper. For the time-of-flight measurements, the chopper disk employed had a diameter of 17.8 cm and 9 slits, 0.13 cm wide, evenly spaced on the perimeter. The chopper was operated at 130 rps, giving a gating time of 18 μsec. The flight path was 65 cm. The time-of-flight detector was a miniature open ionization gage, with a buffer-amplifier circuit built directly onto the collector element. For mixtures, we attempted to use an Extranuclear quadrupole for detection, but the results were never satisfactory when calibrated against the ion-gage for pure gases. This is in contrast to our previous studies with an Electronics Associates, Inc. quadrupole,[3] which did agree with ion-gage results. We have since learned that it is necessary to reduce the emission current below 1.0 ma to prevent excessive ion storage time in the Extranuclear ionizer for the time-of-flight experiment. The time-of-flight data were fit to the ellipsoidal velocity distribution by nonlinear least-squares analysis, which included the convolution of the gating functions, to obtain T_{11} and V.[7]

The fluorescence measurements of emission from the first negative system of N_2^+ were made in the manner of Scott et al.[8,9] Instead of probing a freejet expansion with a high-energy (25 keV) electron beam, we probe the molecular beam downstream of the skimmer with a low-energy electron gun (300-500 eV). The advantages of this procedure are the absence of secondary electron effects in the analysis and an easily constructed electron gun. Our gun utilized an indirectly heated cathode and formed a two-dimensional electron beam 0.3 cm wide x 1 cm long, in the direction of the beam, with a typical beam current of 40 mA. The optical arrangement consisted of a simple converging lens, which focused the collected emission onto a 0.3 m, f/5.3 McPherson monochromator. A slit width of 30 μ allowed sufficient separation of the rotational transition lines. The emission spectrum was detected by an EMI 8850 photomultiplier tube; the signal was processed by an SSR model 1110 dual channel counter, and was recorded by a strip chart recorder. The spectra were analyzed by digital computer in the manner originally suggested by Muntz[10] with no additional corrections for secondary electron effects; our source chamber pressure is low enough ($< 5 \times 10^{-5}$ Torr) that background effects should be negligible. Under our conditions, we had sufficient intensity to analyze the first 7 ± 1 rotational levels. The relative populations of the rotational energy levels up to the 7th level was sufficiently close to a Boltzmann distribution to allow the assignment of a rotational temperature.

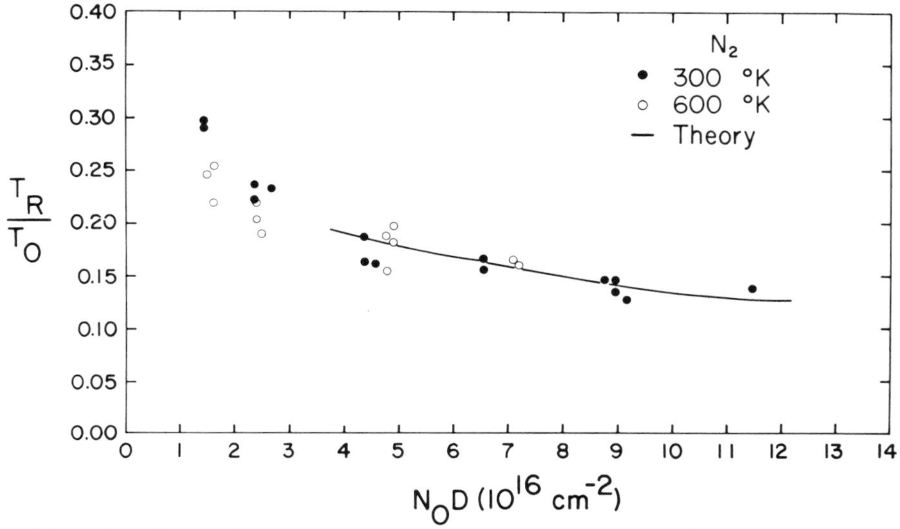

Fig. 2 The reduced rotational temperature T_R/To vs N_oD for pure N_2 expansions.

Two sets of runs were made for pure N_2 expansions; one set at $\sim 300^oK$ and one set at $\sim 600^oK$. The rotational temperatures obtained are shown in Fig. 2. The temperature is normalized by the stagnation temperature, and the stagnation conditions are expressed by the product of source density and orifice diameter $N_oD_n(cm^{-2})$, the expected correlating parameter for expansion processes dominated by bimolecular collisions with velocity independent cross-sections. These normalizations scale out the source temperature dependence in the data. The fact that the data points for the two source temperatures lie nearly on the same curve suggests that the rotational collision number Z_r is not a strong function of temperature in this region. Also shown in Fig. 2 are the results of our solution to the linear relaxation equation (discussed below) with a $Z_r = 3$.

By use of these T_r's and the corresponding T_{11}, and V from the time-of-flight measurements, we calculated $T_\perp$ from Eq. (1). The results are shown in Fig. 3. We find that, for reasonable pressures, $T_\perp \neq T_{11}$, but rather $T_\perp \to o$ as T_{11} becomes "frozen." In view of the similar behavior for monatomic gas expansions, this result is not surprising. Conversely, the energy balance can be used successfully for N_2 to predict T_r if we take $T_\perp = 0$, and if $N_oD > 3.3 \times 10^{16}$ cm^{-2} (this corresponds to $P_oD \gtrsim 1$ Torr cm at 300^oK).

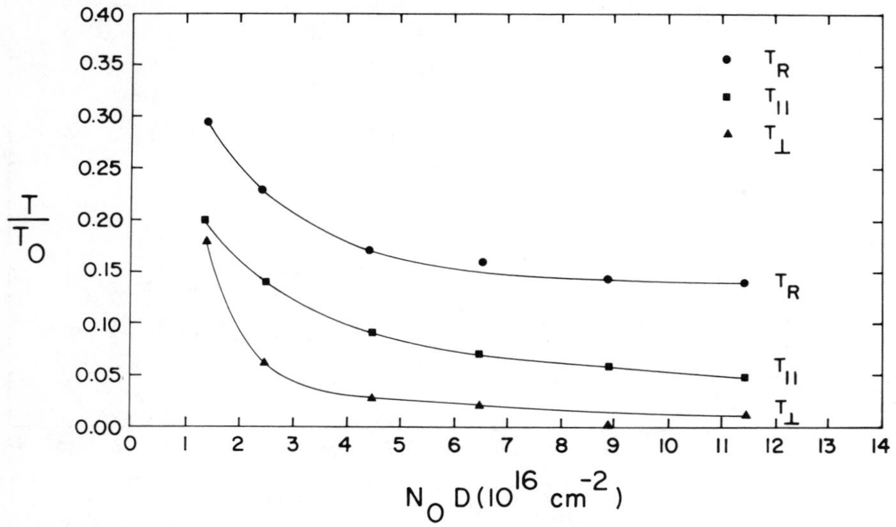

Fig. 3 The temperature ratios T_R/To, T_{11}/To, and $T_{\perp}$/To vs N_oD.

Relaxation in Mixtures

We also measured T_r for N_2 in mixtures of He/N_2 (50%) and Ar/N_2 (50%). These results are shown in Figs. 4 and 5. The

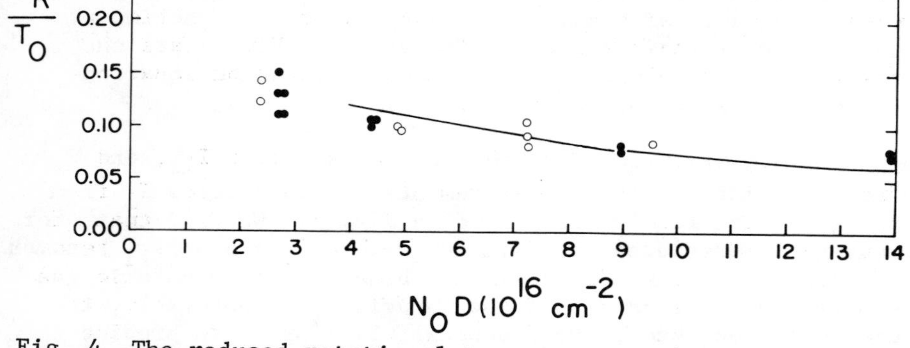

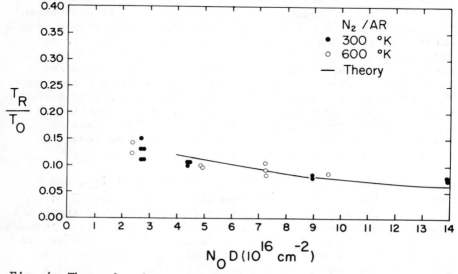

Fig. 4 The reduced rotational temperature T_R/To vs N_oD for 50% molar concentration of Ar in N_2.

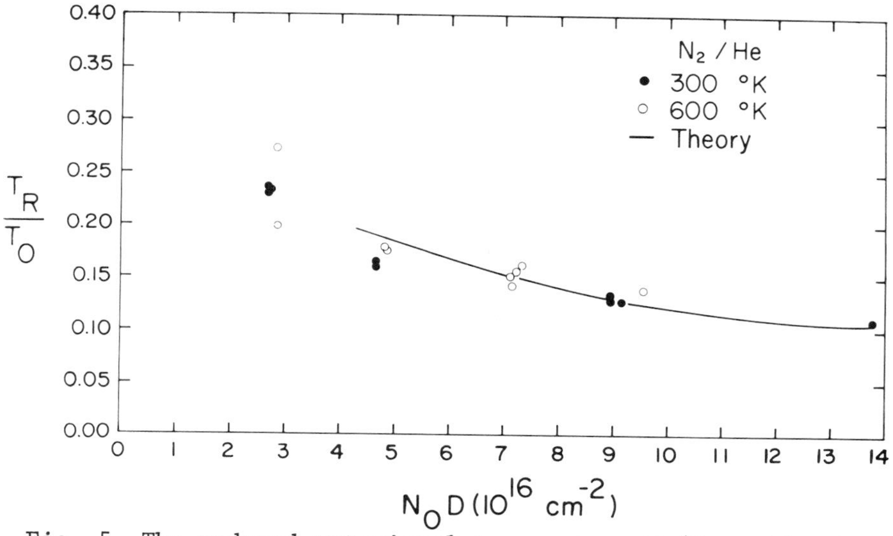

Fig. 5 The reduced rotational temperature T_R/To vs N_oD for 50% molar concentration of He in N_2.

theoretical curves that are shown are based on the following analysis, which is a straightforward extension of earlier work on relaxation of a pure gas.[2] Let $\nu_{ij} = n_j\sigma_{ij}^2(8\pi RT/\mu_{ij})^{\frac{1}{2}}$ be the collision frequency for species i in a bath of species j for a hard sphere of diameter σ_{ij}. The number density is denoted by n_j, and the reduced mass of species i and j are denoted by μ_{ij}. The equation describing the relaxation of rotational temperature in a binary mixture with rotational collision numbers Z_{AA} and Z_{AB}, for like and unlike collisions, respectively, is

$$DT_r/Dt = (\nu_{AA}/Z_{AA} + \nu_{AB}/Z_{AB})(T_t - T_R) \qquad (2)$$

where T_t is the translational temperature. In steady state, $DT_r/Dt = V(dT_r/dX)$, where X is the coordinate along a streamline. Normalizing static properties by the stagnation conditions and the distance by the orifice diameter D, defining the mole fraction of the diatomic molecule A to be α, and the total number density $N = n_i + n_j$, Eq. (2) becomes

$$\frac{d(T_r/T_o)}{d(X/D)} = \left[\frac{\alpha\sigma_{AA}^2}{Z_{AA}\mu_{AA}^{\frac{1}{2}}} + \frac{(1-\alpha)\sigma_{AB}^2}{Z_{AB}\mu_{AB}^{\frac{1}{2}}}\right](N_oD)\left(\frac{N}{N_o}\right)\left(\frac{8\pi RT}{V^2}\right)^{\frac{1}{2}}\left(\frac{T_t}{T_o} - \frac{T_r}{T_o}\right) \qquad (3)$$

For the present mixtures of a monatomic ($C_p = 5/2\ R$) and a diatomic ($C_p = 7/2\ R$) gas, the appropriate mixture heat capacity ratio is $\gamma = (5/2 + \alpha)/(3/2 + \alpha)$, the average molecular weight is $W = \alpha W_A + (1 - \alpha)W_B$, the speed of sound is $a = [\gamma RT/W]^{\frac{1}{2}}$, and the Mach number is $M = V/a$. Following Gallagher and Fenn,[2] the isentropic relations are used to evaluate (N/N_o) and (T_t/T_o). With the proper γ and M, one obtains

$$\frac{d(T_r/T_o)}{d(X/D)} = B\,\frac{1 - (T_r/T_o)\left\{1 + [(\gamma-1)/2]\,M^2\right\}}{\gamma^{\frac{1}{2}} M \left\{1 + [(\gamma-1)/2]\,M^2\right\}^{(\gamma/\gamma-1)}} \tag{4}$$

where

$$B = \left[\frac{\alpha\sigma_{AA}^{\;2}}{Z_{AA}\mu_{AA}^{\frac{1}{2}}} + \frac{(1-\alpha)\sigma_{AB}^{\;2}}{Z_{AB}\mu_{AB}^{\frac{1}{2}}}\right](8\pi W)^{\frac{1}{2}}(N_oD)$$

and

$$\mu_{AA} = W_A/2 \text{ and } \mu_{AB} = W_A W_B/(W_A + W_B)$$

For a given value of γ, isentropic characteristic solutions for M(X/D) are available,[11] and a universal solution of terminal rotational temperature ratio T_r/T_o vs B can be obtained by integrating Eq. (4) until T_r/T_o shows a negligible change (typically to $X/D \sim 20$). We have used a standard Runge-Kutta program to carry this out. Conversely, an experimental measurement of the terminal T_r gives B, from which the cross sections for rotational relaxation $\pi\sigma_{ij}^{\;2}/Z_{ij}$ can be computed. For pure N_2 (Fig. 2) we get best agreement for $\pi\sigma_{AA}^{\;2}/Z_A = 14\text{Å}^2$. Using this value for $\pi\sigma_{AA}^{\;2}/Z_{AA}$ in the analysis of the mixture data of Figs. 4 and 5, we find a best fit with $\pi\sigma_{ij}^{\;2} = 22\text{Å}^2$ for N_2/Ar and 3 Å^2 for N_2/He. The fit of this simple relaxation theory to the data is shown in Figs. 2, 4, and 5. The theoretical curves are not extended below $N_oD \sim 3 \times 10^{16}\ \text{cm}^{-2}$ because the free-jet expansions are very poor below this value and the validity of the characteristic solutions is in doubt. In order to compute the effective collision number, we must select σ_{ij} values. By use of typical hard-sphere values at room temperatures, $\sigma(N_2\text{-}N_2) = 3.68\text{Å}$, $\sigma(N_2\text{-Ar}) = 3.74\text{Å}$, $\sigma(N_2\text{-He}) = 2.7\text{Å}$, we calculate the following effective rotational collision numbers: $Z_{N_2\text{-}N_2} = 3 \pm 1$, $Z_{N_2\text{-Ar}} = 2 \pm 1$,

$Z_{N_2\text{-}He} = 8 \pm 2$. Had we evaluated the σ_{ij} at the temperatures appropriate to the freejet temperatures where rotational freezing occurs ($\sim 60^oK$), the σ_{ij} would be larger by factors up to two, and hence the collision numbers would be made larger artificially. Our value for pure N_2 is within the scatter of most previous data[2,12,13] (typically $Z_r = 4 \pm 1$ at 300^oK). We were not able to find experimental data in the literature with which to compare our mixture data. Theoretical calculations have been carried out for the N_2-Ar system by several authors. Pattengill[14] reports total rotational inelastic cross-sections for the N_2-Ar system of about $30 \pm 3Å^2$ at 300^oK. Further experiments over a larger range of temperatures and concentrations should be carried out before a quantitative comparison is made.

We have been able to correlate our numerical results for T_r as a function of B and γ by the following relation

$$(Tr/To) = (1.25) \left[(\gamma-1)/2\right] (4\gamma^{\frac{1}{2}}/B)^{2(\gamma-1)/\gamma} A^{2/\gamma} \qquad (5)$$

where A is the numerical factor used in correlations of the characteristic solutions for Mach number (e.g., A = 3.65 for $\gamma = 1.4$).[11] The form of Eq. (5) was determined by a sudden freeze approximation, and the coefficient 1.25 was determined by a best fit. Experimenters can use Eq. (5) to estimate T_r to within 10% for the range $3 < N_0D < 20. \times 10^{16}$ cm^{-2}.

Terminal Temperature of Polyatomic Gas

Although it appears that the prediction of rotational temperature is adequately treated by the previous analysis, at least for systems that can be treated with an effective T_r and Z_r, the prediction of T_{11} for polyatomic expansions seems more difficult. We have developed a simple scheme, which takes advantage of the successful correlations for monatomic gases and the relaxation analysis discussed previously.

We first use the preceding analysis, integrating Eq. (4) or using Eq. (5) to find T_r and the $(X/D)_r$, where rotational freezing occurs. We then assume that the flow is isentropic to this $(X/D)_r$, where T_r freezes, and then the flow behaves as a monatomic gas ($\gamma = 5/3$) expansion beyond $(X/D)_r$. Hypothetical N_0' and T_0' are calculated which, for an isentropic

$\gamma = 5/3$ expansion, would yield the same local N and T at $(X/D)_r$ as the polyatomic expansion, i.e. the same collision frequency. These N_0' and T_0' then are used directly in the monatomic gas correlations for T_{11}.

As an example, if we make the approximation that $M \gg 1$ then $M = A(X/D)^{\gamma-1}$ is the form of characteristic solution for the Mach number.[11] It is then straightforward to show that the monatomic gas Mach number, M_2, which would exist at the freeze point $(X/D)_r$, is

$$M_2 = A_2/A_1^{(\gamma_2-1)/(\gamma_1-1)} \left(2T_0/(\gamma_1-1)T_r\right)^{(\gamma_2-1)/2(\gamma_1-1)} \quad (6)$$

where subscript 1 corresponds to the polyatomic case and subscript 2 to the monatomic case. In the isentropic approximation, we also have $N = N_0(T_r/T_0)^{1/\gamma_1-1}$. Knowing $T = T_r$ from Eq. (5), M_2 from Eq. (6), and N at $(X/D)_r$ for a monatomic expansion permits a calculation of the N_0', T_0' from the isentropic relations

$$T_0' = T_r\left(1 + \frac{\gamma_2-1}{2} M_2^2\right) \quad (7a)$$

$$N_0' = N(T_0'/T_r)^{1/(\gamma_2-1)} \quad (7b)$$

We then use the following correlation for T_{11}

$$T_{11} \simeq 0.86\ T_0'/\left[N_0'D\,\sigma^2\right]^{0.90} = 1.43\ T_0'/\left[N_0'D(6C_6/kT_0')^{\frac{1}{3}}\right]^{0.90} \quad (8)$$

Equation (7) comes from the theory of Andres,[3] fit to the monatomic gas data. Although the correlations usually are written in terms of Mach number, we have used them to correlate T_{11} by setting $V \simeq (5RT_0')^{\frac{1}{2}}$ for monatomic gases, since in free-jet expansions with $N_0D > 3 \times 10^{16}$ cm^{-2} the velocity is within a few percent of this terminal value.

In Table 1 we show the results of this method, using Eqs. (5-8). The data are all for pure N_2, with $T_0 \sim 300^oK$ and we have used $\sigma = 3.68$ Å and $Z_r = 3$ to be consistent with our previous results. The data of Buck et al.[15] were obtained

Table 1 Prediction of terminal temperature for pure N_2 at $T_o = 300^oK$ ($\sigma = 3.68$ Å, Z = 3)

P_oD(Torr-cm)	T_{11}(exp)	T_{11}, Eq. (7)	T_r(exp)	T_r, Eq. (5)
2	21°K	18°K	47°K	55°K
2.7	17	15	44	46
3.6	13	12	40	39
6.5[15]	10.7	9		28
10[15]	5.8	7		22

with a velocity selector; the other data were obtained in the present study with time-of-flight; T_r(exp) was obtained spectroscopically.

For this range of data, the agreement seems satisfactory and suggests that this simplified analysis may be useful to experimentalists. As correlations for T_{11} in mixtures of monatomic gases[3] become available, Eq. (8) can be adjusted easily to predict T_{11} for polyatomics in mixtures.

To summarize the recommended procedure: (i) calculate T_r from Eq. (5) or numerical integration of Eq. (4); (ii) calculate T_{11} from Eq. (8), using either Eqs. (6,7a,7b) or the exact characteristic solutions[11] to get N_o' and T_o'; and (iii) calculate V by the energy balance, Eq. (1), with $T_\perp = 0$.

References

[1]Willis, R. D. and Hamel, B. B., "Non-Equilibrium Effects in Spherical Expansions of Polyatomic Gases and Gas Mixtures," Proceedings of the Fifth International Symposium on Rarefied Gas Dynamics, Vol. 1, 1967, pp. 837-860.

[2]Gallagher, R. J. and Fenn, J. B., "A Free-Jet Study of the Relaxation of Molecular Nitrogen from 300 - 1000°K," Proceedings of the Ninth International Symposium on Rarefied Gas Dynamics, Vol. 1, 1974, pp. B19-1 - B19-10.

[3]Miller, D. R. and Andres, R. P., "Translational Relaxation in Low Density Supersonic Jets," Proceedings of the Sixth

International Symposium of Rarefied Gas Dynamics, Vol. 2, 1968, pp. 1385-1397.

[4]Miller, D. R., Toennies, J. P., and Winkelman, K., "Quantum Effects in Highly Expanded Helium Nozzle Beams," Proceedings of the Ninth International Symposium on Rarefied Gas Dynamics, Vol. 2, 1974, pp. C.9-1 - C.9-9.

[5]Cattolica, R., Robben, F., and Talbot, L., "The Ellipsoidal Velocity Distribution Function and Translational Non-Equilibrium," Proceedings of the Ninth International Symposium on Rarefied Gas Dynamics, Vol. 1, 1974, pp. B.16-1 - B.16-11.

[6]Edwards, R. H. and Cheng, H. K., "Distribution Function and Temperatures in a Monatomic Gas Under Steady Expansion into a Vacuum," Proceedings of the Fifth International Symposium on Rarefied Gas Dynamics, Vol. 1, 1967, pp. 819-835.

[7]Subbarao, R. B. and Miller, D. R., "Velocity Distribution Measurements of 0.06 - 1.4 ev Argon and Neon Atoms Scattered from the (111) Plane of a Silver Crystal," The Journal of Chemical Physics, Vol. 58, June 1973, pp. 5247-5257.

[8]Scott, P. B. and Mincer, T., "Measurement of the Rotational State Distribution of a Molecular Beam," Entropie, No. 30, Dec. 1969, pp. 170-173.

[9]Scott, P. B., Mincer, T. R., and Muntz, E. P., "The Electron Beam Fluorescence Method for Molecular Scattering Experiments," The Review of Scientific Instruments, Vol. 45, Feb. 1974, pp. 207-209.

[10]Muntz, E. P., "Measurement of Rotational Temperature, Vibrational Temperature, and Molecule Concentration, in Non-Radiating Flows of Low Density Nitrogen," University of Toronto, UTIA Rep. no. 71, 1961.

[11]Anderson, J. B., "Inviscid Free Jet Flow with Low Specific Heat Ratios," AIAA Journal, Vol. 10, 1972, pp. 112-115.

[12]Miller, D. R. and Andres, R. P., "Rotational Relaxation of Molecular Nitrogen," The Journal of Chemical Physics, Vol. 46, May 1967, pp. 3418-3423.

[13]Lewis, J. W. L., Price, L. L., and Kinslow, M., "Rotational Relaxation of N_2 in Heated Expansion Flow Fields," Proceedings of the Ninth International Symposium on Rarefied Gas Dynamics, Vol. 1, 1974, pp. B.17-1 - B.17-9.

[14]Pattengill, M. D., "An Application of the Semiclassical Approximation of the Generalized Phase Shift Treatment of Rotational Excitation: Ar-N_2," The Journal of Chemical Physics, Vol. 62, 1975, pp. 3137-3142.

[15]Buck, U., Pauly, H., Pust, D., and Schleusener, J., "Molecular Beams from Free Jet Expansion of Molecules and Mixed Gases," Proceedings of the Ninth International Symposium on Rarefied Gas Dynamics, Vol. 2, 1974, pp. C.10-1 - C.10-9.

SHOCK WAVE STRUCTURE IN WATER VAPOR

R. Synofzik,* W. Garen,* and
G. Wortberg†

Technische Hochschule Aachen,
Aachen, Germany

and

A. Frohn≠

Universitaet Stuttgart,
Stuttgart, Germany

Abstract

Vibrational relaxation times of pure water vapor at pressures between 0.1 and 0.5 mm Hg have been derived from measured density profiles of shock waves at Mach numbers between 1.2 and 3.0. The results indicate that approximately 14 to 70 collisions are necessary to reach equilibrium. The results have been compared with experimental and theoretical data found in the literature.

Introduction

The relaxation phenomena in gases after sudden changes of state are influenced strongly by the presence of water vapor.[1,2] In order to clarify this situation, knowledge of the relaxation phenomena in water vapor itself is necessary. A

Presented as Paper 80 at the 10th International Symposium on Rarefied Gas Dynamics, Aspen, Colo., July 19-23, 1976.

*Graduate Assistant, Institute fuer Allgemeine Mechanik.

†Professor, Institute fuer Allgemeine Mechanik.

≠Professor, Institute fuer Thermodynamik der Luft- und Raumfahrt.

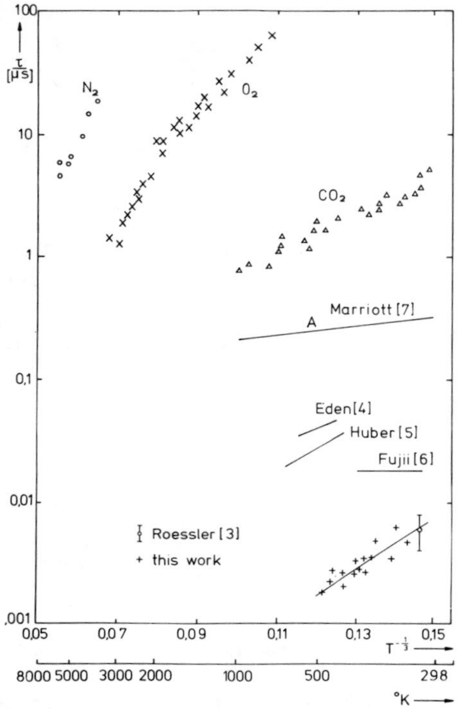

Fig. 1 Vibrational relaxation time vs $T^{-1/3}$.

few results for the relaxation of the vibrational excitation over a small range of temperature have been reported in the literature,[3-1] but the reported relaxation times differ by a factor of up to 100 (Fig. 1). The aim of our investigations is to measure the relaxation of vibrational excitation in pure rarefied water vapor over a large range of temperature. The vibrational relaxation time may be derived from the measured density structure of weak shock waves. The water vapor has to be highly rarefied and the shock Mach number must be rather small in order to resolve the density profile, which is fairly narrow in the case of water vapor because of the short relaxation time.

Experimental Technique

The measurements were performed in a conventional shock tube equipped with a special pneumatic valve, which replaced the commonly used diaphragm.[8] A few details may be found in the paper by Garen et al.[9] Because of the very small initial pressure difference between driver and driven section, the modification of the shock tube just mentioned was necessary to generate weak shock waves in pure rarefied water vapor. The density was measured by means of a laser differential

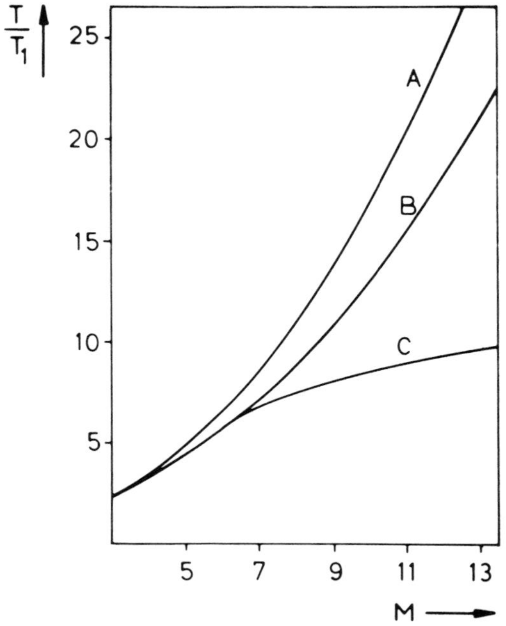

Fig. 2 Nondimensional equilibrium temperature vs shock Mach number; curve A frozen vibration and dissociation; curve B frozen dissociation; curve C all internal degrees of freedom in equilibrium.

interferometer. Again details of this technique and especially of how to avoid boundary-layer effects, may be found in the paper of Garen et al.[9] At all measurements, the initial pressure was less than 0.5 mm Hg. The shock Mach number varied between 1.2 and 3.0. At these Mach numbers, dissociation and ionization may be neglected, as can be seen from our numerical calculations of the equilibrium state (Fig. 2).

Results and Discussion

A typical density profile is depicted in Fig. 3a which shows normalized density vs nondimensional distance. The first steep rise of density, which is due to excitation of translational and rotational degrees of freedom, may be separated clearly from the second rise, which stems from the vibrational excitation. It should be mentioned that Fig. 3a was obtained by digitalizing an oscilloscope trace, followed by smoothing out the data by spline functions. An

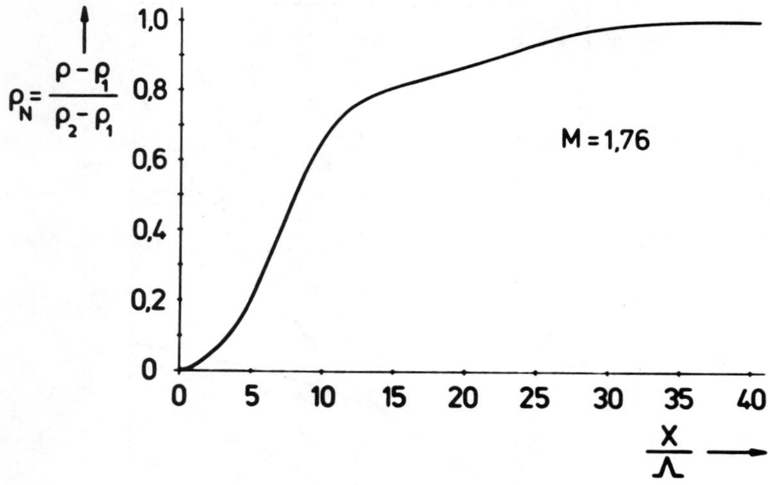

Fig. 3a Normalized density vs nondimensional distance.

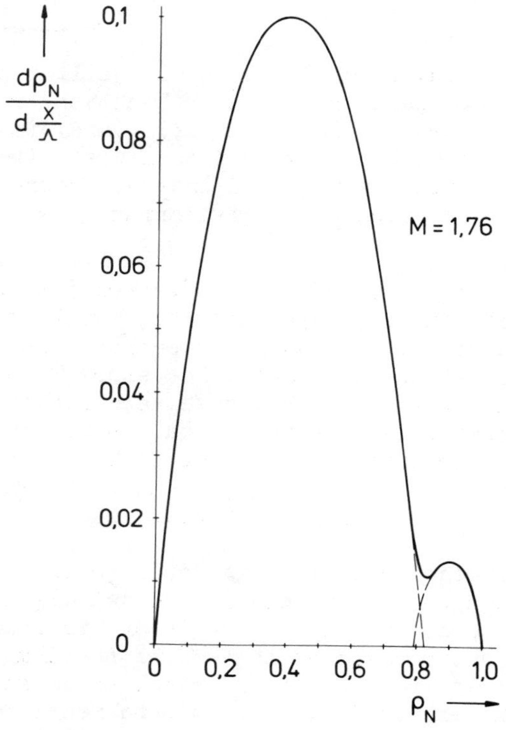

Fig. 3b Density gradient vs normalized density.

even more precise picture of the density rise may be obtained from a plot of density gradient vs normalized density (Fig. 3b). This kind of plot is very sensitive to small variations of the density profile.

The vibrational relaxation time may be derived from Fig. 3a in the following manner. The width of the translational rotational shock was determined, as usual, from the maximum slope of the first density rise. The half width of the rise, which is due to vibrational excitation, may be determined by the equilibrium density, and the line that is given by the maximum slope in the second rise.

The density gradient profile of the first density rise, as depicted in Fig. 3b, is symmetrical as predicted by Mott-Smith theory. Assuming the same symmetry for the second rise, there then exists a simple method to determine the width of the second rise, as may be seen from the figure. Also, the overlapping of both density rises may be detected easily. It should be mentioned that this symmetry of the second density rise was tacitly anticipated for the determination of the width of this rise from Fig. 3a, as described in the preceding.

In Fig. 1, some typical results for the vibrational relaxation time normalized to a pressure of 760 mm Hg are plotted vs $T^{-1/3}$. Our results agree very well with a single value reported by Roesler and Sahm,[3] which was obtained from sound absorption measurements. Large discrepancies exist between older experimental results[4-6] and the theoretical results of Marriott[7].

References

[1]v. Rosenberg, C.W., Bray, K.N.C. and Pratt, N.H., "Shock Tube Vibrational Relaxation Measurements: N_2 Relaxation by H_2O and the CO-N_2 V-V Rate," The Journal of Chemical Physics, Vol. 56, April 1972, p. 3230.

[2]Hodgson, J.P., "Vibrational Relaxation Effects in Weak Shock Waves in Air and the Structure of Sonic Bangs," Journal of Fluid Mechanics, Vol. 58, March 1973, p. 1817.

[3]Roesler, H. and Sahm, K.F., "Vibrational and Rotational Relaxation in Water Vapor," The Journal of the Acoustical Society of America, Vol. 31, February 1965, p. 386.

[4]Eden, R.B., Lindsay, R.B. and Zink, H., "Acoustic attenu-

ation and Relaxation Phenomena in Steam at High Temperature and Pressure," Transactions of the American Society of Mechanical Engineers, Vol. 83, January 1961, p. 137.

[5]Huber, P.W. and Kantrowitz, A., "Heat-Capacity Lag Measurements in Various Gases," The Journal of Chemical Physics, Vol. 15, May 1947, p. 275.

[6]Fujii, Y., Lindsay, R.B., and Urushihara, K., "Ultrasonic Absorption and Relaxation Times in Nitrogen, Oxygen, and Water Vapor," The Journal of the Acoustical Society of America, Vol. 35, July 1963, p. 961.

[7]Marriott, R., "Molecular Collision Cross Sections and the Effect of Hydrogen on Vibrational Relaxation in Water Vapor," Proceedings of the Physical Society, Vol. 88, July 1966, p. 617.

[8]Garen, W., Synofzik, R., and Frohn, A., "Shock Tube for Generating Weak Shock Waves," AIAA Journal, Vol. 12, August 1974, pp. 1132-1134.

[9]Garen, W., Synofzik, R., Wortberg, G., and Frohn, A., "Experimental Investigation of the Structure of Weak Shock Waves in Noble Gases," presented as paper 71 at the International Symposium on Rarefied Gas Dynamics, Aspen, Colo., July 19-23, 1976; published elsewhere in this Volume.

NON-MAXWELLIAN VELOCITY DISTRIBUTIONS IN SUPERSONIC EXPANSIONS OF ARGON

A.H.M. Habets,* H.C.W. Beijerinck,† N.F. Verster,‡ and J.P.L.M.N. de Warrimont*

Eindhoven University of Technology, Eindhoven, the Netherlands

Abstract

Accurate measurements have been performed of the parallel and the perpendicular velocity distributions in supersonic expansions of Argon. Extensive cryopumping is used to eliminate interaction effects. High resolution time-of-flight spectroscopy is used for measuring the parallel distribution. Measurements are done with three converging nozzles of 20, 40, and 80 μm diam. in the pressure range of 50 to 3000 Torr. Temperature is varied from 300° to 400° K. A least-squares analysis with several model functions has been performed to determine the shape of the measured velocity distribution functions. The parallel velocity distribution is fully described by the sum of two Boltzmann distributions, differing both in central velocity u and speed ratio S. Speed ratios typically differ 30%; the central velocities differ considerably less than the thermal width of the distribution. The colder of the two components is normally the more populated, typically by a factor 2. The average speed ratio behavior is compared with theory. The perpendicular velocity distribution is fully described by the sum of two Boltzmann distributions. In terms of the virtual source formalism, this corresponds to the sum of two gaussian virtual source components. The ratio of the virtual source radii typically is five. The radius of the colder distribution is of the order of a few times the nozzle diameter. The colder distribution is several times more densily

Presented as Paper 107a at the 10th International Symposium on Rarefied Gas Dynamics, July 19-23, 1976, Aspen, Colo.

*Ir. of Physics, Physics Department.
†Ph.D. in Physics, Physics Department.
‡Professor of Physics, Physics Department.

populated than the warmer. The mean virtual source radius is compared with theory.

Introduction

Velocity distributions in supersonic expansions into vacuum have been subject to both theoretical and experimental study[1,2] for over 10 years. In an early stage, experiments indicated the existence of notable deviations from a simple Maxwellian distribution in the observed perpendicular velocity distributions.[3] A similar effect, although less pronounced, was observed in detailed measurements of the parallel velocity distributions.[4] Theoretical investigations[5] showed results that agreed with experiment, but at the time no quantitative data could be derived from the latter. Since then, the experimental study of velocity distributions has made considerable progress. Results on Argon are given in Refs. 6-8, showing an increasing accuracy due to the suppression of interaction effects and the perfection of measuring techniques. Still, however, for the description of non-Maxwellian effects, no sufficiently precise data were available. Theoretically, the effect was investigated again by Willis,[9] with the result that, for the perpendicular distribution, the deviation appeared so large as to be essential for a good description, whereas for the parallel distribution it would show only as a small deformation.

We will present new quantitative data on the amount of deviation from simple Maxwellian shape for both the parallel and the perpendicular velocity distributions in supersonic expansions of Argon. For the perpendicular distributions, these are the results of a thorough analysis of existing data, which were published previously.[10] This paper contained only a preliminary analysis in terms of a simple Maxwellian velocity distribution.

The parallel distribution data result from a set of high-precision time-of-flight measurements, carried on as part of a research program on supersonic expansion, incorporating also the correlation of parallel and perpendicular distributions and the effect of condensation on the monomer- and multimer-fraction velocity distribution.

Apparatus

In this section, we review the experimental setups used for the perpendicular and the parallel measurements, respectively. For the perpendicular measurements, all necessary information is to be found in our previous report.[10] We will present here the final analysis of the data preliminarily reported there.

For the parallel velocity measurements, we used a high-resolution time-of-flight spectrometer, which has been described previously.[11,12] It consists of a series of 6 differentially pumped chambers with pressures varying from the 10^{-8} Torr range to 10^{-11} Torr partial Argon pressure at the detector.

The source chamber contains a cryopump at about 20 K, completely surrounding the nozzle, which constitutes the expansion chamber. The high sticking coefficient[13] insures undisturbed expansion. In order to avoid skimmer interaction, no skimmer was placed in the expansion, and the collimation of the beam is postponed until after the exit of the cryopump, 15 cm downstream of the nozzle. The beam then is chopped and passes on to the detector, located 2 m from the nozzle. The detector consists of an electron bombardement ionizer, a quadrupole mass-filter and an electron multiplier. Single ion pulses are further processsed in nuclear physics pulse handling electronics and fed into a PDP-11 with a multiscaling interface.[12,14] Systematic errors in the TOF spectrum, due to memory effects in the ionizer and the detector electronics, have been thoroughly investigated[12,16] and eliminated by a suitable choice of the detector settings.

Special care was taken in the construction of the three nozzles used. In order to insure a well-defined temperature, we used copper, which was electroformed onto a steel needle ground to a well-defined sharp cone. After machining the assembly roughly to the desired shape, the tip was carefully honed, using a microscope to check progress. Finally the steel needle was withdrawn, leaving a sharp-edged circular hole at the end of a conical section, which was exactly the same in all three nozzles. This enabled us to check the dependence of the measured effects on the nozzle diameter, which proved particularly useful in studying condensation effects.

The internal full cone angle is 60°. Nozzle diameters used are 19, 46, and 85 μm. By use of various collimating apertures, we could vary the fraction of the virtual source[10] contributing to the detector signal. For the measurements reported here, the whole virtual source is sampled. The chopper period used is about 7 msec, and the duty cycle is adjustable at 0.0065 or 0.020. Flight time to the detector is typically 3 msec. The multiscaler-interface is set to 20 μsec time channels. The resulting velocity resolution of the apparatus is 1.5 or 4%, mainly due to the width of the chopper slit. The relative accuracy of velocity measurements is 0.1%; the absolute calibration is better than 0.5%. The accuracy in the measured TOF spectra must be better than 0.5% in the top of the distribution to allow the present detailed analysis. Measuring time is then typically 20 min, determined by the maximum counting frequency, which is handled undistorted by the detector electronics. For low-intensity beams, measuring times up to several hours were needed.

Data Analysis

Both perpendicular and parallel measured data were analyzed, using least-squares data-reduction techniques. The measured perpendicular beam intensity profiles are most easily thought of as consisting of the convolution of the virtual source distribution and the transmission function of the skimmer collimator. The latter is a simple rectangular function, as a slit skimmer was used.[10] A typical example of a resulting beam profile is given in Fig. 1.

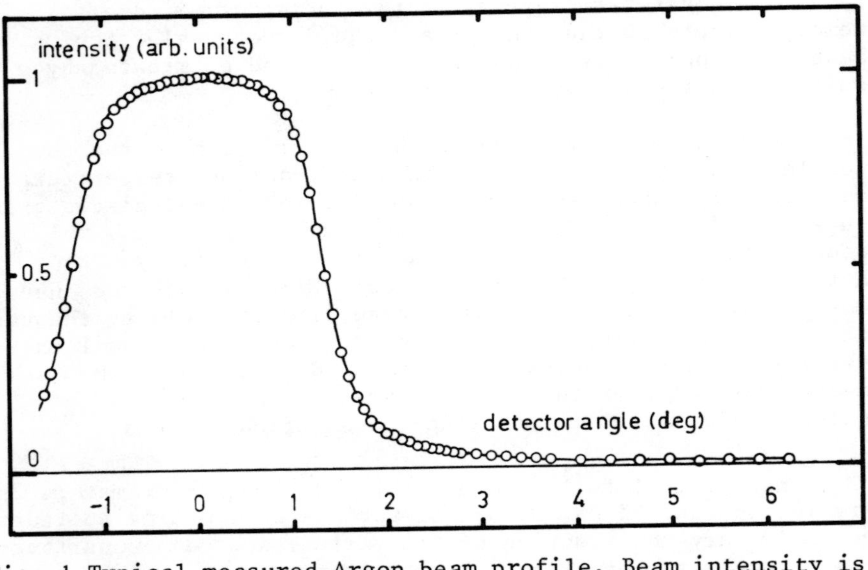

Fig. 1 Typical measured Argon beam profile. Beam intensity is measured as a function of angular detector position. A 53 μm nozzle at 290° K was used. Nozzle pressure is 150 Torr, nozzle skimmer distance is 10.7 mm, skimmer slit width is 0.43 mm. The drawn curve is calculated from a best-fit bimodal distribution. Constant background is about 0.005.

Let $f(r,\theta)$ be the normalized virtual source intensity distribution in the nozzle plane, with (r,θ) plane polar coordinates. We assume f to depend on the radial distance r only. An obvious ansatz for the distribution is

$$f(r) = 2\pi r\ (\pi R^2)^{-1} \exp -\ (r/R)^2 \qquad (1)$$

which corresponds to a Maxwellian distribution of the perpendicular velocity. Here R is the characteristic virtual source radius. It turned out, however, that the data could not be represented by a distribution of the form of Eq.(1) plus a constant background. The resulting R values were considerably higher than those derived from the slope of the profiles at the geometrical beam edges, and the residue after subtraction of the best-fit model function displayed a pronounced structure. Apparently, the broad tails of the measured profiles were not accounted for in the model. We then attempted to extend the model with extra terms, decreasing more slowly with increasing distance. The only one of these giving really satisfactory results is of the form

$$f(r) = 2\pi r\ \{c_1(\pi R_1{}^2)^{-1} \exp -(r/R_1)^2 + c_2(\pi R_2{}^2)^{-1} \exp -(r/R_2)^2\} \qquad (2)$$

with the sum of the relative populations $c_1 + c_2 = 1$. For simplicity we omit in this description the constant background, which was used as an extra parameter in the least-squares fits. We will refer to this type of function as a bimodal distribution.

By using this model in a weighted least-squares procedure, we were able to remove nearly all structure from the postfit residue. The best-fit curve is indicated in the example given in Fig. 1. The results of this analysis of the perpendicular velocity distribution will be given in the next section.

A similar problem arises in the analysis of high-precision parallel velocity distribution measurements. The commonly used form of the normalized velocity distribution of supersonic beam intensity or flux is

$$F(v) = A\ v^3 \exp -\ S^2((v/u) - 1)^2 \qquad (3)$$

Here S is the speed ratio, related to the parallel temperature $T_{/\!/}$ as $S = u/(2kT_{/\!/}/m)^{\frac{1}{2}}$, u is the central or flow velocity, and A is a normalization constant approximately equal to

$$A = \pi^{-\frac{1}{2}}\ S\ u^{-3}\ (1 - \tfrac{1}{2}S^{-2}) \text{ for } S > 3 \qquad (4)$$

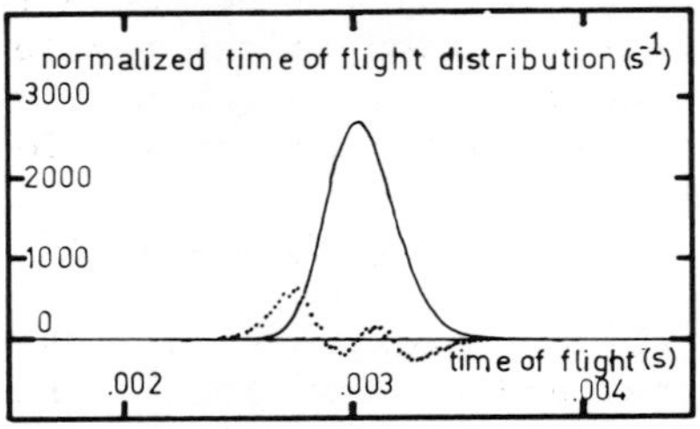

Fig. 2 Typical normalized time-of-flight distribution of Argon. The drawn curve is a best-fit simple Maxwellian distribution. Points represent the five times enlarged residue, showing typical structure. A 85 μm nozzle at 288° K was used. Nozzle pressure is 200 Torr.

The corresponding time-of-flight distribution of the density, as measured with an electron-bombardement detector, is

$$f(t) = a\ t^{-4}\ \exp - S^2((\bar{t}/t) - 1)^2 \tag{5}$$

with $\bar{t}$ the central time-of-flight corresponding to u, and

$$a = \pi^{-\frac{1}{2}}\ S\ \bar{t}^3\ (1 - \tfrac{1}{2}S^{-2})\ \text{for}\ S > 3 \tag{6}$$

This model function, extended by an adjustable constant background, was used in the weighted least-squares fit procedure mentioned previously. The experimental data were first deconvoluted to correct for the effect of the finite chopper gate time.

Although the results were considerably better than in the case of the perpendicular beam profiles, here too a residual structure was observed. This is illustrated by a typical example in Fig. 2. The residue is exaggerated 5 times with respect to the best-fit curve shown.

The form and the amplitude of the residue vary systematically over the range of experimental parameters. The maximum amplitude ranges from 3% of the top height of the fitted curve for very narrow spectra to 10% for low speed ratios. We attempted to describe the phenomenon quantitatively by repeating the fitting procedure for a selection of the available data, using several trial functions to extend the model of Eq. (5). Among these were functions with powers of t different from -4 in the preexponential factor, functions with the exponential factor replaced by the sum of two exponential terms, and combinations of the two former. Of these, only the second set gave satisfactory results. Eventually, the model function chosen was the time-of-flight distribution corresponding to

$$F(v) = v^3\{c_1 A_1 \exp - S_1^2((v/u_1) -1)^2 +$$
$$c_2 A_2 \exp -S_2^2((v/u_2)-1)^2\} \quad (7)$$

Mostly the data reduction procedure converged satisfactorily for sufficiently accurate data, difficulties arising only for very narrow spectra of high speed ratio, where the number of effective measured points became too small. The curve fitting process as a whole, however, was considerably more complicated than in the perpendicular data case, as could be seen from the sometimes quite high correlations of the fitted parameters. The difference $u_1 - u_2$ always proved small as compared to the width u/S of the distribution.

Results and Discussion

As remarked in the preceding section, the analysis of perpendicular data was much easier than in the parallel case, since the non-Maxwellian character of the velocity distribution is much stronger in the perpendicular case. This has been pointed out in several theoretical investigations.[5,9] The analysis of the perpendicular distribution data yields the individual virtual source values of the two component distributions, and the relative population given by the ratio c_1/c_2.

Figure 3 gives the reduced virtual source values $R_i^* = R_i/d$ as a function of the dimensionless quantity ξ^*

$$\xi^* = \xi / \xi_{ref} \quad (8)$$
$$\xi = n\, d\, T^{-1/3} \quad (9)$$

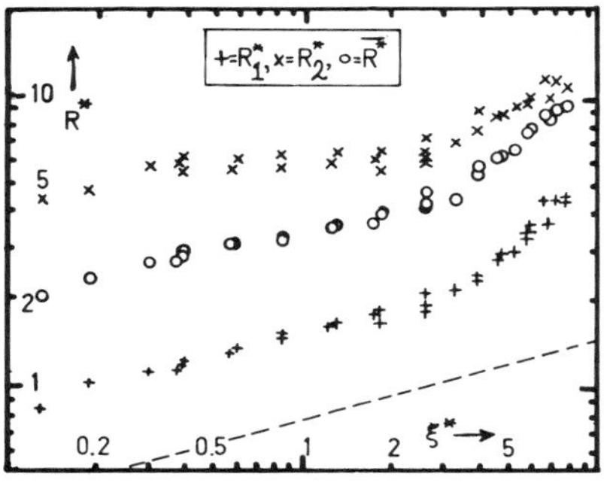

Fig. 3 Characteristic radii of the two virtual source components as a function of ξ^* : R_1^* and R_2^* are the colder and warmer component radii respectively, $\overline{R}^*$ is the average radius. A theoretical line is indicated. Measurements were done using a 53 μm nozzle at 290° K. Nozzle pressures are in the range 50-3000 Torr. Nozzle skimmer distance is 10.7 mm.

$$\xi_{ref} = 10^{20}\ m^{-2}\ K^{-1/3} \tag{10}$$

n, T, and d are reservoir density, temperature, and nozzle diameter, respectively. Here the combination ξ of source parameters is chosen, since dimensional analysis shows this to be the quantity governing the transition to free-molecular flow when the intermolecular potential is effectively an inverse sixth-power only. For the given choice of the reference value ξ_{ref}, the value of ξ^* becomes unity at a pressure of 400 Torr for a 50 μm nozzle at 300° K. The mean virtual source radius $\overline{R}^*$ is obtained by averaging the two component R_i^*, weighing with the relative populations c_i. We choose the following definition of $\overline{R}^*$

$$\overline{R}^* = (c_1\ R_1^{*\,2} + c_2\ R_2^{*\,2})^{\frac{1}{2}} \tag{11}$$

which gives the correct correspondence to the mean perpendicular temperature.

These values also are indicated in Fig. 3. The theoretical line represents the result of a sudden freeze argument. A sudden freeze surface is determined from the known continuum behavior and from the terminal parallel temperature, as follows from the thermal conduction model discussed in Ref. 15. The apparent absolute value discrepancy should be due to the crudeness of this argument. On the other hand, the general trend, as is apparent from the roughly equal slopes of average and theoretical lines, shows a gratifying correspondence.

The discontinuity in behavior, visible at about $\xi^* = 2.5$, is ascribed to condensation. This is in agreement with an experimentally found criterium for the onset of condensation for Argon, given by

$$p\ d^{0.88}\ T^{-2.28} \approx 4.10^{-7}\ \text{Torr}\ m^{0.88}\ K^{-2.28} \tag{12}$$

Here p is the reservoir pressure.

The relative population ratio c_1/c_2 of cold to warm distribution is plotted as a function of ξ^* in Fig. 4. Over the noncondensing range, the component with the higher virtual source width has a population varying from 20 to 50% with increasing ξ^*. It is interesting to compare the present virtual source width data, given in Fig. 3, to the preliminary analysis given in Ref. 10. The latter, calculated from the slope of the

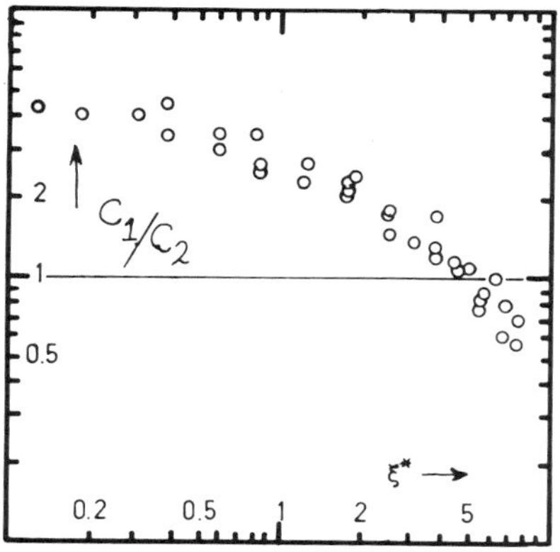

Fig. 4 Population ratio of the two components of the virtual source distribution as a function of ξ^*. Source conditions are those of Fig. 3.

beam profile at the geometrical beam edge, practically coincide with the present set of narrow R_1^* values. As expected, such a simplified analysis is sensitive only to the narrow central part of the virtual source. As to the magnitude of the bimodality effect, we see roughly a factor 5 difference in virtual source radius corresponding to a factor 25 in perpendicular temperature. In view of the population ratio of the warmer distribution, the effect is large enough that it must be taken into account when calculating supersonic beam profiles or intensity as a function of skimmer width. At the moment, the data shown here are the only ones available. The variation of this bimodality with nozzle diameter and temperature has not been measured as yet, the only parameter varied being nozzle pressure. Data on the influence of the nozzle skimmer distance x are very limited. The few available measured points indicate an x^{-1} dependence of the mean virtual source radius $\bar{R}^*$, whereas the primitive analysis[10] clearly resulted in an x^{-2} dependence. We feel that the illustration of conclusions, which was formulated in Ref. 9, is an interesting one.

The parallel velocity distribution data are characterized by the speed ratio, central velocity, and relative population for each of the component distributions. Speed ratio values are given in Fig. 5 as a function of ξ^*. The experimental points represent six different combinations of nozzle diameter and nozzle temperature. Accuracy can be estimated from the spread in the measured points. For each T, d combination, the high pressure end of the experimental region showed condensation effects. These points were omitted from this plot, according to the criterium given in Eq. (12). At the moment there are no theoretical predictions for the individual parameters. Since the scaling of S_i to the quantity

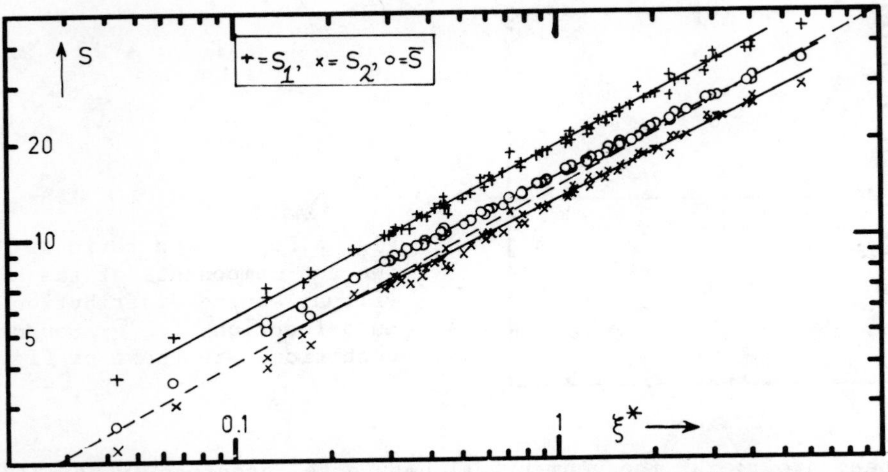

Fig. 5 Speed ratio values of the two components of the parallel velocity distribution as a function of ξ^*. S_1 and S_2 are the colder and warmer component speed ratio, respectively, $\overline{S}$ is the average speed ratio. Full lines are the best-fit representations given in the text. Broken line is a theoretical relation. Measurements were done using three nozzles of 19, 46, and 85 μm at 288° and 374° K.

ξ^* appears to be quite satisfactory, we represent the two component speed ratios by the following best-fit expressions

$$S_1 = 19.0\ (\xi^*)^{0.525 \pm 0.01} \tag{13}$$

$$S_2 = 13.2\ (\xi^*)^{0.495 \pm 0.01} \tag{14}$$

resulting in a relative difference of about 30% in the experimental range. In defining the average speed ratio $\overline{S}$ of the distribution, which we want to compare to available theoretical values, extra care is needed as compared to the perpendicular distributions. We must account for the fact that the two components have different central velocities. The definition based on the energy-moment is

$$\overline{S} = \overline{u} \,/\, (2\ \overline{(v - \overline{u})^2}\ )^{\frac{1}{2}} \tag{15}$$

in which $\overline{u}$ is the flow velocity

$$\overline{u} = c_1\, u_1 + c_2\, u_2 \tag{16}$$

with c_i the relative population of the normalized component distributions ($c_1 + c_2 = 1$). This is equivalent to calculating $\overline{S}$ as

$$\overline{S} = \overline{u} / (2k\overline{T}_{/\!/} /m)^{\frac{1}{2}} \qquad (17)$$

in which the mean temperature $\overline{T}_{/\!/}$ is given by

$$\overline{T}_{/\!/} = c_1 T_1 + c_2 T_2 + c_1 c_2 (m/k) (u_1 - u_2)^2 \qquad (18)$$

The $\overline{S}$ values determined in this way are also plotted in Fig. 5. The best-fit description of the $\overline{S}$ behavior is

$$\overline{S} = 15.2 (\xi^*)^{0.495 \pm 0.01} \qquad (19)$$

which is to be compared to the theoretical result

$$S = 14.0 (\xi^*)^{6/11} \qquad (20)$$

of the thermal conduction model.[15] The theoretical line also is indicated in Fig. 5.

Whereas the speed ratios $S_{1,2}$ and $\overline{S}$ scale quite well with ξ^*, the behavior of the parameters u_1, u_2, and c_1/c_2 is less clear. Figure 6 shows $u_{1,2}$ and $\overline{u}$ values of one T, d combination. The average velocity shows a steady increase with nozzle pressure, typically of the order of 1% over a decade of pressure. This behavior is clearly distinguished from the region of beginning condensation, which is characterized by a much steeper velocity increase.

The two-component central velocities are spaced on both sides of the averaged points. The typical distance of the two velocities is of the order of 0.5% in the high-pressure region,

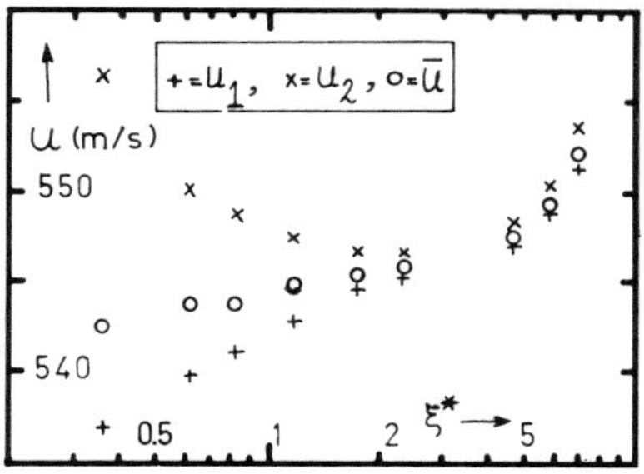

Fig. 6 Central velocities of the two components of the parallel velocity distribution as a function of ξ^*. u_1 and u_2 belong to the component with the higher and the lower speed ratio, respectively, $\overline{u}$ is the average value. Measurements shown were done using a 46 μm nozzle at 288° K. Behavior is typical for all nozzles and temperatures used.

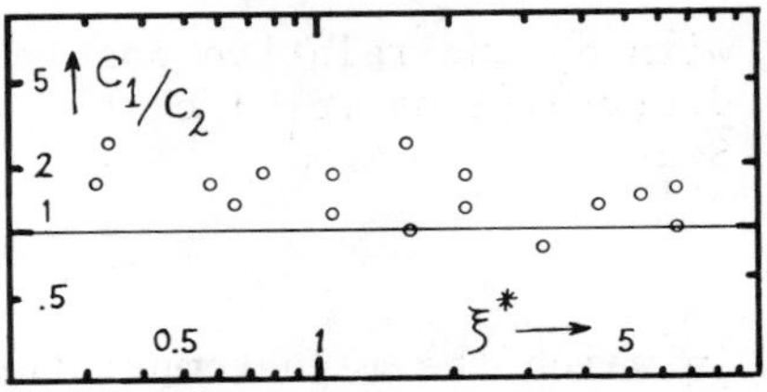

Fig. 7 Population ratio of the two components of the parallel distribution as a function of ξ^*. c_1 belongs to the higher speed ratio. Data shown are of 46 µm nozzle at 288° K.

increasing to 5% with decreasing pressure. In fact, the distance between the two curves increases so quickly that the steady decrease of the average velocity with decreasing pressure is overcompensated, resulting in a rise of the higher of the two velocities. It is interesting to note that the higher velocity always belongs to the component with the lower speed ratio.

In Fig. 7 we give, again, for one T, d combination, the values of the population ratios c_1/c_2 of the colder over the warmer components. As in the case of the central velocities, we did not yet succeed in scaling these universally to a combination of source parameters. An extra difficulty here is the excessive spread of the calculated values, because of the high correlation of the estimates of c_1 and c_2. Here it is shown that, especially for the higher speed ratio values, the two-component distributions become more or less indistinguishable for the least-squares-fit process. In relation to this, we stress the fact that the choice of the bimodal distribution used in the analysis was made on technical rather than physical grounds. Consequently, it would be unwise to attach too much physical importance to the individual parameters used in the description, or to their precise dependence on source conditions. The small difference in shape of the components, and the practically equal population, point in our view, to the need for a more objective description of the deviations from Maxwellian shape, e.g., by approximating the spectra by a series expansion in Hermite polynominals with an exponential factor using the known average values of $\overline{S}$ and $\overline{u}$. We intend to reanalyze the data in this way, in order to obtain a more systematic insight into the physical processes affecting the velocity distribution.

In conclusion, inspecting the observed amount of bimodality in the perpendicular and parallel distributions, we can state that the effect is by far stronger in the former case. Although the perpendicular distributions show a secondary component that is much less populated than in the parallel distributions, its effect is still the larger, because of the enormous temperature difference of a factor 25 between the two components. The two parallel temperatures, on the other hand, differ only by about a factor of 2. Contrary to the perpendicular measurements, the parallel data therefore seem to have little practical importance on the subject of supersonic beam production. Both sets of data, however, should prove useful as a check on more detailed theoretical treatments of the transition from continuum to free-molecular flow.

References

[1] Anderson, J.B., and Fenn, J.B., "Velocity Distributions in Molecular Beams from Nozzle Sources", Physics of Fluids, Vol. 8, May 1965, p. 780.

[2] Hamel, B.B., and Willis, D.R., "Kinetic Theory of Source Flow Expansion with Application to the Free Jet", Physics of Fluids, Vol. 9, May 1966, p. 829.

[3] Abuaf, N., et al., "Studies of Low Density Supsersonic Jets", Rarified Gas Dynamics, edited by C.L. Brundin, Academic Press, New York, 1967, Vol. 2, p. 1317.

[4] Scott, J.E., and Phipps, J.A., "Translational Freezing in Freely Expanding Jets", Rarified Gas Dynamics, edited by C.L. Brundin, Academic Press, New York, 1967, Vol. 2, p.1337.

[5] Edwards, R.H., and Cheng, H.K., "Distribution Function and Temperatures in a Monatomic Gas under Steady Expansion into a Vacuum", Rarified Gas Dynamics, edited by C.L.Brundin, Academic Press, New York, 1967, Vol. 1, p.819.

[6] Miller, D.R., and Andres, R.P., "Translational Relaxation in Low Density Supsersonic Jets", Rarified Gas Dynamics, edited by R. Trilling and H.Y. Wachman, Academic Press, New York, 1969, Vol. 2, p. 1385.

[7] Buck, U., Pauly, H., Pust, D., and Schleusener, J., "Molecular Beams from Free Jet Expansions of Molecules and Mixed Gases", Rarified Gas Dynamics, edited by M.Becker and M. Fiebig, DFVLR Press, Porz-Wahn, Germany, 1974, Vol. 2, p. C10-1.

[8] Cattolica, R., Robben, F., and Talbot, L., "The Ellipsoidal Velocity Distribution Function and Translational Non-Equilibrium", Rarified Gas Dynamics, edited by M. Becker and M. Fiebig, DFVLR Press, Porz-Wahn, Germany, 1974, Vol. 1, p. B16-1.

[9] Willis, D.R., Hamel, B.B., and Lin, J.T., "Development of the Distribution Function on the Centerline of a Free Jet Expansion", Physics of Fluids, Vol. 15, April 1972, p. 573.

[10] Habets, A.H.M., Beijerinck, H.C.W., and Verster, N.F., "Perpendicular Temperature Measurements on a Cryopumped Supersonic Argon Jet", Rarified Gas Dynamics, edited by M. Becker and M. Fiebig, DFVLR Press, Porz-Wahn, Germany, 1974, Vol. 1, p. B6-1.

[11] Beijerinck, H.C.W., Habets, A.H.M., and Verster, N.F., "Crossed Beam Measurements of Ar-Ar Total Cross-sections with a Time-of-Flight Machine", Rarified Gas Dynamics, edited by M. Becker and M. Fiebig, DFVLR Press, Porz-Wahn, Germany, 1974, Vol. 1, p. B34-1.

[12] Beijerinck, H.C.W., "Molecular Beam Studies with a Time-of-Flight Machine", Thesis, Eindhoven University of Technology, 1975.

[13] Habets, A.H.M., Verster, N.F., and Vlimmeren, Q.A.G. van, "Simple Device for Measurements of the Sticking Coefficient of Cryopumps", Rev. Sci. Instr., Vol. 46, May 1975, p. 613.

[14] Geel, J. and Verster, N.F., "A Fast Multiscaler Interface for a PDP-11 Computer", J. Phys. E.: Sci. Instr., Vol. 6, July 1973, p. 644.

[15] Habets, A.H.M., Verster, N.F., de Warrimont, J.P.L.M.N., and Beijerinck, H.C.W., to be published.

[16] Beijerinck, H.C.W., Moonen, R.G.J.M., and Verster, N.F., "Calibration of a Time-of-Flight Machine for Molecular Beam Studies", J. Phys. E.: Sci. Instr., Vol. 7, January 1974, p.31.

EXCITED STATE DISTRIBUTION FUNCTION OF ARGON IN A FREELY EXPANDING ARCJET PLUME

C. C. Limbaugh*

Arnold Engineering Development Center,
Arnold Air Force Station, Tenn.

Abstract

The excited state distribution function of argon is determined at two axial positions in the nonequilibrium environment of the free expansion into vacuum of an argon arcjet operating at a total pressure and temperature of 50 Torr and 4460°K. Radial inversion of narrow spectral line absorption data is used to determine the centerline density of the first four excited states, including two resonance and two metastable states. Abel inversion of the absolute radiance of selected spectral lines is used to determine the centerline density of highly excited states in equilibrium with the free electron density. The first four excited states were markedly elevated above the equilibrium condition at the free electron temperature and showed no evident tendency to approach equilibrium. Calculations using a system of eigenstate rate equations were made to predict the nonequilibrium distribution function at the downstream axial position using the upstream distribution function as initial conditions. The calculations showed a much stronger collisional coupling between the lower excited states and a much faster relaxation toward the equilibrium relationship than was actually observed. The difference in the predicted and observed relaxation rates is consistent with over-

---

Presented as Paper 99 at the 10th International Symposium on Rarefied Gas Dynamics, July 19-23, 1976, Aspen, Colo. Research reported in this paper was conducted by the Arnold Engineering Development Center, Air Force Systems Command. Research results were obtained by personnel of ARO, Inc., contract operator of AEDC.

*Physicist, Physics Section, Technology Applications Branch, Engine Test Facility.

estimation of the excitation and de-excitation collision cross sections by the semiclassical methods used to calculate the cross sections.

Introduction

The diagnostics of partially ionized gases was put on firm analytic grounds with the presentation of the collisional-radiative recombination (CRR) Model.[1] However, that model requires the quasisteady state (QSS) assumption, in which the time rate of change of each excited internal state available to the gas species is negligible relative to the net rate of recombination. Although the QSS assumption may be appropriate in some applications, there are other cases in which the validity, and hence result, of the assumption is questionable. An example of such a phenomenon is the free expansion into vacuum of a plasma plume. In this case, excited state density depletion within a control volume due to the gas expansion can become competitive with collisional and radiative effects, and it is necessary to resort to a system of time-dependent coupled differential equations for correct description of the excited state distribution function. The results of calculations with this parent system of differential equations, termed the eigenstate rate equations (ERE), have been examined previously, independent of gasdynamical effects for hydrogen plasmas[2] and helium plasmas.[3,4] The results of that earlier work suggested that the electronic state density relaxation times could be significant with respect to typical plume expansion times.[5] Other investigators have pursued direct comparisons between experiment and theory.[6,7] However, the experimental conditions were such that the QSS assumption was appropriate and thus the CRR analytic approach that was used was valid. Other recent theoretical investigations[8,9] have reinforced the view that transient deviations from QSS predictions are to be expected and confirmed the importance of including the quantum state structure in correctly assessing equilibrium criteria for plasmas.

The present work is concerned with the experimental determination and subsequent comparisons to theoretical predictions of the excited state distribution function in the free expansion into vacuum of a low density argon plasma. The arc-jet was operated at a total pressure of 50 Torr and a total temperature of 4460°K. Experimental determination of excited state densities was accomplished by inversion of narrow spectral line absorption and emission data. The theoretical predictions are made without resort to the simplifying QSS assumption, and includes accounting of gas expansion effects, as well as the transient coupling between atomic states.

Eigenstate Rate Equations

The ERE provides a complete transient description of each quantum state available to the atoms in the plume, including collisional coupling to atoms in other quantum states. The earlier work for the helium atom[3,4] was extended to argon and modeling of the plume gasdynamic expansion was included. One of the coupled equations of the ERE, as used here, can be written

$$\frac{dn(p)}{dt} = \{n_e \sum_q n(q)\, K(q;p) + \sum_{q>p} n(q)\, A(q;p) + n_e^2 n^+ K(c;p) + n_e n^+ \beta(p)\} - \{n_e n(p) \sum_q K(p;q) + n_e n(p)\, K(p;c) + n(p) \sum_{q<p} A(p;q)\} + \{\frac{n(p)}{n_s} \frac{\partial n_s}{\partial M} \frac{\partial M}{\partial X} \sqrt{\gamma R T_s}\} \quad (1)$$

where n is the population density, p and q symbolically represent those quantum numbers required to represent each appropriate quantum state, the subscript "e" and superscript "+" identify the free electron and singly ionized ion densities, the subscript "s" identifies the static density, K represents the collisional rate coefficient, M the Mach number, x the displacement coordinate, and the T_s the heavy body static temperature.

In Eq. (1) the terms in the first set of braces represent the rate with which state p is being populated due to: 1) electron collisions with atoms in state q resulting in the transition of the bound electron to state p; 2) radiative transitions from higher quantum states into state p; 3) three body recombination (two electrons and one ion); and 4) radiative recombination, respectively. The second set of braces represents the rate of depopulation of state p due to: 1) electron-atom collisions resulting in an excitation or de-excitation from state p to state q; 2) electron-atom ionizing collisions; and 3) radiative de-excitation of state p to some lower state q. For the problem studied here, these gas dynamic processes correspond to the $1/r^2$ density decay normally associated with free-jet expansions. In the formulation of Eq. (1), all electron-atom collisional transitions are allowed, including direct as well as exchange transitions. Those radiative transitions violating the dipole selection rules have the appropriate transition probability set to zero. Radiative absorption is approximated by reducing or zeroing the appropriate transition probability. Heavy body collisional transitions are assumed to be insignificant in comparison to the electron-atom collisional transitions because the collision frequencies differ by three orders of magnitude whereas the more important cross sections are comparable in magnitude. A two-temperature model[10] in which the electron temperature is much higher than the heavy body temperature is assumed. The free electrons are assumed to have a Maxwellian energy distribution. The rate coefficients are obtained by integrating the energy dependent cross section for the transition over the free electron distribution function. For the work reported here, the semi-classical Gryzinski[11] formulation for the collisional cross sections was used, Seaton's[12] expressions for the radiative recombination rates were modified for Argon, radiative transition probabilities of prominent lines were taken from Wiese,[13] and transition probabilities of other lines were calculated by making a hydrogenic approximation using Green, Rush, and Chandler's matrix elements.[14] Energy levels of the argon atom were taken from Moore.[15]

For the gasdynamic contribution to the rate of decay of the excited state densities along the axis of the freejet plume, perfect gas and isentropic relations are assumed. This is justified on the basis that, for reasonable ionization levels, the gasdynamical effects provide the dominant forces on the heavy bodies, and the assumption is adequate for a first approximation. The expression

$$n_s/n_o = \{1 + [(\gamma - 1)/2]\, M^2\}^{-1/\gamma-1} \tag{2}$$

provides the Mach number dependent static density and the Ashkenas-Sherman model[16] is used to provide the model for the plume centerline Mach number distribution.

An equation similar to Eq. (1) can be written for each quantum state available to the plasma. This system, along with the expression for the conservation of atoms, obtained by assuming single ionization

$$\Sigma\, n(p) + n_e = n_s \tag{3}$$

provides a square, and hence determinate system of coupled, nonlinear, first order differential equations. The system of equations is truncated at some upper excited state which will assuredly be in equilibrium with the free electron density. Closely adjacent states which will maintain an equilibrium configuration are averaged and treated as single states at the appropriate energy. For the conditions studied here, after averaging, Eqs. (1) and (3) represent a system of 46 nonlinear coupled differential equations and one algebraic equation which were solved numerically for the time dependent distribution function of the excited argon species along the arcjet plume centerline. It is an initial value problem and the experimental observations at one position provide the starting conditions for the calculations of the distribution function at the next position.

Experiment and Analysis

The plasma plume was generated by the free expansion into vacuum from a 0.635-cm-diam straight nozzle of an argon arcjet operating at a total pressure of 50 Torr and total temperature of 4460°K. Power was supplied to the arcjet by a dc rectifier operating at 150 A and a voltage of 20 V.

Spatial profiles of absolute radiance and absorption of selected isolated spectral lines, listed in Table 1, were ob-

Table 1 Spectral lines used in experiment

State no.	Wavelength (nm)	Spectroscopic notation	Energy (cm^{-1})	Type of measurement
1	763.5	$4s[3/2]_2$	93144	Absorption
2	738.4 751.4	$4s[3/2]_1$	93751	↓
3	772.4 794.8	$4s'[1/2]_0$	94554	↓
4	750.3 840.8 922.4	$4s'[1/2]_1$	95400	↓
5	430.0	$5p[5/2]_2$	116999	Emission
6	416.4	$5p[3/2]_1$	117151	↓
7	415.8 426.6	$5p[3/2]_2$	117184	↓
8	425.9	$5p'[1/2]_0$	118871	↓
9	603.2	$5d[7/2]_4$	122036	↓
10	604.3	$5d[7/2]_3$	122160	↓

tained at two and four nozzle diameters downstream of the nozzle exit plane. These data were obtained by traversing the 1.6 mm parallel beam field of view of a 3/4-m spectrometer across the plasma transverse to the plume axis. A diagram of the freely expanding plume and the relative measurement position are shown schematically in Fig. 1.

The absorption measurements were used to determine spatially resolved number densities of the first four excited states of argon, two of which are metastable ($4s[3/2]_2$ and $4s'[1/2]_0$) and two of which have resonance transitions to the ground state ($4s[3/2]_1$ and $4s'[1/2]_1$). The absorption data were radially inverted by assuming Doppler broadened profiles and cylindrical plume geometry. At the conditions of this work, the other broadening mechanisms are negligible compared to the Doppler mechanism. The static gas temperature, required for the inversion, is determined from a method of char-

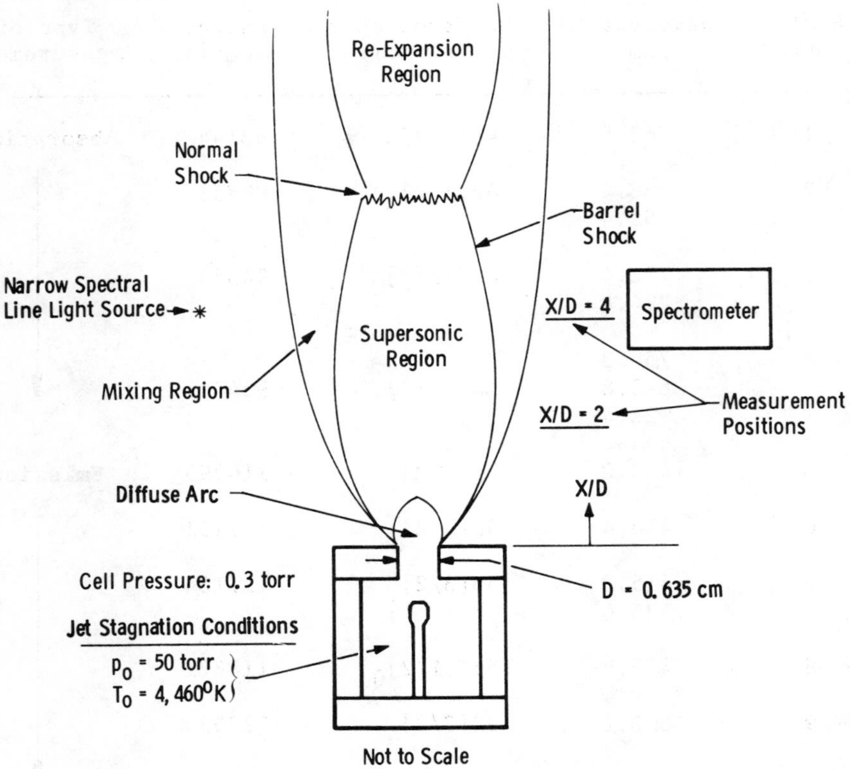

Fig. 1 Arcjet plume schematic showing measurement positions.

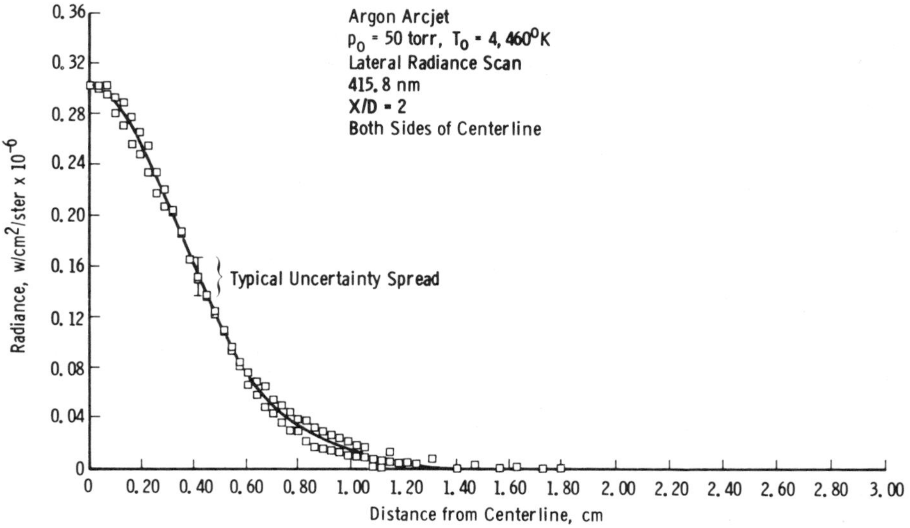

Fig. 2 Typical lateral radiance scan from argon arcjet at X/D = 2, 415.8 nm.

acteristics[17] solution for the plume expansion while the state density is determined iteratively from the data to yield the profile of absorber species which yield the measured transmission.

The emission data were inverted using the Abel inversion and provide spatially resolved number densities of highly excited states of the argon plasma. The specific technique used here, based on fitting the raw data to a series of least squares spline fit polynomials, is analytic and with proper arrangement of terms the accuracy of the resultant emission coefficient is determined as a function of the accuracy of the experimental data, with the effects of the numerical smoothing included.[18]

Results

Experimental Determination of Excited State Densities

Spatial profiles of the absolute radiance were obtained at each of the wavelengths from 415.8 to 604.3 nm, listed in Table 1 at axial positions of two and four nozzle diameters downstream of the nozzle exit plane. A sample of the data obtained at 415.8 nm at X/D = 2 is shown in Fig. 2. The abcissa is the distance from the plume centerline, and the ordinate is measured in W/cm^2-sr. The symbols represent raw data taken from either side of the observed plasma centerline. As is seen, plume symmetry is good, with the slight asymmetry being well within experimental uncertainty. The data shown in Fig. 2 are typical of the profiles at either axial station, although the plume was somewhat larger at X/D = 4, and showed evidence of the presence of the barrel shock. The bars show, typically, two standard deviation error bounds, and the solid line is the result of performing the least squares fit to the raw data. This curve is used for subsequent Abel inversion of the radiance values of local values of emission coefficients.

The emission coefficient is directly proportional to the number density of the emitting specie. Hence, if the number densities n(p) fit into an equilibrium configuration, a Boltzmann plot of the logarithm of the emission coefficient divided by the statistical weight vs. the excitation energy will yield the excitation temperature from the slope of the resulting straight line. The assumption of equilibrium with the free electron density causes the excitation temperature to be identified with the electron temperature and the Law of Mass Action, yields the electron density. Table 2 lists the centerline electron temperature and electron density determined from the emission data in this manner at each of the axial positions X/D = 2 and X/D = 4. Included are the estimates of the 2σ uncertainties of the results based on the 2σ uncertainty of the raw data.

Spatial profiles of the absorption were obtained at each of the wavelengths from 738.4 to 922.4 nm, listed in Table 1. A sample of the data for 763.5 nm at X/D = 2 is shown in Fig. 3. The ordinate is percent transmission and the abcissa is displacement from the observed centerline of the plume. As with the emission data, points from both sides of the observed centerline are plotted, and symmetry is again observed to be good. The line is the result of a polynomial curve fit to the data. The data shown in Fig. 3 are typical of the absorption data obtained at the other wavelengths and the other axial position.

Table 2 Centerline electron temperature and number density

X/D	Electron temperature (°K)	Electron density (cm^{-3})
2	7760 ± 763	$(6.34 \pm 0.45)\ 10^{14}$
4	7200 ± 834	$(2.19 \pm 0.32)\ 10^{14}$

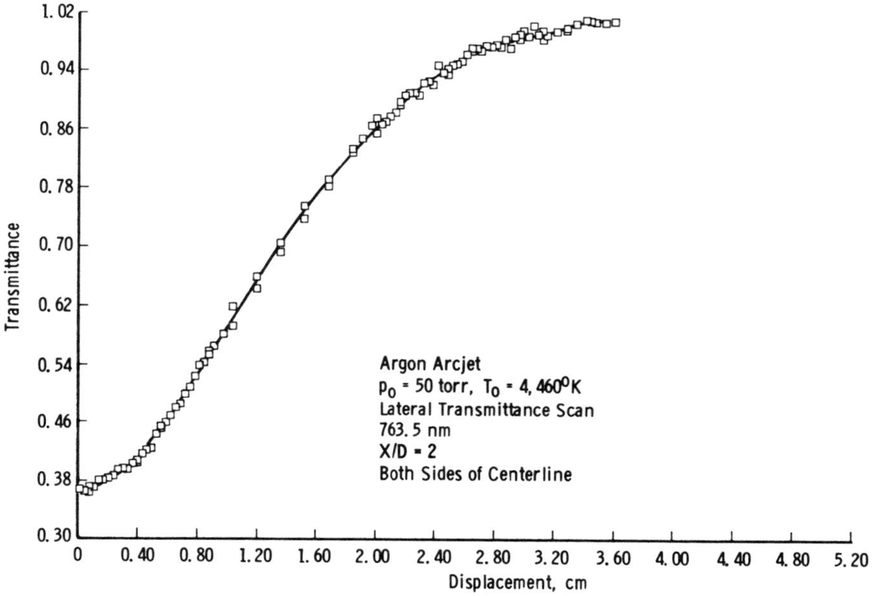

Fig. 3 Typical lateral transmittance scan from argon arcjet, X/D = 2,763.5 nm.

It is interesting to note, comparing Figs. 2 and 3, that the apparent diameter of the plume is larger in absorption than in emission. This is attributed to the presence of a higher than normal concentration of metastable (quantum states $4s[3/2]_2$ and $4s'[1/2]_0$) atoms in the ambient exhaust gases surrounding the plume. The other low lying states, the resonance $4s[3/2]_1$ and $4s'[1/2]_1$ levels, have very strong radiative transitions in the vacuum ultraviolet and the plasma and ambient gases are considered to be optically opaque to these wavelengths, and thus also effectively metastable. The long lifetime of these low-lying states increases their relative concentration in the gases surrounding the plume, and entrainment and mixing of the exhaust gases into the outer periphery of the plume cause it to have a larger diameter in absorption than in emission.

The number density profile for each of the first four excited states as a function of the plume radius was obtained from the absorption data. Comparison on a Boltzmann plot, Fig. 4, of the centerline densities of these states with the centerline densities of the highly excited states, determined from the emission data, illustrates graphically the deviations from equilibrium of the individual states. The data from both

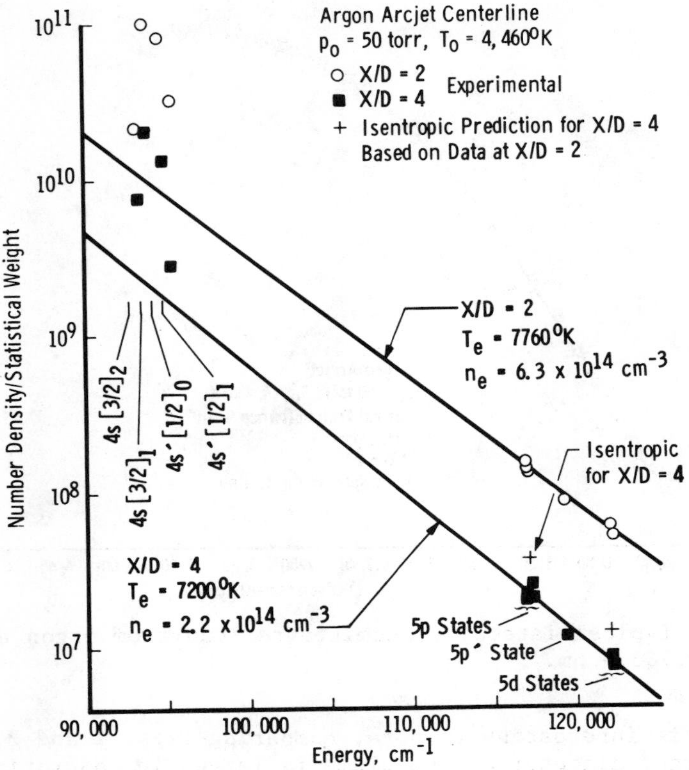

Fig. 4 Boltzmann plot of argon excited state distribution function, arcjet centerline.

axial positions are shown, with the higher density results corresponding to an axial position of X/D = 2. The lines correspond to the least squares straight line curve fit to the upper excited state densities in the semilog plane and have been extended to the lower energies for comparison purposes. As is seen the densities of the first four excited states are markedly above the densities required for equilibrium with the upper excited states. Further, since on a Boltzmann plot those states in collisional equilibrium with each other will fall on the same straight line, there is no apparent single line that would connect more than two of the lower excited states densities. Thus, there is no apparent collective equilibrium description that is satisfactory for these lower states.

Comparison of Experiment with Theory

It is of interest to compare the observed densities at X/D = 4 to the density predicted by an isentropic expansion based upon the distribution at X/D = 2. This is shown by the crosses in Fig. 4 for two of the upper excited states. The observed densities are noticeably less than the computed densities, thus showing that the nonisentropic recombination effects, included by Eq. (1), are necessary for prediction or interpretation of measurements in such an environment.

It is also of interest to compare the energy contained in the excited internal states to the energy of random translational motion of the heavy bodies, characterized by the static temperature. These results are summarized in Table 3. The static properties are determined by a method of characteristics[17] solution based upon stagnation conditions in the arcjet, T_o = 4460°K, p_o = 50 Torr, determined from a Rayleigh heating analysis[19] (constant area heat addition) modified to account for nozzle losses. The losses are obtained by comparing measured and predicted arcjet thrust. As is seen the energy stored in internal energy is approximately 2% of the energy of the random translational motion at X/D = 2 and approximately 5% at X/D = 4. It is to be noted from Fig. 4 that the excited states comprise less than 0.01% of the flow species.

Calculations of the predicted distribution function of excited states at X/D = 4 were made using the ERE system of equations represented by Eqs. (1) and (3). The observed distribution function at X/D = 2 was used for the initial conditions. Those state densities above the $4s[1/2]_1$ state, not measured at X/D = 2, were assumed to be in equilibrium with the free electron density for their initial value. The results of these calculations are summarized in Fig. 5. The abcissa of Fig. 5 is the state excitation energy, and the ordinate is the calculated or experimentally determined density of a particular state ratioed to that density in equilibrium with the free electron density. The experimental results at both X/D = 2 and X/D = 4, denoted by the closed symbols, as well as the calculated densities at X/D = 4, denoted by the open symbols, are

Table 3 Comparison of internal and kinetic energy of random motion

X/D	T_s (°K)	n_s (1/cm^3)	Kinetic energy (erg/cm^3)	Internal energy (erg/cm^3)
2	520	4.3×10^{15}	464	10.8
4	210	1.1×10^{15}	45	2.2

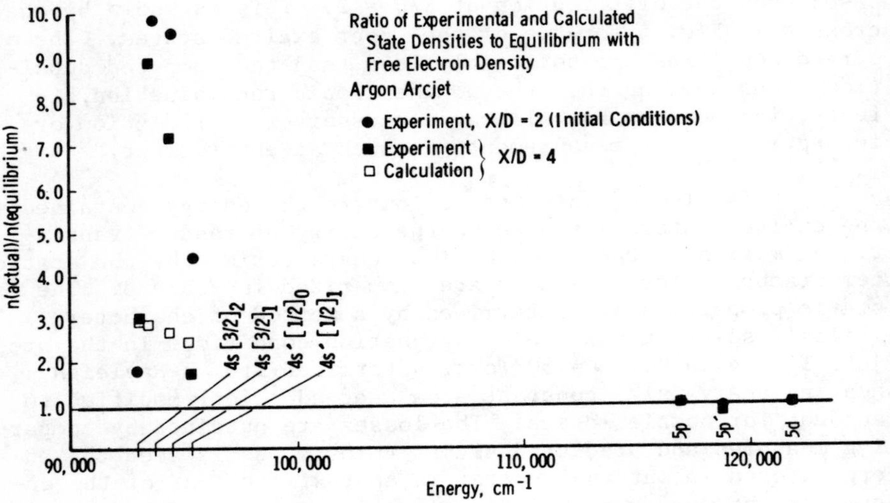

Fig. 5 Ratio of experimental and calculated number densities to Saha equilibrium density.

shown. As is seen, the experimentally determined densities at X/D = 4 show a much more pronounced nonequilibrium character than the ERE calculations, and the densities determined by the calculations relax toward the equilibrium value much faster than was actually observed.

The calculated distribution function for the four lowest excited states shows that the states are near to an equilibrium configuration among themselves although the departure from equilibrium with the other states is significant. This is evidenced in Fig. 5 by the ratios n(actual)/n(equilibrium) for these four levels all being near the same value. This implies that the Gryzinski formulation of the collision cross sections, used for the rate coefficient determinations here, predicts a much larger value for these lower states than is actually the case. Thus the collisional coupling between the states is overestimated so that the distribution drives toward the equilibrium configuration faster in the model than in the actual case. This is in spite of the fact that quasisteady state calculations of recombination rates using Gryzinski cross sections appear to agree with experimentally observed recombination rates.[20]

Summary and Conclusion

The distribution of excited state densities determined in the freely expanding plume of an argon arcjet shows decidedly nonequilibrium characteristics. The distribution of upper excited states indicates that these states are in equilibrium with the free electron density. The four lowest excited state densities are elevated above this equilibrium configuration by factors ranging from two to ten. The observed distribution does not lend itself to description by any equilibrium based theory since the states evidencing nonequilibrium did not show any distinguishing relationships among themselves. Calculations of the detailed distribution function based upon a complete, coupled, transient modeling of the time development of each state showed nonequilibrium results also, although the calculated relaxation toward the equilibrium distribution was much faster than was observed experimentally. The calculated densities for the nonequilibrium states showed a tendency to establish a collisional equilibrium among themselves, in contrast to the experimentally measured densities. These observations imply that the semiclassical Gryzinski cross sections, used to determine the collisional rate coefficient, overestimate the physical cross section and lead to stronger than actual collisional coupling between the states. Thus, the more correct (in principle) quantum mechanical cross section calculation should be used. The nature of the observed distribution function suggests strongly that detailed coupled calculations, such as were attempted here, are necessary in order to predict the excited state distribution function, or, conversely, to interpret correctly the results of measurements from such environments. Finally, the energy content of the observed distribution function is approximately an order of magnitude above that predicted by the Law of Mass Action, and, as the plume expands, the energy content of the excited state distribution function increases relative to the kinetic energy of the random translational motion of the gas, characterized by the static temperature.

References

[1]Bates, D. R., Kingston, A. E., and McWhirter, R. W. P., "Recombination Between Electrons and Atomic Ions." Proceedings of the Royal Society of London, Series A, Vol. 267, May 1962, pp. 297-312.

[2]Limbaugh, C. C., McGregor, W. K., and Mason, A. A., "Numerical Study of the Early Population Density Relaxation of Thermal Hydrogen Plasmas," Arnold Engineering Development Center, Arnold Air Force Station, Tenn., AEDC-TR-69-156, October 1969.

[3]Limbaugh, C. C., "The Transient Behavior of Collisional Radiative Recombination in Atomic Helium Plasmas," Ph.d. Dissertation, University of Tennessee, Knoxville, Tenn., December 1971.

[4]Limbaugh, C. C. and Mason, A. A., "Validity of the Quasisteady State and Collisional-Radiative Recombination for Helium Plasmas. I. Pure Afterglows," Physical Review A., Vol. 4, No. 6, December 1971, pp. 2368-2377.

[5]Limbaugh, C. C. and McGregor, W. K., "The Transient Behavior of Quantum State Density Distributions in Atomic Helium Plasmas," Eighth Symposium (International) on Rarefied Gas Dynamics, Stanford, Calif., July 1972.

[6]Park, C., "Comparison of Electron and Electronic Temperature in Recombining Nozzle Flow of Ionized Nitrogen-Hydrogen Mixture. Part 1. Theory," Journal of Plasma Physics, Vol. 9, Part 2, April 1973, pp. 187-216.

[7]Park, C., "Comparison of Electron and Electronic Temperature in Recombining Nozzle Flows of Ionized Nitrogen-Hydrogen Mixture. Part 2. Experiment," Journal of Plasma Physics, Vol. 9, Part 2, April 1973, pp. 217-234.

[8]Hogarth, W. L. and McElwain, D. L. S., "Internal Relaxation, Ionization, and Recombination in a Dense Hydrogen Plasma. II. Population Distributions and Rate Coefficients," Proceedings of the Royal Society, Series A, Vol. 345, August 1975, pp. 265-276.

[9]Hey, J. D., "Criteria for Local Thermal Equilibrium in Non-Hydrogenic Plasmas," Journal of Quantitative Spectroscopy and Radiative Transfer, Vol. 16, No. 1, 1976, pp. 69-75.

[10]McGregor, W. K. and Brewer, L. E., "Spectroscopy of Supersonic Plasma. III. Electron Temperature Measurements in Argon Plasma by Two Independent Methods," Arnold Engineering Development Center, Arnold Air Force Station, Tenn., AEDC-TR-65-131, November 1965.

[11]Gryzinski, M., "Classical Theory of Atomic Collisions. I. Theory of Inelastic Collisions," Physical Review, Vol. 138, No. 2a, April 1965, pp. A336-A358.

[12]Seaton, J. J., "Radiative Recombination of Hydrogenic Ions," Monthly Notices of the Royal Astronomical Society, Vol. 119, No. 2, 1959, pp. 81-89.

[13]Wiese, W. L., Smith, M. W., and Miles, B. M., Atomic Transition Probabilities, Vol. II., Sodium Through Calcium, NSRDS-NBS 22, National Bureau of Standards, Washington, D. C., 1969.

[14]Green, L. F., Rush, R. P., and Chandler, C. D., "Oscillator Strengths and Matrix Elements for the Electric Dipole Moment for Hydrogen," Astrophysical Journal, Vol. 125, No. 3, Supplement 26, 1957, pp. 835.

[15]Moore, C. E., Atomic Energy Levels, Vol. I, National Bureau of Standards, Washington, D. C., 1949.

[16]Ashkenas, H. and Sherman, F. S., "The Structure and Utilization of Supersonic Freejets in Low Density Wind Tunnels," Rarefied Gas Dynamics, edited by J. H. de Leeuw, Vol. II, Supplement 3, Academic Press, New York, 1966, pp. 84-105.

[17]Fox, J. H., "An Axially Symmetric, Inviscid, Real Gas, Non-Isoenergetic Solution by the Method of Characteristics," Masters Thesis, University of Tennessee, Knoxville, Tenn., 1968.

[18]Shelby, R. T., "Abel Inversion Error Propagation Analysis," Masters Thesis, University of Tennessee, Knoxville, Tenn., 1976.

[19]Bryson, R. J., "A Method for Determining the Bulk Properties of Arc-Heated Argon," Arnold Engineering Development Center, Arnold Air Force Station, Tenn., AEDC-TR-69-125, June 1969.

[20]Wanless, D., "Electron-Ion Recombination in Argon," Journal of Physics B: Atomic and Molecular Physics, Vol. 4, April 1971, pp. 522-527.

RELAXATION OF HOT ATOMS WITH CHEMICAL REACTION

Katsuhisa Koura *

National Aerospace Laboratory, Chofu, Tokyo, Japan

Abstract

The relaxation of the hot-atom velocity distribution in the hot-atom reaction is studied by solving the Boltzmann equation with the Monte Carlo simulation. The explicit time-dependent hot-atom velocity distribution, temperature, and rate constant from initial to steady states are obtained for a simple model system corresponding to the $F-H_2$ system with the hot-atom reaction $F + H_2 \rightarrow HF + H$, where hot atoms are dispersed in the heat bath of surrounding molecules without internal degrees of freedom and the cross sections are chosen as a simple hard-sphere model. The high-temperature steady state (HTS), where the hot-atom velocity distribution is different from the Maxwell distribution, exists for the ratio of reactive to elastic cross sections $\sigma_r/\sigma_e \gtrsim 5$ in consistence with the prediction of Keizer with the Maxwell distribution approximation, which indicates that HTS may not exist for the $F-H_2$ system with $\sigma_r/\sigma_e \simeq 2$ at HTS.

Introduction

A chemical reaction yields the nonequilibrium velocity distribution and reaction rate, which was studied originally by Curtiss[1] and Prigogine and Xhrouet[2] with the Boltzmann equation. In the hot-atom reaction, where hot atoms produced by nuclear or photochemical processes and atomic beam

---

Presented as Paper 48 at the 10th International Symposium on Rarefied Gas Dynamics, Aspen, Colo., July 19-23, 1976.
The author wishes to thank Professor N. Oshima and Dr. I. Wada for helpful discussions.

*Principal Research Officer.

equipments are cooled by surrounding gases and enter into chemical reactions with surrounding molecules, Keizer[3] developed a steady-state theory based on the Boltzmann equation and indicated that the low- (LTS) or high-temperature nonequilibrium steady state (HTS) exists in consequence of a balancing between the heating and cooling effects due to reactive and nonreactive collisions, where the hot-atom velocity distribution is approximated by the Maxwell distribution. In an earlier paper,[4] it was shown that LTS or HTS, where the hot-atom velocity distribution is different from the Maxwell distribution, exists in consistence with the prediction of Keizer.

In this paper, the relaxation of the hot-atom velocity distribution in the hot-atom reaction corresponding to $F + H_2 \rightarrow HF + H$[5] is studied by solving the time-dependent Boltzmann equation with the Monte Carlo simulation.[4] The explicit time-dependent hot-atom velocity distribution, temperature, and rate constant from initial to steady states are obtained.

Model System

In a spatially homogeneous system where hot atoms A are dispersed in the heat bath of surrounding molecules R without internal degrees of freedom and undergo the hot-atom reaction

$$A + R \rightarrow \text{products} \tag{1}$$

the velocity distribution function $f(\mathbf{v},t)$ of A with velocity $\mathbf{v}$ at time t obeys the Boltzmann equation[3]

$$\partial f/\partial t = \int (f'f'_R - ff_R) g I_e(g,\Omega)\, d\Omega\, d\mathbf{v}_R - \int ff_R g\sigma_r(g)\, d\mathbf{v}_R \tag{2}$$

where $I_e(g,\Omega)$ is the differential elastic cross section for the A-R scattering at a solid angle Ω with the relative velocity $g = |\mathbf{v} - \mathbf{v}_R|$, $\mathbf{v}_R$ is the velocity of R, and $\sigma_r(g)$ is the total reaction cross section for the hot-atom reaction (1).

The velocity distribution function $f_R(\mathbf{v}_R)$ of R is assumed to be the Maxwell equilibrium distribution with the constant ambient temperature T_R and the number density n_R

$$f_R = n_R(m_R/2\pi kT_R)^{3/2} \exp(-m_R v_R^2/2kT_R) \tag{3}$$

where $v_R = |\mathbf{v}_R|$, m_R is the mass of R, and k is the Boltzmann constant. The hot-atom velocity distribution function $f(\mathbf{v},t)$

described by Eq. (2) is obtained by the Monte Carlo simulation.[4] The cross sections are chosen as a simple hard-sphere model[3]

$$I_e(g,\Omega) = \sigma_e/4\pi \tag{4}$$

$$\begin{aligned} \sigma_r(g) &= \sigma_r, \; E_1 < E < E_2 \\ &= 0, \text{ elsewhere} \end{aligned} \tag{5}$$

where $E = \mu g^2/2$, $\mu = mm_R/(m + m_R)$, and m is the mass of A. The initial hot-atom velocity distribution is taken to be the δ function distribution (δ) with the temperature T(0) and the number density n(0)

$$f(0) = n(0)\ \delta\{v - [3kT(0)/m]^{1/2}\}/4\pi v^2 \tag{6}$$

or the Maxwell distribution (MD)

$$f(0) = n(0)[m/2\pi kT(0)]^{3/2}\ \exp[-mv^2/2kT(0)] \tag{7}$$

where $v = |\mathbf{v}|$.

Results and Discussion

For the hot-atom reaction $1 \ll E_2/kT_R \ll T(0)/T_R$, the energy ratios E_1/kT_R and E_2/kT_R, the temperature ratio $T(0)/T_R$, and the mass ratio m/m_R are taken to be 0, 100, 1000, and 10, respectively, in correspondence to the F-H_2 system[5] with the hot-atom reaction $F + H_2 \rightarrow HF + H$ ($E_2 \simeq 10$ eV and $T_R \simeq 10^3$ °K).

The time evolution of the hot-atom temperature, $T = (m/3kn)\int fv^2\, d\mathbf{v}$, is presented in Fig. 1 for the initial δ, where n is the number density of hot atoms and t_0 is the mean elastic collision time, defined by $t_0 = [n_R\sigma_e(8kT_R/\pi\mu)^{1/2}]^{-1}$. The temperature ratio T/T_R decreases to the steady-state value, which is much greater than unity $[T/T_R \gtrsim (2m/3\mu)(E_2/kT_R)]$ for $\sigma_r/\sigma_e \gtrsim 5$ (HTS) or less than unity for $\sigma_r/\sigma_e \lesssim 4$ (LTS). T/T_R at HTS increases with the increase of σ_r/σ_e, since the heating effect[3] increases. For LTS, a high-temperature quasisteady state exists, which is indicated at $t/t_0 \simeq 2$ for $\sigma_r/\sigma_e = 3$. For the initial MD, the evolution of T/T_R is shown for $\sigma_r/\sigma_e = 5$. T/T_R increases to a maximum value and approaches the same steady-state value with δ, since the heating is stronger than the cooling at the initial stage.

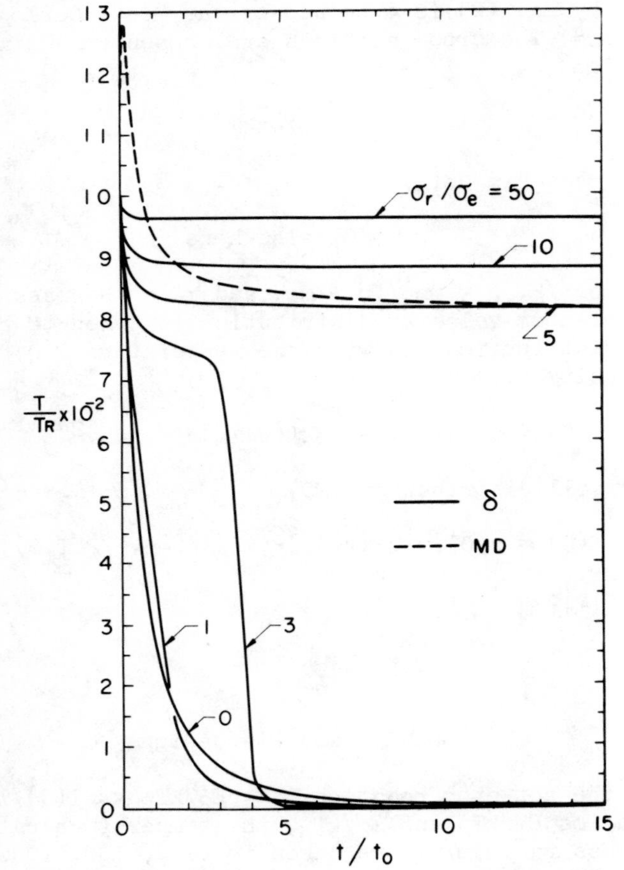

Fig. 1 Time evolution of hot-atom temperature.

The ratio of the hot-atom energy kT at HTS to the upper energy limit E_2 is presented in Fig. 2. The results[4] for $m/m_R \lesssim 1$ are also shown in comparison with that of Keizer. The lower limit of σ_r/σ_e for the existence of HTS for $m/m_R = 10$ is about 5, which is nearly equal to that for $m/m_R = 1$ and that of Keizer for $m/m_R \ll 1$. Therefore, HTS may not exist for the F-H_2 system[5,6] with $\sigma_r/\sigma_e \simeq 2$ at HTS, which is evaluated by $\sigma_r \simeq 6$ (Å^2) and $\sigma_e = 5.894E(\mathrm{eV})^{-0.325}$ (Å^2) at $E \simeq E_2 \simeq 10$ eV. The minimum value of kT/E_2 is estimated to be $kT/E_2 \simeq 2m/3\mu$ by taking account of a near δ function shape of the hot-atom velocity distribution (Fig. 4), which leads to $kT/E_2 \simeq 2m/3m_R$ for $m/m_R \gg 1$. The sensitivity of kT/E_2 to σ_r/σ_e decreases with the increase of E_2/kT_R.

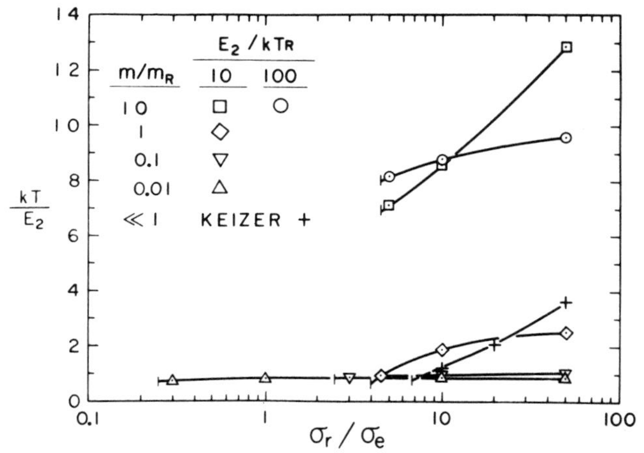

Fig. 2 Hot-atom temperature at high-temperature steady state in comparison with that of Maxwell distribution approximation by Keizer.

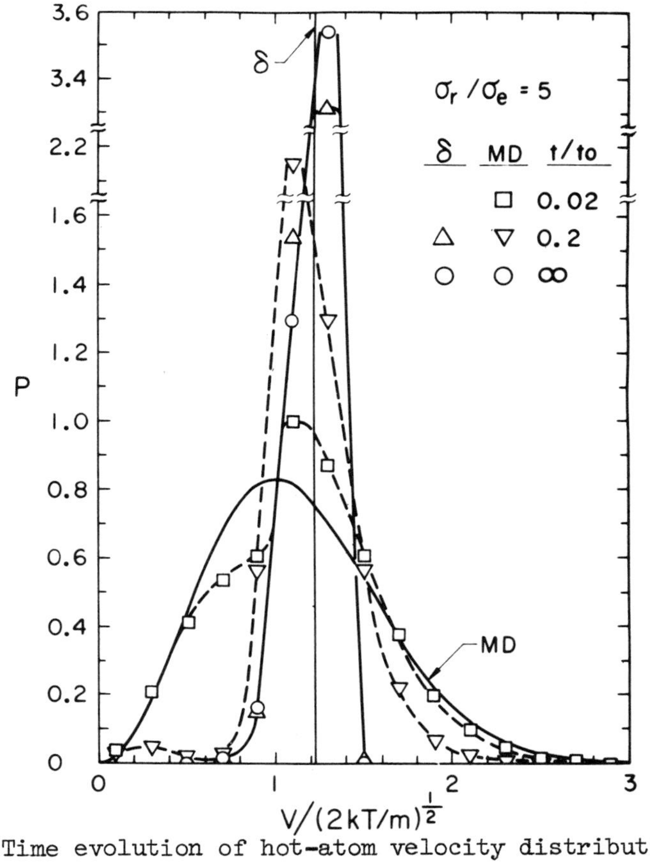

Fig. 3 Time evolution of hot-atom velocity distribution.

The typical time evolution of the hot-atom velocity distribution function, $P = 4\pi v^2 f(v)/n$, is presented in Fig. 3 for $\sigma_r/\sigma_e = 5$ (HTS). The initial δ function peak decreases in consequence of the cooling before the reaction begins ($t/t_0 \lesssim 0.2$) and, after the reaction begins effectively, increases to the steady-state hot peak (∞) just above the velocity $v_{E2} \equiv (2E_2/\mu)^{1/2}$ in consequence of the removing of cooled atoms with $v < v_{E2}$ by the reaction (heating) and the cooling of hot atoms to $v \simeq v_{E2}$. For the initial MD, the hot and cool peaks appear at the initial stage ($t/t_0 = 0.02$). The hot peak increases to the same steady-state value with δ in consequence of the heating even at the initial stage and the cooling to $v \simeq v_{E2}$. The cool peak decreases and disappears at the steady state in consequence of the removing of low-energy atoms with $v < v_{E2}$ by the reaction.

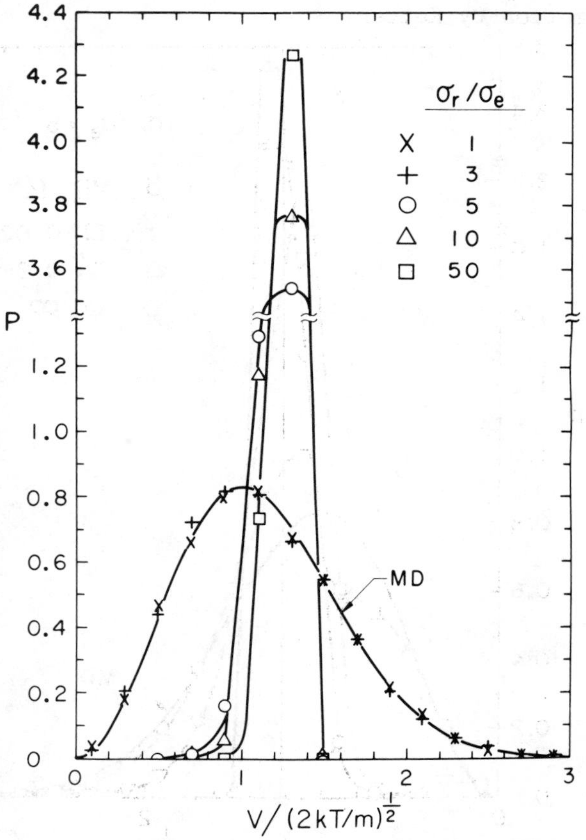

Fig. 4 Steady-state hot-atom velocity distribution.

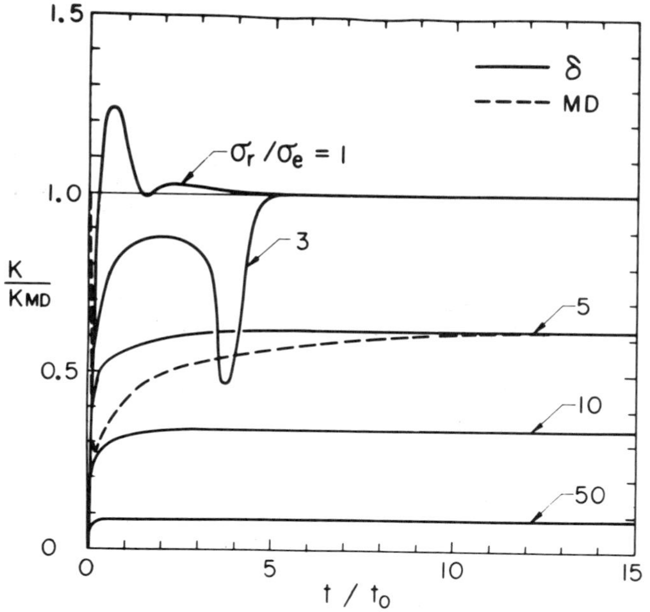

Fig. 5 Time evolution of ratio of rate constant to that of Maxwell distribution approximation by Keizer.

The steady-state P is presented in Fig. 4. For LTS ($\sigma_r/\sigma_e \lesssim 4$), P indicates a near Maxwell distribution (MD). For HTS ($\sigma_r/\sigma_e \gtrsim 5$), P indicates a near δ function shape with the hot peak just above v_{E2}. The hot peak increases with the increase of σ_r/σ_e, since the heating effect increases.

The time evolution of the ratio of the rate constant, $K = (1/nn_R)\int ff_R g\sigma_r(g)\, dv\, dv_R$, to that of the MD approximation by Keizer, $K_{MD} = \sigma_r(8kT_M/\pi\mu)^{1/2}[(1 + E_1/kT_M)\exp(-E_1/kT_M) - (1 + E_2/kT_M)\exp(-E_2/kT_M)]$, where $T_M = (mT_R + m_RT)/(m + m_R)$, is presented in Fig. 5 for the initial δ. The evolution of K/K_{MD} with the initial value of zero indicates a rather complicated behavior for LTS. The steady-state K/K_{MD} for HTS is less than unity and decreases with the increase of σ_r/σ_e, since the population of cooled atoms with $v < v_{E2}$ decreases (Fig. 4). The decrease of K from K_{MD} is more than about 40%. For the initial MD, the evolution of K/K_{MD} is shown for $\sigma_r/\sigma_e = 5$. K/K_{MD} decreases from the initial value of unity to a minimum value and approaches the same steady-state value with δ.

The simple hard-sphere model yields the lower limit $\sigma_r/\sigma_e \simeq 5$ for the existence of HTS, which indicates that HTS may not exist for the F-H_2 system with $\sigma_r/\sigma_e \simeq 2$ at HTS.

Work on the $F-H_2$ system with more realistic cross sections[5,6] is in progress.

References

[1]Curtiss, C. F., "The Equilibrium Assumption in the Theory of Absolute Reaction Rates," University of Wisconsin, Naval Research Laboratory Report CM-476, June 1948.

[2]Prigogine, I. and Xhrouet, E., "On the Perturbation of Maxwell Distribution Function by Chemical Reactions in Gases," Physica, Vol. XV, Dec. 1949, pp. 913-932.

[3]Keizer, J., "Steady State Theory of Hot Atom Reactions," The Journal of Chemical Physics, Vol. 58, May 1973, pp. 4524 -4535.

[4]Koura, K., "Nonequilibrium Velocity Distribution and Reaction Rate in Hot-Atom Reaction," The Journal of Chemical Physics (to be published).

[5]Feng, D. F., Grant, E. R., and Root, J. W., "Hydrogen Abstraction Reactions by Atomic Fluorine. V. Time-Independent Nonthermal Rate Constants for the $^{18}F + H_2$ and $^{18}F + D_2$ Reactions," The Journal of Chemical Physics, Vol. 64, April 1976, pp. 3450-3456.

[6]Riley, M. E. and Matzen, M. K., "Non-Maxwellian H and F Velocity Distributions in an H_2-F_2 Reaction," The Journal of Chemical Physics, Vol. 63, Dec. 1975, pp. 4787-4799.

CHAPTER IX—IONIZED GASES

PREDICTION OF POSITIVE ION COLLECTION BY A ROCKET-BORNE MASS SPECTROMETER

T. Sugimura*

TRW Defense and Space Systems Group,
Redondo Beach, Calif.

Abstract

The Monte Carlo direct simulation method has been used to describe the flow about a rocket borne mass spectrometer through the ionosphere. In order to predict the flux of positive ions to the probe the flowfield due to neutrals, ions, and electrons has been determined with a consistent Poisson solution for the electric potential. A systematic parametric variation for the positive ion collection has been completed for the following parameters: speed ratio 3.10, nondimensional front face potential -200, zero sidewall potential, Debye numbers from 0.01 to 1000, and Knudsen numbers from 0.007 to 1000. The smallest value of the Knudsen number, 0.007, corresponds to an altitude of 70 km, the lowest altitude of practical interest. It is found that at 70 km the flowfield is nearly continuum, so that the Knudsen numbers considered in the present study include the entire range of rarefaction.

Nomenclature

e = charge on electron
k = Boltzmann constant

Presented as Paper 59 at the 10th International Symposium on Rarefied Gas Cynamics, Aspen, Colo., July 19-23, 1976. This research was sponsored by the Air Force Geophysics Laboratories, Air Force Systems Command, under contract F19628-74-C-0128. The author is grateful to C. Sherman of AFGL for many helpful discussions and to M. E. Gardner for assistance in carrying out the numerical computations.

*Member Technical Staff, Engineering Science Lab.

m, m_i = mass of neutral molecules, ions
n, n_i, n_e = number density of neutrals, ions, electrons
rd = disk radius
v_m = most probable molecular speed = $\sqrt{2kT/m}$
x = axial coordinate
A_D = disk area
C_I = ion flux coefficient = $\dot{N}/n_{i\infty}U_\infty A_D$
D = orifice plate diameter
Kn = Knudsen number = λ_∞/D
$\dot{N}$ = flux of ions
R = orifice plate radius
S = speed ratio = U_∞/V_m
T = temperature
U = mean gas velocity in x direction
λ = neutral mean free path
λ_D = ion Debye length = $\sqrt{kT_\infty/4\pi e^2 n_{i\infty}}$
ϕ = electrical potential
Φ = nondimensional electric potential = $e\phi/kT$

Subscripts

∞ = conditions in freestream
o = orifice plate value

Introduction

A vehicle carrying aeronometric measuring instruments through the ionosphere will travel through the entire spectrum of flow rarefaction, at low altitudes, through gas that is so dense that the flow about the vehicle may be considered to be that of a continuum; i.e., one need not take explicit account of the molecular structure of the atmosphere. At higher altitudes, where the molecule mean free paths are not negligibly small compared to the vehicle dimension, the flow becomes rarefied; i.e., the molecular velocity distribution departs significantly from the nearly Boltzmann distribution that characterizes near-equilibrium continuum flows. At extremely high altitudes, the flow is well described by the concept of a collisionless flow, the well-known free-molecule limit.

The proper theoretical description of the flow is a matter of considerable importance, especially for sampling using mass spectrometers, since a connection must be established between the properties of the ambient atmosphere and the strength and chemical state of the flux entering the mass spectrometer proper. The influences that distort the sample from the ambient state will depend on the character of the external flow about the body. Although it recognizes the importance of internal distorting influences, this work is directed solely to the external disturbances.

In order to enhance the flux of charged species to the measuring instrument, experimenters often maintain the instrument housing at an electrostatic potential above or below the potential of the ambient gas. This introduces additional parameters that affect the collection process. The Monte Carlo direct simulation method may be considered a numerical experiment carried out on a digital computer, where the paths of molecules flowing past a body are followed, and collisions in the gas are computed by statistical sampling. A complete description of the flowfield and surface fluxes is obtained. The simulation method produces a solution of the Boltzmann equation[1]; hence, the solution is valid at all density levels in the atmosphere.

In the work discussed here, the simulation technique was extended to describe the flow of positive ions about a flat-faced cylinder traveling through a weakly ionized gas. This geometry was chosen to represent the shapes currently in extensive use to house rocket-borne mass spectrometers. A typical arrangement is shown in Fig. 1. The front face of the

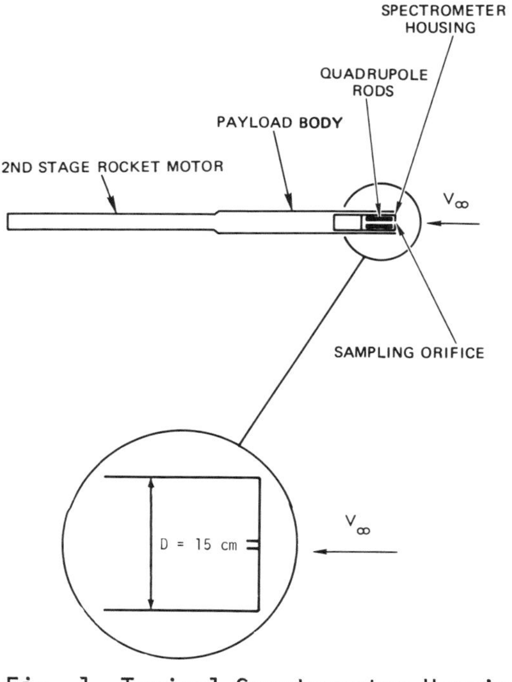

Fig. 1 Typical Spectrometer Housing.

cylinder, hereinafter called the orifice plate, was assumed to be held at a negative potential with respect to the ambient gas, and the positive ion flux induced by the resulting electric field, as well as the density, temperature, and velocity of the ions in the flowfield about the cylinder were computed.

As discussed by Bird[1], the technique produces a solution of the Boltzmann equation which is, in conventional notation

$$\vec{U}_n \cdot \frac{\partial f_n}{\partial \vec{x}} = \int (f'_n f' - f_n f)\, v_f b \, db \, d\varepsilon \, dc$$

where the left-hand side of the equation represents the substantial derivative of the neutral distribution function f_n, and the right-hand side represents the collision term.

Since the ionosphere contains charged species, some additional features were required to use the technique to determine ion collection. The ambient gas was assumed to be composed of neutral molecules, positive ions, and electrons. The concentration of ions and electrons relative to neutrals is actually so low in the ionosphere that the former may be considered trace species. Thus, the flow of the charged species will be affected by the neutrals, but not vice versa. Likewise, binary encounters between charged particles can be neglected: all charged particle interaction arises through the electric field. Thus, the equation that governs the positive ion behavior is the Boltzmann equation with the body force term for electric field added is

$$\vec{U}_i \cdot \frac{\partial f_i}{\partial \vec{x}} + \frac{e\vec{E}}{m} \cdot \frac{\partial f_i}{\partial \vec{c}_i} = \int (f'_i f'_n - f_i f_n)\, v_r b \, db \, d\varepsilon \, dc$$

The right-hand collision term now accounts for only ion-neutral encounters.

The quantity of primary concern is the flux of positive ions to the stagnation point of the cylinder orifice plate. The net current is not of interest; hence the electrons in the gas are of interest insofar as their spatial distribution affects the net space charge. The electrons were prescribed to be an inviscid gas in Boltzmann equilibrium under the electric field.

The body geometry considered in this study was a right circular cylinder at zero angle of attack. The electric potential of the face of the cylinder with respect to the ambient gas was specified as an input parameter, whereas the side of the cylinder was assumed to be at zero potential with respect to the ambient gas. This means that the small charge buildup at zero current on the actual flight article (the floating potential) was assumed to be zero. Only that portion of the cylinder side was simulated which was required to compute accurately ion current to the front face.

The hard-sphere interaction law was used for neutral-neutral and neutral-charged particle interactions. The numerical magnitudes of collision cross sections for neutrals and ions were taken to be identical in the present study and the ion and electron temperatures in the ambient gas at infinity were assumed equal to the neutral temperature at infinity for these calculations. The method, however, could treat arbitrary values of these parameters which are input variables. The effects of varying the speed ratio, front face potential and the effects of different cross sections were discussed by Sugimura and Vogenitz[2].

To complete the solution procedure for ions, the Poisson equation

$$\nabla^2\phi = 4\pi e\ (n_i - n_e)$$

which governs the electric potential, must be solved in conjunction with the Boltzmann equation in such a way that the electric field solution is consistent with the space-charge distribution.

The present study is a continuation of work reported previously,[2-4] where the details of the Monte Carlo simulation procedure and the assumptions made to perform the calculations have been described in detail and will not be repeated here. The results presented include a detailed examination of the transition flow regime through a systematic study of the positive ion collection for the following range of parameters: speed ratio S = 3.10, nondimensional front face potential Φ_0 = -200, Debye numbers λ_D/D = 0.01 through 1000, and Knudsen numbers λ_∞/D = 0.007 through 1000. The smallest value of the Knudsen number corresponds to 70 km, the lowest altitude of practical interest. In Fig. 2, the range of Debye numbers shown as a function of Knudsen number or altitude was taken from Narcisi.[5] This figure shows the maximum and minimum Debye numbers expected for a given altitude.

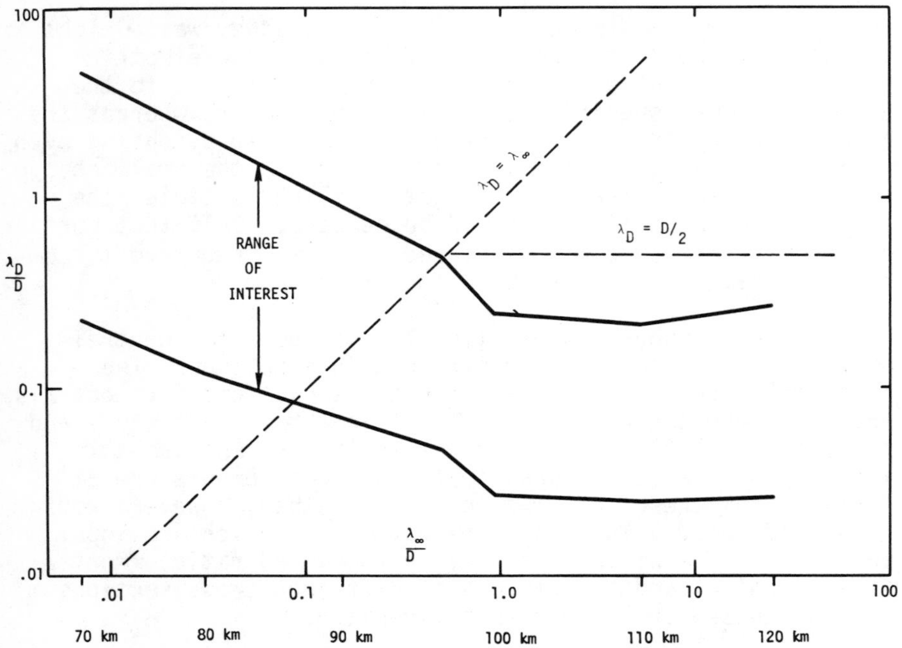

Fig. 2 Limits for Debye Number with Altitude.

Results

Free Molecule Flow

The ion flux variation across the front face of the collecting disk is shown in Fig. 3 for a Knudsen number of $\lambda_\infty/D = 10$ for Debye numbers from 0.10 to 1000. Also shown in this figure is a result for a Knudsen number and Debye number both equal to 1000. It can be seen that the two results for a Debye number of 1000 are very similar and represent the collisionless and nearly collisionless cases. The flux for $\lambda_\infty/D = 10$ is slightly lower than the flux for $\lambda_\infty/D = 1000$, which shows the effect of a finite number of collisions. However, the decrease in the flux coefficient at a value of $r_D/R = 0.07$ is due to a small sample size, at least a factor of ten less than the sample size over the entire front face, and the stagnation point flux is found by extrapolating through the other points. A typical sample size at the stagnation point was approximately 3000 particles. The results for a Debye number of 1.0 and 1000 show little difference, indicating again that the Debye number range between 1.0 and 1000 results in a Laplace solution of Poisson's equation. As the Debye number is reduced further, the ion flux coefficient

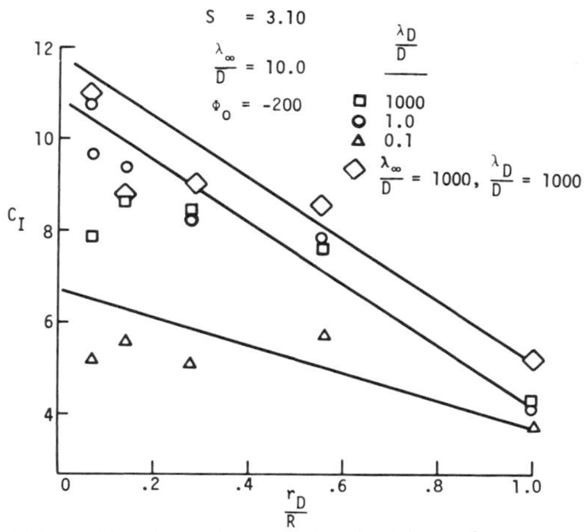

Fig. 3 Ion Flux Variation Across Front Face for Near Free Molecule Flow

also decreases. For all of these results, the linear variation of the ion flux with radius which characterizes the near-free molecule flow is found again.

Near Free Molecule Flow

The ion flux distribution with radius for a Knudsen number of 1.0 is shown in Fig. 4. By comparing these results with those for a Knudsen number of 10, it can be seen that the effect of collisions has reduced the ion flux by approximately one-third. Again for large Debye numbers the flux variation across the plate is about a factor of 2, whereas for the small Debye numbers, e.g., $\lambda_D/D = 0.05$, the flux is approximately constant with radius. This indicates that the sheath is very close to the body surface for small Debye numbers and the electric field is approximately constant.

Transition Flow (90 km)

The density variation for both ions and neutrals for a Knudsen number of 0.20 is shown in Fig. 5 for Debye numbers 0.02 and 1000. It is seen that there is a considerable separation between the ions and neutrals in contrast to the result at 70 km discussed later. There is also no indication of a gasdynamic shock in the neutral particles which is expected in the continuum case.

Fig. 6 shows the flux variation across the front face, where again the stagnation point flux is determined by extrap-

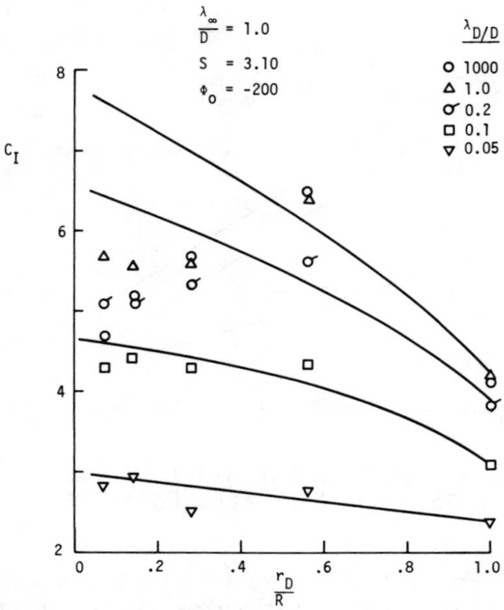

Fig. 4 Ion Flux Distribution Along Front Face for Knudsen Number 1.0

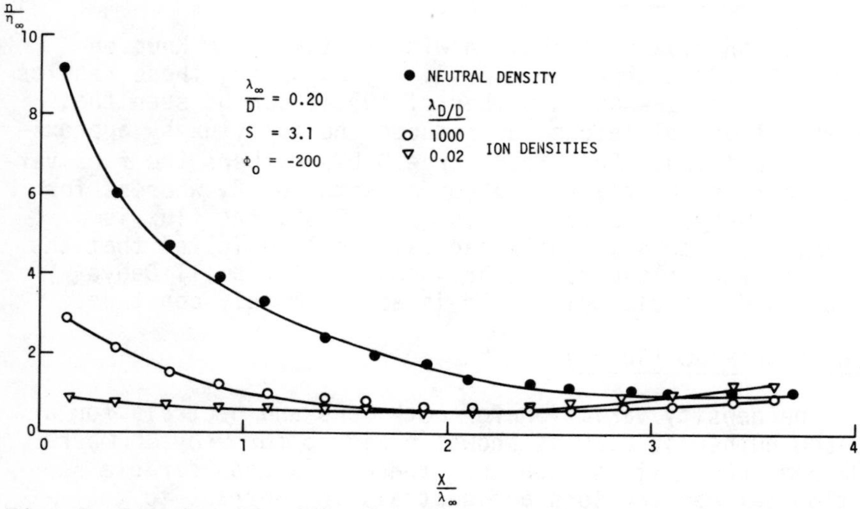

Fig. 5 Density Distribution Along Stagnation Line for 90 km

olation. In extrapolating by eye, the values near the stagnation have less weight because of the smaller number of samples at these interior points, Similar to other Knudsen numbers, the flux becomes uniform as the Debye number decreases.

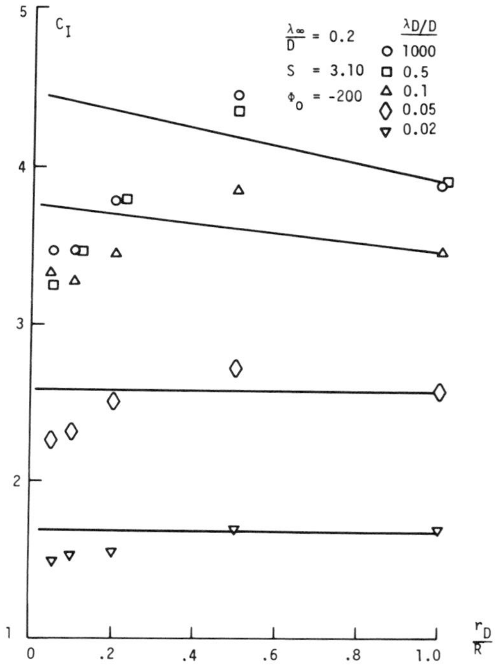

Fig. 6 Ion Flux Distribution along Front Face at 90 km.

The variation of the electric potential along the stagnation line is shown in Fig. 7. It is seen that for Debye numbers greater than 0.5 the potential distribution does not change and is given by the Laplace solution. For smaller Debye numbers, the potential decreases faster and the sheath becomes smaller. For a Debye number of 0.02, the potential becomes positive at approximately $X/\lambda_\infty \simeq 2.5$.

Near Continuum Flow (70 km)

Fig. 8 shows the density distributions along the stagnation line for both ions and neutral particles for two extreme values of the Debye numbers, 0.03 and 1000, for a Knudsen number of $\lambda_\infty/D = 0.007$. It should be emphasized that the abscissa is measured in mean free path units X/λ_∞ so that a value of $X/\lambda_\infty = 20$ is in terms of the diameter

$$X/D = (X/\lambda_\infty)(\lambda_\infty/D) = (20)(0.007) = 0.14$$

It is seen from Fig. 8 that the first effects of the body on the neutral gas particles occurs at approximately $X/\lambda_\infty \simeq 20$ or $X \simeq 0.14D$.

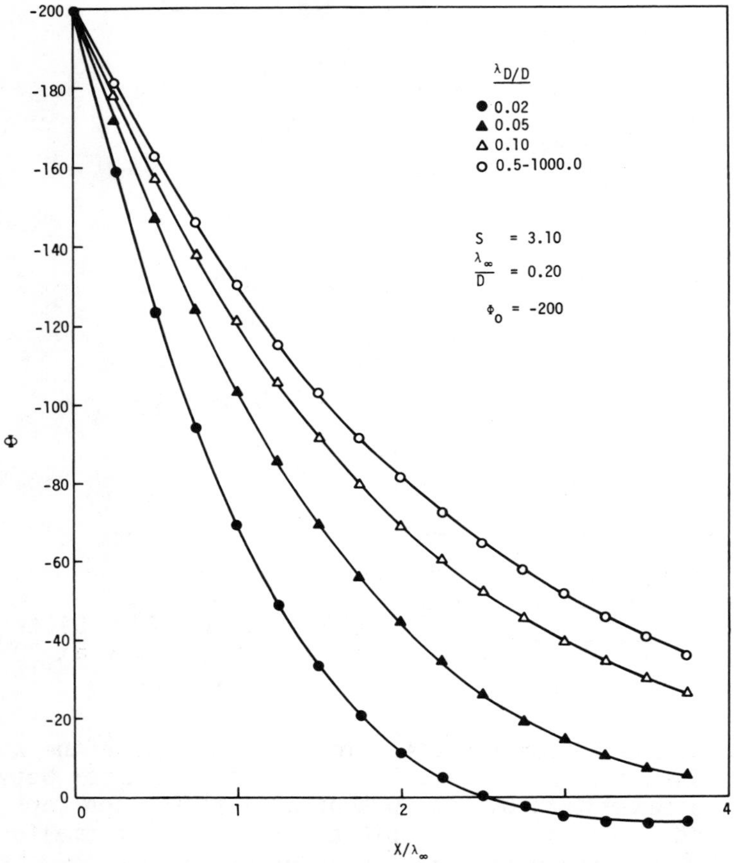

Fig. 7 Electric Potential Along Stagnation Line.

In Fig. 9 the potential field variation along the stagnation line are shown for the two Debye numbers. It is seen that in both cases the electric field penetrates far upstream to accelerate the positive ions toward the collecting plate. The ions are, thus, affected far upstream of the neutral particle gasdynamic shock identified by the rise in density from $X/\lambda_\infty \cong 20$ to $X/\lambda_\infty \cong 10$. In Fig. 9, large differences in electric potential are shown for Debye numbers of 0.03 and 1000. The linear behavior of the potential for $\lambda_D/D = 1000$ is because the upstream boundary is located at approximately, $X/\lambda_\infty \cong 60$ which corresponds to $X \cong 0.42D$. For this case, the solution of the Poisson equation is the Laplace solution along the stagnation line determined for a configuration similar to the field between two parallel plates which gives a linear variation. Although the ion density is higher near the body for the larger Debye number the electric field is lower and

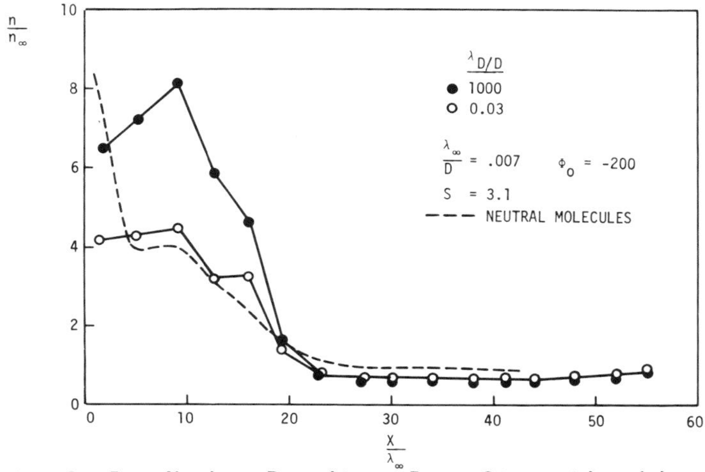

Fig. 8 Ion Number Density Along Stagnation Line.

the net effect on the ion flux at the stagnation point is that it remains approximately constant, $C_I \simeq 2.7$, for the Debye number range; $0.03 \leq \lambda_D/D \leq 1000$.

In Fig. 10 the ion flux variation across the front face shows two effects. First, the variation with Debye number in the range $0.03 \leq \lambda_D/D \leq 1000$ is very small, and the flux at the stagnation is approximately constant. This result is shown later in Fig. 12. Secondly, the ion flux peaks sharply at the stagnation point reflecting the very high densities for this near-continuum flow field.

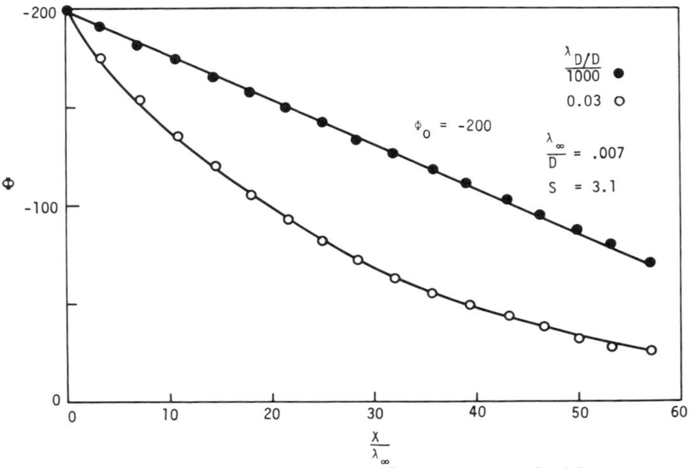

Fig. 9 Non-Dimensional Potential Along Stagnation Line.

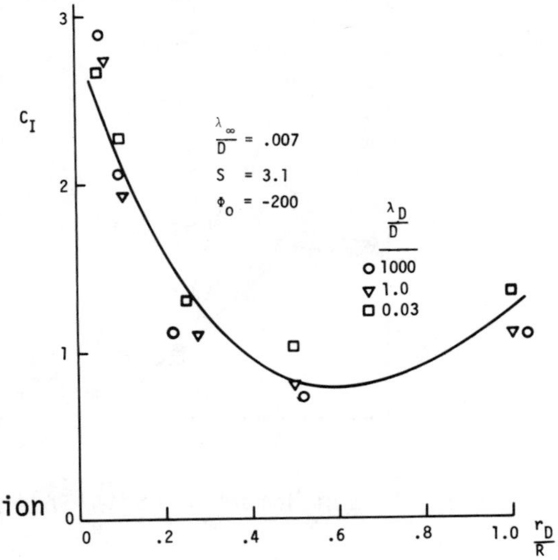

Fig. 10 Ion Flux Distribution Along Front Face at 70 km.

Fig. 11 shows velocity field along the stagnation line for both ions and neutrals. Again, as with the density, the effect of the electric field upstream of the disturbance in the neutral molecules is apparent, the field corresponding to the larger Debye number again penetrating further from the body. It should be pointed out that the ion velocity at the body surface ($X/\lambda_\infty = 0$) is not zero, unlike the velocity for the neutrals, since the ions are assumed to be converted to neutrals when they strike the body. These calculations at

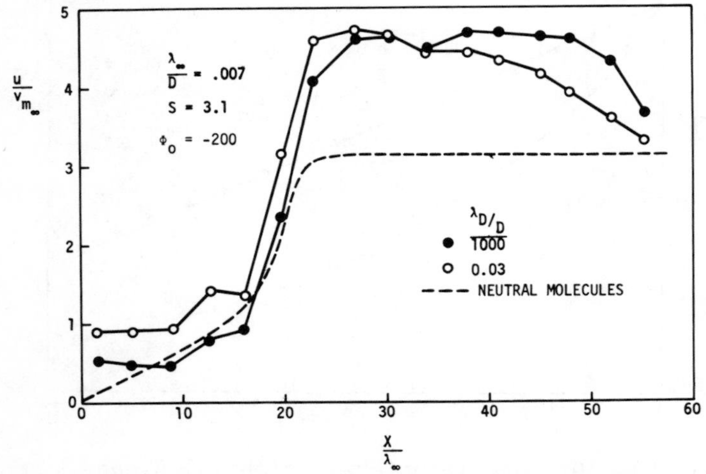

Fig. 11 Ion Velocity Along Stagnation Line.

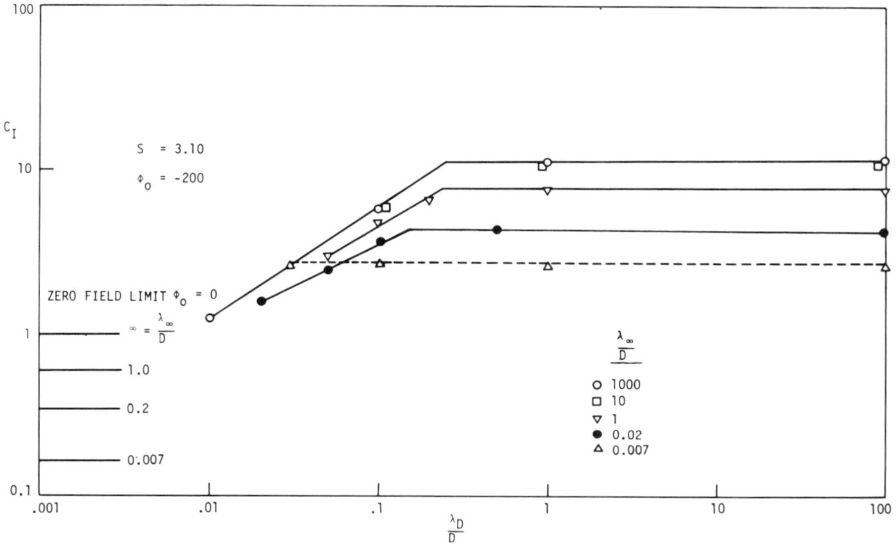

Fig. 12 Stagnation Point Flux with Debye Number.

70 km required approximately 25 minutes on the CDC 6600 computer for the most time consuming calculations at small Debye number. The ion flux samples for a typical case at small Debye number were about 400 for the stagnation point and 4000 over the entire front face.

Summary

A summary of the results for the parametric study is shown in Fig. 12, where the stagnation ion flux coefficient is shown as a function of Debye number for Knudsen numbers equal to λ_∞/D = 0.007, 0.20, 1.0, 10., 1000, a speed ratio of 3.10, and a nondimensional front face potential of -200. Also included on this figure on the left hand side are the values of the flux coefficient in the zero-field limit (Φ_0 = 0), which represent the minimum value for a given Knudsen number. Note that the value of the zero-field limit for the free molecule case ($\lambda_\infty/D = \infty$) is unity as required.

It can be seen that the variation of the flux coefficient C_I with Debye number exhibits a common trend for Knudsen numbers $\geq$0.20. That is the flux coefficient is constant as the Debye number is reduced until a "break-point" is reached, after which the flux coefficient decreases with decreasing Debye number. A further observation is that the break-point occurs at a Debye number approximately equal to the Knudsen number for transitional flows; e.g. λ_∞/D = 0.20. For the collisionless flow $\lambda_\infty/D \geq 10$, the break point occurs at a value of $\lambda_D/D \cong 0.5$ or $\lambda_D \sim$ body radius. The effects of collisions can be seen in the results for Knudsen number unity as seen by the reduction in the flux coefficient below the collisionless limit. However, the "break-point" remains at approximately 0.5. The results for the near-continuum flow with Knudsen number 0.007 exhibit no "break-point" for the range of Debye numbers considered. If the "break-point" is assumed to occur at a Debye number approximately equal to the Knudsen number the transition should occur at a Debye number near 0.007. Calculations for Debye numbers less than 0.03 for these conditions have not been successful to date because of numerical instabilities.

Although these results do not represent a complete parametric study the following trends seem reasonable.

1) The flux coefficient decreases with decreasing Debye number for a fixed Knudsen number.

2) The flux coefficient decreases with decreasing Knudsen numbers for fixed Debye numbers greater than the body radius, i.e. $\lambda_D/D > 0.5$.

3) The "break point" for the variation of the ion flux coefficient with Debye number can be estimated by

$$(\lambda_D)_{BP} = \text{minimum}\ (\lambda_\infty, R)$$

i.e. the "break point" is given by $(\lambda_D)_{BP} = \lambda_\infty$ or $(\lambda_D)_{BP} = R$ whichever is less. Therefore, for $\lambda_D > (\lambda_D)_{BP}$ the ion flux is determined by taking $\lambda_D \to \infty$.

These limits for the "break point" are shown on Fig. 2. It can be seen that in the "range of interest" below 90 km the ion flux can be determined by the infinite Debye number limit.

In many applications the quality of the flux entering the instrument is just as important as the number of particles. Fig. 13 gives the percentage of free stream ions in the stagnation point flux as a function of Knudsen number. These results are approximately independent of Debye number for the range $0.01 \leq \lambda_D/D \leq 1000$. This shows, for example, that below an altitude of 90 km ($\lambda_\infty/D = 0.2$) the entire flux has been affected by the body and only for Knudsen numbers greater than 10 is the flux completely undisturbed.

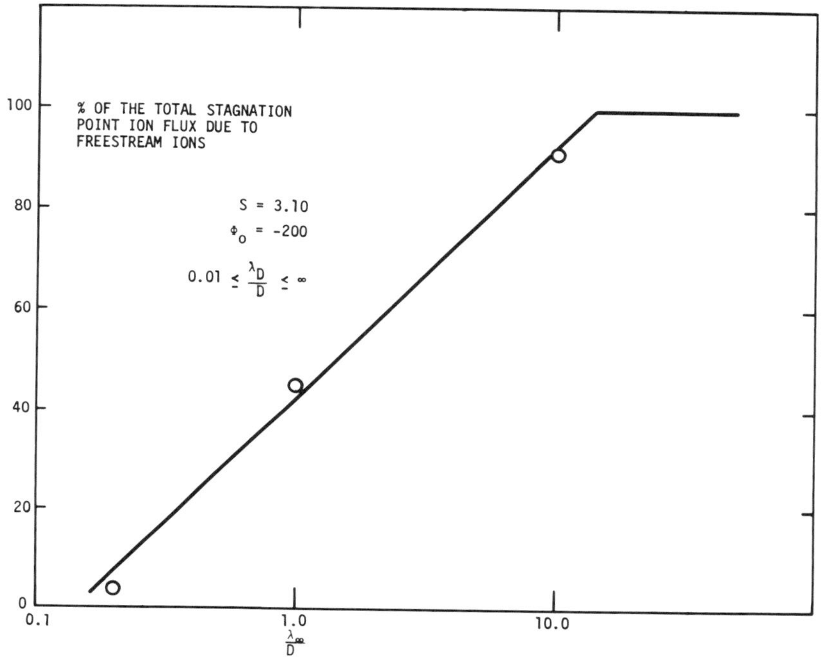

Fig. 13 Fraction of Freestream Ions in Stagnation Flux.

Conclusions

From the results of the present study and a previous study[2], a functional relation of the form

$$C_I = f(S_\infty, \Phi_0, \lambda_\infty/D, \lambda_D/D)$$

can be determined for the altitudes which are characterized by Knudsen numbers in the transitional regime or larger, e.g. 0.20 to ∞ in the present study. For Knudsen numbers less than transitional the results shown in Fig. 12 indicate that the correlational function will be complicated by the location of the "break point." This problem becomes unimportant, however, if the range of parameters is restricted to the "range of interest" shown in Fig. 2.

Furthermore, in the collection of large positive (or negative) ions the number of collisions that the ions experience before entering the mass spectrometer is important because of perturbations to the ambient states, i.e., chemical, internal, etc. The result indicated here that the fraction of free stream ions is independent of the Debye number and electric field may be of some use to instrument designers.

References

[1]Bird, G. A., "Direct Simulation and the Boltzmann Equation," The Physics of Fluids, Vol. 13, Nov. 1970, pp. 2676-2681.

[2]Sugimura, T. and Vogenitz, F. W., "Monte Carlo Simulation of Ion Collection by a Rocket-Borne Mass Spectrometer," Journal of Geophysical Research, Vol. 80, Feb. 1975, pp. 673-684.

[3]Sugimura, T. and Vogenitz, F. W., "Monte Carlo Simulation of Ion Collection by a Rocket-Borne Mass Spectrometer for Collisionless and Transitional Flowfields," TR-73-0448, July 1973, Air Force Cambridge Research Labs., Bedford, Mass.

[4]Sugimura, T., "Monte Carlo Simulation of Ion Collection by a Rocket-Borne Mass Spectrometer, Scientific Rept. No. 1" AFCRL-TR-75-0240, Feb. 1975, Air Force Cambridge Research Labs., Bedford, Mass.

[5]Narcisi, "Composition Studies of the Lower Atmosphere, Four Lectures Presented at the International School of Atmospheric Physics," June 1970, Air Force Cambridge Research Labs., Bedford, Mass.

IONOSPHERIC AERODYNAMICS AND SPACE SIMULATION STUDIES UTILIZING THE SHUTTLE/SPACELAB PLATFORM

U. Samir*

University of Michigan, Ann Arbor, Mich.

and

Tel Aviv University, Ramat Aviv, Israel

N. H. Stone† and W. A. Oran†

NASA Marshall Space Flight Center, Huntsville, Ala.

Abstract

A new philosophy for the second generation of in situ experiments in the area of plasma flows over bodies is discussed. A brief review of the present state of the theory and terrestrial laboratory and in situ studies is given, which points toward the need for further in situ investigations which utilize independently maneuverable test bodies and diagnostic instrument packages on board the Shuttle/Spacelab. The possibility of utilizing the Space Shuttle/Spacelab in the broader context of a general in situ plasma laboratory is considered, as well as the possibility of addressing questions of geo/astrophysical interest if the principle of qualitative scaling is used in conjunction with such a facility.

1. Introduction

Aspects of the interaction between a body and a flowing gas have been investigated since the early days of aerodynamics. With the advent of space flight, the range of interest was ex-

---

Presented at Paper 128 at the 10th International Symposium on Rarefied Gas Dynamics, Aspen, Colo., July 19-23, 1976.

*Research Scientist and Associate Professor, Space Physics Research Laboratory.

†Physicist, Magnetospheric and Plasma Physics Branch.

tended to include the interaction between an electrically charged body moving rapidly in the near-Earth ionospheric rarefield plasma. This new aspect of aerodynamics is sometimes called "ionospheric aerodynamics." Extending the range of investigation to problems involved in the flow interactions occurring in deep space leads to studies that are within the general areas of "cosmic-gas dynamics" and "plasma astrophysics."

An extensive in situ effort to study these space flow interactions was begun with the launch of satellites. This has centered mainly on studying the solar-wind interaction with the Earth, although the recent success of the planetary flybys has increased the interest in various other space flow interaction problems.

A particular case of plasma flows over bodies in space is the interaction between orbiting spacecraft and the near Earth environmental plasma. The complex of phenomena involved in this interaction also has been studied since the early days of spaceflight, primarily for three reasons: 1) it is a basic scientific (or academic) problem in rarefied plasma physics or rarefied plasma dynamics, 2) it has applications to the better understanding of the reliability of in situ observations and to experiment planning, and 3) there is the possibility of the spacecraft acting as a kind of model to explore and understand aspects of plasma flow interactions occurring with bodies in deep space which may be of interest to plasma-astrophysics, and cosmic-gas dynamics.

In this paper we shall review briefly the present state of knowledge of the spacecraft/space plasma interaction in the context of ionospheric aerodynamics. The review will include the state of both theoretical and experimental (in situ and laboratory) work, detailing some of the reasons why the results of previous in situ investigations are of a fragmentary and sometimes rather qualitative nature. This will clarify why a continuation of the research program using multibody in situ studies is required to properly understand the complex of phenomena occurring in the interaction. Then we will discuss some of the fundamental questions that may be answered by conducting multibody in situ experiments. We also consider utilizations of the Shuttle as a general near-Earth plasma laboratory where active, controlled experiments can be performed which, using the principle of qualitative scaling, may address questions of geo/astrophysical interest. Finally, we shall outline, in general terms, an example program to study flow interactions with bodies in space utilizing the Shuttle/Spacelab as an experimental platform.

2. Spacecraft-Ionosphere Interactions: Present State of Knowledge

2.1 About the Interaction

A spacecraft moving hypersonically or supersonically through a space plasma interacts with it and produces significant disturbances. The interaction is mutual and the phenomena involved are coupled by the effects of both the spacecraft and the spaceplasma. Therefore, the characteristics of the disturbances depend upon the plasma flow conditions and body properties. The spacecraft acquires a potential due to the accretion of the surrounding charged particles and due to other charging mechanisms such as photoemission, particle emission, magnetic field effects (both via the $\underline{V} \times \underline{B}$ effect and the influence of the magnetic field on particle collection), energetic particle bombardment, etc. (e.g., Ref., 1).

The surrounding space plasma is perturbed strongly by the satellite motion, and a wake zone, depleted unequally of ions and electrons is formed behind the spacecraft. Charged particle current enhancements as well as enhancements in electron temperature have been observed in this zone. In addition to the creation of the wake, the steep gradient in electron and ion densities at the boundaries of the wake zone may give rise to some kind of plasma instability and/or weak turbulence. It also is possible that a potential well exists behind the spacecraft at some distance downstream from its surface. One might anticipate that plasma oscillations and/or possibly particle energizations are associated with the potential well. Finally, the hypersonic/supersonic characteristic of the satellite may lead to the creation of shocks, ahead and/or behind the satellite.

Besides the basic science aspects, the interaction has a direct application to in situ measurements. There can be little doubt that the disturbances created via the interaction have profound relevance to the reliability and quality of particle and field measurements from satellites and rockets. Although there is an awareness of this problem, very little is known of practical utility about the influence of the disturbances on in situ measurements. The lack of knowledge of the latter may be very critical in the planning and design of future Shuttle-borne experiments utilizing common plasma diagnostics such as Langmuir probes, retarding potential analyzers, etc.

Preliminary studies to answer such questions as: 1) how extensive is the disturbed zone around the Shuttle itself, and 2) how should the operations be planned to best overcome the effects of the disturbances, will have to be undertaken before in situ experiments can be conducted.

2.2 Theoretical Studies

Theoretically, we seek the self-consisten solution of the Vlasov-Poisson equations, given in the form

$$\frac{\partial f_+}{\partial t} + \underline{V}\frac{\partial f_+}{\underline{r}} + \frac{e\underline{E}}{M_+} + \frac{e}{M_+ c}[(\underline{V} + \underline{V}_S) \times \underline{B}_E]\frac{\partial f_+}{\partial t} = 0 \quad (1)$$

$$\frac{\partial f_e}{\partial t} + \underline{V}\frac{\partial f_e}{\underline{r}} - \frac{e\underline{E}}{M_e} + \frac{e}{M_e c}[(\underline{V} + \underline{V}_S) \times \underline{B}_E]\frac{\partial f_e}{\partial t} = 0 \quad (2)$$

$$\nabla^2\phi = -4\pi e\{\int f_+ d\,V - \int f_e d\,V\} = -4\pi e[N_+ - N_e] \quad (3)$$

$$E = -\text{grad}\ \phi$$

where $f_+(t,\underline{r},\underline{v})$ and $f_e(t,\underline{r},\underline{v})$ are the distribution functions of the charged particles (ions and electrons) and N_+ and N_e are their densities. The electric field $\underline{E}$ is caused by both the potential field of the spacecraft and the space charge. $\underline{B}_E$ is the magnetic field of the Earth and $\underline{V}_S$ is the spacecraft velocity.

The boundary conditions for the potential ϕ, are: $\phi = \phi_0$ at $r = R_0$ and $\phi = o$ for $r \to \infty$ where R_0 is the radius of the spacecraft. The value of ϕ_0 is determined from the condition that the total current to the spacecraft equals zero. The current includes the collection of positive ions and electrons as well as contributions from photoemission, magnetic field effects, r.f. radiation, secondary emission, etc. In most theoretical studies presently available Eq. (2) has been replaced by an assumed electron density of the form $N_e(r) = N_{eo} \exp(e\phi/kT_{eo})$, where N_{eo} is the ambient electron density and T_{eo} the ambient electron temperature. The method of calculating N_+ is still a subject of controversy. Most studies ignore magnetic field effects and limit themselves to solving Eqs. (1 - 3) for $(\partial f/\partial t) = 0$. There are also differences involving the inclusion or neglect of the ion thermal motion with respect to the spacecraft velocity ($\underline{V}_S$) in solving Eqs. (1 - 3).

As far as the plasma flow regimes are concerned, most of the theoretical studies assume

$$L(n),\ L(+e) \gg R_0 \gg \lambda_D \quad (4a)$$

$$V_T(+) << V_S << V_T(e) \quad (4b)$$

$$R_L(e) << R_0 < R_L(+) \quad (4c)$$

Here, L(n) and L(+e) are the mean free paths of neutral and charged particles respectively, $V_T(+)$ and $V_T(e)$ the thermal velocities of ions and electrons respectively, and $R_L(+)$ and $R_L(e)$ the Larmor radii of ions and electrons respectively. In the most general way (4) is valid although with reservations for (4b) and sometimes even for (4c).

Despite the extensive theoretical work done for the case of spacecraft-ionosphere interactions, there is at present no unified theory or model that can account for the various structural characteristics and the physical processes involved. Among the most relevant paper we cite Ref. 2-9.

2.3 Laboratory Studies of Spacecraft/Space Plasma Interaction

In recent years ground-based laboratory studies have been used to investigate aspects of the spacecraft/space plasma interaction. Although these studies can simulate the interaction only partially, they nevertheless offer capabilities that are difficult to obtain with single satellite systems. For example, parts of the zones of disturbance can, at least in principle, be studied using a variety of disagnostic instruments and some of the flow parameters can be varied independently over a wide range of plasma parameters. Table 1 presents a partial summary of the range of flow parameters and target bodies used in some of the facilities. The principle results of these laboratory systems include observations of: 1) an ion void region in the near wake zone, 2) ion current enhancements in the mid-wake region in the absence of a significant transverse component of the random ion motion, 3) wave-like structure in the far wake zone, and 4) electron temperature enhancement in the near wake region. Also, in some instances partial parametric studies were conducted. Additional details are given in Ref. 10. However, these laboratory systems suffer from certain inherent limitations, among which are the inability to duplicate charged particle distributions accurately and the adverse influence of wall effects. There is little doubt that a complete understanding of the physical processes occurring in the entire interaction region can be obtained only when a detailed systematic study is made in situ.

Table 1. Summary of plasma flow parameters and body geometry used in simulation experiments[a]

Group	Body geometry	$e\phi_S/kT_e$	$S = \frac{V_{flow}}{[2kT_e/m_+]^{1/2}}$	R_o/λ_D	Downstream extent of profiles in R_o	T_i/T_e
MIT	Sphere	-2.75 to -44	3.6 to 8.2	1.8 to 46	0 to 58	<<1
	Long cylinder	+3.5 to -50	3.2 to 28	0.005 to 1.6	0 to 57000	<<1
Cal Tech	Disk	-2.5 to -30	5.4 to 7.0	4 to 32	0 to 20	<<1, ~1
	Sphere	-1.6 to -22	5.1 to 6.2	6 to 33	0 to 20	<<1, ~1
Skv. & Now.	Sphere	0 to 67	5.0 to 11.5	18 to 30	0 to 12	<<1
ONERA	Cylinder Sphere	float to -100	10 to 14.3	0.7 to 24	0 to 180	<<1, ~1
ESTEC	Disk/cylinder	0 to -97	6.4 to 9.0	10 to 18	~2.0	<<1
MSFC	Sphere	-3.5 to -48	4 to 17	1 to 25	0 to 24	<<1
	Cylinder	-4 to -48	7.3 to 12	12 to 25	0 to 9.5	<<1
	Square sheet	-3.5 to -46	~7.5	~14	0 to 9.5	<<1
TRW	Sphere	$0 \leq \|\phi_S\| \leq 400$	V_{flow} = 7.7 11 km/s	R_o = 1 cm 2.5 cm	4 to 17.5	<<1
RARDE	Sphere/cone	+25 to -360	5.7 to 12	~5	2 to 20	<<1
Levedev Inst.	Disk	~11	1 to 2.6	25 to 35	1 to 30	~1

[a]After Ref. 10, including results of Ref. 11.

2.4 Present State of In Situ Experimental Studies

In situ measurements that were used in the initial studies of spacecraft/space plasma interactions came mainly from: 1) the Explorer 8 satellite, 2) the Ariel I satellite, 3) the Explorer 31 satellite, and 4) the Gemini-Agena 10 manned spacecraft. The measurements were made by various types of current collectors (e.g., Langmuir probes, retarding potential analyzers, guarded and unquarded spherical probes). The satellite configurations and probe location for the Explorer 8, the Ariel I, and the Explorer 31 satellites are given in Fig. 1a and the Gemini-Agena flight configuration and probe location in Fig. 1b.

The measurements used from a, b, and c of Fig. 1a provided for 1) several examples showing the variation of electron current (I_e) with the angle of attack (θ) between 0 and 180° and in the altitude range 400-2300 km, at $r \approx R_o$, i.e., at the closest vicinity to the satellite surface - $I_e = f_1(\theta)_{r = R_o}$; 2) one example showing $N_+/N_e = f_2(\theta)_{r \approx 2R_o}$ in the angular range: $120° \leq \theta \leq 180°$; 3) several examples showing $I_e = f_e(\theta)_{r = 5R_o}$ for various orientations of the spacecraft with respect to the velocity vector (see Fig. 1b for the boom probe); 4) several examples showing $I_+ = f_4(\theta)_{r \approx R_o}$ in the angular range $90° \leq \theta \leq 165°$.

In some cases, these measurements did not furnish all of the information required for computing the values of all of the relevant plasma parameters. This latter limitation will be reduced largely when measurements from the Atmosphere Explorer (AE) satellites will be used. However, the severe limitations of single-spacecraft observations still will remain; namely, the measurements are made only over a small region of the zone of disturbance.

From the Gemini-Agena 10 experiment (see Fig. 1b for probe locations), which adopted a different experimental philosophy (i.e., using a two-body system), a single axial profile of electron and ion current was obtained. This profile extended to distances downstream from the surface of the spacecraft to about $r = M \times R_o$, where M is the ionic Mach number and R_o is the radius of the spacecraft.[12] A review paper that presents and discusses the in situ observations prior to 1973 is given.[13] More recent in situ studies, again from single-body satellites are given in Ref. 14-16.

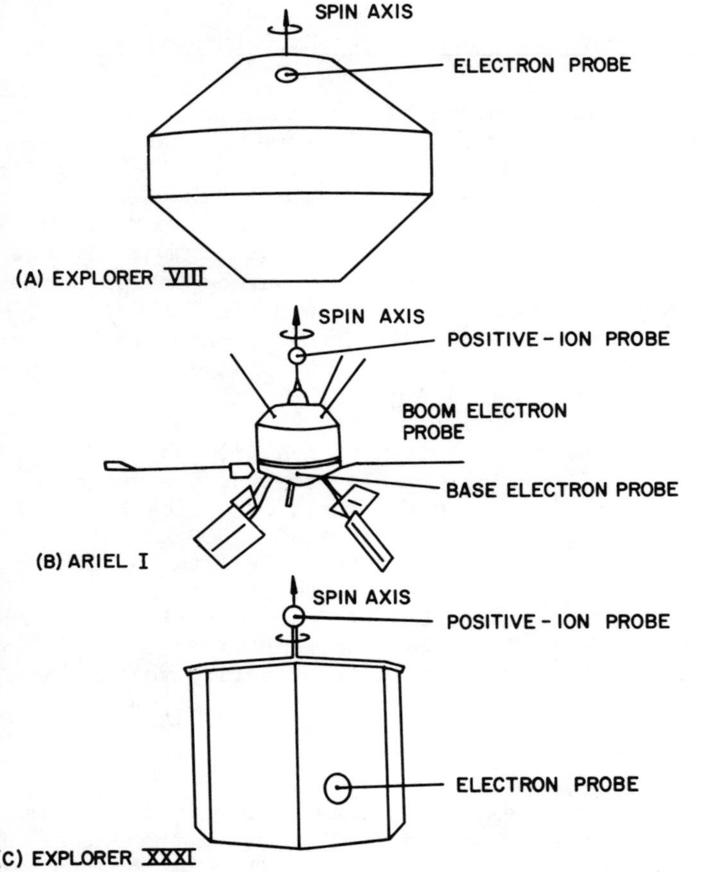

Fig. 1a A diagrammatic drawing showing satellite configuration and probe locations (a) Explorer 8, (b) Ariel I, (c) Explorer 31.

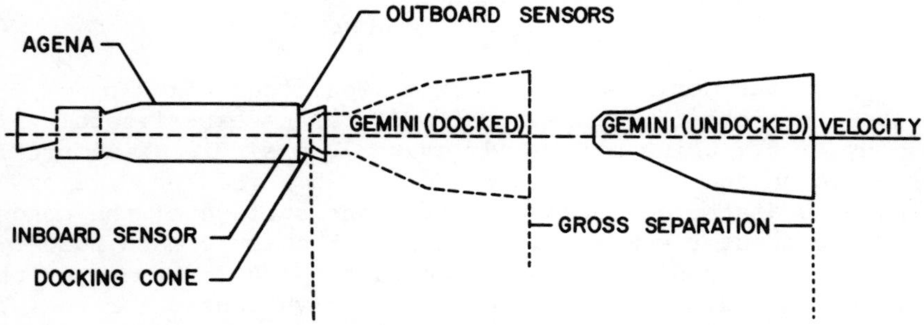

Fig. 1b Diagrammatic representation of the Gemini-Agena flight configuration and probe location.

3. Utilization of the Shuttle/Spacelab

3.1 The Space Shuttle/Spacelab as an In Situ Near-Earth Plasma Laboratory

Since the concept of a general, in situ, near-Earth laboratory implies a new mode of experimentation which will be made possible for the first time by the Shuttle/Spacelab system, perhaps the best indication of the utility of this experimental philosophy can be obtained from analogous types of studies which have been carried out in the past in terrestrial laboratories.

Before the advent of the space era, several laboratories throughout the world engaged in terrestrial laboratory experiments that were of magnetospheric, astrophysic and cosmic interest. Extensive discussions are given in Ref. 17-21.

As early as 70 years ago, such studies were aimed at simulating the near-Earth space environment.[22] At that time, the Earth's environment was poorly understood and the technical capabilities were very limited.[17] However, those investigations opened up and stimulated theoretical studies that are of current interest and validity. One example is the concept of magnetic-field aligned currents recognized by Birkeland which recently came to be recognized generally as an important in situ phenomena. The study of such currents in the Earth's environment is now a topic of active research.

It should be realized that traditionally when "simulation" of cosmic, astrophysic, or planetary phenomena was attempted, the efforts were directed toward the simulation of the entire system configurations.

More recently, experiments in terrestrial laboratories have made use of the principle of "qualitative scaling." In general, this principle states that: 1) the non-dimensional parameters characterizing the phenomena which are of the order of unity in nature should remain so in the laboratory, 2) non-dimensional parameters, which in nature are oders of magnitude smaller or larger than unity, should accordingly remain so in the laboratory although there is no need to retain the same order of magnitude. We submit that another type of laboratory simulation experiment which is called "process simulation"[17] may not require more than utilizing the concept of "qualitative scaling." Falthammar and others[17-19], have demonstrated the degrees of success achieved studying various problems of interest in the interaction of the solar wind with the Earth, utilizing this principle. In other words, the principle of qualitative scaling is powerful if the main interest and scientific objective is focused on the understanding of physical processes, rather than on the morphology of "systems" in space.

In view of the past success of terrestrial laboratory studies which applied the principle of "qualitative scaling" to certain geophysical phenomena, utilizing this philosophy in similar types of experiments conducted in the natural, ionospheric plasma may be of utmost interest and promise: a new insight into geo/astrophysical phenomena which could not have been investigated before. For example, those types of studies should indicate what type of interactions develop when a body possesses a magnetic field; indicate the validity of a quasi-continuum formulation of collisionless plasma flow problems; and provide the opportunity to examine aspects of the hypersonic-analogue often used to describe the interaction of the solar wind with the planets.

One question that becomes immediately obvious, and therefore should be dealt with at this point, concerns the relationship between laboratory and in situ experimental efforts. It should be pointed out that the authors do not view laboratory and in situ investigations as competing modes of experimentation. A brief discussion should be sufficient to show how the two complement each other.

In the laboratory, competing scaling requirements often are encountered. For example, when attempting to model large-scale geo/astrophysical phenomena, it may be required to preserve the inequalities

$$(R_{L(+)}/R_o) \ll 1,\ (L_{(+e)}/R_o \gg 1,\ (R_o/\lambda_D) \gg 1,\ R_{me} \gg 1$$

among others, where R_{me} is the magnetic Reynolds number. It will be seen that all of these cannot be maintained in the laboratory since a frozen in magnetic field ($R_{me} \gg 1$) generally is attained in conjunction with relatively high charged particle densities (10^{13} - $10^{14}/cm^3$). This produces a very small mean free path and Debye shielding distance, therefore, making the above set of dimensionless parameters mutually exclusive.

Although geo/astrophysical phenomena can be studied "piecewise" as in the case of the Earth's bow shock (which has a small characteristic thickness), in general, large scale phenomena cannot be modeled correctly (in total) in the laboratory. In the ionosphere, however, particle densities are low enough and scale size can be made large enough, so that the preceding problems should not be encountered in a number of experiments.

A further difficulty which arises when large scale phenomena are reduced to the laboratory scale, is the effect of diagnostic instrumentation on the model. Simply put, it is not (or to date, has not been) possible to fabricate instruments that are small compared to the scale size of such experiments. Experiments of this type can be conducted on such a scale in the ionosphere (~10 to 10^2 m body radii) as to make instrument effects truly insignificant. In addition, the in situ experiments, being conducted in a natural, essentially unbounded, plasma have none of the problems of non-Maxwellian charged particle distributions, residual gas and slow ions, or wall effects, which are inherent to laboratory plasma facilities. It therefore appears that the optimal approach to studies of complex plasma flow problems involves a union of both experimental modes, as well as sound, theoretical support. Such a philosophy would seem to be in line with the natural, scientific evolution and development of a space sciences program.

3.2 Example of a Program to Study Plasma Flow Interactions with Bodies in Space Using the Shuttle/Spacelab

In this section we outline, in general, the philosophy of experimental programs that could use the Shuttle/Spacelab facility as a near-Earth plasma laboratory to conduct flow interaction studies, e.g., to carry out experimental investigations in ionospheric aerodynamics as well as "simulation" experiments of geo/astrophysical interest. In this approach the Shuttle/Spacelab system would be used as a carrier of a variety of target bodies, the appropriate diagnostic packages, and the subsystems required to maneuver a target body independently and a set of mapping diagnostic instruments, to determine and control their relative position and their altitude with respect to the Shuttle velocity vector, and to record, reduce, display, and transmit experimental data.

Test bodies are objects used to generate the disturbances to be studied in the ambient plasma flowfield. These bodies should range in complexity, depending on specific experimental objectives, from simple planar, cylindrical, or spherical conductors to more sophisticated bodies with controllable electric and magnetic characteristics. The size of the test bodies also will vary with experimental objectives from radii of ~1 meter (with applications to probe theory) to several tens or even ~100 m (required fcr $R_o/\lambda_D >> 1$, $R_L/R_o << 1$, etc).

Insulating as well as conducting body materials may be used to understand better some aspects of differential charging. It will be of interest and importance to understand better the flowfield around the bodies as a function of the surface properties. In more advanced studies, the test bodies should include dipole magnetic fields of controllable orientation and strength. Processes involved in oblique shocks could be reviewed. The comparison between flowfield around magnetized and unmagnetized bodies would be very interesting and relevant in its wide range of applications to geo/astrophysics.

The characteristics of the disturbances generated in the ambient plasma flow by test bodies would be measured by a "mapping" diagnostic instrument package. This is a package of appropriate diagnostic instruments that can be maneuvered independently of the test body in such a way that it scans (or maps) the various zones of disturbance around the body. In addition, a second instrument package consisting of essentially the same set of instruments, which remains continuously in the undisturbed flow, would monitor the ambient plasma in order to distinguish between the temporal and spatial fluctuations of the ionosphere and those generated by the test body.

It is clear that, in order to understand the complex of phenomena involved in the flow interaction of a moving body with a space plasma, the studies should be conducted in a parametric manner; that is, the behavior of various processes, structure, and effects should be determined as a function of the appropriate governing parameters.

In all cases, the test body, as well as the two diagnostic instrument packages, must be removed from the Shuttle generated plasma disturbances and electromagnetic interference. This, together with the required maneuverability, can be attained by a variety of combinations of fixed or maneuverable booms, free flying or maneuverable subsatellites, and free flying or tethered balloons.

One mode involved both the test body and the instrument package mounted on booms (see Fig. 2). In this case, the booms must be maneuverable. However, if the experiment is performed, for example, using a simple ejectable body with a boom mounted instrument package, then the boom may be fixed. If a maneuverable subsatellite is used, then an ejectable free flying or tethered body may be used, and no boom would be required, with the possible exception of a small fixed boom for the ambient plasma-monitoring instrument package.

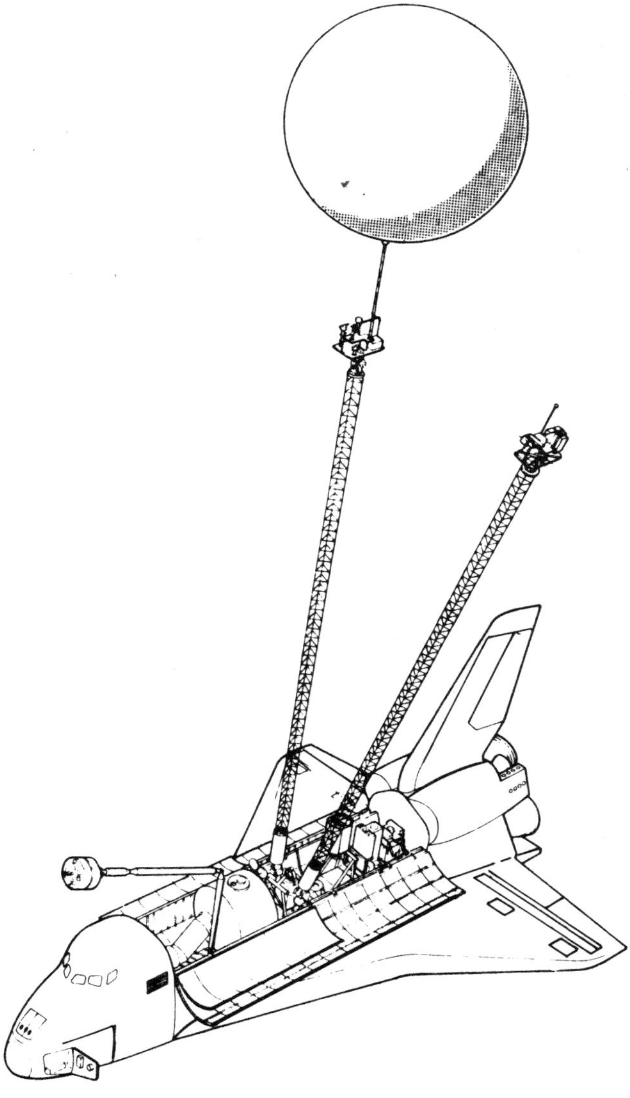

Fig. 2 Boom-mounted Shuttle/Spacelab plasma flow experiment (artist conception).

Obviously, all modes for conducting the experiments will not produce the same scientific return. For example, the use of boom mounted test bodies will limit the maximum size of the body, depending on the length of the boom. The use of boom mounted instruments with a free flying body, on the other hand, will limit the extent to which the disturbance generated in the plasma flow can be mapped, depending on the allowable maneuverability of the Shuttle.

The use of maneuverable subsatellites to carry diagnostic instruments and/or as test bodies will increase the scope of the investigation very significantly, and hence the scientific return. In this mode, the experiment could be made with large inflatable-balloon type target bodies, which would allow, for example, attainment of large ratios of ($R_o/R_{L(+)}$ and R_o/λ_D). In addition, it also will permit investigation of the "far wake" characteristics of the disturbance.

4. Summary

We have briefly described the overall state of knowledge regarding the spacecraft/space plasma interaction. From this description the limitations of presently available single satellite systems and of laboratory systems simulating aspects of the interaction became obvious. The advantages of the multibody facility provided by the Shuttle/Spacelab was discussed in the context of an evolutionary program starting from simple flight configuration with restricted objectives and progressing towards more sophisticated experiments which provide a comprehensive investigation of the complex relationship between various physical mechanisms and the plasma parameters governing the interaction. If the principle of "qualitative scaling" is applied, these sophisticated experiments may also answer questions about physical processes of geo/astrophysical interest.

References

[1]Kasha, M. A., The Ionosphere and Its Interaction with Satellites, Gordon and Breach, New York, 1969.

[2]Call, S. M., "The Interaction of a Satellite with the Ionosphere," Rept. 46, 1969, Plasma Laboratory, Columbia University.

[3]Liu, V. C., "Ionospheric Gas Dynamics of Satellite and Diagnostic Probes," Space Science Reviews, Vol. 9, June 1969, p. 423.

[4]Gurevich, A. V., Pitaevski, L. P., and Smirnova, V. V., "Ionospheric Aerodynamics," Soviet Physics-Uspekhi, Vol. 92, March-April 1970, p. 595.

[5]Grabowski, R., and Fischer, T., "Theoretical Density Distribution of Plasma Streaming Around a Cylinder," Planetary and Space Science, Vol. 23, February 1975, p. 287.

[6]Gurevich, A. V., and Pitaevski, L. P., "Non-linear Dynamics of a Rarefied Ionized Gas," Progress in Aerospace Science, Vol. 16, September-November 1975, p. 227.

[7]Parker, L. W., "Computation of Collisionless Steady-State Plasma Flow Past a Charged Disc," CR-144159, February 1976, NASA.

[8]Liu, V. C., "On Ionospheric Aerodynamics," Progress in Aero-Space Science, Vol. 16, September-November 1975, p. 273.

[9]Martin, A. R., "Numerical Solutions to the Problem of Charged Particle Flow Around an Ionospheric Spacecraft," Planetary and Space Science, Vol. 22, January 1974, p. 121.

[10]Samir, U., Oran, W. A., and Stone, N. H., "Laboratory Simulation of Space Aerodynamic Phenomena: Satellite Wake Studies," Rarefied Gas Dynamics, Vol. 11, 1974, p. D.11.1.

[11]Fournier, G., and Pigache, D., "Wakes in Collisionless Plasmas," The Physics of Fluids, Vol. 18, November 1975, p. 1443.

[12]Troy, B. E., and Samir, U., and Medved, D. B., "Some Wake Observations on the Gemini/Agena Two-body System," Journal of Astronautical Science, Vol. 18, December 1969, p. 173.

[13]Samir, U., "Charged Particle Distribution in the Nearest Vicinity of Ionospheric Satellites - Comparison of the Main Results from the Ariel I, Explorer 31, and Gemini-Agena 10 Spacecraft," Photon and Particle Interactions with Surfaces in Space, edited by R.J.L. Grard, Reidel, Dordrecht, Netherlands, 1973, p. 193.

[14]Samir, U., First, M., Maier, E. J., and Troy, B. E., "A Comparison of the Gurevich et al. and the Liu-Jew Wake Models for the Ion Flux Around a Satellite," Journal of Atmospheric and Terrestrial Physics, Vol. 37, April 1975, pp. 557-586.

[15]Troy, Jr., B. E., Maier, E. J., and Samir, U., "Electron Temperatures in the Wake of an Ionospheric Satellite," Journal of Geophysical Research, Vol. 80, March 1975, p. 993.

[16]Vernet-Mayer, N., "Rocket Spin Effects on the Current Collected by a Cylindrical Probe in the Ionosphere," Journal of Geophysical Research, Vol. 81, January 1976, p. 450.

[17]Fälthammar, C. G., "Laboratory Experiments of Magnetospheric Interest," Space Science Reviews, Vol. 15, May 1974, p. 803.

[18]Kawashima, N., "Characteristics of Intruding Plasma in a Simulated Magnetosphere," Cosmic Electrodynamics, Vol. 1, February 1971, p. 415.

[19]Podgorny, I. M., and Dubinin, E. M., "Laboratory Experiments Directed Toward the Investigation of Magnetospheric Phenomena," Space Science Reviews, Vol. 15, May 1974, p. 287.

[20]Podgorny, I. M., and Sagdeev, R. Z., "Physics of the Interplanetary Plasma and Laboratory Experiments," Soviet Physics-Uspekhi, Vol. 98, January-February 1976, p. 445.

[21]Alfven, H., and Fälthammar, C. G., Cosmical Electro-Dynamics, The Claredon Press, Oxford, 1963.

[22]Birkeland, K., The Norwegian Aurora Polar's Expedition of 1902-1903, Aschenjoug and Co., Oslo, Norway, 1908.

TRANSPORT PROPERTIES IN NONISOTHERMAL RAREFIED PLASMAS

Wallace F. Walters*
University of Delaware, Newark, Del.

Abstract

A modified Chapman-Enskog method of solution of the Lenard-Balescu equation is presented. This method yields zeroth-order electron and ion solutions that are Maxwellian, having the same macroscopic flow velocity, with electron and ion temperatures that need not be equal. Transport coefficients for fully ionized hydrogen are computed for ratios of electron temperature to ion temperature of 1 - 100. These coefficients are found to exhibit corrections of up to 20% because of the inclusion of collective phenomena in the analysis. Corrections arising from the modified method of solution also appear. In the limit of equal electron temperature and ion temperature with collective contributions set equal to zero, the values of the transport coefficients are found to agree with previous results.

I. Introduction

Fully ionized laboratory plasmas frequently are found to exist in a state where the electrons and ions of the plasma are at different temperatures. This situation occurs when either plasma component is heated preferentially. Because of the large difference in mass, the energy exchange between electrons and ions is ineffective and leads to a long-lived state in which the temperatures of these two components are

---

Presented as Paper 65 at the 10th International Symposium on Rarefied Gas Dynamics, Aspen, Colo., July 19-23, 1976.
This paper is based on work performed at Sandia Laboratories and supported by the U.S. Energy Research and Development Administration.

* Assistant Professor, Mechanical and Aerospace Engineering.

not equal. This paper will deal with plasmas in which the ratio of electron to ion temperature T_e/T_i ranges from 1 to 100.

The determination of transport properties in such a plasma is complicated by the fact that long-range collective interactions, specifically the emission and absorption of ion acoustic waves by electrons,[1-3] are important, necessitating the use of the Lenard-Balescu kinetic equation.[4-6] Complications also arise because the classical Chapman-Enskog method[7,8] of solving the kinetic equation is adequate only when both like and unlike particle interactions dominate the behavior of the plasma; this is clearly not the case for systems far from thermal equilibrium.

Several investigators[2,3,9,10] have obtained transport coefficients for a hot-electron cold-ion plasma. In Refs. 9 and 10, a modified Chapman-Enskog method is applied to the Landau and Boltzmann kinetic equations, respectively. In the fully ionized limit, the results of these two works are in agreement. The Lenard-Balescu kinetic equation is used in Refs. 2 and 3; however, neither of these works treats the ion equation. In Ref. 2, there appears to be no rationale for ordering the various terms in the kinetic equation, and, as a result, the classical Chapman-Enskog method is used to solve the kinetic equation. A numerical solution is obtained from the electron equation in Ref. 3, but complete transport relations are not exhibited.

In this paper, a rationale will be established for judging the order of magnitude of the various terms in the Lenard-Balescu equation. This rationale is based upon test-particle transfer frequencies. By using this rationale, a modified Chapman-Enskog method for solving the Lenard-Balescu equation is developed. Since the Lenard-Balescu equation is used, the effect of collective phenomena on the transport properties is retained.

In Sec. II, test-particle transfer frequencies obtained using the linearized Lenard-Balescu equation are exhibited. These frequencies then are used to order the interaction terms in the kinetic equation, and formal solutions are obtained for the electron and ion one-particle distribution functions. Both electron and ion transport coefficients are computed in Sec. III, and the effect of the inclusion of collective phenomena on these coefficients is indicated. The effect of the modified method of solution on the transport coefficients and transport relations also is indicated.

II. Solution of the Kinetic Equation

A. Transfer Frequencies

The Lenard-Balescu equation can be written as

$$\frac{\partial f_b}{\partial t} + \vec{c} \cdot \frac{\partial f_b}{\partial \vec{r}} + \frac{\vec{F}_b}{m_b} \cdot \frac{\partial f_b}{\partial \vec{c}} = \sum_{a=e,i} \frac{8\pi^4 e^4}{m_b} \int_0^{\ell_o} d^3\ell \; d^e c' \vec{\ell} \cdot \frac{\partial}{\partial \vec{c}}$$

$$\left| \frac{V_\ell}{\varepsilon^+(\vec{\ell},U)} \right|^2 \delta\,(\vec{\ell}\cdot g)\,\left\{ \frac{\vec{\ell}}{m_b} \cdot \frac{\partial}{\partial \vec{c}} - \frac{\vec{\ell}}{m_a} \cdot \frac{\partial}{\partial \vec{c}'} f_b(\vec{c}) f_a(\vec{c}') \right\} \qquad (1)$$

where

$$\varepsilon^+(\vec{\ell},U) = 1 + i \sum_{a=e,i} \frac{4\pi^2 e^2}{m_a \ell^2} \int d^3c' \delta \; (\vec{\ell}\cdot g)\vec{\ell} \cdot \frac{\partial f_a(\vec{c})}{\partial \vec{c}'}$$

$$\vec{g} = \vec{c} - \vec{c}',$$

$$\delta\_(y) = \delta(y) - (i/\pi)\mathrm{Pr}(1/y)$$

$$U = \vec{\ell} \cdot \vec{c}/\ell, \qquad \ell_o = kT_e/e^2, \qquad eV_\ell = e/2\pi^2\ell^2$$

The quantity $e/2\pi^2\ell^2_{\rightarrow}$ is the Fourier transform of the e/r coulomb potential, ℓ is the wave vector, δ is the Dirac delta function, and the symbol Pr indicates the principal value when placed under the integral sign. The upper limit of the integration is ℓ_o, which is the inverse of the distance of closest approach of thermal electrons. The value of the integral depends very weakly on the cutoff value ℓ_o, since wave numbers larger than ℓ_o correspond to large angle deflections, which have negligible effect in fully ionized systems. The function $\varepsilon^+(\vec{\ell},U)$ is the dielectric function, which can be obtained from the linearized Vlasov equation. A comparison of the Landau interaction integral[11] with the Lenard-Balescu interaction integral reveals that, if $\varepsilon^+(\vec{\ell},U) = 1$ in the latter, the two integrals are identical. It is the inclusion of the dielectric function which allows one to account for long-range collective effects properly. Hence, the Landau integral cannot account adequately for long-range interactions.

Using the Lenard-Balescu interaction integral (J_{LB}), both the momentum transfer and energy transfer frequencies for both electron and ion test particles scattering within a background plasma, in which the electron temperature T_e is greater than or equal to the ion temperature T_i, have been calculated and are exhibited in this section. The test-particle method of obtaining energy and momentum relaxation times is described in Ref. 12.

The notation ν_{Eta} refers to the energy transfer frequency of the test-particle stream t interacting with the plasma component a. A similar notation is used for the momentum frequencies, with E replaced by M. All frequencies obtained using the Lenard-Balescu interaction integral are found to be the sum of two terms. One term contains the coulomb logarithm $\ln\Lambda$ and is the collisional contribution obtained when the Landau interaction integral is used to model interactions. The other term is due entirely to the presence of collective effects in the nonisothermal plasma. The subscript L denotes the Landau or collisional contribution to a frequency whereas the subscript C denotes the collective contribution.

In order to simplify the presentation of the collective component of the frequencies, the following relations are introduced

$$(\nu_{Eta})_C = |A_{Eta}/\ln\Lambda|\ (\nu_{Eta})_L$$

$$(\nu_{Mta})_C = |A_{Mta}/\ln\Lambda|\ (\nu_{Mta})_L$$

Table 1 exhibits the values of A_{Mta} and A_{Eta} for each frequency for five temperature ratios. Notice that as the temperature ratio increases the collective contribution grows, as expected. Assuming that $\ln\Lambda = 10$, one finds, by referring to Table 1, that the collective term is a substantial fraction of the collisional term for the transfer frequencies ν_{Mee} ν_{Eii}, ν_{Eee}, and ν_{Eei}.

Test-particle transfer frequencies or transfer times have been calculated by several authors.[12-14] By making use of a temperature relaxation model, other authors[1,15,16] have obtained relaxation times of frequencies closely related to some of the test-particle frequencies exhibited in this section. However, none of the preceding works contain the complete set of 16 frequencies, both collective and collisional, which are necessary to determine correctly the order of the

Table 1 Energy transfer and momentum transfer frequencies

	Collective coefficients T_e/T_i				
	1	30	60	80	100
A_{Mee}	-0.03	1.67	2.75	3.43	4.11
A_{Mei}	-0.349	-0.068	-0.045	-0.040	-0.036
A_{Mie}	-0.24	-0.20	-0.20	-0.20	-0.20
A_{Mii}	-0.44	-0.38	-0.38	-0.38	-0.38
A_{Eee}	-2.31	-0.080	0.472	0.708	0.884
A_{Eei}	-0.156	1.09	1.27	1.31	1.35
A_{Eie}	0.22	-0.15	-0.16	-.016	-0.16
A_{Eii}	-1.51	-2.80	-2.84	-2.85	-2.85

various interaction terms appearing in the Lenard-Balescu equation.

B. Modified Chapman-Enskog Formalism

The modified Chapman-Enskog solutions of both the electron and ion equations are obtained when the terms in each equation are ordered correctly. The Lenard-Balescu interaction integral J_{LB} is written as $J_{LB} = J_L + J_C$ where J_L is the Landau collision integral and J_C is the collective interaction integral. Using dimensionless variables and splitting the Lenard-Balescu interaction, the ion and electron equations take the form

$$\omega_t\ (\partial f_a/\partial t) + \omega_r\ \vec{c} \cdot (\partial f_a/\partial \vec{r}) + \omega_c\ \vec{A}_a \cdot (\partial f_a/\partial \vec{c}) =$$
$$(\omega_{aa})_L\ J_L\ (f_a,f_a) + (\omega_{ab})_L\ J_L\ (f_a,f_b)$$
$$+ (\omega_{aa})_C\ J_C\ (f_a,f_a) + (\omega_{ab})_C\ J_C\ (f_a,f_b) \qquad (2)$$

The ω's represent the frequencies associated with each term in this scaled equation.

At this point, it is assumed that the test-particle frequencies computed in the previous section are representative of the frequencies associated with the interaction terms in the equation. Hence, the interaction frequency is replaced by either the momentum or energy frequency corresponding to that interaction, depending on which is larger. The ion equation is divided by the frequency $(\nu_{Mie})_L$, and the electron equation is divided by $(\nu_{Mei})_C$. These steps yield the following equations

$$\frac{\omega_t}{(\nu_{Mie})_L}\frac{\partial f_i}{\partial t} \quad \frac{\omega_r}{(\nu_{Mie})_L}\vec{c}\cdot\frac{\partial f_i}{\partial \vec{r}} + \frac{\omega_c}{(\nu_{Mie})_L} \rightarrow \cdot \frac{\partial f_i}{\partial \vec{c}} =$$

$$\frac{1}{\varepsilon} J_L\,(f_i,f_i) + J_L\,(f_i,f_e) + J_C\,(f_i,f_i) + \varepsilon\, J_C\,(f_i,f_e) \qquad (3)$$

$$\frac{\omega_r}{(\nu_{Mei})_C}\frac{\partial f_e}{\partial t} + \frac{\omega_r}{(\nu_{Mei})_C}\vec{c}'\cdot\frac{\partial f_e}{\partial \vec{r}} + \frac{\omega_{c'}}{(\nu_{Mei})_C}\vec{A}'\cdot\frac{\partial f_e}{\partial \vec{c}'} =$$

$$\frac{1}{\varepsilon} J_L\,(f_e,f_e) + \frac{1}{\varepsilon} J_M\,(f_e,f_i) + \frac{1}{\varepsilon} J_C\,(f_e,f_e) +$$

$$J_C\,(f_e,f_i) \qquad (4)$$

Primes will be used throughout to indicate electron variables when the subscript e is impractical.

In both equations, the parameter ε is identified as

$$\varepsilon = (m_e/m_i)^{\frac{1}{2}} = (\nu_{Mie})_L/(\nu_{Mii})_L = (\nu_{Mei})_C/(\nu_{Mee})_L$$

Since $(\nu_{Eei})_L/(\nu_{Mei})_L$ is $O(m_e/m_i)$, the interaction integral $J_L(f_e,f_i)$ has been replaced by a simplified integral that accounts for momentum transfer only and yields $(\nu_{Mei})_L$ accurate to $O(m_e/m_i)$. This simplified integral $J_M(f_e,f_i)$ is given by

$$J_M\,(f_e,f_i) = \frac{2e^4\pi}{m_e^2}\ln\Lambda\,\frac{\partial}{\partial u'_\alpha}\,\frac{(u')^2\delta_{\alpha\beta} - u'_\alpha u'_\beta}{(u')^3}\,\frac{\partial f_e}{\partial u'_\beta}\int d^3u\, f_i(\vec{u}') \qquad (5)$$

where

$\vec{u} = \vec{c}-\vec{w}$ = ion peculiar velocity

$\vec{u}' = \vec{c}'-\vec{w}$ = electron peculiar velocity

$\vec{w}$ = hydrodynamic or mass flow velocity

This result is obtained by using the following approximation in the Landau interaction integral $J_L(f_e, f_i)$

$$\delta(\mu' - u\mu/u') \simeq \delta(\mu') \tag{6}$$

where

$$\mu = \vec{\ell} \cdot \vec{u}/u, \qquad \mu' = \vec{\ell} \quad \vec{u}'/u'$$

As in the classical Chapman-Enskog development, it now is assumed that the terms on the left-hand side of Eqs. (3) and (4) are of order unity and that f_a can be written as

$$f_a = f_a^o (1 + \varepsilon \phi_a)$$

With these assumptions the solutions of the zeroth-order (ε^o) ion equation and zeroth-order electron equation are

$$f_a^o(\vec{c}) = n_a (m_a/2\pi kT_a)^{3/2} \quad \exp - [(m_a/2kT_a)(\vec{c}-\vec{w})^2]$$

where n_a, m_a, and T_a are the density, mass, and temperature of the constituent of type a (a=i,e). These solutions are Maxwellian about the hydrodynamic velocity w with a temperature T_a. In contrast to the zeroth-order "classical" solutions, the solutions just exhibited do not require that the constituents be at the same temperature in zeroth order. In other words, these zeroth-order solutions allow the treatment of two-temperature plasmas.

The first-order equation for each constituent is of the same form. Both equations are inhomogeneous linear integral equations for ϕ_a. The conditions for the existence of a solution of an equation of this type are that the solutions of the homogeneous equation be orthogonal to the inhomogeneity. Since there are three collisional invariants for ions ϕ_i^H, then three equations are generated. These equations are the ion Euler equations. There are only two collisional invariants for electrons ϕ_i^H; therefore, only two electron Euler equations are obtained. Using the peculiar velocity $\vec{u} = \vec{c} - \vec{w}$, the two Euler equations that hold for both electrons and ions are

$$(Dn_a/Dt) + n_a\,(\partial/\partial\vec{r}) \cdot \vec{w} = 0 \tag{7}$$

$$(D/Dt)[(3/2)n_a kT_a] + (5/2)n_a kT_a\,(\partial/\partial\vec{r}) \cdot \vec{w} = E^o_{ab} \tag{8}$$

$$(a,b = e,i;\ a \neq b)$$

$$D/Dt = (\partial/\partial t) + \vec{w}\,(\partial/\partial\vec{w})$$

In addition, for ions there is a third Euler equation coore-sponding to $\phi^H_i = \vec{u}$. This Euler equation is

$$(D\vec{w}/Dt)\,(\vec{F}_i/m_i) + (1/n_i m_i)(\partial/\partial\vec{r})(n_i kT_i) = 0 \tag{9}$$

where

$$\int u^2\, J_L\,(f^o_i,f^o_e)\, d^3u = \frac{2E^o_{ei}}{m_i}$$

$$\int u'^2\, J_C\,(f^o_e,f^o_i)\, d^3u' = \frac{2E^o_{ie}}{m_e}$$

The three ion Euler equations are used to eliminate the time derivatives on the left-hand side of the first-order ion equation. Having done this, the first-order ion equation becomes

$$f^o_i\,\{(W^2 - \tfrac{5}{2})\,\frac{u_\alpha}{T_i}\frac{\partial T_i}{\partial r_\alpha} + 2\,\overset{\circ}{W}_{\alpha\beta}\,\frac{\partial w_\beta}{\partial r_\alpha} + \frac{2}{3n_i kT_i}(W^2-\tfrac{3}{2})E^o_{ie}\} - J_L(f^o_i,f^o_e)$$

$$= J_L\,(f^o_i,f^o_i\phi_i) + J_L\,(f^o_i\phi_i,f^o_i) \tag{10}$$

In a similar fashion, the two electron Euler equations and the third ion Euler equation, Eq. (9), are used to cast the first-order electron equation in the following form

$$f_e^o \{(W'^2 - \frac{5}{2}) \frac{u'_\alpha}{T_e} \frac{\partial T_e}{\partial r_\alpha} + 2 \overset{o}{W}'_{\alpha\beta} \frac{\partial w_\beta}{\partial r_\alpha}$$

$$+ \frac{2}{3n_e kT_e} (W'^2 - \frac{3}{2}) E^o_{ei} + u'_\alpha d^e_\alpha\} - J_C (f_e^o, f_i^o)$$

$$= J_{LB} (f_e^o, f_e^o \phi_e) + J_{LB} (f_e^o \phi_e, f_e^o) + J_M (f_e^o \phi_e, f_i^o) \quad (11)$$

In the preceding equations

$$W = u(m_i/2kT_i)^{\frac{1}{2}}, \quad \overset{o}{W}_{\alpha\beta} = W_\alpha W_\beta - (W^2/3)\delta_{\alpha\beta}, \quad W' = u'(m_e/2kT_e)^{\frac{1}{2}}$$

In Eq. (11)

$$\vec{d}^e = \frac{1}{n_e} \frac{\partial n_e}{\partial \vec{r}} - (\frac{T_i m_e}{T_e m_i}) \frac{1}{n_i} - \frac{1}{kT_e} \{\vec{F}_e - (\frac{m_e}{m_i}) \vec{F}_i\}$$

$$+ \frac{1}{T_e} \{\frac{\partial T_e}{\partial \vec{r}} - (\frac{m_e}{m_i}) \frac{\partial T_i}{\partial \vec{r}}\}$$

Following the classical Chapman-Enskog development, conditions are imposed to insure a unique solution ϕ_a. The first five velocity moments of the perturbation $f_i \phi_i$ must be identically zero. For the electron perturbation only, the zeroth and scalar second moment must vanish. Since E^o_{ab} are scalar and are not functions of velocity and since $J_L (f_i^o, f_e^o)$ and $J_C (f_e^o, f_i^o)$ are isotropic functions of the velocity, the unique solutions of the first-order ion and electron equations are

$$\phi_i(W) = -A_i (W^2) W_\alpha (\partial/\partial r_\phi) (\ln T_i) - B_i (W^2) \overset{o}{W}_{\alpha\beta} (\partial w_\beta/\partial r_\alpha)$$

$$+ \Psi (W^2) \quad (12)$$

$$\phi_e = -A_e (W'^2) W'_\alpha (\partial/\partial r_\alpha) (\ln T_e) - B_e (W'^2) \overset{o}{W}'_{\alpha\beta} (\partial w_\beta/\partial r_\alpha)$$

$$- D_e (W'^2) W'_\alpha d^e_\alpha + \Psi_e (W'^2) \quad (13)$$

Each of these solutions is the sum of the homogeneous solution plus the particular integral. The particular solution in the electron case, for example, has been obtained by noting that ϕ_e must be a scalar that is a linear combination of $\partial \ln T_e/\partial \vec{r}$, $\partial w_\beta/\partial r_\alpha$, $\vec{d}^e$, and an arbitrary scalar.

If ϕ_i and ϕ_e are substituted in Eqs. (10) and (11), respectively, then Eq. (10) yields two equations in A_i and B_i and Eq. (11) yields three equations in A_e, D_e, and B_e. These equations are solved in the next section by using the following expansion in terms of the generalized Laguerre polynomials

$$A_j\,(W^2) = \sum_{r=0}^{\infty} a_r^j\, L_r^{3/2}\,(W^2) \tag{14a}$$

$$B_j\,(W^2) = \sum_{r=0}^{\infty} b_r^j\, L_r^{5/2}\,(W^2) \quad (j=i,e) \tag{14b}$$

$$D_e(W'^2) = \sum_{r=0}^{\infty} d_n\, L_r^{3/2}\,(W'^2) \tag{14c}$$

III. Transport Coefficients

In this section, expressions for electron current, heat flux, and pressure tensor are exhibited. The various transport coefficients such as electrical conductivity are computed as a function of temperature ratio. Since collective interaction directly affects only the electron transport properties, the detailed expressions are obtained for only the electron component of the plasma.

Using Eq. (13) for ϕ_e and the relation for $\vec{d}_e$, it is found that

$$\vec{J}_e = \sigma\vec{E} + \alpha_e\,\vec{\nabla}\,T_e - \alpha_i\,\vec{\nabla}\,T_i + \gamma_e\,\vec{\nabla}\,n_e + \gamma_i\,\vec{\nabla}\,n_i \tag{15}$$

where

$$\sigma = [n_e e^2/(2\,m_e k T_e)^{\frac{1}{2}}]\,[1 + (m_e/m_i)]\,d_o$$

$$\alpha_e = n_e e\,(k/2\,m_e\,T_e)^{\frac{1}{2}}\,(a_o^e + d_o)$$

$$\alpha_i = n_e e\,(k/2\,m_e\,T_e)^{\frac{1}{2}}\,d_o\,(m_e/m_i)\,,\quad \gamma_e = e\,(kT_e/2m_e)^{\frac{1}{2}}(d_o)$$

$$\gamma_i = (n_e\,m_e\,T_i e/n_i m_i T_e)(kT_e/2m_e)^{\frac{1}{2}}\,d_o$$

In a similar fashion, the electron heat flux $\vec{q}_e$ is found to be

$$\vec{q}_e = - \beta_e \vec{E} - \kappa_e \vec{\nabla} T_e - \kappa_i^e \vec{\nabla} T_e - \delta_e \vec{\nabla} n_e - \delta_i \vec{\nabla} n_i \quad (16)$$

where

$$\beta_e = - 5/4(n_e e\ k)(2kT_e/m_e)^{1/2} (d_o - d_i)(1 + m_e/m_i)$$

$$\kappa_e = + 5/4(n_e\ k)(2kT_e/m_e)^{1/2} (a_o^e - a_1^e + d_o^e - d_1^e)$$

$$\kappa_i^e = 5/4(n_e\ k)(2kT_e/m_e)^{1/2}\ m_e/m_i\ (d_1 - d_o)$$

$$\delta_e = 5/4(k\ T_e)(2kT_e/m_e)^{1/2}\ (d_o - d_1)$$

$$\delta_i = 5/4\ (n_e k/n_i)\ (2kT_e/m_e)^{1/2}\ (m_e T_i/m_i T_e)\ (d_1 - d_o)$$

The electron viscosity μ_e is given by

$$\mu_e = (n_e kT_e/2)\ b_o^e \quad (17)$$

The ion thermal conductivity and viscosity both have the same value as in the classical solution. If $\vec{\nabla}P_e = \vec{\nabla}P_i = 0$, then the expressions for electron current and heat flux become

$$\vec{j}_e = \sigma \vec{E} + \alpha_e^* \vec{\nabla} T_e \quad (18)$$

$$\vec{q}_e = - \beta_e \vec{E} - \kappa_e^* \vec{\nabla} T_e \quad (19)$$

where

$$^*_e = n_e e\ (k/2\ m_e\ T_e)^{\pi}$$

$$\alpha_e^* = (n_e e)(k/2\ m_e\ T_e)^{1/2}\ a_o^e$$

$$\kappa_e^* = (5/4)(n_e k)(2kT_e/m_e)^{1/2}\ (a_o^e - a_1^e)$$

The Laguerre polynomial expansion coefficients that appear in the expressions for current, heat flux, and viscosity can be determined by solving Eqs. 10-11. The third-order approximation is used to determine a_o^e, a_1^e, d_o, and d_1, whereas the second-order approximation is used to determine b_o^e.

If collective contributions are treated as perturbations to the usual collisional contributions, then all of the coefficients may be written in the form shown below for a_o^e

$$a_o^e = a_{oL}^e + (a_{o1}^e/\ln\Lambda) + \ldots \tag{20}$$

The leading term in these expansions is the classical value obtained by using only the Landau interaction integral.

The coefficients in the second term of the expressions for a_o^e, a_1^e, d_o, and d_1 are shown in Table 2. These coefficients are the first-order correction arising from collective interactions. The electron transport coefficients obtained, appearing in the electron heat flux and current equation, exhibit corrections at T_e/T_i = 100 of up to 20% where compared to the accepted values of Spitzer and Harm.[17] The electron viscosity exhibits a similar correction, and, in addition, it is seen that the electron viscosity becomes equal to the ion viscosity at $T_e/T_i \simeq 6$ and is greater than the ion viscosity for large values of T_e/T_i. For a temperature ratio of one and zero pressure gradient, it was found that the transport coefficients agreed with those of Spitzer and Harm[17] and DeVoto.[18]

An attempt was made to compare the results of this paper with those of Ref. 2. It was found that only the

Table 2 Collective contributions to expansion coefficients as a function of T_e/T_i [multiply by $(k\,T_e)^2/n_e e^4 \ln\Lambda$]

T_e/T_i	a_{o1}	a_{11}	d_{o1}	d_{11}	b_{o1}^e
1	-0.1447	0.3375	-0.0450	0.0579	-0.0804
3	-0.2803	0.5256	-0.1226	0.1121	-0.1671
10	-0.4807	0.8024	-0.2384	0.1923	-0.2963
40	-0.8011	1.2109	-0.4432	0.3204	-0.5213
100	-1.1778	1.6186	-0.7265	0.4711	-0.8259

values of electrical conductivity were in good agreement at $T_e/T_i = 100$. There appears to be an error in Ref. 2, however, because the coefficients in the equations corresponding to Eqs. (43) and (45) do not satisfy the Onsager relation.[19] This relation is

$$\beta_e = \alpha_e^* + (5kT_e/2e)\,\sigma$$

The coefficients obtained in this paper satisfy the Onsager relation to one part in 10^5 for all values of T_e/T_i with collective effects included.

IV. Conclusions

The results obtained in this paper have demonstrated that the use of test-particle frequencies to estimate the order of magnitude of terms appearing in the kinetic equation is a valu

valid procedure. This procedure was used to obtain transport coefficients for a fully ionized hydrogen plasma; however, the procedure is sufficiently general that it can be applied to any fully ionized plasma.

The transport coefficients that have been obtained exhibit two distinct corrections to be accepted values. The first type of correction is due to the modified Chapman-Enskog method of solution, whereas the second type is due to the inclusion of long-range collective phenomena that exist in two-temperature plasmas. In the limit of equal constituent temperature and neglecting collective effects, the coefficients reduce to those computed by Spitzer and Harm.[17]

The results of this paper are valid so long as $\ln\Lambda \geq 8$. One should take care, however, to note the following points concerning extensions of this work to larger values of T_e/T_i and smaller values of $\ln\Lambda$. If too large a value of T_e/T_i is chosen, the total energy of the fluctuations (in this case, ion acoustic waves) may not be much smaller than the total kinetic energy of the plasma. If this should occur, the Lenard-Balescu equation is no longer valid.[20]

The expansion parameter Λ^{-1} appears in the derivation of the Lenard-Balescu equation, and, if Λ^{-1} becomes too large (i.e., $\ln\Lambda$ too small), the equation once again is invalid. Lastly, the effects of collisional damping are ignored in the Lenard-Balescu equation. The collisional damping rate can be ignored only if the Landau damping decrement of the ion acoustic wave is much greater than (ν_{Mii}).

References

[1] Wu, C.S., Klevans, E.H., and Prinack, J.R., "Temperature Relaxation in a Fully Ionized Plasma", The Physics of Fluids, Vol. 8, June 1965, pp. 1126-1133.

[2] Gorbunov, L.M. and Silin, V.P., "Theory of Transport Effects in a Nonisothermal Fully Ionized Plasma", Zhurnal Tekhnicheskoi Fiziki, Vol. 34, March 1964, pp. 385-394; English Translation: Soviet Physics-Technical Physics, Vol. 9, Sept. 1964, pp. 305-311.

[3] Pearson, G.A., "Nonlinear Effects of Fluctuations in a Current-Carrying Plasma," The Physics of Fluids, Vol. 10, April 1967, pp. 685-695.

[4] Lenard, A., "On Bogoliuhov's Kinetic Equation for a Spatially Homogeneous Plasma," Anals of Physics, Vol. 10, July 1960, pp. 390-400.

[5] Guernsey, R., "The Kinetic Theory of Fully Ionized Gases", Ph.D. Dissertation, 1960, University of Michigan.

[6] Balescu, R., "Irreversible Processes in Ionized Gases, "The Physics of Fluids, Vol. 3, Jan. 1960, pp. 52-63.

[7] Chapman, S. and Cowling, T.G., The Mathematical Theory of Non-Uniform Gases, Cambridge University Press, New York, 1958 Chap. 7.

[8] Hirschfelder, J.O., Curtiss, C.E., and Bird, R.B., Molecular Theory of Gases and Liquids, Wiley, New York, 1954.

[9] Braginskil, S.I., "Transport Phenomena in a Completely Ionized Two-Temperature Plasma", Journal of Experimental and Theoretical Physics, (U.S.S.R.), Vol. 33, Aug. 1957, pp. 459-472; English translation: Soviet Physics-JETP, Vol. 6, Feb. 1958, pp. 358-368.

[10] Chmieleski, R.M. "Transport Properties of a Non-equilibrium Partially Ionized Gas", Ph.D. Dissertation, 1966, Stanford University.

[11] Balescu, R., Statistical Mechanics of Charged Particles, Wiley, New York, 1963, Chap. 3.

[12] Montogmery, D.C. and Tidman, D.A., Plasma Kinetic Theory, McGraw-Hill, New York, 1964, Chap. 3.

[13] Tidman, D.A. and Eviatar, A., "Scattering of a Test Particle by Enhanced Plasma Fluctuations," The Physics of Fluids, Vol. 8, Nov. 1965, pp. 2059-2065.

[14] Spitzer, L., Physics of Fully Ionized Gases, Interscience, New York, 1956, pp. 76-81.

[15] Ramazashvili, R.R., Rukhadze, A.A., and Silin, V.P., "Temperature Equilibration Rate for Charged Particles in a Plasma", Journal of Experimental and Theoretical Physics (U.S.S.R.). Vol. 43, Oct. 1962, pp. 1323-1330; English translation: Soviet Physics-JETP, Vol. 16, April 1963, pp. 939-944.

[16] Kihara, T., Aono, O., and Itikawa, Y., "Unified Theory of Relaxations in Plasmas, II. Applications", Journal of the Physical Society of Japan, Vol. 18, July 1963, pp. 1043-1050.

[17] Spitzer, L. and Harm, R., "Transport Phenomena in a Completely Ionized Gas", Physical Review, Vol. 89, March 1953, pp. 977-981.

[18] Devoto, R.S., "Transport Properties of Ionized Monatomic Gases," The Physics of Fluids, Vol. 9, June 1966, pp. 1230-1240.

[19] DeGrout, Thermodynamics of Irreversible Processes, North Holland Publishing Co., Amsterdam, 1952, Chap. 2.

[20] Rostoker, M., "Superposition of Dressed Test Particles," The Physics of Fluids, Vol. 7, April 1964, pp. 479-490.

COMPARATIVE STUDY OF SPHERICAL AND CYLINDRICAL DRIFT PROBES

H. Makita,* and K. Kuriki+

Institute of Space and Aeronautical Science, University of Tokyo, Tokyo, Japan

Abstract

Current-voltage characteristics of drifting spherical and cylindrical Langmuir probes has been studied experimentally in collisionless plasmas. In order to single out the effect of probe speed on the ion current collection, the experiment was carried out by a rotating arm facility. With this device the probe speed as well as properties of the undisturbed plasma was accurately determined. The ion current of the spherical probe showed good agreements with the theory of Godard at very large or small R_p/λ_D, the ratio of probe radius to Debye length. Discrepancies were found at the intermediate R_p/λ_D, where the spherically asymmetric potential distribution was important. At some R_p/λ_D, the ion current collection by the cylindrical probe was dominated by the end effect, which was explained in termes of the spherical probe characteristics.

Introduction

Recently, interest in drift probe problems has been greatly increased in relation to the ionospheric plasma measurement by the satellite born probes or the probe diagnostics in the plasma wind tunnels. On this problem Godard[1] calculated the velocity-dependent current collection of the spherical and cylindrical probes. The calculation covers wide

---

Presented as Paper 33 at the 10th International Sym-Posium on Rarefied Gas Dynamics, Aspen, Colo., July 19-23, 1976.

*Research Associate.
+Associate Professor.

ranges of parameters such as the probe radius to Debye length ratio R_p/λ_D, nondimensional probe potential $\chi = eV/kT_e$, and ion-to-electron-temperature ratio T_i/T_e, where e and k denote the electronic charge and the Boltzmann constant. In his calculation, the sheath distortion has not been taken into account.

So far, few experimental works have succeeded in demonstrating the direct velocity dependence of the ion current collection characteristics of the drift probes, espacially in the case of spherical probe. It has been very difficult in the plasma wind tunnel experiment to control the velocity alone without altering the plasma properties and to measure the velocity accurately. This is why the experimental results have failed to show the pure effects of drift velocity on the probe characteristic and, moreover, why the theoretical development of the drift probe problem has been retarded.

The rotating arm facility, shown in Figure 1, enables us to specify the drift velocity U_d to any value between 0 and 290 m/sec with an accuracy of 0.1%. Current signals picked up by the usual probe circuit as the probe passed through the test section at first were amplified with 80-110 dB gain. An instantaneous signal was sampled at the moment when the probe arrived at the center of the test section in every rotation, and then held steady until the next arrival. Meanwhile, the probe bias was changed gradually. By the stationary probe located near the drift probe orbit, the disturbences produced by one probe passage in the test section were sure to be extinguished completely before the next probe arrival. Accordingly, we obtained the probe characteristics at any speed ratio on an X-Y recorder as if it were obtained in a uniform plasma[2]. With these devices, we can detect any minute change of the probe characteristics caused by the velocity change. The speed ratio S_d is as large as about 1.7 in the present experiment. This intermediate speed range is important, since, at hypersonic velocity, the ion current collection simply depends linearly on the drift velocity.

In the low number density plasma, the spherical drift probe has rather simple experimental configuration. On the other hand, the cylindrical probe has very complicated one because of its finite length. Therefore, the results of the cylindrical probe are always affected by the three-dimensional sheath structure. In the present paper, the three-dimensional behaviours of the cylindrical drift probe are explained in connection with the results of spherical drift probe.

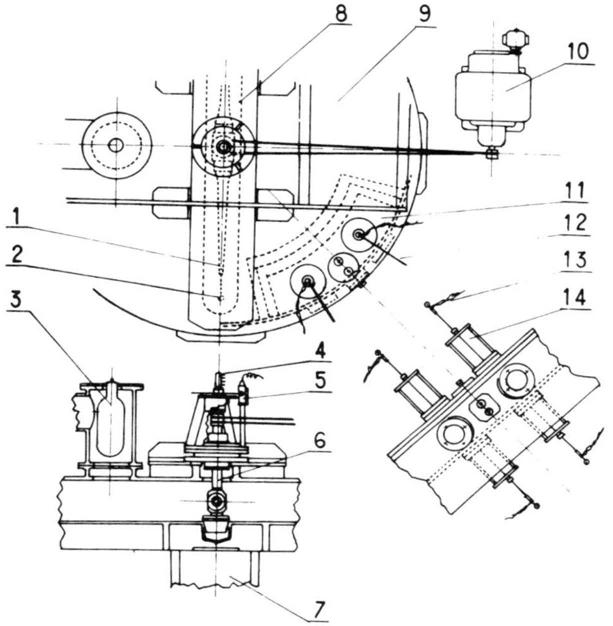

Fig. 1 Rotating arm facility: 1 rotating arm, 2 spherical probe, 3 cold trap, 4 mercury slip ring, 5 probe position detector, 6, rotating shaft, 7 concrete base, 8 arm installation port, 9 vacuum tank, 10 electric motor, 11 test section, 12 gas feeding pipe, 13 discharge power lines, 14 plasma source container.

Experimental Results and Discussion

The present experiments were made in the plasma of $n_i = 10^5 \sim 10^8/cm^3$ and $T_e = 500° \sim 2000°K$, which gives the R_p/λ_D values between about 0.3 ~ 20 for the spherical probes and 0.03 ~ 2 for the cylindrical probes. Ion temperature T_i was determined by the two probe method newly developed by the authors. This method utilizes the large difference in T_i/T_e dependence between the spherical and cylindrical probes in the stationary collision-free plasma. T_i determination is indispensable to normalize the drift velocity as $S_d = U_d/(kT_i/m_i)^{1/2}$, where m_i denotes the ion mass. As a result, we obtained T_i almost equal to T_e. The detailes of T_i measurement will be reported elsewhere[3]. T_e, n_i and the floating potential distribution were assured to be sufficiently uniform at the central part of the test section compared with the largest probe radius of 5 mm. We employed spherical probes of 3.82 and 10- mm diam and cylindrical probes of 0.30 -mm diam and 75-mm length and 1.25 - mm diam and 25 - mm length. By these differences in probe sizes, we can check the size dependence of the results.

Normalized ion current I_i at the nondimensional probe potential $\chi = -25$ is plotted against the probe speed ratio S_d for several R_p/λ_D in the cases of spherical probes in Fig. 2 (a) and of cylindrical probes in Fig. 2 (b), where I_i is defined as $J_i/en_i(kT_e/m_i)^{1/2}$ and χ as $e(V-V_p)kT_e$, and J_i is the ion current collected by the probe, and V_p the plasma potential. In Fig. 2 (a) for the the spherical probe, I_i for $R_p/\lambda_D = 0.550$, which is the smallest for the spherical probe, decreases with the increase in S_d and that for the largest value of $R_p/\lambda_D = 15.1$, and increases to the contrary. These two cases show good agreements with the theoretical results of Godard , shown by dashed lines. For the intermediate values of R_p/λ_D, the experimental plots differ from the theoretical prediction. Especially around $R_p/\lambda_D \approx 0.8$ and at higher drift velocities, the difference is marked by the abrupt increase in I_i of the experimental plots.

Then we can observe the bump in I_i-S_d curves at $R_p/\lambda_D = 1 \sim 3$. S_d corresponding to the bump shifts to the smaller value as R_p/λ_D is increased. This is because the potential enhancement[4,5] in the wake, which is caused by the ions strongly deflected toward the probe surface, tends to move downstream out of the rear sheath more easily as the sheath becomes thinner.

In the case of the cylindrical drift probes, discrepancies were also found between the experimental and theoretical results. This is due partly to the end effect inherent with the cylindrical probe of finite length and partly due to the theoretical assumptions of the sheath structure to be unaffected by the drift motion. Two bumps are observed in these I_i-S_d curves; the first bump at the lower drift velocity and the second one at the higher drift velocity. These bumps show the similar R_p/λ_D dependence to that of the spherical drift probe. The first bump is considered to have resulted from the two-dimensional sheath deformation by the localized potential enhancement due to the ion concentration in the wake just like the spherical drift probe. It occurs at almost the same S_d as the bump of the spherical drift probe at the same R_p/λ_s, where λ_s is the sheath thickness estimated from the results of Laframboise[6]. The same R_p/λ_s means that these two proves have similar sheath structures in spite of the large difference in

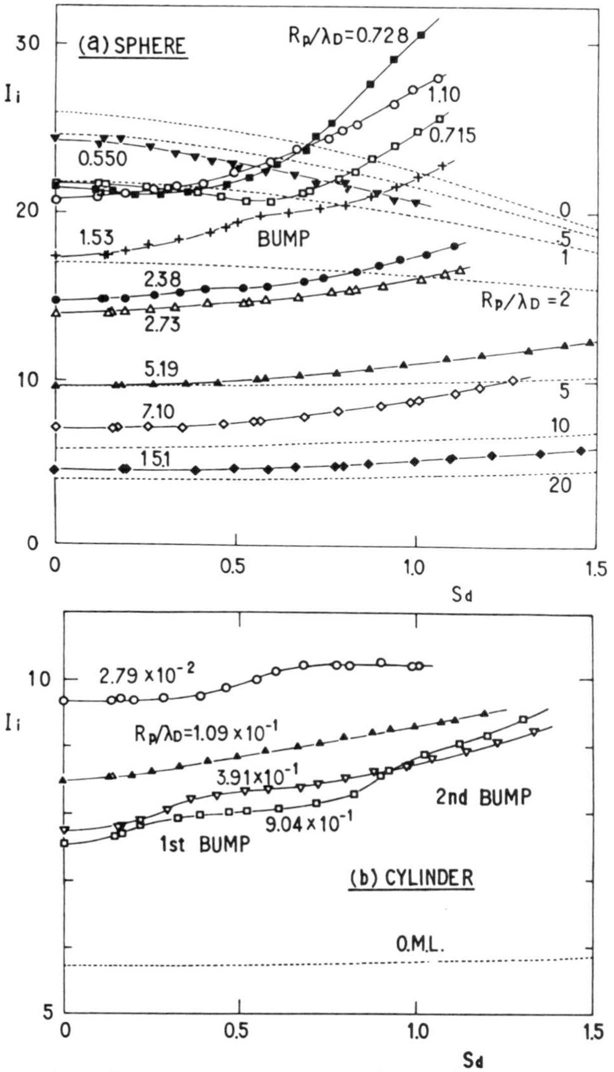

Fig. 2 Ion current at χ = -25 vs speed ratio for various R_p/λ_D. Theoretical values of Godard are plotted by broken lines. (a) Spherical probe, (b) Cylindrical probe.

R_p/λ_D. On the other hand, the second bump is considered to be caused by the end effect. At the probe end, the sheath structure resembles closely that of the spherical probe of the same R_p/λ_D. The second bump of the cylindrical probe occurs almost

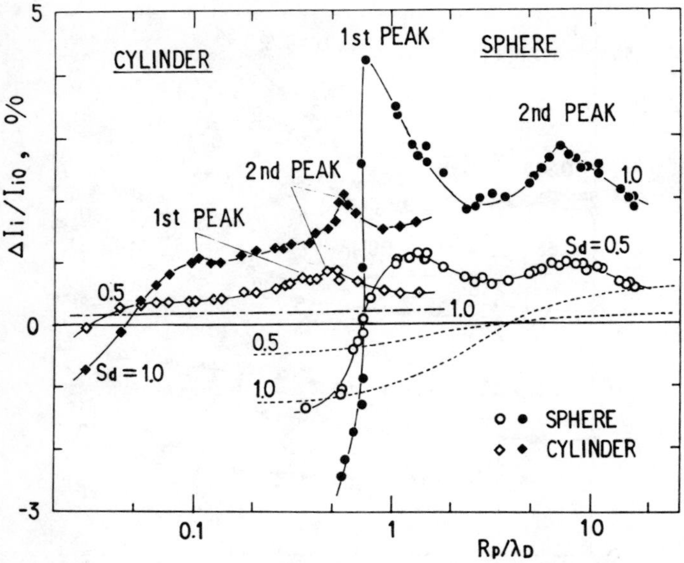

Fig. 3 Fractional variation of ion current $\Delta I_i/I_{io}$ of the spherical and cylindrical probes at $\chi = -15$ vs R_p/λ_D for various speed ratios. Theoretical curves of Godard are plotted by broken lines.

at the same R_p/λ_D as the bump is found in the spherical drift probe characteristics. Figure 3 shows the variations of the change in ion current of the spherical and cylindrical probes normalized by the stationary ion current of the same probe $\Delta I_i/I_{io} = (I_i - I_{io})/I_{io}$, where suffix 0 denotes the value at $S_d = 0$. Spherical probe shows two peaks in the $\Delta I_i/I_{io}$ vs R_p/λ_D curve. We call the peaks the first and second peaks in order from the smaller R_p/λ_D. The second peak is presumed to be formed by the effective increase in the influx of ions into the elongated rear sheath.

As R_p/λ_D decreases from 7, the increment in ion current becomes smaller except around the first peak, and eventually $\Delta I_i/I_{io}$ becomes negative. This decrease in ΔI_i is caused by the effective decrease in the frontal sheath area due to the ion flow. Around $R_p/\lambda_D = 0.8$, the ion current collection has a sharp enhancement, and we call it the first peak. The first peak is very high, and its position moves to the smaller R_p/λ_D with the increase in drift velocity. The first peak originates in the bumps shown in Figure 2 (a).

We also see the two kind of peaks in the case of cylindrical drift probe, and we call them the first and second peaks of cylindrical probe in order from smaller R_p/λ_D. The first peak is rather small and shifts to the smaller R_p/λ_D with the increase in drift velocity. The first peak originate from the first bump in Fig. 2 (b). The peak is less prominent than the first peak of the spherical probe because the inward deflection of the ions toward the wake axis is only two-dimensional in the case of a cylindrical probe. The second peak lies at $R_p/\lambda_D \sim 0.7$ and is caused by the three-dimensional current collection around the end of the cylinder. This peak corresponds to the second bump shown in Fig. 2 (b).

Difference between the experimental and theoretical results is greater than the spherical probe. This is due partly to the cylindrically symmetric sheath, assumed in the theory, and partly to the end effect. The end effect is considered to be not so severe except for around the second peak, if R_p/λ_D is large enough.

In the region of $R_p/\lambda_D \lesssim 0.05$, however, the cylindrical probe is not long enough to have a two-dimensional sheath. In this region, the probe-length-to-sheath-thickness ratio ℓ_p/λ_s is smaller than about 3. The sheath becomes similar to the spherical one, and therefore the ion current tends to decrease with drift velocity. From these reasons, the theoretical prediction, shown by dashed lines, might differ greatly from the experimental results. In conclusion, when the three-dimensional sheath structure is essential to the cylindrical probe characteristics, its complicated current collection can be explained in terms of that by the spherical probe.

References

[1] Godard, R., "Systematical Model for Cylindrical and Spherical Collectors in a Flowing Collisionless Plasma," Ph. D. Thesis, 1975, York University, Toronto, Ontario.

[2] Makita, H., and Kuriki, K., "Current Collection by Spherical Probe Drifting in a Collisionless Plasma," submitted to Physics of Fluids.

[3] Makita, H., and Kuriki, K., "Ion Temperature Determination by Two-Probe Method," submitted to Reviews of Scientific Instruments.

[4] Fournier, G., "Ecoulement de Plasma sans Collisions Autour d'eu Cylindre en vue d'Applications aux Sondes Ionspheriques," Rept. 137, 1971, Office National d'Etudes et de Recherches Aerospatials.

[5] Hester, S. D., and Sonin, A. A., "A Laboratory Study of the Wake of Ionospheric Satellite," AIAA Journal, Vol. 8, June, 1970, pp. 1090-1098.

[6] Laframboise, J. G., "Theory of Spherical and Cylindrical Langmuir Probes in a Collisionless Maxwellian Plasma at Rest," UTIAS Rept. 100, 1966, Univ. of Toronto.

PHYSICS OF IONIZING SHOCK WAVES IN MONATOMIC GASES AT LARGE MACH NUMBERS

M. Krishnamurthy* and S. M. Ramachandra+

Hindustan Aeronautics Ltd., Bangalore, India

Abstract

This paper describes a study of the physical parameters of ionizing shock waves in monatomic gases at large Mach numbers when a simultaneous study of the gasdynamic shock and the relaxation zone is necessary. Assuming a trimodal distribution for the atoms and Maxwellian distribution for the electrons, the flow equations are obtained from the Boltzmann equation using the moment technique. Experimental cross sections are used for the ionizing collisions, and inverse fifth-power law has been assumed for the weak elastic interactions. Using the appropriate boundary conditions, the set of equations has been reduced to obtain the jump conditions across an ionizing shock. The flow equations have been solved numerically for argon gas to obtain the shock structure profiles. The variations across the shock of the electrical conductivity and the refractive index to sodium yellow line radiation and of the electric potential due to electron-ion polarization have been computed.

Nomenclature

B_{aa} = constant of proportionality in the atom-atom elastic interaction force law, dyn-cm^5

B_{ae} = constant of proportionality in the atom-electron elastic interaction force law, dyn-cm^5

C_{aa} = constant of the atom-atom collisional excitation cross section, cm^2/eV

Presented as Paper 16 at the 10th International Symposium on Rarefied Gas Dynamics, Aspen, Colo., July 19-23, 1976.

*Deputy Design Engineer, Design Bureau.

+Deputy Chief Designer, Design Bureau.

C_{ae} = constant of the atom-electron collisional excitation cross section, cm^2/eV
$\underset{\sim}{c}$ = molecular velocity vector, cm/sec
e = electron charge, esu
g_a = electronic partition function for the neutral atom
g_+ = electronic partition function for the singly ionized atom
h = Planck's constant
I = ionization potential, eV
I* = excitation potential, eV
$\hat{i}$ = unit vector in the direction of the flow
k = Boltzmann constant
m = mass of the particle, g
n = number density, cm^{-3}
T = translational temperature, °K
U = flow velocity, cm/sec
x = shock coordinate, cm
α = degree of ionization, polarizability, cm^3
ϵ = ratio of electron mass to atom mass
ρ = density, g/cm^3
λ = characteristic length, cm
σ = electrical conductivity, mho/cm
μ = refractive index

Introduction

The structural aspects of a strong ionizing shock wave for Mach numbers ~ 30, when the gasdynamic shock and the relaxation zone are comparable in size, have been studied by Chubb.[1] He studied the two regions separately. Inside the gasdynamic shock, he assumed a bimodal distribution function[2] for atoms and Maxwellian distribution functions for electron and ion gases. He included the inelastic collisional effects within the gasdynamic shock, in addition to the predominant elastic collisional dissipative mechanisms. In the relaxation zone, all of the three species were assumed to have Maxwellian distribution. The plasma equations along with the electron equations describe the flow completely. A study of this type is incomplete when the identification of the two regions becomes difficult. Identification of the regions is ambiguous for large values of upstream Mach number, temperature, etc. It is particularly so when the atom-atom collisional excitation cross section is large. This difficulty can be overcome by resorting to a unified study of the entire nonequilibrium region. Krishnamurthy and Ramachandra[3] have studied the problem using the Boltzmann equation to describe the flow in the entire region and a moment method for its solution. In the present study, a trimodal distribution function characterizing the three possible equilibrium states has been assumed for the atom gas, whereas Maxwellian distributions for the electron and ion gases have been postulated. In addition to the usual moment equations of the Boltzmann equation, namely, mass, momentum, and energy, a c_x^2 moment equation[2] has been used to supplement the set of equations. This study includes the effect of elastic collisional dissipative mechanisms that may be active inside the relaxation zone, especially toward the end of it, where steep gradients in the temperature and velocity profiles are noticed. A characteristic length based on the upstream velocity has been chosen in place of the commonly used upstream mean free paths to nondimensionalise the x coordinate. The set of equations will be solved numerically in two phases: first for the downstream equilibrium conditions, and then for the structure profiles.

The shock thickness and the thickness parameter as defined by Petschek and Byron[4] always have been used for the purpose of comparison with either experiments or parallel theories. These are only gross parameters and do not reveal a detailed comparison. Furthermore, with the increase of upstream Mach number, the shock thickness and the thickness parameter asymptotically tend to constant values[3], and hence the choice of the shock thickness or the thickness parameter will be inappropriate for the comparison purpose. Under this circumstance, comparison of the shock structure profiles definitely would be better and yields more reliable information regarding various collision rates, their cross sections, and their relative effects on the shock structure profile. The electrical and optical properties of the fluid within the shock are measurable with comparatively less difficulty than the gasdynamic variables of state. Consequently, the present investigation deals with a theoretical estimate of the profiles of the refractive index, the electrical conductivity, and the electric potential due to electron-ion polarization for argon gas. The study also includes the effect of initial pressure on the shock structure profiles. Radiation and precursor effects and the effects of multiple ionization have not been included in the present studies, even though these are important at higher Mach numbers and lower initial pressures.

Flow Equations

The Boltzmann equation is used to describe the distribution functions of the three species, namely, atoms, ions, and electrons using the Debye length as the cutoff parameter for the divergent integrals. The distribution functions for the three species for a one-dimensional flow can be assumed in the form

$$f_a = \sum_{j=1}^{3} n_j (m_a/2\pi kT_j)^{3/2} \exp\left[-m_a(\underline{c}_a - U_j\hat{\imath})^2/2kT_j\right] \tag{1a}$$

$$f_e = n_e (m_e/2\pi kT_e)^{3/2} \exp\left[-m_e(\underline{c}_e - U_e\hat{\imath})^2/2kT_e\right] \tag{1b}$$

$$f_i = n_i (m_i/2\pi kT_i)^{3/2} \exp\left[-m_i(\underline{c}_i - U_i\hat{\imath})^2/2kT_i\right] \tag{1c}$$

where f is the velocity distribution function, and the subscripts a, e, i represent atom, electron, and ion, respectively; n_1, n_2, n_3, n_e, n_i, U_e, U_i, T_e, and T_i are functions of x. The subscripts j = 1, 2, 3 represent, respectively, the upstream equilibrium state, a fictitious intermediate state with fully excited translational degrees, and the ionized downstream equilibrium state. The atom density, velocity, and temperature at any point are defined, respectively, by

$$n_a = \sum_{j=1}^{3} n_j \tag{2a}$$

$$U_a = (1/n_a) \sum_{j=1}^{3} n_j U_j \tag{2b}$$

$$T_a = (1/n_a) \sum_{j=1}^{3} n_j \left[T_j + (m_a/6kn_a) \sum_{k=1}^{3} n_k (U_j - U_k)^2\right] \tag{2c}$$

The equations of conservation of mass, momentum, and energy can be written for the three species separately by taking the appropriate moments of the Boltzmann equation with the distribution functions (1). The overall plasma conservation equations obtained by adding the corresponding set of equations replace the set of ion equations in the present study.[5] An additional c_x^2 moment equation has been included for the completion of the set of equations. Since the atom-ion elastic interaction cross section is large compared to other collision cross sections, it is possible to assume that the atom and ion gases have the same velocity and temperature. In addition, with the quasineutrality approximation it can be written that

$$U_a = U_i = U_e\,; \qquad T_a = T_i \tag{3}$$

The various quantities are nondimensionalised using the up-stream atom density n_o and the temperature T_1 as follows

$$N_j = n_j/n_o, \bar{U}_j = U_j/\sqrt{2kT_1/m_a}, \bar{T}_j = T_j/T_1 \; ; \quad j = 1,2,3,e,i \tag{4}$$

The characteristic length λ used for nondimensionalising x is defined by

$$\lambda = (U_1/\pi n_o)(m_a/2B_{aa})^{1/2} \tag{5}$$

For $M_1 \to \infty$, λ can be shown[5] to vary as the mean free path on the hot side of the inert gas shock so that $\lambda < \infty$. Consequently, λ can be used with the same advantages as the mean free path of the hot side of the inert gas shock,[6] whereas λ still characterizes the upstream flow. Table 1 gives the values of λ for different upstream Mach numbers. The non-dimensional x and the electric field are defined by

$$\xi = x/\lambda \; , \quad \bar{E} = E/E_o \tag{6}$$

where E is the electric field and

$$E_o = (\pi n_o m_e/e)(2B_{aa}/m_a)^{1/2}(2kT_1/m_a)^{1/2} \tag{7}$$

The boundary conditions are

$$\xi \to -\infty, N_1 \to 1, N_j \to 0, \bar{E} \to 0; \quad j = 2,3,e,i \tag{8a}$$

$$\xi \to +\infty, N_1 \to 0, N_2 \to 0, \bar{E} \to 0, U_e = U_i = U_a \to U_3, T_e = T_i = T_a \to T_3 \tag{8b}$$

Table 1 Characteristic length λ (p_1 = 1.0 Torr, T_1 = 300°K)

M_1	15	20	25	30	35	40
λ, mm	0.2753	0.3671	0.4589	0.5506	0.6424	0.7342

Using the boundary condition (8a), the nondimensional governing equations can be written as[3]

$$\frac{d}{d\xi}\Big[\sum_j N_j \bar{U}_j\Big(\bar{U}_j^2+\frac{3}{2}\bar{\beta}_j\Big)\Big]= \bar{U}_1\sum_j\Bigg[-\frac{1}{4}A_2(5)\sum_k N_j N_k \bar{U}_{jk}^2$$

$$+\left(\frac{2B_{ae}}{B_{aa}}\right)^{1/2}\epsilon^{1/2} N_e N_j \bar{\beta}_{je}\, P_1\left(\frac{\bar{U}_j}{\sqrt{\bar{\beta}_{je}}},\frac{\bar{\beta}_j}{\bar{\beta}_{je}},\frac{\bar{U}_{je}}{\sqrt{\bar{\beta}_{je}}},\epsilon\right)$$

$$+\left(\frac{B_{ai}}{B_{aa}}\right)^{1/2} N_e N_j \bar{\beta}_{ji}\, P_1\left(\frac{\bar{U}_j}{\sqrt{\bar{\beta}_{ji}}},\frac{\bar{\beta}_j}{\bar{\beta}_{ji}},\frac{\bar{U}_{ji}}{\sqrt{\bar{\beta}_{ji}}},\frac{1}{2}\right)$$

$$-D_1\Bigg\{\frac{1}{8\epsilon}\frac{C_{aa}}{C_{ae}}\sum_k N_j N_k \bar{\beta}_{jk}^{5/2} F_2\left(\frac{\bar{U}_j}{\sqrt{\bar{\beta}_{jk}}},\frac{\bar{\beta}_j}{\bar{\beta}_{jk}},\frac{\bar{U}_{jk}}{\sqrt{\bar{\beta}_{jk}}},\frac{\bar{g}_{oa}}{\sqrt{\bar{\beta}_{jk}}}\right)$$

$$+\frac{1}{2}N_e N_j \bar{\beta}_{je}^{5/2} F_2\left(\frac{\bar{U}_j}{\sqrt{\bar{\beta}_{je}}},\frac{\bar{\beta}_e}{\bar{\beta}_{je}},\frac{\bar{U}_{je}}{\sqrt{\bar{\beta}_{je}}},\frac{\bar{g}_{oe}}{\sqrt{\bar{\beta}_{je}}}\right)$$

$$-\frac{1}{3}D_2 N_e^3\left(\bar{U}_e^2+\frac{1}{2}\bar{\beta}_i\right)\left(2+\frac{\bar{g}_{oe}^2}{\bar{\beta}_e}\right)\Bigg\}\Bigg] \tag{9}$$

$$\frac{d}{d\xi}\left[N_e\bar{U}_e\right]=\frac{1}{4}D_1\bar{U}_1\sum_j\Bigg[\frac{1}{4\epsilon}\frac{C_{aa}}{C_{ae}}(1-R_1)\sum_k N_j N_k \bar{\beta}_{jk}^{3/2}F_1\left(\frac{\bar{U}_{jk}}{\sqrt{\bar{\beta}_{jk}}},\frac{\bar{g}_{oa}}{\sqrt{\bar{\beta}_{jk}}}\right)$$

$$+(1-R_2)N_e N_j \bar{\beta}_{je}^{3/2} F_1\left(\frac{\bar{U}_{je}}{\sqrt{\bar{\beta}_{je}}},\frac{\bar{g}_{oe}}{\sqrt{\bar{\beta}_{je}}}\right)\Bigg] \tag{10}$$

$$\frac{d}{d\xi}\left[N_e\bar{\beta}_e\right]=-2N_e\bar{U}_1\bar{E} \tag{11}$$

$$\frac{d\bar{\beta}_e}{d\xi}=\frac{2}{3}\frac{\bar{U}_1}{N_e\bar{U}_e}\Bigg[D_3 N_e\sum_j N_j\left\{3(\bar{\beta}_j-\epsilon\bar{\beta}_e)+2\bar{U}_j\bar{U}_{je}\right\}$$

$$+2D_4 A_1(2) N_e^2 \bar{\beta}_{ie}^{-3/2}(\bar{\beta}_i-\epsilon\bar{\beta}_e)$$

$$-\frac{1}{4}D_1\bar{g}_I^2 N_e(1-R_2)\sum_j N_j \bar{\beta}_{je}^{3/2} F_1\left(\frac{\bar{U}_j}{\sqrt{\bar{\beta}_{je}}},\frac{\bar{g}_{oe}}{\sqrt{\bar{\beta}_{je}}}\right)\Bigg]$$

$$-\frac{2}{3}\frac{\bar{\beta}_e}{\bar{U}_e}\frac{d\bar{U}_e}{d\xi}-\frac{\bar{\beta}_e}{N_e\bar{U}_e}\frac{d}{d\xi}(N_e\bar{U}_e) \tag{12}$$

$$\sum_j N_j\bar{U}_j+N_e\bar{U}_e=\bar{U}_1 \tag{13}$$

$$\sum_j N_j(\bar{U}_j^2+\frac{1}{2}\bar{\beta}_j)+N_e\bar{U}_e^2+\frac{1}{2}N_e\{(1-\epsilon)\bar{\beta}_i+\epsilon\bar{\beta}_e\}-\frac{\pi}{4}\epsilon^2 n_o\frac{B_{aa}}{e^2}\bar{E}^2=\bar{U}_1^2+\frac{1}{2} \tag{14}$$

$$\sum_j N_j\bar{U}_j(\bar{U}_j^2+\frac{5}{2}\bar{\beta}_j)+N_e\bar{U}_e\{\epsilon(\frac{5}{2}\bar{\beta}_e+\bar{g}_I^2)+(\bar{U}_e^2+\frac{5}{2}\bar{\beta}_i)\}=\bar{U}_1(\bar{U}_1^2+\frac{5}{2}) \tag{15}$$

where

$$\beta=2kT/m\ ,\ \bar{U}_{mn}=\bar{U}_m-\bar{U}_n\ ,\ \bar{\beta}_{mn}=\bar{\beta}_m+\bar{\beta}_n$$

$$g_I=(2I/m_e)^{1/2},\ g_{oa}=(4I^*/m_a)^{1/2},\ g_{oe}=(2I^*/m_e)^{1/2},\ \bar{g}=g/\sqrt{\beta_1}$$

$$D_1=(1/\sqrt{2}\pi^{3/2})(C_{ae}/\sqrt{B_{aa}})m_e m_a^{1/2}\beta_1^{3/2}$$

$$D_2=4\pi^{3/2}(g_a/g_+)(h/2\pi m_e)^3(n_o/\beta_1^{3/2})$$

$$D_3=A_1(5)(2B_{ae}/\epsilon B_{aa})^{1/2}$$

$$D_4=(8m_a/\pi B_{aa})^{1/2}(e^2/m_e)^2/\beta_1^{3/2}$$

$$R_1=D_2(N_e^2/N_a)(\bar{\beta}_a/\epsilon)^{-3/2}exp(I/kT_a)$$

$$R_2=D_2(N_e^2/N_a)\bar{\beta}_e^{-3/2}exp(I/kT_e)$$

$$A_1(5)=0.422\ ,A_2(5)=0.436\ ,A_1(2)=\ln(1+9kT_e^3/4\pi e^6 n_e) \tag{16}$$

The functions F_1, F_2, and P_1 are defined in Ref.3. In obtaining Eqs. (9-15), only atom-atom and atom-electron two-step

Table 2 Properties of argon

Atomic weight	39.95
I, eV	15.7
I*, eV	11.5
α, cm^3	1.64×10^{-24}
B_{aa}, dyn-cm^5	1.124×10^{-42}
C_{aa}, cm^2/eV	2.5×10^{-20}
C_{ae}, cm^2/eV	7.0×10^{-18}
g_a	1.0
g_+	5.6

collisional ionization have been considered. The recombination rates are given by the Saha equation. The properties of the gas and the interaction constants used in the present analysis are summarized in Table 2.

Study of the Equilibrium Region

Introducing the boundary condition (8b) in Eqs. (13-15) yields the jump equations in the flow parameters across the shockwave. Thus, defining the equilibrium degree of ionization as

$$\alpha_{eq} = (N_e)_{eq} / [(N_e)_{eq} + (N_a)_{eq}] \tag{17}$$

the jump equations across the shock may be written in terms of α_{eq} and the upstream parameters as follows

$$(N_a)_{eq} = 8(1-\alpha_{eq})/3\,G(M_1, \alpha_{eq}) \tag{18}$$

$$\bar{U}_{eq} = (3/8)\bar{U}_1\,G(M_1, \alpha_{eq}) \tag{19}$$

$$\bar{\beta}_{eq} = \frac{1}{2(1+\alpha_{eq})}\left[\frac{5}{2}\left(1+\frac{1}{3}M_1^2\right) - \frac{3}{16}\left(1+\frac{5}{3}M_1^2\right)\left\{G(M_1,\alpha_{eq}) - \left(\frac{I}{kT_1}\right)\alpha_{eq}\right\}\right] \qquad (20)$$

where

$$G(M_1,\alpha_{eq}) = \left[\left(\frac{5}{3}+\frac{1}{M_1^2}\right) - \sqrt{\left(1-\frac{1}{M_1^2}\right)^2 + \frac{32}{15}\frac{\alpha_{eq}}{M_1^2}\left(\frac{I}{kT_1}\right)}\right]$$

Equations (18-20) reduce to the usual Rankine-Hugoniot relations when $\alpha_{eq} = 0$. The Saha equation may be written appropriately as

$$\alpha_{eq}^2/(1-\alpha_{eq}) = (3K\bar{\beta}_{eq}^{3/2}/8n_o)G(M_1,\alpha_{eq})\exp(-I/kT_1\bar{\beta}_{eq}) \qquad (21)$$

where $K = 2(g_+/g_a)(2\pi m_e kT_1/h^2)^{3/2}$. In the preceding equations, M_1 is the upstream Mach number defined by $M_1 = \sqrt{(6/5)}\bar{U}_1$. Equations (20) and (21) have been solved numerically for argon gas for different upstream Mach numbers, and pressures to obtain α_{eq} and $\bar{\beta}_{eq}$. Equations (18) and (19) directly give the atom density $(N_a)_{eq}$ and velocity $\bar{U}_{eq}$. The results are presented in Table 3, in which for comparison the corresponding jump values for the case of an inert monatomic gas also are given.

Shock Structure Profiles

In obtaining the structure profiles, the plasma momentum equation has been omitted. The set of six equations (9-13 and 15) has been solved numerically for the six variables N_e, $\bar{\beta}_e$, E, N_1, N_2, and N_3 for different initial conditions. An initial value $N_1 = 0.99$ and the corresponding value of N_2 obtained from the inert gas solution of Mott-Smith have been used. The electron gas is assumed to be in thermal equilibrium with the heavy gas at this initial point. A value of 10^{-6} has been assumed for the initial degree of ionization. Knowing the variation of N_1, N_2, and N_3, the atom properties can be obtained using the Eqs.(2). The electrical conductivity and refractive index variations across the shock wave have been computed knowing the variations of the atom and electron properties. The contribution to the electrical conductivity of an ionized gas arises from the close encounters between the electrons and neutral particles, and also

Table 3 Downstream equilibrium properties across an ionizing normal shock for argon (T_1 = 300°K, p_1 = 1.0 Torr)

M_1	α_{eq}	M_3	M_2	ρ_3/ρ_1	ρ_2/ρ_1	T_3/T_1	T_2/T_1
10	0.0123	0.4791	0.4543	4.375	3.8835	29.26	32.12
11	0.0285	0.4563	0.4531	4.939	3.9032	31.86	38.69
12	0.0490	0.4290	0.4522	5.568	3.9184	33.88	45.87
13	0.0729	0.4039	0.4514	6.219	3.9302	35.58	53.69
14	0.0995	0.3821	0.4509	6.872	3.9340	37.08	62.12
15	0.1287	0.3635	0.4504	7.515	3.9474	38.44	71.19
16	0.1603	0.3474	0.4500	8.142	3.9537	39.72	80.87
17	0.1943	0.3337	0.4497	8.749	3.9589	40.93	91.19
18	0.2304	0.3218	0.4494	9.329	3.9633	42.09	102.12
19	0.2687	0.3116	0.4492	9.882	3.9670	43.22	113.69
20	0.3090	0.3026	0.4490	10.404	3.9702	44.32	125.87
21	0.3515	0.2948	0.4488	10.893	3.9730	45.41	138.69
22	0.3958	0.2880	0.4487	11.347	3.9750	46.50	152.12
23	0.4421	0.2820	0.4486	11.766	3.9774	67.61	166.19
24	0.4902	0.2767	0.4485	12.147	3.9793	48.73	180.87
25	0.5399	0.2723	0.4484	12.488	3.9809	49.88	196.19
26	0.5913	0.2681	0.4483	12.787	3.9823	51.09	212.12
27	0.6440	0.2646	0.4482	13.041	3.9836	52.37	228.69
28	0.6979	0.2615	0.4481	13.243	3.9847	53.77	245.87
29	0.7525	0.2590	0.4481	13.387	3.9858	55.32	263.69
30	0.8072	0.2565	0.4480	13.459	3.9867	57.13	282.12
31	0.8608	0.2545	0.4480	13.434	3.9875	59.34	301.19
32	0.9113	0.2518	0.4479	13.270	3.9883	62.26	320.87
33	0.9542	0.2485	0.4479	12.887	3.9890	66.52	341.19
34	0.9826	0.2441	0.4478	12.203	3.9896	73.15	362.12
35	1.0000	0.2401	0.4478	11.417	3.9902	81.62	383.69
36	1.0000	0.2513	0.4478	10.505	3.9907	93.07	405.87
37	1.0000	0.2612	0.4477	9.777	3.9912	104.80	428.69
38	1.0000	0.2706	0.4477	9.182	3.9917	116.84	452.12
39	1.0000	0.2793	0.4477	8.687	3.9921	129.16	476.19
40	1.0000	0.2872	0.4477	8.270	3.9925	141.79	500.87

from the long-range interactions between electrons and ions. Lin et al.[7] have pointed out that, for a high degree of ionization ($\alpha > 10^{-3}$ for argon), the diffusion of electrons is limited primarily by long-range Coulomb interaction of the gas, and the contribution arising from the electron interaction with the neutral particles is negligible. Spitzer and Härm[8] have obtained the electrical conductivity by considering the effect of electron-ion encounters as a problem of diffusion in the velocity space. According to Spitzer and Härm, the electrical conductivity σ is given by

$$\sigma = 0.591\,(kT_e)^{3/2} / m_e^{1/2} e^2 \ln(\lambda_D/\lambda_L) \tag{22}$$

where λ_D and λ_L are, respectively, the Debye shielding length and the Landau distance. Equation (22) has a relative inaccuracy of the order of ln (λ_D/λ_L) arising from the uncertainty in the choice of the cutoff distance. In the present study, since $\alpha > 10^{-3}$ over a large part of the shock, the electrical conductivity computed from Eq.(22) is quite accurate, and the contribution to the electrical conductivity arising from electron-atom encounters can be ignored. The refractive index μ for a partially ionized gas composed of ground state atoms, ions, and free electrons can be expressed as[9]

$$\mu = 1 + 2\pi(\alpha_a n_a + \alpha_i n_i) - (1/2)(\omega_p/\omega)^2 \tag{23}$$

where $\omega_p = (4\pi n_e e^2/m_e)^{1/2}$, the plasma frequency; ω is the impressed frequency; and α_a and α_i are the polarizabilities of the ground state atom and ion, respectively. Equation (23) is applicable when the plasma frequency is much smaller than the applied frequency and when the number density of the excited atoms is small compared to the density of the atoms in the ground state. In the present case, $\omega_p/\omega \sim 10^{-10}$, and Eq.(23) can be applied conveniently. The refractive index across the shock wave has been computed to sodium yellow line (λ = 5890 Å) taking a value of α_i/α_a = 0.67.[9] Furthermore, the electric potential due to the field E is given by

$$V = -\int_{-\infty}^{x} E\,dx \tag{24}$$

Equation (24) can be written in dimensionless form as

$$\overline{V} = eV/kT_1 = 2\epsilon\overline{U}_1 \int_{-\infty}^{\xi} \overline{E}\, d\xi \tag{25}$$

The variation of the dimensionless electric potential within the shock has been studied.

Discussion and Conclusions

It can be seen from Table 3 that, for an upstream temperature of 300°K and pressure of 1 Torr, the effect of ionization is significant only for Mach numbers greater than 10. For this set of upstream conditions, complete ionization occurs at an initial Mach number of 35. The downstream temperature for this range of upstream Mach numbers increases at a much smaller rate due to the effect of ionization, and, once the ionization is complete, it starts rising very steeply. The downstream density ratio increases very fast to reach a maximum of about 13.5 and then falls to reach an asymptotic value of 4, corresponding to the monatomic case. However, long before this happens the second ionization will start, resulting in a rise of the density again. For the upstream Mach numbers for which the ionization just starts, the downstream Mach number M_3 takes a value slightly higher than the corresponding inert gas value, as indicated in Table 3 for $M_1 = 10$. For these Mach numbers, the decrease of speed of sound overweighs the reduction in the mass velocity with ionization. The downstream equilibrium Mach number decreases first, then reaches a minimum value corresponding to $\alpha_{eq} = 1$, and increases further with increase of M_1.

The variation of the electrical conductivity across the shock wave is plotted in Fig. 1 for different initial pressures. The curves are given only for the region where $\alpha > 10^{-3}$. Qualitatively, the σ profiles exhibit variation similar to the electron temperature profiles. This is because of the strong dependence of σ on the electron temperature and a weak logarithmic dependence on the electron density. In the initial region, a steep rise in the electron temperature gives rise to a corresponding sudden increase of the electrical conductivity. Over a large part of the relaxation zone where the electron temperature varies very slowly, the electrical conductivity also gradually increases, and toward the end it rises steeply because of an increase in the electron density. Toward the end of the relaxation zone, a fall in the electron

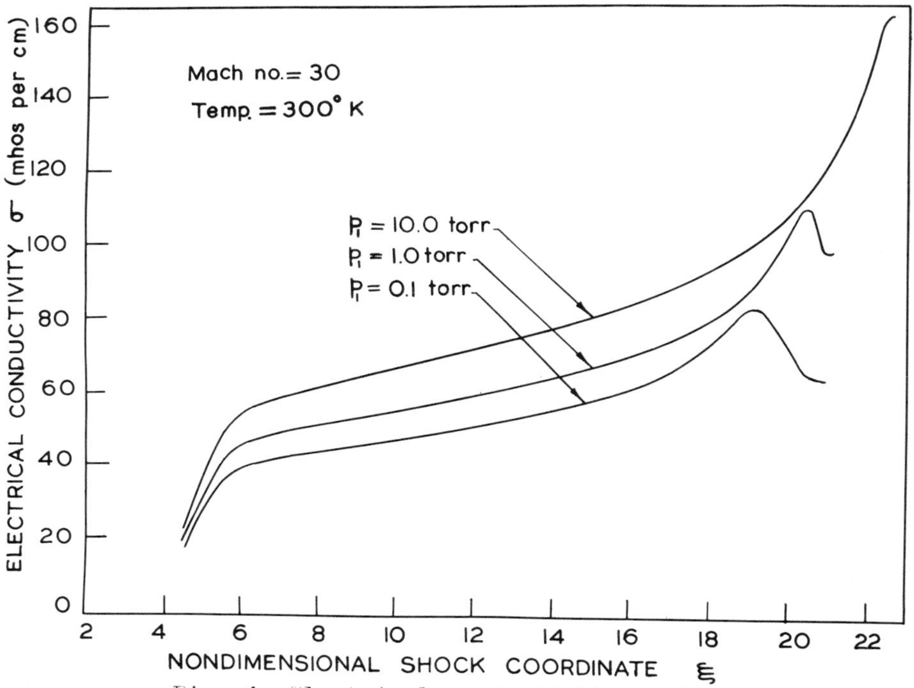

Fig. 1 Electrical conductivity profiles.

temperature influences σ, resulting in a peak in the σ profile before attaining the equilibrium value. It has been found in the present investigation that this peak behavior disappears for larger upstream Mach numbers. For $M_1 = 40$, toward the end of the relaxation zone, the electrical conductivity rises steeply and attains its equilibrium value. The variation of the refractive index μ across the shock wave to sodium yellow radiation is presented in Fig. 2. Since the refractive index depends upon the density of the gas, for a given initial temperature, $(\mu-1)/p_1$ is independent of initial density in the case of an inert gas, and hence $|(\mu-1)/p_1|$ is chosen as the ordinate. In the gasdynamic shock where the electron density is small, the contribution to μ arises mainly from the atom density and increases with the atom density. Later, as the electron density builds up, the third term in the Eq.(23) becomes significant, and μ starts decreasing and reaches the equilibrium value. In Fig. 3, the electric potential variation across the shock is plotted. A continuously increasing electric potential throughout the shock indicates that the electrons move upstream relative to the ions in order to smooth out the gradients. The steep rise in the electric potential in the initial part of the shock is

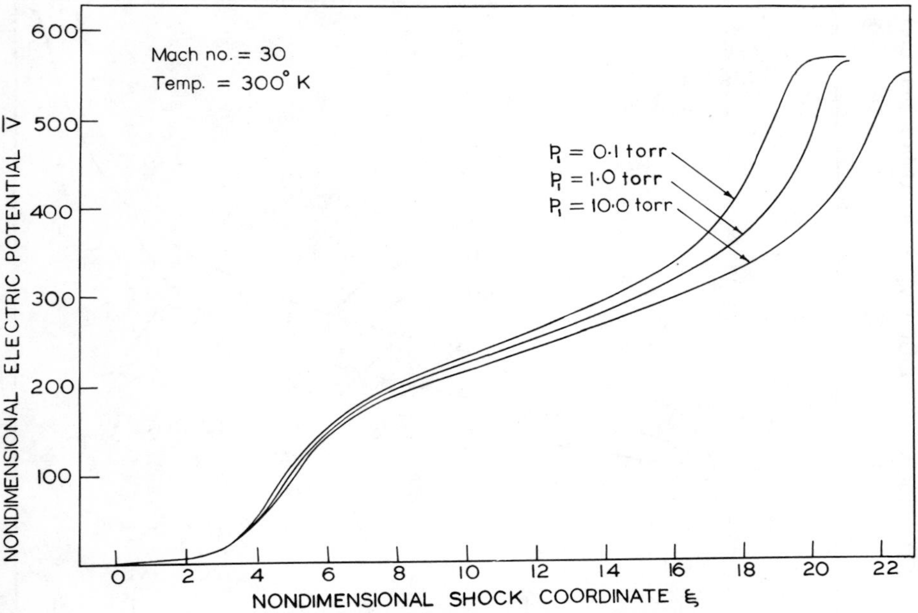

Fig. 2 Refractive index profiles.

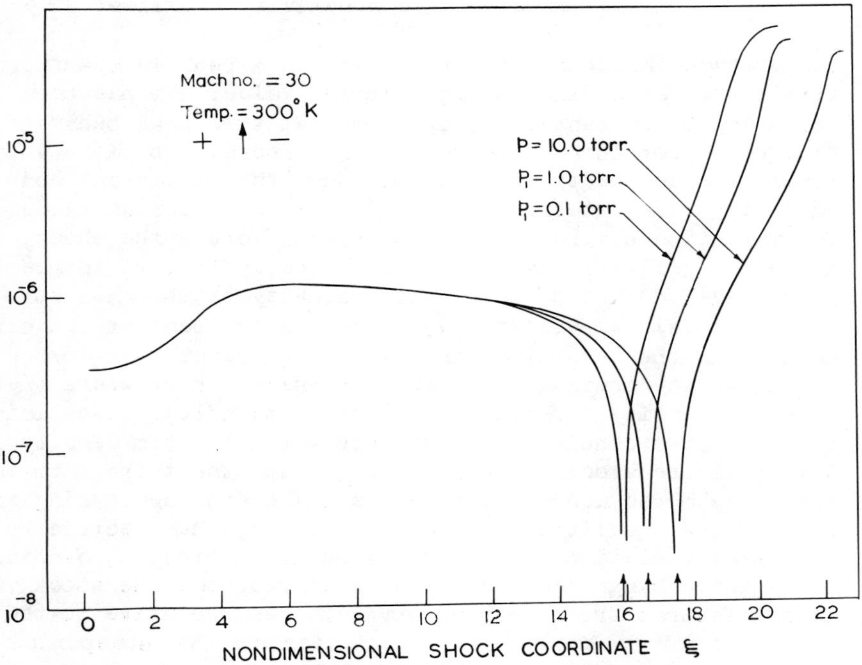

Fig. 3 Electric potential profiles.

due to thermal diffusion of the electrons resulting from a steep gradient in the electron temperature in this zone. In the middle of the relaxation zone, where the electron temperature gradient is very low, the electron diffusion resulting from the electron density gradient results in an overall rise in the electric potential. Toward the end of the zone, large electron density gradients give rise to strong electron diffusion, which results in a steep rise of the electric potential before reaching the equilibrium value.

Nonavailability of the measured shock structure profiles for the Mach numbers considered here precludes any attempt to compare the results of the present investigation with them. Nevertheless, emphasis is laid, in the present work, on the development of an approach for the unified description of the complete nonequilibrium region and an estimate of the physical quantities amenable to measurement.

References

[1]Chubb, D. L., "Ionizing Shock Structure in a Monatomic Gas," The Physics of Fluids, Vol. 11, November 1968, pp. 2363-2376.

[2]Mott-Smith, H. M., "The Solution of the Boltzmann Equation for a Shockwave," The Physical Review, Vol. 82, June 1951, pp. 885-892.

[3]Krishnamurthy, M. and Ramachandra, S. M., "Ionizing Shock-waves in Monatomic Gases Through a Kinetic Theory Approach," Acta Astronautica, Vol. 2, May-June 1975, pp. 367-389.

[4]Petschek, H. E. and Byron, S. R., "Approach to Equilibrium Ionization Behind Strong Shockwaves in Argon," Annals of Physics, Vol. 1, June 1957, pp. 270-315.

[5]Krishnamurthy, M., "A Kinetic Theory Approach to Shock Structure in Monatomic Gases with Ionization," Ph.D. Thesis, 1971, Indian Institute of Technology, Kanpur, India.

[6]Narasimha, R. and Deshpande, S. M., "Minimum Error Solution of the Boltzmann Equation for Shock Structure," Journal of Fluid Mechanics, Vol. 36, May 1969, pp. 555-570.

[7] Lin, S. C., Resler, E. L., and Kantrowitz, A., "Electrical Conductivity of Highly Ionized Argon Produced by Shockwaves," Journal of Applied Physics, Vol. 26, January 1955, pp. 95-109.

[8] Spitzer, L., Jr. and Härm, R., "Transport Phenomena in a Completely Ionized Gas," The Physical Review, Vol. 89, March 1953, pp. 977-981.

[9] Horn, K. P., Wong, H., and Bershader, D., "Radiative Behavior of Shock Heated Argon Plasma Flow," Journal of Plasma Physics, Vol. 1, May 1967, pp. 157-170.

CHAPTER X—ATOMIC & MOLECULAR BEAMS

NOZZLE BEAM SPEED RATIOS ABOVE 300 SKIMMED IN A ZONE OF SILENCE OF He FREEJETS

R. Campargue,* A. Lebéhot,+ and J.C. Lemonnier‡

Division de Chimie, Service de Chimie Physique
Centre d'Etudes Nucléaires de Saclay, Gif-sur-Yvette, France.

Abstract

The nozzle-beam technique developed at Saclay is based on the possibility of actual skimming in a "zone of silence" of a freejet overexpanded in a relatively high-pressure environment ($10^{-2} \lesssim P_1 \lesssim 1$ Torr). This principle enables one to operate at very high P_oD^* (P_o = inlet pressure, D^* = nozzle diameter) and consequently very high terminal Mach numbers enhanced by quantum effects at least in He freejets. Nevertheless, such a nozzle skimmer system must be optimized carefully for achieving the present He beam speed ratios which seem to be the highest up to now, i.e., $S_{//} \simeq 340$ corresponding to velocity spreads as low as 0.5% or temperatures as low as 6.5×10^{-3}°K. Recent improvements in skimming, time-of-flight resolution and speed ratio evaluation are promising for going to higher $S_{//}$ and even to the quantum mechanical predictions by Miller, Toennies, and Winkelmann. The absence of condensation in such helium beams is established. These experiments are extended to hydrogen ($S_{//} \simeq 70$) and seeded light-gas freejets ($S_{//} \lesssim 60$ with Ar or Xe) at 293°K and to an available large range of source temperature (80°-3000°K).

I. Introduction

In previous investigations performed with our first minibeam generator,[1] the experimental conditions, for maximum beam

---

Presented as Paper 134 at the 10th International Symposium on Rarefied Gas Dynamics, Aspen, Colo., July 19-23, 1976.

*Dr. ès Sciences, Chef du Laboratoire des Jets Moléculaires
+Candidate Dr. ès Sciences
‡Member of Professional Staff.

intensity, were found to be in agreement with a simple relation[2]

$$X_s/D^* = 0.125\ \left[(1/Kn_o)(P_o/P_1)\right]^{1/3} \qquad (1)$$

with

X_s/D^* = nozzle skimmer distance, nozzle diameter D^*
$Kn_o = \lambda_o/D^*$ = nozzle Knudsen number (λ = mean free path)
P_o = nozzle stagnation pressure
$0.01 \lesssim P_1 \lesssim 1$ Torr = expansion chamber background pressure

This semiempirical relation, which recently has been confirmed[3] with a remarkable precision in large ranges of the preceding experimental parameters, has shown different possibilities and mainly a new way of improving the nozzle-beam technique.[4] The beam intensity increasing with the skimmer Mach number and this latter being an increasing function of X_s/D^*, the quality of the molecular beam is correspondingly better as the right-hand member of (1) is increased. This quantity can be enhanced in ways considered separately below.

A. Increase at Constant D^* of the Expansion Ratio P_o/P_1

Very high values of P_o/P_1 are attainable only by means of large diffusion pumps or cryopumps having a high pumping speed

$$\mathcal{S}_1 \propto P_o D^{*2}/P_1 \propto X_M^2 \qquad (2)$$

in the nozzle chamber (X_M = nozzle Mach disk distance). This conventional solution allows one to operate at low background pressure ($10^{-5} \lesssim P_1 \lesssim 10^{-3}$ Torr) and thus attenuate easily the background disturbance of the freejet. Nevertheless, the expansion ratio is used very uncompletely for obtaining high terminal Mach numbers because of translational freezing[5] resulting from the relatively low $P_o D^*$ and P_1 acceptable with such pumping systems. Thus, the optimal distance X_s/D^* is considerably smaller than predicted from (1), this relation not being valid at such low densities.[2]

B. Reduction of D^* at Constant Values of $\mathcal{S}_1$ and P_1

For a given gas, at a given source temperature T_o, the relation (1) can be rewritten

$$X_s/D^* \propto \left[P_o D^*(P_o/P_1)\right]^{1/3} \propto D^{*-1} \qquad (3)$$

The last simplification obtained for fixed $\mathcal{S}_1$, P_1 and consequently fixed $P_o D^{*2}$, suggests an improvement by going to small nozzle diameters as expected from theoretical freejet studies[6] and already shown in our early experimental nozzle-beam work.[1]

C. Increase of P_o and P_1 at Constant Values of δ_1 and D^*

The advantage of this possibility, which is the basis of an alternative for improving the nozzle-beam technique,[4] appears clearly in the following expression of (1) and (3)

$$X_s/D^* \propto (P_o^2/P_1)^{1/3} \qquad (4)$$

The improvement obtained by increasing the pressure levels results in part from the production, by this method, of a "zone of silence"[6] (at least in the streamtube subtended by the skimmer orifice) unaffected by the background gas[7] and equivalent to the centerline of a freejet expanding into an ideal vacuum. Furthermore, the very large values acceptable for P_oD^* enable one to produce very high terminal Mach numbers[5] even with pumps of modest size, because of the relatively high pressure P_1 used (for instance, 0.25 Torr instead of 10^{-4} Torr with the conventional method). In spite of such a relatively high-pressure environment, actual skimming is possible within the shock barrel, thanks to the overexpansion occurring upstream of the Mach disk.[6] Nevertheless, it should be emphasized that the nozzle-skimmer system must be optimized carefully[7] for operating in these conditions and mainly for obtaining the extremely high performance achieved at Saclay from 1974 on.

The best experimental results recently obtained by using miniature nozzles and skimming in a "zone of silence" of a free-jet initially were estimated from measurements carried out with a time-of-flight (TOF) system having insufficient resolution.[3] In spite of this limitation, the very high Mach numbers published for helium beams[3] ($\mathcal{M}_{//} > 140$) were about an order of magnitude above all of those obtained in previous experimental studies[8] and also higher than expected from theoretical predictions.[5] According to recent calculations by Miller et al.[9] such deviations were attributed to quantum effects in helium free-jets. Finally, our experimental findings seem to have encouraged Toennies and Winkelmann to go further in the development of their quantum theory and even its experimental verification.[10] During the same period, a first improvement of our TOF system allowed us to measure $S_{//} \simeq 200$ in helium beams and to present a preliminary verification of the theoretical work.[11] After new improvements concerning mainly the time response of the electronics, these experiments were refined, extended to hydrogen and seeded beams, and finally reported in the present paper. In spite of the very high values used for $P_oD^* \lesssim 450$ Torr-cm, the absence of condensation is established at least for helium beams at room temperature.

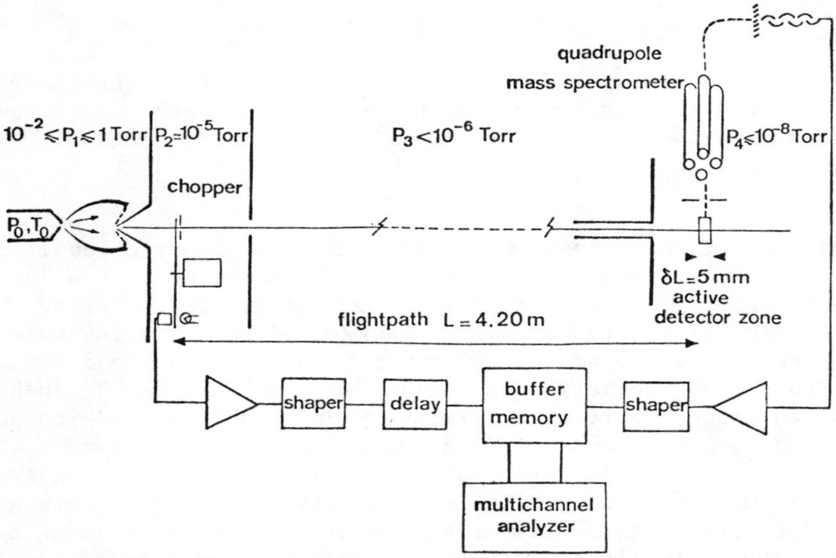

Fig. 1 Schematic view of the apparatus used for the production and analysis of nozzle beams with extremely narrow velocity distributions.

II. Experimental

The description and main characteristics of the nozzle-beam generators based on the principle developed at Saclay were reported previously.[1-4,7,11] The experimental system used for the production and analysis of the high-performance molecular beams presented in this paper is shown schematically in Fig.1. The freejets are produced through miniature nozzles ($D^* = 29.5$ and 53 μm) from a high stagnation pressure $P_0 \lesssim 200$ bars into an environment where the relatively high background pressure($10^{-2} \lesssim P_1 \lesssim 1$ Torr) is maintained by means of two Roots pumps in series backed by a mechanical pump (for helium $\mathcal{S}_1 \simeq 350$ l-sec^{-1}). It has been observed that the efficiency of a nozzle may be reduced by the existence of asperities or defects of circularity at the throat. However, the skimming is the key problem now conveniently solved with conical skimmers having the following characteristics: orifice edge thickness of 1 to 2 μm, orifice diameter of about 0.5 mm, inner and outer angles of 50° and 60°, respectively, and outer length of 19 mm. The beam formation being terminated in the postskimmer region, a 24,000-rpm chopper is able to operate conveniently in the collimating chamber.

In order to resolve velocity distributions now lower than 1%, several changes were made in the preceding TOF system,[3] leading to the following improvements: 1)the half-width of the triangular shutter function has been decreased from 10 to 5 μsec; 2) the flight-path L has been increased from 1.56 to 4.20m ; 3) the thickness δL of the active zone of the detector has been reduced from 20 to about 5 mm by replacing the Bayard-Alpert gage previously used by a quadrupole mass spectrometer (Riber, QMM 17, of reduced size), placed perpendicular to the beam axis in an ultravacuum region ($P_4 \lesssim 10^{-8}$ Torr). In these new conditions, the shutter function half-width (SFHW = 5 μsec) seems to be small enough compared to the full width at half maximum of the TOF curves (FWHM $\lesssim$ 12.5 μsec for helium at $T_o \simeq 293°K$). Indeed, the resolution thus achieved and given by $R = FWHM/SFHW = 12.5/5 = 2.5$ should lead to the correct FWHM with a broadening of less than 10% according to Scott et al.[12], and Hagena and Varma.[12] Furthermore, the time spent in the active detector zone also being relatively small (∿ 2.8 μsec for helium at $T_o \sim 293°K$) the convolution product of the incoming TOF distribution by the detector function leads to negligible broadenings much smaller than those due to the beam shutter. However, other broadenings can result from space-charge effects in the ionizer of the quadrupole (see He results). Finally, the transit time of the ions to the multiplier results in a simple time displacement of the signal which is negligible compared to the total time of flight (< 10^{-2} for helium at $T_o \simeq 293°K$). In conclusion, the measured half-widths of the extremely narrow TOF distributions achieved can be regarded only as upper limits. For the experimental results presented below, the parallel speed ratios $S_{/\!/}$ are derived directly from the TOF curves by using a simple relationship previously established[7] for Maxwellian distributions

$$S_{/\!/} = 1.665 \times TOF/FWHM \qquad (5)$$

In spite of the very high beam intensity achieved (10^{20} to 10^{21} mol-ster^{-1}-sec^{-1}), signal-averaging instruments are necessary with the long flight path used (L = 4.20m). Such devices are useful mainly with seeded beams at low concentrations, for species or even isotope discriminations. Because of the insufficient time resolution of the multichannel analyzer (10 μsec/channel for measuring FWHM $\gtrsim$ 12.5 μsec), the signal now is first stored in a buffer memory having a convenient response time (from 10 μsec down to 100 nsec/channel) and afterwards transferred slowly into the multichannel analyzer (Fig.1).

III. Results and Discussion

A. Helium Beams at Room Temperature

The experiments on helium beams were performed with P_oD^* varying in the large range of 1 to 450 Torr-cm. The un-

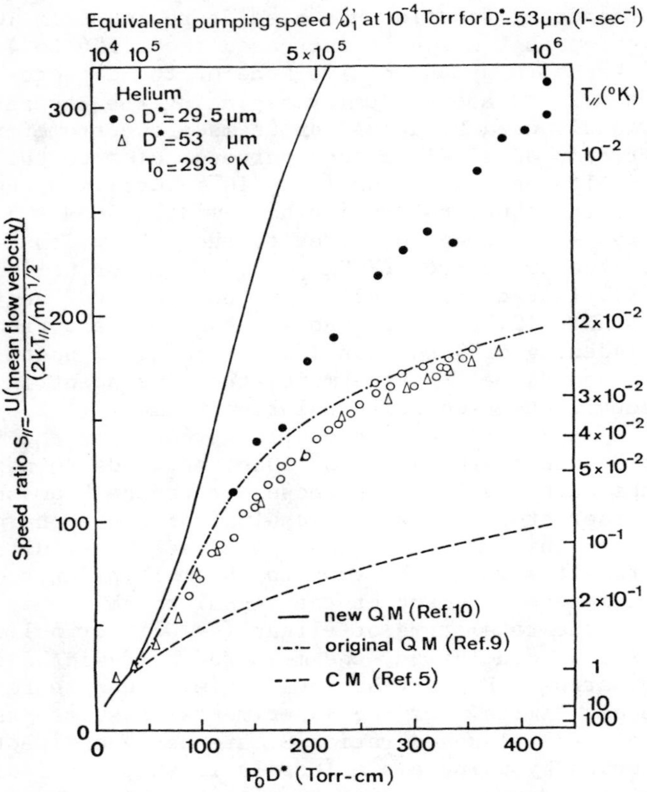

Fig. 2 Original (Δ, ○) and recent (●) experimental results on helium beams at room temperature, compared to classical (CM) and quantum mechanical (QM) predictions by Miller, Toennies, and Winkelmann.

certainty on this product is of the order of 2%, since P_o (high precision manometer) and D^* (flow-rate measurements) can be measured both with an error of less than 1%. It is interesting to notice (Fig.2) the equivalent pumping speeds ($\mathcal{S}_1' \lesssim 10^6$ l-sec^{-1}) that would be required to maintain $P_1 \simeq 10^{-4}$ Torr by means of a diffusion pump allowing one to operate in such a P_oD^* range, with D^* = 53 µm. The values of the parallel speed ratios $S_{//}$ determined directly from the TOF distributions and (5) are plotted versus P_oD^* in Fig.2 without any correction. The equivalent translational temperatures $T_{//}$ are also reported. These experimental data, which appear as a confirmation and even an improvement of previous results,[3,11] are compared with the available theoretical predictions, based on a treatment using moments of the Boltzmann-equation and collision integrals calculated both for a Lennard-Jones potential with classical theory[5] as well as quantum theory.[9,10]

At low P_0D^*, the experimental points agree well with the predictions of both theories and even with measurements made for helium in other laboratories[8] (see Fig.5, Ref.3). This low P_0D^* range corresponds to the conditions used mainly in our early molecular-beam work.[1] The measured speed ratios $S_{//}$ deviate from the classical theory (CM in Fig.2) and seem to follow quantum mechanical predictions for P_0D^* higher than about 30 Torr-cm. In a first series of experiments,[11] which were even refined by using our new device (Fig.1), the results appeared with an excellent reproductibility and on a single curve with D^* = 29.5 and 53 μm (see open circles and triangles in Fig.2). Furthermore, the measured values were nearly the same as predicted from the original quantum theory (original QM in Fig.2) where the expansion process was assumed to become "frozen" at the skimmer entrance. Nevertheless, this preliminary agreement must be considered as a simple coincidence since the experimental and theoretical speed ratios are now higher than originally.

The rise of about 50% obtained recently for the highest values of $S_{//}$ (see filled circles in Fig.2) results from improvements concerning the beam production and mainly the TOF analysis. The performance of the beam generator has been increased by using a new skimmer having an edge of better quality. The enhancement attributed previously to the use of helium of higher purety (99.9995% instead of 99.997%) was in fact amplified by the elimination of other concurrent effects linked to the TOF system. For instance, broadenings in the velocity distributions were due to imperfect alignments of the narrow slits defining the triangular shutter function. Additional spreads accompanying the variations of the heating filament current of the detector ionizer, seem to result from space-charge effects and to be uncompletely eliminated until now. Such a disturbance could be avoided by using our metastable detector.[3] The rise of the theoretical speed ratios results from stopping the expansion, not at the skimmer entrance as originally, but downstream of the skimmer. This quantum theory[10] (new QM in Fig.2) seems to be much more realistic than the original QM, mainly in our conditions where the significant skimmer Knudsen number is in the range of 10^{-1} to 10^{-2}, because of the large quantum mechanical cross sections He-He. Among the new results, our measured $S_{//}$ are smaller than expected theoretically (Fig.2). They are also smaller than those obtained in Gottingen[10] with the conventional method (P_1 in the range of 10^{-4} to 10^{-3} Torr) for the same relatively low P_0D^* in the range of 30 to about 100 Torr-cm. These deviations are

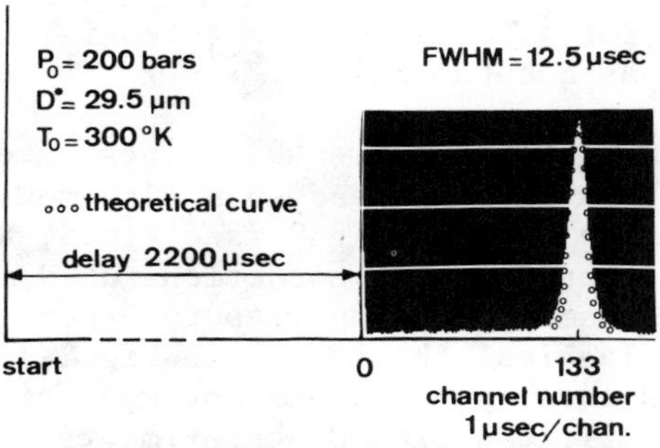

Fig. 3 Narrowest He beam oscillogram compared to Maxwellian distribution
$S_{//} \simeq 310$ or $T_{//} \simeq 8 \times 10^{-3}$°K (without correction)
$S_{//} \simeq 340$ or $T_{//} \simeq 6.5 \times 10^{-3}$°K (with correction[12]).

attributed to the very different conditions used at Saclay ($P_1 \sim 0.25$ Torr), the difficulty in analysing velocity spreads smaller than 1% and the different methods employed for evaluating $S_{//}$ and $T_{//}$. Nevertheless, the recent improvements in skimming and TOF resolution and also possible refinements in speed ratio evaluations are promising for going to higher $S_{//}$ and even to the theoretical values.

In spite of the present limitations, extremely narrow velocity distributions with spreads as low as 0.5% now are achieved directly. For instance, a speed ratio $S_{//} \simeq 310$ can be derived directly from one of the best oscillogram (Fig.3) and relation (5). A translational temperature $T_{//} \simeq 8 \times 10^{-3}$°K, in agreement with the value corresponding to this $S_{//}$, has been calculated directly from this symmetrical and nearly Maxwellian curve, reflecting the extremely high quality of the beam and the absence of condensation. If the correction due to the finite open time of the beam shutter is taken into account,[12] these values become $S_{//} \simeq 340$ and $T_{//} \simeq 6.5 \times 10^{-3}$°K. One can summarize the main reasons allowing us to conclude that condensation effects are not present in our highly expanded helium beams from room temperature: 1) the symmetry and the nearly Maxwellian form of the TOF distributions (see Fig.3); 2) the agreement at least qualitative with the predictions from a quantum theory[9,10]; 3) the fixed location (with a precision better than 1%) of the TOF curves, corresponding exactly to the theoretical velocities for uncondensed He beams, i.e., 1770 $msec^{-1}$ at $T_o = 293$°K; 4) the absence of ion signals He_n^+ in the available mass range (from 1 to 500) of the quadrupole; and 5) the absence of discontinuities in all the experimental results and the validity of the relation (1) for P_oD^* varying from 1 to 450 Torr-cm.[3]

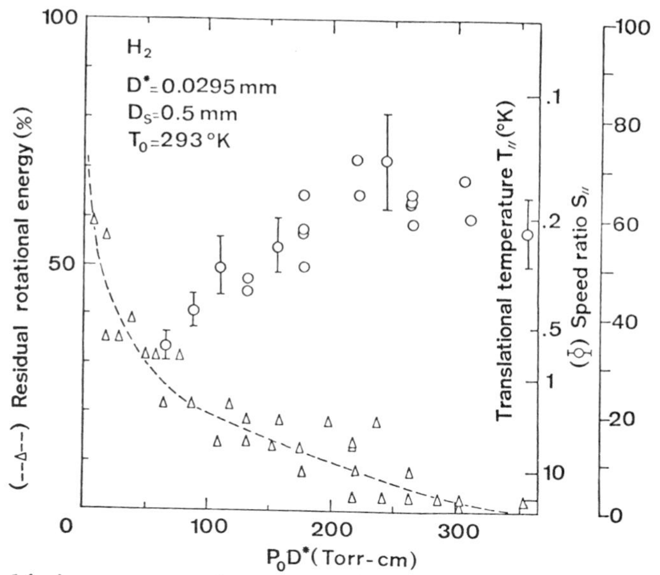

Fig. 4 Preliminary experimental results on Hydrogen beams at room temperature.

B. Hydrogen Beams at Room Temperature

Experiments currently under way also are performed with hydrogen freejets expanded from a 293°K source having an orifice diameter $D^* \simeq 29.5$ μm. The preliminary results (Fig.4), obtained in a large range of P_o (up to 180 bars), lead to the following remarks:

1) The speed ratios are not so high as those achieved for helium and seem to have a maximum value ($S_{//} \simeq 70$) for $P_oD^* \simeq 225$ Torr-cm.

2) The levelling-off is attributed to condensation because extremely small but significant amounts of H_3^+ and H_5^+ signals ($< 10^{-3}$ of the H_2^+ signal) were detected for the highest P_oD^*. Similar observations (maximum for $S_{//}$ and detection of H_3^+) were made previously by Buck et al.[8] at lower P_oD^* and $T_o = 85$°K

3) The residual rotational energy derived from TOF analysis and energy balance, appears to be negligible at large values of P_oD^*. Nevertheless, the maximum kinetic energy is only about 20% higher than the theoretical predictions for $\gamma = 5/3$, or about 20% below the $\gamma = 7/5$ predictions, because of the relatively high (85°K) rotational characteristic temperature of hydrogen.

As shown in Fig.4, these preliminary measurements are not very well reproductible and must be refined. The main disturbance seems also to be due to space-charge effects in the detector zone, which appear even more important than those previously observed in analyzing helium beams.

C. Seeded light gas (He, H_2) molecular beams

Other experiments also were carried out for heavy gases seeded in helium or hydrogen freejets produced in the conditions of Fig. 2 or 4, respectively. With a convenient dilution ($\lesssim$ 1%) and/or a sufficiently high source temperature T_0, in the available range of 80° to 3000°K, this method enables one to obtain narrow velocity distributions even for uncondensed heavy molecules. Typical TOF curves were published previously[3] for such beams generated in a large energy range extended to about 40 eV. The speed ratios $S_{//} \simeq 60$ seem to be attainable without condensation for argon and xenon, respectively, seeded in light gas freejets. The He speed ratios are reduced by the presence of the heavy molecules, even at low concentration (< 0.1%), and always smaller (from 1/3 to 1/2) than that measured for Ar or Xe. These results are in agreement with the previous findings of Miller and Andres[5] and even with our observations concerning the gas purety. The condensation effects detected by using the quadrupole mass spectrometer seem also to be indicated by a remarkable asymmetry, with a low velocity tail, in the TOF curves obtained, for heavy molecules, in pure gases or gas mixtures.

Low rotational temperatures also are attainable for complicated polyatomic molecules. Thus helium freejets, exploiting quantum effects (Fig.2), were used recently at the University of Chicago[13] to obtain in the gas phase, without condensation, ultracold polyatomic molecules (NO_2, I_2, tetrazine). These molecules were seeded in helium freejets produced according to our principle and observed by laser-induced fluorescence. Mainly because of the substantial reduction of the rotational structure (0.5°K $\lesssim T_{rot} \lesssim$ 0.8°K), this new method called "molecular-jet-spectroscopy", allows a remarkable simplification of the absorption spectra of complicated molecules. Furthermore the great selectivity of excitation attainable by this technique may augment the potential of laser methods in isotope separation.

IV. Conclusion

In spite of present limitations mentioned previously, the lower limits already achieved in the measurements appear as extremely high speed ratios ($S_{//} \simeq 340$ for helium). In addition to the fundamental aspect of the properties of highly expanded freejets, the possibility has been shown of actual skimming in a "zone of silence", allowing the extraction of even extremely low temperatures ($T_{//} \simeq 6.5 \times 10^{-3}$°K). Thus, the extremely high performance achieved offers new possibilities in scattering experiments (gas phase or gas surface) and also in isotope separation processes involving jets and/or beams, such as the crossed-jet or beam methods and the expansion-induced separation (differential slip effect). The potential of such beams results mainly from the very narrow velocity spreads obtained (∿ 0.5%), the large energy range available [3,11] from 10^{-2} to 40 eV (by using a new 50 bar-3000°K heated nozzle), and the limitation of the expansion into the nozzle chamber (0.1 to 0.5 Torr).

References

[1] Campargue, R., "High Intensity Supersonic Molecular Beam Apparatus," Review of Scientific Instruments, Vol. 35, Jan. 1964, pp. 111-112; also Rarefied Gas Dynamics, Vol. 2, edited by J.H. de Leeuw, Academic Press, New York, 1966, pp. 279-298.

[2] Campargue, R., "Dimensionless Number Linked to Background and Skimmer Jet Interaction in Nozzle Beam Generation," Rarefied Gas Dynamics, Vol. 2, edited by L. Trilling and H.Y. Wachman, Academic Press, New York, 1969, pp. 1003-1007.

[3] Campargue, R., and Lebéhot, A., "High Intensity Supersonic Molecular Beams with Extremely Narrow Energy Spreads in the 0 - 37 eV Range," Rarefied Gas Dynamics, Vol. 2, C. 11, edited by M. Becker and M. Fiebig, DFVLR Press, Porz-Wahn, West Germany, 1974, pp. 1-12.

[4] Campargue, R., "Perfectionnements aux Procédés et Dispositifs pour Produire des Jets par Détente Libre d'un Gaz," Patent 1,579,570 filed, May 1968 (France) and Foreign Patents (U.S.A., West Germany, Great Britain, Canada, etc.).

[5] Miller, D.R., and Andres, R.P., "Translational Relaxation in Low Density Supersonic Jets," Rarefied Gas Dynamics, Vol. 2, edited by L. Trilling and H.Y. Wachman, Academic Press, New York, 1969, pp. 1385-1402; also references reported in this paper e.g. Anderson, J.B., and Fenn, J.B.; Hamel, B.B., and Willis, D.R.; Edwards, R.H., and Cheng, H.K.; Muntz, E.P.; Scott, J.E., and Phipps, J.A.

[6]Sherman, F.S., "A survey of Experimental Results and Methods for the Transition Regime of Rarefied Gas Dynamics," Rarefied Gas Dynamics, Vol. 2, edited by J.A. Laurmann, Academic Press, New York, 1963, pp. 228-260; also Ashkenas, H., and Sherman, F.S., "The Structure and Utilization of Supersonic Freejets in Low Density Wind Tunnels," Rarefied Gas Dynamics, Vol. 2, edited by J.H. de Leeuw, Academic Press, New York, 1966, pp. 84-105.

[7]Campargue, R., "Aerodynamic Separation Effect on Gas and Isotope Mixtures Induced by Invasion of the Freejet Shock Wave Structure," Journal of Chemical Physics, Vol. 52, Feb. 1970, pp. 1795-1802; also "Etude, par Simple et Double Extraction de Jets Supersoniques Purs ou Dopés, des Effets Intervenant dans la Formation d'un Faisceau Moléculaire de Haute Intensité et d'Energie comprise entre 0 et 25 eV." Thesis, 1970, University of Paris; also Report R-4213, 1972, Commissariat à l'Energie Atomique, France, pp. 1-246; also Campargue, R., and Breton, J.P., "Amélioration de la Méthode des Jets Moléculaires Supersoniques par Augmentation Simultanée des Pressions Génératrice et Résiduelle," Entropie, Vol. 42, Nov.-Dec. 1971, pp. 18-28.

[8]Abuaf, N., Anderson, J.B., Andres, R.P., Fenn, J.B., and Miller, D.R., "Studies of Low Density Supersonic Jets," Rarefied Gas Dynamics, Vol. 2, edited by C.L. Brundin, Academic Press, New York, 1967, pp. 1317-1336; also Eckelt, W.R., Hose, H., and Schügerl, K., "Investigation of Low Density Supersonic Jets by Means of a Molecular Beam Sampling Technique," Zeitschrift fur Naturforschung, Vol. 24a, 1969, pp. 1365-1373; also Buck, U., Düker, M., Pauly, J., and Pust, D., "Molecular Beams with Extremely Narrow Velocity Spreads From Freejets Expansions," Proceedings of the 4th International Symposium on Molecular Beams, 1973, Cannes, France, pp. 70-82.

[9]Miller, D.R., Toennies, J.P., and Winkelmann, K., "Quantum Effects in Highly Expanded Helium Nozzle Beams," Rarefied Gas Dynamics, Vol. 2, C.9, edited by M. Becker and M. Fiebig, DFVLR Press, Porz-Wahn, West Germany, 1974, pp. 1-9.

[10]Toennies, J.P., and Winkelmann, K., private communication, 1975; also Winkelmann, K., private communication, 1976; also Brusdeylins, G., Meyer, J.D., Toennies, J.P., and Winkelmann, K., Proceedings of the 10th International Symposium on Rarefied Gas Dynamics, Aspen, Colo., 1976, this volume.

[11] Campargue, R., Lebéhot, A., Lemonnier, J.C., Marette, D., and Pebay, J., "Générateur de Jet Moléculaire Supersonique Fonctionnant en Monochromateur dans le Domaine de 10^{-2} à 40 eV, "Proceedings of the 5th International Symposium on Molecular Beams, 1975, Nice, France, A.7, pp. 1-15.

[12] Scott, P.B., Bauer, P.H., Wachmann, H.Y. and Trilling, L., "Velocity Distribution Measurements by a Sensitive Time-of-Flight Method," Rarefied Gas Dynamics, Vol. 2, edited by C.L. Brundin, Academic Press, New York, 1973, pp. 1353-1368; also Hagena, O.F. and Varma, A.K., "Time-of-Flight Velocity Analysis of Atomic and Molecular Beams," Review of Scientific Instruments, Vol. 39, Jan. 1968, pp. 47-52.

[13] Wharton, L., private communication, 1975; also Smalley, R.E., Levy, D.H., and Wharton, L., "Molecular-Jet Spectroscopy," Laser Focus, Nov. 1975, pp. 40-43; also Levy, D.H., private communication, 1976.

PRODUCTION OF HELIUM NOZZLE BEAMS WITH VERY HIGH SPEED RATIOS

G. Brusdeylins,* H.-D. Meyer*
J.P. Toennies,† and K. Winkelmann*

Max-Planck-Institut für Strömungsforschung,
Göttingen, Germany

Abstract

Earlier measurements of large speed ratios in He have been extended to speed ratios of 225 in a new, improved apparatus with greater time-of-flight resolving power. The high resolution has made it possible to observe significant deviations in the velocity distribution from the simple Maxwellian form observed in previous work. The speed ratios evaluated assuming a single Maxwellian have been found to be in good agreement with a moments solution of the Boltzmann equation in which a Maxwellian form also is assumed. This simplified theory and the observed deviation in the velocity distribution are both in reasonable agreement with a recent Monte Carlo simulation calculation by Chatwani under otherwise identical conditions. These new results demonstrate that temperatures as low 8.4×10^{-3} °K and beams with velocity spreads (FWHM) as low as 0.7% can be achieved. The theory suggests that temperatures as low as 2×10^{-3} °K and relative velocity spreads as low as 0.275% should be achievable with our apparatus.

Presented as Paper 69 at the 10th International Symposium on Rarefied Gas Dynamics, Aspen, Colo., July 19-23, 1976.

* Diplom-Physiker.

† Professor, Director of the Division "Molekulare Wechselwirkungen".

1. Introduction

Nozzle beams offer a unique source of very intense, nearly monoenergetic beams. They are ideally suited for the study of inelastic scattering from other molecules or from surfaces.[1,2] In addition, in nozzle beams extremely low temperatures can be achieved without condensation. This feature has been used to cool the internal degrees of polyatomic molecules and thereby simplify their spectra.[3,4] Nozzle beams frequently are characterized by their speed ratio

$$S = [(1/2\ mu^2)/kT]^{1/2}$$

where u is the average velocity of the flowing gas and T is the temperature of the gas in the moving coordinate system. For a monatomic gas ($\gamma = 5/3$) and large speed ratios ($S > 10$), the narrowing of velocity distribution in the beam and the temperature reduction are given by

$$\Delta v/u \simeq 1.65/S \quad , \quad T/T_o \simeq 2.5/S^2$$

where Δv is the FWHM of the velocity distribution and T_o is the stagnation temperature.

In a previous paper, we were able to show for the first time that it is possible to achieve He nozzle beams with speed ratios as high as S = 43.[5] These results were confirmed by Campargue and co-workers.[6] In the work reported here, we have succeeded in achieving even higher speed ratios up to S = 225. The results are compared with a moments solution of the Boltzmann equation using an ellipsoidal Maxwellian velocity distribution[7] and using quantum-mechanical collision integrals.[5,8] A careful examination of the measured velocity distribution has revealed a deviation from the expected Maxwellian form. Such deviations, which of course cannot be explained by the preceding theory, also have been found in recent theoretical calculation based on the Bird Monte Carlo simulation technique[9] and carried out by Chatwani in our laboratory.[10] The measured beam velocities show a slight shift to higher velocities with increasing stagnation pressure. This shift could be explained by an increase in the enthalpy and is described by the use of the Benedict-Webb-Rubin equation of state.

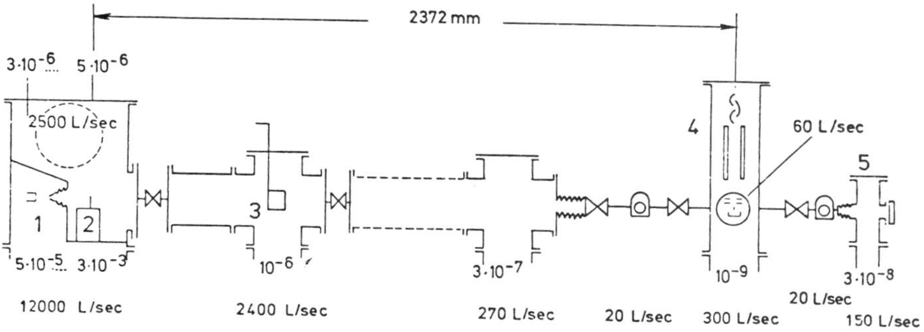

Fig. 1 Schematic diagram of the nozzle beam apparatus: 1 = skimmer chamber, 2 = chopper motor, 3 = differential pumping chamber, 4 = detector, 5 = beam catching chamber.

2. Apparatus

Figure 1 shows a schematic diagram of the apparatus. The nozzle is located at the left in the chamber marked 1. The nozzle skimmer region is pumped by a Leybold-Heraeus DI-12000 diffusion pump, which is backed by an Alcatel MIV 1000 Root's pump. In this way, a helium gas throughput of up to 20 Torr-liter/sec could be achieved at about 3×10^{-3} Torr. The beam is chopped by a rotating disk (two 2-mm slots on a 16-cm-diam disk rotating at 600 Hz). The next chamber (3) provides for differential pumping. After a total flight path of 2.37 m, the beam particles are ionized in a commercial quadrupole mass spectrometer. The pulses produced by the arrival of individual ions are stored in a multichannel analyzer (6 μsec channel time). The beam then passes into the last chamber (5), where a small pump serves to remove the gas from the beam. Below we describe in more detail some of the more important apparatus components.

2.1 The Nozzle

The nozzle with hole diameters of 20 and 50 μm was used. Small molybdenum plates, containing the small holes, were obtained from Siemens,[‡] where they are used as aperture lenses in electron microscopes. The effective diameter, which is an important quantity for comparison with theory, was deter-

‡ Siemens AG. 3000 Hannover, Postfach 5329, Germany

mined by flux measurements.[11] The basic equation for flow in a throat is

$$d(p_o V) = p_o \; E \; C^* \; F_{eff} \; dt \qquad (1)$$

where

$$E = [2/(\gamma+1)]^{1/(\gamma-1)} \; ; \; C^* = [2\gamma/(\gamma+1)] \; (kT_o/m)$$

F_{eff} is the effective area of the throat and was used to determine the diameter $[F_{eff} = (\pi d^2_{eff})/4]$. No dependence of the effective diameter on the Reynolds number Re, which varies with pressure, was observed. For two nozzles of nominal diameters of 20 and 50 μm, the flux measurements yielded values of 14.5 and 45.5, μm, respectively. Microscopic examination of the holes gave diameters of 19 ± 1 and 47 ± 1 μm.

2.2 The Skimmer

The long, slender, conical skimmers ($32^o/25^o$, full angles) first suggested by Bossel et al.[12] were used in all of these studies. The overall length was l = 25 mm, and the diameter of the opening was 0.6 mm. Recent theoretical work[13] has revealed that even more slender skimmers, which, according to Bossel,[14] should be used at high speed ratios, will not perform satisfactorily because of "choking" produced by collisions with the inner surface. The skimmers were produced by a galvanic process in our shop by a technique similar to that since described in the literature.[15]

2.3 The Detector and Time-of-Flight (TOF) Electronics

The commercial quadrupole mass spectrometer (Extranuclear Laboratory) is mounted so that its axis is perpendicular to the beam axis. In this way, the beam can pass through the ionization region without wall collisions. The length of the ionization region is about 5 mm, corresponding to a spatial resolution of $\Delta l/l \simeq 0.25\%$. Care was taken to keep the electron emission current sufficiently small to prevent ion storage by space charges in the ionizer.[§] Space-charge effects could be detected by long tails to longer times in the TOF distributions. The pulses produced by each ion were

---

[§] For the used ionization arrangement an emission current less than 5 mA and an extractor voltage of 50 V had been chosen to work well in our case.

stored in a commercial multichannel analyzer (Northern Scientific ECON IA), which was modified to give an effective channel width of 6 μsec.

The overall response time of the detector was determined by an absolute calibration of the entire detection system. This was achieved by using a high-resolution Fizeau-type velocity selector ($\Delta v/v \simeq 1.5 \times 10^{-2}$) in the primary beam, in addition to the beam chopper. A comparison of the measured TOF distributions at two different stagnation temperatures (different values of u) yields the absolute velocity transmission function of the velocity selector, as well as the overall response time. This was found to 10 μsec and results in an overall resolution of $\Delta t/t \simeq 0.7\%$ for He at $T_o = 300^\circ K$. A typical measurement of a time-of-flight spectrum took about 10 min. Changes in the stagnation temperature during a run can cause a smearing of the TOF distribution. To avoid such an effect, the stagnation temperature of the nozzle itself was held constant at $\pm 0.1^\circ C$. An electronically switched heater cooler arrangement regulates the temperature in a water circulation line which is in contact with the nozzle wall.

3. Experimental Results

3.1 Velocity Distribution

In order to determine a speed ratio from the data, a temperature has to be extracted from the TOF distribution. In almost all previous work, measured velocity distribution could be fitted satisfactorily to a modified Maxwellian distribution[16]

$$f_M(v_{\parallel}) \propto v_{\parallel}^2 \exp\left[-m\,(v_{\parallel} - u)^2/2\,kT_{\parallel}^M\right] \qquad (2)$$

where the index $\parallel$ denotes the parallel component (with respect to the beam centerline), which is measured in the experiment. In our experiments with He, we found for $30 \lesssim S \lesssim 100$ a significant deviation from the simple Maxwellian form. This is illustrated in Fig. 2, where a best-fit Maxwellian distribution is compared to the experiments. Although providing a good fit in the central portion, the Maxwellian cannot fit the broad wings of the distribution. As shown in Fig. 3, two Maxwellian distributions with different temperatures $T_{\parallel}^1$ and $T_{\parallel}^2$ ($T_{\parallel}^2 > T_{\parallel}^1$) do provide a satis-

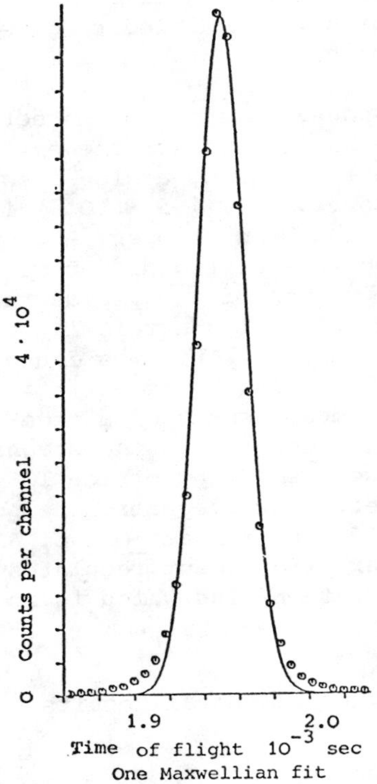

Fig. 2 Comparison of a single Maxwellian and the measured time-of-flight distribution. Channel width is 6 μsec. The speed ratio determined from this fit is $S_{\parallel} = 102.1$.

factory fit. Thus the overall distribution could be assigned a mean temperature defined by

$$\overline{T_{\parallel}} = T_{\parallel}^{1}\,(1-\alpha) + \alpha\, T_{\parallel}^{2}, \quad (0<\alpha<1) \tag{3}$$

where $\alpha/(1-\alpha)$ is the ratio of the area of the $T_{\parallel}^{2}$ distribution to that of the $T_{\parallel}^{1}$ distribution. Typically, α was found to lie between 0.16 and 0.5, and the ratio of temperatures between 4 and 9. A more satisfactory way of assigning a temperature is to use the concept of a mean kinetic energy within the beam

$$1/2\; kT_{\parallel}^{k} = 1/2\; m \int (v_{\parallel} - u)^2\, f(v_{\parallel})\, d\,v_{\parallel} \tag{4}$$

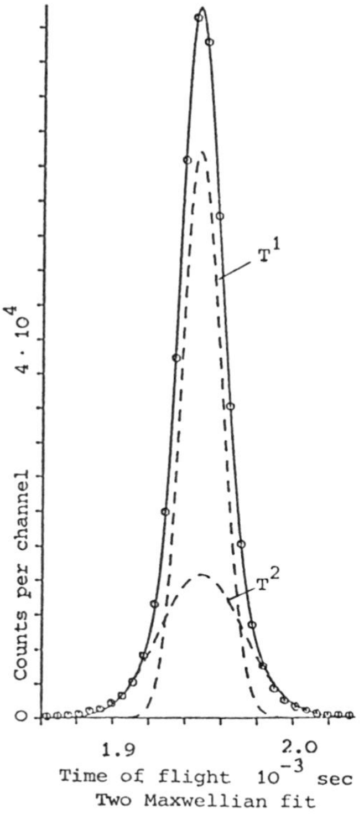

Fig. 3 Comparison of the sum of two Maxwellian distributions and the measured time-of-flight distribution. The measured distribution was the same as in Fig. 2. The speed ratio determined from this weighted average temperature (see text) is $S_{\parallel} = 85.1$.

where $f(v_{\parallel})$ is the measured velocity distribution. The temperatures $T_{\parallel}^{k}$ do depend very sensitively on the value of u used in Eq. (4). u was determined in each case from the best-fit Maxwellian distribution [Eq. (2)]. In general, values of $T_{\parallel}^{k}$ differed only by about 3% from values of $\overline{T}_{\parallel}$, based on Eq. (3).

A convenient measure of the deviation in the shapes of the distributions as well as of the differences in the speed ratio resulting from the two different fit procedures is the quantity

$$(T_{\parallel}^{M}/T_{\parallel}^{k})^{1/2} \equiv S_{\parallel}^{k}/S_{\parallel}^{M} \tag{5}$$

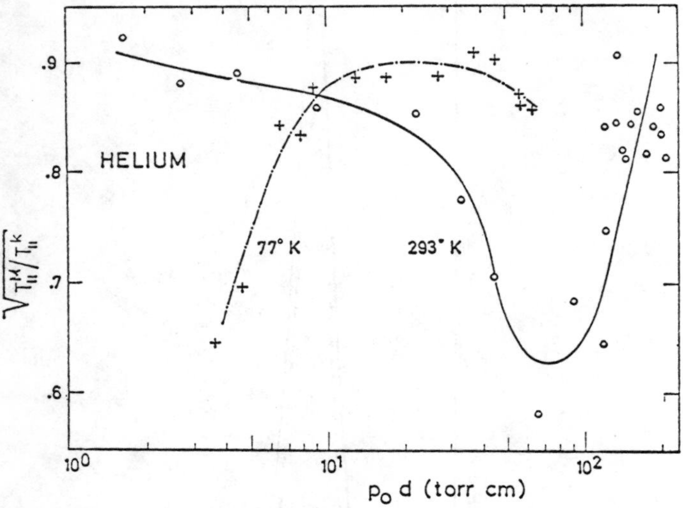

Fig. 4 The ratio of the square roots of the parallel temperatures calculated by two different procedures vs. $p_o d$. The temperature $T_{\parallel}^{M}$ is obtained from the best fit of the measured time-of-flight distribution by a single Maxwellian. The temperature $T_{\parallel}^{k}$ in the denominator is the kinetic integral over the measured distribution.

This quantity is plotted in Fig. 4 as a function of $p_o d$ for measurements at stagnation temperatures of 77° and 293° ^{o}K. The smallest temperature ratios, indicating extended wings, are found for T_o = 293°K at $p_o d \simeq 80$ and for T_o = 77°K at $p_o d \simeq 4$ Torr-cm. The data for T_o = 293°K show that with increasing $p_o d$ the deviations from the Maxwellian form disappear. This also is found for T_o = 77°K. The slight dropoff at large $p_o d$ observed in this curve correlates with large deviations from the theoretical curve (see Fig. 5). A simple explanation of these deviations will be presented in the last section.

3.2 Terminal Speed Ratios

Figure 5 shows a comparison of the measured speed ratios with theory. For the sake of comparison with the theoretical curve, to be discussed below, the speed ratios shown here are based on a fit using a single Maxwellian distribution. The speed ratios corresponding to $T_{\parallel}^{k}$ would be considerably smaller and can be obtained with the aid of Fig. 4. The errors shown in Fig. 5 were obtained from a conservative esti-

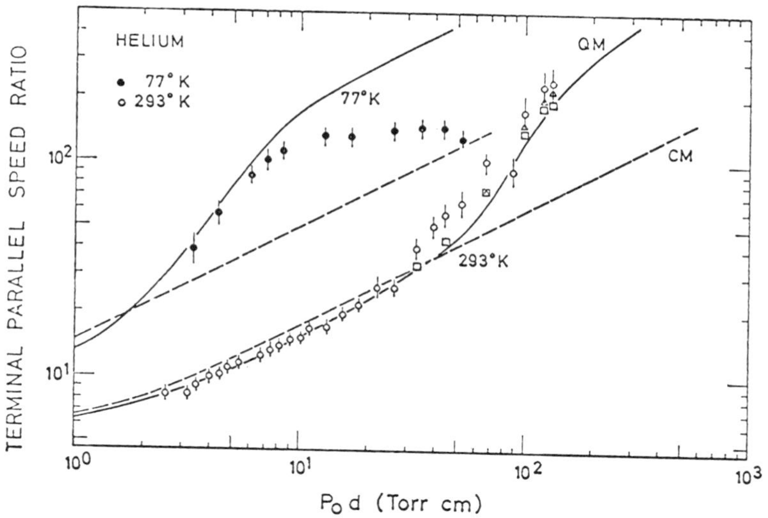

Fig. 5 Comparison of the measured and calculated terminal parallel speed ratios vs. $p_0 d$. The nozzle diameters were d = 0.00145 cm for 293°K and d = 0.00455 cm for 77°K. The CM curve is calculated for classical mechanical collision integrals and QM for quantum-mechanical collision integrals. The data were taken from the one Maxwellian fit, because the theory[7] assumes an ellipsoidal Maxwellian velocity distribution. In addition some measured $S_{\parallel}^{K}$ (Δ) and $\overline{S}_{\parallel}$ (□ , 2-Maxwellian-fit) values are plotted.

mate of statistical errors, as well as of systematic errors. The increase in the overall error with increasing speed ratio is largely due to systematic effects.

The theoretical curves are based on a theory, which is described in detail in Ref. 8 and is essentially an extension of earlier work by Miller and Andres.[7,5] In the theory, the Boltzmann equation is solved by a method of moments. The solution is made possible by assuming Maxwellian distribution functions for $v_{\parallel}$ and $v_{\perp}$. This is the reason why we compare with experiments fitted to $T_{\parallel}^{M}$. In the theory, the collisional interaction comes in by way of collision integrals similiar to those encountered in the Chapman-Enskog transport theory.

These collision integrals customarily are calculated using classical mechanics for an assumed potential. However, it is well known that for light particles quantum effects become important at low temperatures (< 50°K). In the case of

He, quantum effects lead to a dramatic increase in the collision cross sections at small collision energies and correspondingly to an increase in the collision integrals at small $T(< 1^{\circ}K)$. This effect can be attributed to the fact that the only bound state of He, if existent, has an energy that is exceedingly small ($\approx 2 \times 10^{-2}$ °K).

This increase in the collision integrals leads to the observed large increase in the speed ratios for $p_o d \gtrsim 50$ Torr-cm over that predicted by classical mechanics. The calculations and experiments show that the large differences in speed ratios set in at smaller $p_o d$ for the 77°K measurements. Here the low temperatures at which the classical and quantum cross sections diverge are reached for smaller $p_o d$.

In view of the deviation of the observed velocity distribution from the pure Maxwellian form, one might justly suspect a theory based on the assumption of a Maxwellian distribution. The validity of the theory has been established by a careful comparison with independent calculations of speed ratios using the Monte Carlo simulation technique of Bird. These calculations, which were performed by Chatwani[10] in our laboratory, do not require an assumed velocity distribution but actually predict one. These were found to agree well with the results presented in Fig. 4. Calculations carried out for He using the same potential, the same definition of the beam temperature, and termination condition for the expansion yielded results for S which are in reasonable agreement with the theory used in Fig. 5.

Extensive theoretical studies [8] have revealed that the speed ratio is a sensitive function of the assumed intermolecular potential and the point at which the integration of the moment equations is terminated. The results shown in Fig. 5 are for a simple Lennard-Jones (12-6) potential with $\epsilon = 0.94$ meV and $R_m = 2.99$ A. The integration was terminated at a point in the expansion where the perpendicular temperature $T_{\perp}$ was 5% of the parallel temperature $T_{\parallel}$. At this point, which is at large $p_o d$ values well downstream of the skimmer opening, the expansion nearly has ceased, as indicated by the fact that the results are fairly insensitive to the ratio of $T_{\perp}/T_{\parallel}$ chosen as termination condition.

For the results at $T_o = 77$°K, there is a significant deviation between experiment and theory for $p_o d > 8$ Torr-cm. The reason for this deviation is not understood. Monte Carlo simulation studies [13] indicate that the effect may be due in part to skimmer interference. Also, there is the possibility that condensation may heat up the beam and lead to a reduction in S.

4. Discussion

The new measurements reported here have extended the maximum experimental speed ratio in He to S = 225. The measurements reported here have been carried out in a new apparatus, where considerable care has been taken to obtain reliable absolute data. The results show a significant deviation from the usual Maxwellian velocity distributions.

This effect also can be understood in terms of the quantum-mechanical collision cross sections entering into the collision integrals. This cross section Q rises sharply at a collision energy E of $1^{\circ}K$ and has a slope in a logarithmic scale of about $\partial \ln Q/\partial \ln E = -2.5$. When the gas is cooled into this region, a separation of molecules into two species with different temperatures occurs. The cold atoms experience a strong expansion, since they interact by way of a large cross section and thereby become even colder. The atoms in the wings of the distribution have a higher kinetic energy and interact, therefore, by way of a much smaller cross section and, as a consequence, experience a smaller cooling. This separation disappears again at very high speed ratios, since the cross section is nearly constant at a point where the expansion becomes nonisentropic ("freezes"). As a result of this separation into different components, the interpretation of the TOF distribution in terms of speed ratios is no longer straightforward. A comparison of different measurements requires a careful specification of the temperature definition used. In view of this complication, the results can be compared only with a theory capable of including such nonequilibrium distributions. Nevertheless, by neglecting these nonequilibrium effects, we were able to get good agreement for moderate and high $p_0 d$ with a theory, assuming Maxwellian distributions for an assumed potential model. Although the results are sensitive to the potential model used, we cannot determine uniquely a potential from the results, since they are sensitive only to an integral property (the "scattering length") of the potential.

The present results show that very low temperatures and beams with very narrow velocity distributions can be achieved in He. For S = 225 and $T_0 = 293^{\circ}K$, the final temperature in the beam is $15 \times 10^{-3}\ ^{\circ}K$, and for $T_0 = 77^{\circ}K$, $8.4 \times 10^{-3}\ ^{\circ}K$. The corresponding velocity spreads $\Delta v/v$ (Δv at FWHM) are 0.7% and 1.1%, respectively.

Work is now in progress to extend these measurements to even higher speed ratio. By going to pressures of 1500 bar and using d = 5 μm holes, speed ratios of 600 should be achievable with the same pumping system.

References

[1] Toennies, J.P., "Molecular Beam Scattering Experiments on Elastic, Inelastic and Reactive Collisions," Physical Chemistry, an Advanced Treatise: Kinetics of Gas Reactions, Vol. VIA, edited by W. Jost, Academic Press, New York, 1974, pp. 227-381; also Bericht Nr. 134/72, 1972, Max-Planck-Institut für Strömungsforschung, Göttingen, West Germany.

[2] Toennies, J.P., "Scattering of Molecular Beams from Surfaces," Applied Physics, Vol. 3, March 1974, pp. 91-114; also Bericht Nr. 117/73, 1973, Max-Planck-Institut für Strömungsforschung, Göttingen, West Germany.

[3] Malthan, H. and Toennies, J.P., "Direct Molecular Beam Measurements of Rotational State Distributions in Nozzle Beams," Rarefied Gas Dynamics, Vol. II, edited by M. Becker and M. Fiebig, DFVLR Press, Porz-Wahn, West Germany, 1974, pp. C. 14-1 - C. 14-12.

[4] Smalley, R.E., Levy, D.H., and Wharton, L., "The Fluorescence Excitation Spectrum of the He I_2 van der Waals Complex," Journal of Chemical Physics, Vol. 64, April 1976, pp. 3266-3276.

[5] Miller, D.R., Toennies, J.P., and Winkelmann, K., "Quantum Effects in Highly Expanded Nozzle Beams," Rarefied Gas Dynamics, Vol. II, edited by M. Becker and M. Fiebig, DFVLR Press, Porz-Wahn, West Germany, 1974, pp. C 9-1 - C. 9-9.

[6] Campargue, R., Lebêhot, A., Lemonnier, J.C., Marette, D., and Pebay, J., "Generateur de Jet Moléculaires Supersonique Fonctionnant en Monochromateur dans le Domaine de 10^{-2} à 40 eV," Proceedings of the Fifth International Symposium on Molecular Jets, April 7-11, 1975, Nice, France, p. A7.

[7] Miller, D.R. and Andres, R., "Translation Relaxation in Low Density Supersonic Jets," Rarefied Gas Dynamics, Vol. II, edited by L. Trilling and H.Y. Wachman, Academic Press, New York, 1969, pp. 1385-1402.

[8] Toennies, J.P. and Winkelmann, K. (to be submitted to Journal of Chemical Physics).

[9] Bird, G.A., "Direct Simulation and the Boltzmann Equation," The Physics of Fluids, Vol. 13, Oct. 1970, pp. 2676-2681.

[10] Chatwani, A., "Monte Carlo Simulation of Nozzle Beam Expansions," Rarefied Gas Dynamics, 1976, to be published.

[11] Liepmann, H.W., "Gaskinetics and Gasdynamics of Orifice Flow," Journal of Fluid Mechanics, Vol. 10, Febr. 1961, pp. 65-79.

[12] Bossel, U., Hurlbut, F.C., and Sherman, F.S., "Extraction of Molecular Beams from Nearly-Inviscid Hypersonic Free Jets," Rarefied Gas Dynamics, Vol. II, edited by L. Trilling and H.Y. Wachman, Academic Press, New York, 1969, pp. 945-964.

[13] Bird, G.A., "Transition Regime Behaviour of Supersonic Beam Skimmers," Max-Planck-Institut für Strömungsforschung, Göttingen, West Germany (to be published).

[14] Bossel, U., "On the Optimization of Skimmer Geometries," Entropie, No. 42, Nov.-Dec. 1971, pp. 12-17.

[15] Gentry, W.R. and Giese, C.F., "High-Precision Skimmers for Supersonic Molecular Beams," Review of Scientific Instruments, Vol. 46, Jan. 1975, p. 104.

[16] Abuaf, N., Anderson, J.B., Andres, R.P., Fenn, J.B., and Miller, D.R., "Studies of Low Density Supersonic Jets," Rarefied Gas Dynamics, Vol. II, edited by C.L. Brundin, Academic Press, New York, 1967, pp. 1317-1336.

STUDIES OF LASER-GENERATED ATOMIC BEAMS AND METAL-OXIDE CHEMILUMINESCENT REACTIONS

S. P. Tang* and J. F. Friichtenicht†

TRW Defense and Space Systems Group, Redondo Beach, Calif.

Abstract

The velocity distribution of atoms within atomic beam pulses produced by the rapid vaporization of thin metallic films with a pulsed laser have been measured as a function of laser bombardment conditions and the angle of observation. The results show that the average energy of the atoms is a function of the laser beam energy density and the angle of observation as referenced from the normal to the thin-film target. Measurements of the integrated beam flux show that the metallic vapor forms a narrow plume centered about the normal to the target. The velocity distributions of both atomic species from Aℓ-Sn alloy targets are consistent with a Maxwellian distribution superimposed on a center-of-mass motion of the vapor cloud as a whole. However, analysis indicates that the two species equilibrate at different temperatures. Chemiluminescence resulting from metal-oxide formation reactions has been observed.

Introduction

The pulsed laser vaporization of thin solid films deposited on transparent substrates to produce an energetic pulsed atomic beam was first described by Friichtenicht.[1] Marmar et al.,[2] have used this technique to inject metal atoms into plasma for diagnostic purposes.[3] Tang et al.[4] utilized this

Presented as Paper 138 at the 10th International Symposium on Rarefied Gas Dynamics, July 19-23, 1976, Aspen, Colo.
*Member of Professional Staff.
†Manager, Chemical Physics Group.

technique in a crossed-beam configuration to study the chemiluminescent reactions of Ho and B atoms with N_2O. Both Friichtenicht and Marmar et al. report that the characteristics of the beam pulse are strongly dependent upon the laser irradiation conditions and the material in question. At the extreme of a strongly focused laser beam (on the order of 100 J cm^{-2} for a 0.1-μsec pulse) a large fraction of the metal vapor is ionized. At lesser values of energy density, Marmar et al. report the existence of excited neutrals in the beam pulse. For small values of laser beam energy density, the pulse contains predominantly neutral atoms with an average kinetic energy that is a function of the laser energy density.

Since the presence of internally excited atoms within the beam could distort experimental measurements of such quantities as chemiexcitation cross sections, Utterback et al.[5] presented a first-order model of the expansion process, which enables the "temperature" of the expanding vapor cloud to be determined. Comparison of the predictions of this model with experimental measurements are in reasonably good agreement for the range of bombardment conditions, in which the observed average kinetic energy is no more than a few electron volts. For higher average energies, the velocity distribution is skewed in the direction of velocities higher than those consistent with the model. Within the range of validity, the temperature of the vapor cloud is significantly less than might be expected, simply on the basis of the most probable velocity of the atom beam pulse. For the chemiluminescence experiments of Tang et al., the findings of the model were used indirectly to support the contention that the atom beam pulse was "cold" enough to preclude the existence of a large excited state population.

Experiments designed to gain further insight into the beam formation process have been conducted and are described herein. The integrated atom flux as a function of angle has been determined, and additional data relating atom energy to laser energy and laser energy density have been acquired. Velocity distributions of the components of two-element alloys (composed of various mixtures of tin and aluminum) have been measured and compared with the model. It is found that the velocity distributions of the two atomic species are individually consistent with the model; however, the apparent temperatures differ substantially.

Experimental Apparatus

The experimental apparatus is depicted schematically in Fig. 1. A Q-switched ruby laser capable of delivering up to

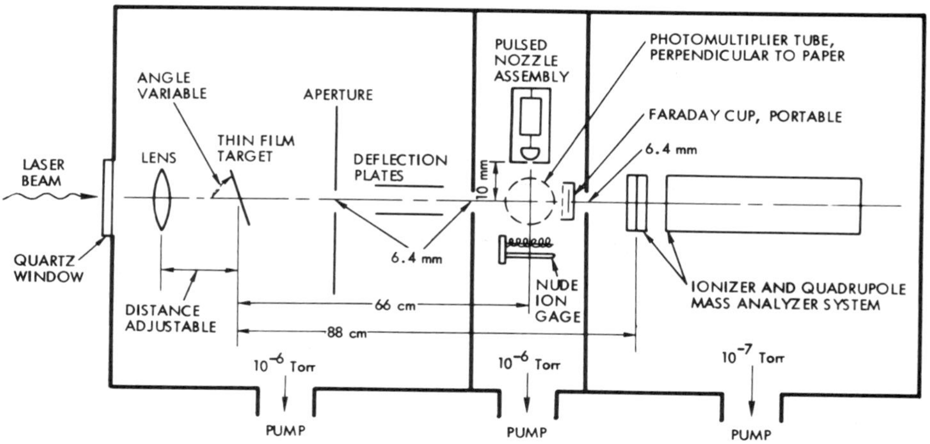

Fig. 1 Schematic diagram of the experimental apparatus.

3 J of radiant energy in approximately 80 nsec was used for these measurements. The collimated laser beam is directed to the target surface through a 12.5-cm focal length lens, mounted within the vacuum system. The laser energy density at the target may be varied by moving the lens to change the lens to target distance. Targets consisting of vapor-deposited metal films on 2.5-by 7.5-cm glass microscope slides are mounted on a rod extending through the vacuum chamber wall. The angle of incidence of the laser beam is varied by rotation of the mounting rod. The mounting rod can be translated along its axis to expose fresh target material to the laser beam. The atomic beam axis is defined by a series of collimating apertures. For the present set of experiments, the atom beam axis and the extended axis of the laser beam coincide.

The interaction region is equipped with a pulsed nozzle beam source. In operation, the pulsed nozzle is actuated prior to the laser Q-switch such that steady-state conditions pertain during the period of interaction with the metal atom pulse. The effective target gas pressure is in the 10^{-4}-to 10^{-2}-Torr range. For the measurements on chemiluminescent reactions discussed below, the interaction region was equipped with a broad band photomultiplier tube, collimated such as to view only the interaction volume. For the velocity distribution measurements, the target gas system was not used leaving the region at a pressure of about 10^{-6}-Torr.

The atom beam detection system consists of an electron impact ionizer (Extranuclear Laboratories, Inc., type II high efficiency ionizer), an rf quadrupole mass filter (Extranuclear Laboratories, Inc., model 324-9 used with either the C or E high-Q head), and a 13-stage electron multiplier tube with Be-Cu dynodes (RCA box and grid type).

Experimental Procedures and Results

Experimental measurements have been performed to determine the dependence of the atom energy on the laser pulse energy, the angular distribution of the vapor cloud, the atom energy as a function of angle of observation, and the characteristics of two-element alloy targets. Additionally, an application of this atomic beam technique has been demonstrated by the measurement of the relative chemiluminescence yield of several metal oxide formation reactions.

Atom Energy Dependence on Laser Beam Energy

The functional dependence of atom energy on the laser beam energy was determined by varying the laser energy for a fixed bombardment geometry and observing the resultant atom beam pulse with the quadrupole mass filter (QMF). Vapor deposited tin films, nominally 1-μ thick, were used for these tests. Several target samples were used, but they were all prepared at the same time and are virtually identical. The plane of the target was normal to the laser beam axis for the entire sequence of shots. The laser energy was varied between approximately 0.06 and 0.18 J. Since the lens location was not changed, the laser energy density was also a function of the laser energy. The effective laser energy density was determined by taking the ratio of the laser beam energy to the area of the hole produced in the film.

The atoms were detected with the QMF assembly located 88-cm from the laser target. The QMF yields a signal whose instantaneous amplitude is proportional to the number density of atoms within the ionizer volume. The velocity of the atoms is specified by their transit time from the laser target to the ionizer. A fraction of the incident atoms were ionized in a region maintained at approximately 70 V positive. It was assumed that the ions traverse the distance from the ionizer to the electron multiplier at a constant velocity, corresponding to the 70-eV ion energy. A correction was made to the observed arrival time allowing for the transit time through the QMF to obtain the free stream velocity of the atoms within the pulse.

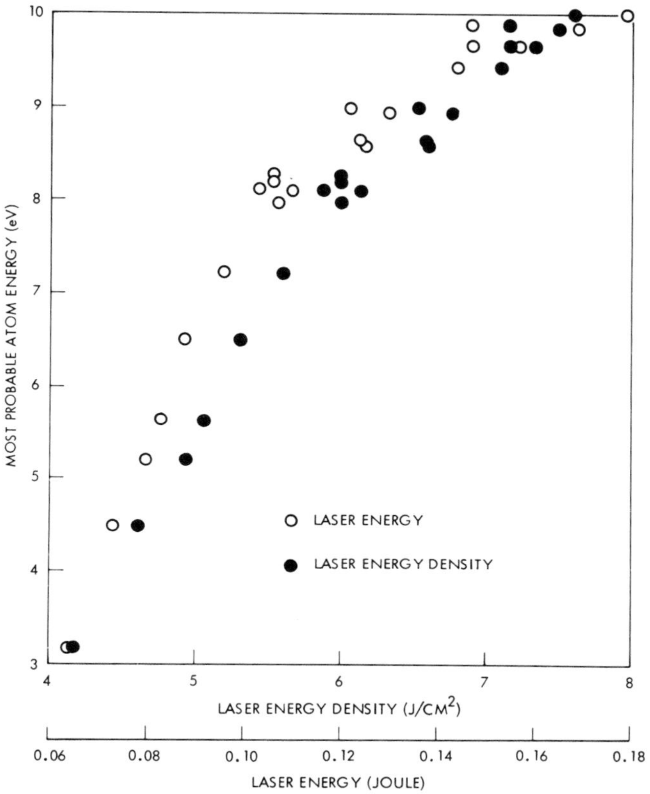

Fig. 2 Most probable atom energy as a function of laser energy and laser energy density obtained with a 1-μ tin target under fixed bombardment conditions.

The results of these measurements are shown in Fig. 2, where the most probable energy of the atom beam pulse is plotted as a function of both laser pulse energy and laser beam energy density. Here, the most probable energy is defined at that corresponding to the signal maximum of the observed atom beam pulse. The results show that the most probable atom energy increases systematically with laser energy and/or laser energy density. It is impossible to separate these two effects on the basis of these data alone.

It seems likely that the atom energy depends more strongly on laser energy density, whereas the magnitude of the pulse is determined by the total energy available. The results graphically illustrate the point made by earlier authors that the energy characteristics of the atomic beam can be varied simply through variation of the laser beam energy.

Angular Distribution Characteristics of the Pulsed Atom Beam

The angular distribution measurements utilized the same basic apparatus, except in this case the angular position of the target plane was systematically varied while the laser energy was maintained at a constant value. Atomic beam pulse signals were obtained from the QMF system in the manner described previously. It should be noted that changing the angular position of the target changes both the angle of incidence of the laser beam on the target surface and the angle of observation. Defining θ as the angle between the laser beam axis and the target normal, the area of the beam spot is given by $A_0/\cos\theta$, where A_0 is the area of the spot at $\theta = 0°$. Thus, changing the angle to other than normal incidence decreases the average laser energy density at the target surface. The angle of observation ϕ is also referenced from the target normal. Specifying the angle of observation also specifies the angle of incidence, and they are numerically equal for the particular geometry used in these experiments.

The results from these measurements are presented in Figs. 3 and 4. Figure 3 gives the most probable energy as previously defined as a function of angle of observation (and angle of incidence). It can be seen that the most probable energy drops by nearly an order of magnitude in going from 0° to 45°; however, the average energy density decreases by less than 30%. It seems clear that the most probable atom energy decreases with increasing angle over and above that expected, simply on the basis of the decrease in laser energy density. Figure 4 gives the integrated atom flux in the beam pulse as a function of the angle of observation. The signal from the QMF system at any instant of time is proportional to the number density of atoms in the ionizer, whereas the elapsed time between the laser Q-switch and the instant of observation specifies the velocity of the atoms in the ionizer at that time. Given ρ and u, the flux J can be determined as a function of time. Graphical integration of J over the duration of the pulse yields of quantity directly proportional to the number of atoms reaching the detector for any burst. It is this quantity that is plotted in Fig. 4. The integrated flux drops dramatically in going from 0° to 45°, showing that atoms are evaporated from the surface in a fairly narrow plume.

Measurements on Aℓ-Sn Alloy Targets

To examine the expansion characteristics of laser-produced vapor clouds containing atomic species with differing

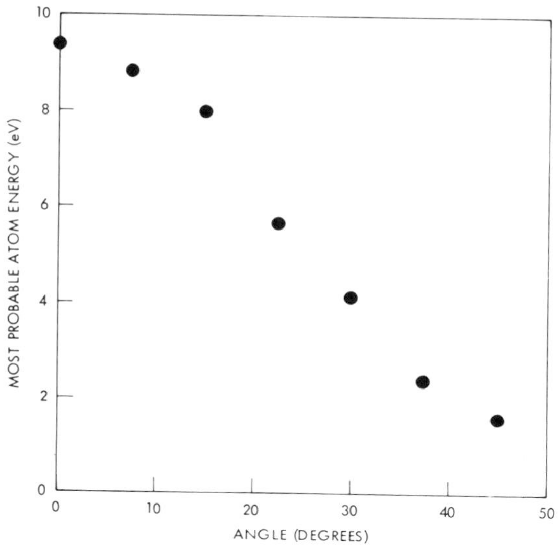

Fig. 3 Most probable atom energy as a function of angle of observation (equal to laser incidence angle) obtained with a 1-μ tin target.

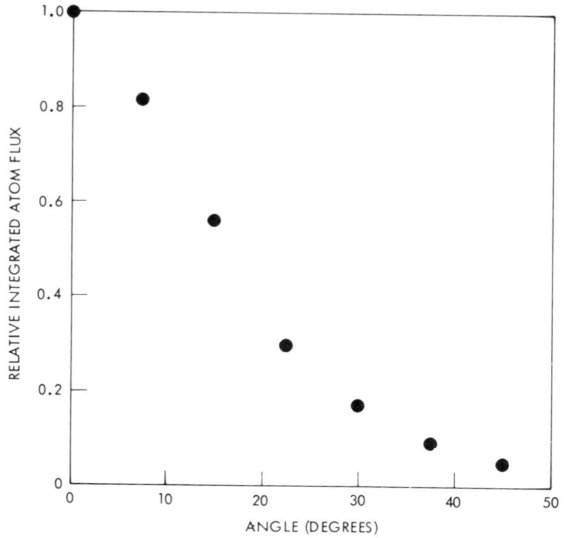

Fig. 4 Relative integrated atom flux as a function of angle of observation (equal to laser incidence angle) obtained with a 1-μ tin target.

Table 1 Alloy compositions used in experimental tests

	Atom %		Weight %	
Alloy	Aℓ	Sn	Aℓ	Sn
1	90	10	67.17	32.83
2	50	50	18.52	81.48
3	10	90	2.46	97.54

atomic weight, thin-film targets of three Aℓ-Sn alloys were prepared in accordance with Table 1. These alloys were prepared from 99.78% weight purity aluminum and 99.97% weight purity tin. Weighed amounts of these metals, having a total weight of 10 g for each alloy, were heated in a graphite crucible under an argon atmosphere in a metallurgical furnace. The temperature of the furnace was set at 800°C, which is considerably above the melting temperature of these alloys predicted by the phase diagram. The aluminum (melting point of 600°C) was melted first and the tin (melting point 232°C) was added. The melts were mechanically stirred to insure good mixing. The melts were kept under argon throughout the cooling and solidification of the alloys. The alloy billets were subsequently vapor deposited on glass slides in the usual manner. No postevaporation analysis of the relative concentrations of the two metals was conducted; however, the phase diagram predicts that the concentrations should remain unchanged during the evaporation process.

The experimental system described in the preceding was used to measure the velocity distributions of the two atomic species under several sets of bombardment conditions. Typical data are shown in Fig. 5. These two curves basically are reproductions of the photographic records of signals obtained with the QMF system normalized to the peak amplitude and with the time base modified such as to yield the arrival time at the ionizer. The QMF resolution $\Delta m/m$ was set at about 0.1, which is more than adequate to separate the two atomic species. The curves shown in Fig. 5 clearly illustrate the point that the most probable velocities of the two distributions differ only slightly, even though the atomic mass ratio of the two species is between four and five.

Chemiluminescent Reactions

In order to illustrate a practical application of this atomic beam technique, preliminary data on chemiluminescent metal-oxide formation reactions were acquired. For these

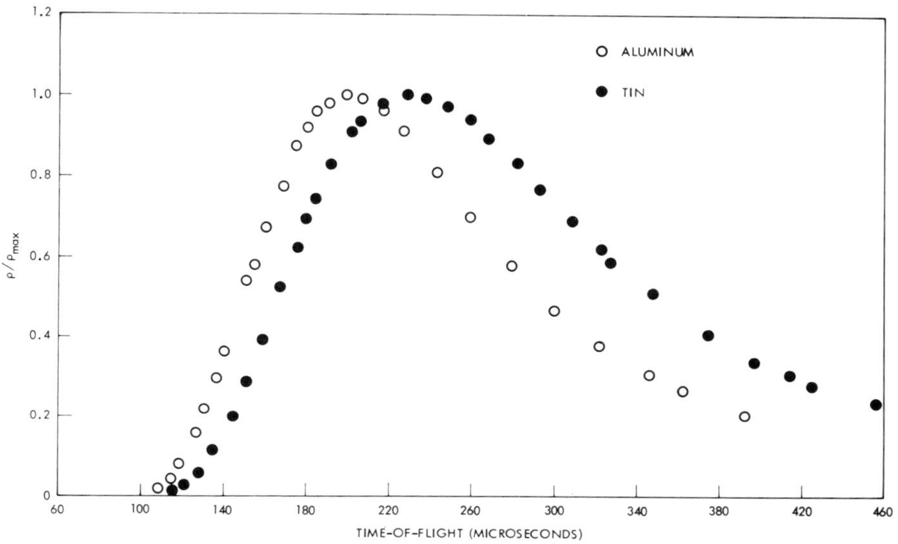

Fig. 5 Comparison of normalized atom number density as a function of time-of-flight in real time of aluminum and tin from a 1-μ alloy target with 50%-Aℓ and 50%-Sn concentration.

measurements, the interaction region shown schematically in Fig. 1 was equipped with a broad-band photomultiplier tube, sensitive to radiation in the visible portion of the spectrum. The experiments consisted of simultaneous measurements of the radiant emission from the target gas region and the intensity of the atom beam pulse for a fixed target gas pressure. Both signals were integrated to yield quantities proportional to the total number of photons produced and the number of atoms in the beam pulse. The ratio of these two quantities is related to the cross section for photon-producing reactions. The results obtained for a series of metal atoms and N_2O and NO_2 target gases are summarized in Table 2. Measurements of this type are but the first of a series required to quantitatively determine specific reaction rates or cross sections.

Discussion of Results

The results presented previously on the expansion characteristics of the laser produced metal vapor clouds as a function of bombardment conditions have been presented primarily to demonstrate the range of beam parameters that can be covered by simply implemented procedures. The qualitative data on chemiluminescent reactions have demonstrated

Table 2 Relative chemiluminescence yields from the reaction of various metal atoms with N_2O and NO_2 target gases

Metal	Target gas N_2O	NO_2
Ba	1.02	1.13
Ce	0.14	0.14
Co	0.11	0.06
Cr	0.09	0.05
Fe	1.19	0.82
Ge	0.76	0.39
Ho	1.23	0.30
In	0.61	...
Mn	∿ 0	...
Mo	1.24	0.54
Nd	0.90	0.60
Ni	∿ 0	...
Si	0.04	0.04
Sm	0.15	0.01
Sn	0.07	0.02
Ti	0.57	0.94

one practical application of this technique. The data on the Aℓ-Sn alloy targets, however, provides a method of exploring some of the more subtle aspects of the atomic beam formation process.

It has been observed consistently (see Refs. 1 and 2) that the distribution of atom velocities observed at a point remote from the laser beam target is narrower than would be expected from the expansion of a thermally equilibrated gas bubble. Utterback et al.[5] presented a simple model of the expansion process wherein it was assumed that the vapor cloud may be described as a hot gas "bubble" containing N_o particles moving with a center of mass velocity perpendicular to the substrate surface. Initially the collision frequency within the gas bubble is high enough that the particles equilibrate to some temperature such that an observer located at the center of mass would measure a Maxwellian velocity distribution. The unconstrained gas bubble expands symmetrically about the center of mass until it becomes collisionless, at which point the particle velocity distribution corresponding to the final temperature becomes "frozen" in the center of mass system. The model mathematically describes the time-dependent variation in the number density of particles that would be observed in the laboratory reference frame for these initial conditions. It is assumed that the dimensions of the bubble and the distance traveled before it becomes collisionless is infinitesmally small as compared to the distance to the detector.

The analysis utilizes the parameter λ defined as the ratio of the center of mass velocity to the root-mean-square thermal velocity, i.e.

$$\lambda = v_{cm}/v_{rms} = v_{cm}/(3kT/m)^{1/2}$$

If both the number density ρ and the time of arrival at the ionizer are nondimensionalized by substitution of the quantities ρ/ρ_{max} and $\eta_\rho = t/t_{max\rho}$, where $t_{max\rho}$ is the time at which the number density reaches its maximum value ρ_{max}, specification of the parameter λ defines a universal curve with ρ/ρ_{max} and η_ρ as the coordinates. Experimentally obtained signals are converted readily to the same format. The experimental curves then can be characterized by the value of λ that gives the best fit to the experimental data. Examples of this procedure are shown in Figs. 6 and 7 for Sn and Aℓ atom pulses obtained from a 50% Sn, 50% Aℓ alloy target. For these data, the laser beam was at normal incidence to the target and the QMF system was located on the target normal ($\phi = 0°$ according to our nomenclature). It can be seen that values of λ equal to 1.2 for the Aℓ atom pulse and 0.9 for the Sn atom pulse provide an excellent fit to the experimental data, which are represented by the points. Curves for $\lambda = 0$ (i.e., $v_{cm} = 0$) also are shown for reference. It might be noted that reasonably good fits were obtained for tin atom pulses observed under a variety of conditions. Under certain conditions, however, the model does not adequately describe the leading edge of the aluminum atom velocity distribution. In these cases, the value of λ that provided the best fit to the low velocity (from the peak downward) part of the distribution was taken to be representative of those particular pulses.

Given a value of λ for a particular experimental curve, and the laboratory velocity distribution as determined by the arrival time at the ionizer, sufficient information is available to determine absolute values of the equilibrium temperature T and the velocity of the center of mass. The findings of this investigation are summarized in Table 3. The most striking conclusion is that, although the velocity distributions of the two atomic species are reasonably

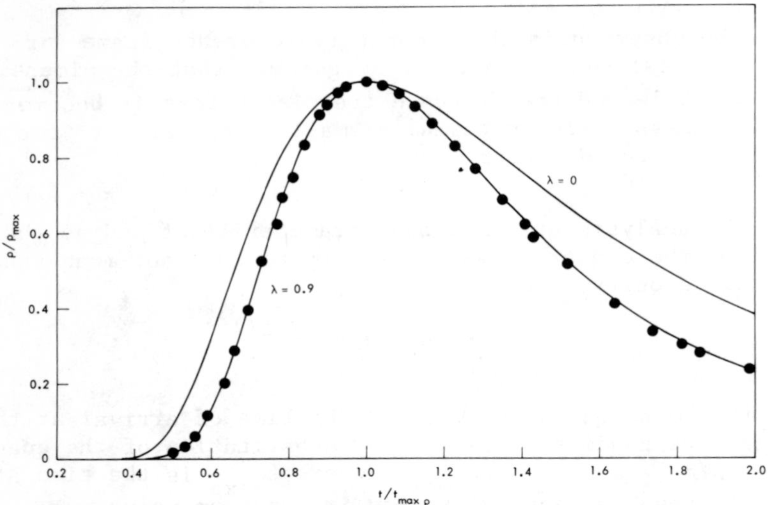

Fig. 6 Normalized number density vs time-of-flight plots. Dots represent the velocity distribution of tin atoms vaporized from a 1-μ alloy target with 50%-Aℓ and 50%-Sn concentration. The λ = 0.9 curve is the best fit according to a theoretical model by Utterback et al. The λ = 0 curve represents a Maxwellian distribution corresponding to a source temperature of the vapor cloud with zero center-of-mass velocity.

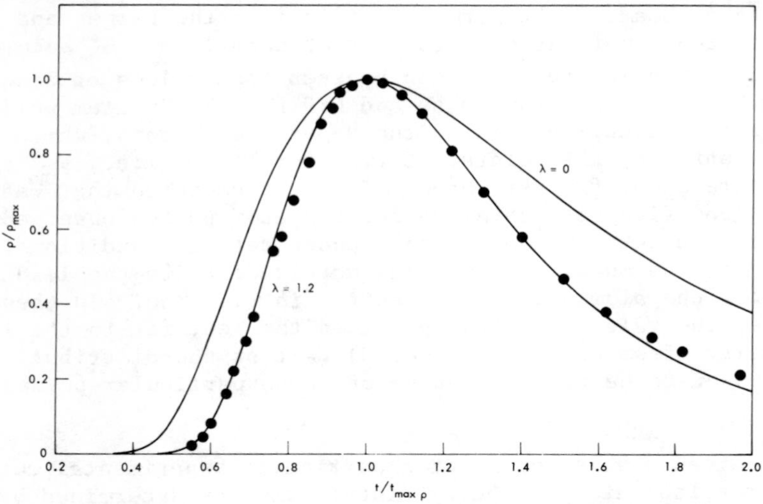

Fig. 7 Normalized number density vs time-of-flight plots. Dots represent the velocity distribution of aluminum atoms vaporized from a 1-μ alloy target with 50%-Aℓ and 50%-Sn concentration. The λ = 1.2 curve is the best fit according to a theoretical model by Utterback et al. The λ = 0 curve represents a Maxwellian distribution corresponding to a source temperature of the vapor cloud with zero center-of-mass velocity.

Table 3 Summary of the results of velocity distributions of atoms from alloy targets

Alloy	Concentration		λ	T, °K	V_{cm}, cm/sec	V_{mp}, cm/sec
1	Aℓ	90%	0.45	13,900	1.6×10^5	4.5×10^5
	Sn	10%	0.85	34,600	2.3×10^5	4.0×10^5
2	Aℓ	50%	1.20	6,800	3.0×10^5	4.4×10^5
	Sn	50%	0.90	29,600	2.2×10^5	3.8×10^5
3	Aℓ	10%	1.20	6,600	3.0×10^5	4.3×10^5
	Sn	90%	0.90	27,900	2.2×10^5	3.7×10^5

well-described by the model, the temperatures and center of mass velocities determined by the model are consistently different for the two atomic species. The aluminum pulse is invariably "colder" than the tin. It might be argued that the fitting process is too imprecise to specify the temperature accurately; however, simply comparing the most probable velocities for the two distributions leads to the same general conclusion. The results are clearly inconsistent with the "zeroth"-order model, where it might be assumed that the expansion is a strictly thermal process and that both velocity distributions are Maxwellian. If this were the case, the most probable velocities of the two distributions would differ by the square root of the mass ratios, or a factor of 2.1. The data in Table 3 show, however, that the most probable Sn atom velocity is only about 10% less than the most probable Aℓ atom velocity.

It is unlikely that the generally good agreement between the model and the measurements is fortuitous. However, as pointed out in the preceding, data reduction through application of the model leads to different apparent "equilibrium" kinetic temperatures for the Sn and Aℓ components in the expanding bubble. This might imply the presence of a dominant inelastic process in Sn-Aℓ atom collisions, tending to differentially cool the lighter component.

References

[1] Friichtenicht, J. F., "Laser-Generated Pulsed Atomic Beams," The Review of Scientific Instruments, Vol. 45, January 1974, pp. 51-56.

[2] Marmar, E. S., Cecchi, J. L., and Cohen, S. A., "System for Rapid Injection of Metal Atoms into Plasmas," The Review of Scientific Instruments, Vol. 46, September 1975, pp. 1149-1154.

[3] Cohen, S. A., Cecchi, J. L., and Marmar, E. S., "Impurity Transport in a Quiescent Tokamak Plasma," Physical Review Letters, Vol. 35, December 1975, pp. 1507-1510.

[4] Tang, S. P., Utterback, N. G., and Friichtenicht, J. F., "Measurement of Chemiluminescent Reaction Cross Sections for $B + N_2O \rightarrow BO^* + N_2$ and $Ho + N_2O \rightarrow HoO^* + N_2$," The Journal of Chemical Physics, Vol. 64, May 1976, pp. 3833-3839.

[5] Utterback, N. G., Tang, S. P., and Friichtenicht, J. F., "Atomic and Ionic Beam Source Utilizing Pulsed Laser Blow Off," The Physics of Fluids, Vol. 19, June 1976, pp. 900-905.

LOW ENERGY TOTAL CROSS SECTIONS FOR THE ARGON-KRYPTON SYSTEM

R. W. York,* W. L. Taylor,† P. T. Pickett≠
Monsanto Research Corporation,
Miamisburg, Ohio

and

R. E. Miers§
Purdue University at Fort Wayne,
Fort Wayne, Ind.

Abstract

Total scattering cross sections have been measured for the argon-krypton system in the low relative velocity range of 3 to 7 (x 10^4) cm/sec. A well-collimated supersonic molecular beam was passed through a target cell containing the scattering gas. The attenuation was measured with an ionization detector. A volumetric flow calibration method was used to obtain accurate pressures in the target cell. Both Ar and Kr were used as target gases. Agreement was obtained between the absolute cross sections upon inversion of beam/target at the same relative velocities to within 7%. These data overlap the measurements of previous workers at the higher velocities, and are in good agreement with those results. Several glory extrema were observed which, with those previously reported,

---

Presented as paper 156 at the 10th International Symposium on Rarefied Gas Dynamics, July 19-23, 1976, Aspen, Colorado.
Mound Laboratory is operated by Monsanto Research Corporation for the U. S. Energy Research and Development Administration under Contract E-33-1-GEN-53.

*Senior Research Physicist, Mound Laboratory.
†Science Fellow, Mound Laboratory.
≠Development Technician, Mound Laboratory.
§Associate Professor of Physics.

were used to determine the product $\varepsilon\sigma$ for the argon-krypton interaction potential. Theoretical calculations of total cross sections using the WKB approximation were made for several interatomic potentials satisfying the $\varepsilon\sigma$ criterion. The results are compared to the experimental data.

Introduction

The elastic scattering of a neutral beam from a target gas has, for many years, provided a direct method for the determination of the energy dependence of the total cross section $Q(v_r)$, at relative velocity v_r, of the beam/target ground state collision.

In the "low energy" region of relative velocities there appear characteristic undulations in $Q(v_r)$ superimposed on the monotonic $(v_r)^{-2/5}$ power dependence of the Schiff-Landau-Lifshitz (SLL) approximation.[1] These undulations, whose maxima and minima are the "glory extrema," occur because of interferences between particles of different angular momenta interacting with the effective potential field. The location of the glory extrema[2] can be used to provide an estimate of the product $\varepsilon\sigma$. The well depth ε and the potential zero location, σ, are the commonly used descriptive parameters for the pair interaction potential. Provided with some guidelines regarding the general characteristics of the potential, one can thereby hopefully reduce the number of potentials that must be considered in detail.

In the present work we have extended the measurements of Bredewout[3] and Rothe, et al.[4] on the argon-krypton system to lower relative velocities and have determined additional extrema for comparison to appropriate potentials for this energy region. Absolute cross section measurements were made by passing a supersonic primary beam through a cell containing the scattering gas and measuring the attenuation of the beam as a function of cell pressure. A dynamic flow calibration procedure for the remote measurement of the absolute target density enabled us to make absolute measurements of the cross section. The glory extrema obtained, along with those given by Bredewout, were used to determine $\varepsilon\sigma$ as given in Ref. 2.

We have compared our experimental data with theoretical calculations of the total cross sections for three Ar-Kr potentials using the WKB approximation. These potentials were the Barker-Fisher-Watts potential[5] using the parameters suggested to Bredewout[3] by Barker for the Ar-Kr system, the Exp-6 potential and the Lennard-Jones potential. Parameters for the latter two potentials were obtained from thermal diffusion data since thermal diffusion is a property quite sensitive to the potential parameters - in particular the potential well depth.

Experimental Procedure

A schematic of the apparatus is shown in Fig. 1. The vacuum chamber is of all stainless steel construction separa-

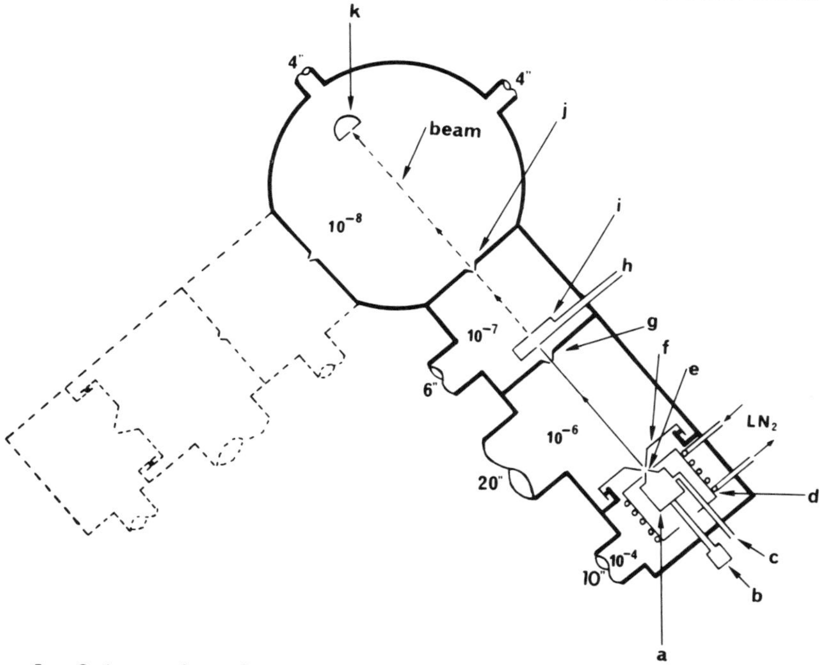

Fig. 1 Schematic of apparatus. The components are as follows: a) source; b) Cryo-tip refrigerator; c) beam gas feed; d) LN_2-cooled radiation shield; e) nozzle; f) adjustable skimmer; g) first collimator; h) target gas feed; i) scattering cell; j) second collimator; and k) the detector. The numbers given in each section are the approximate operating pressures in torr during an experiment and the numbers in the pump openings are the sizes in inches of the diffusion pumps used. The identical second arm of the chamber is shown dotted. It was not used in the present work, although it is open to the center section and the pumps were operating.

ted by bulkheads into individual sections, each of which is pumped by its own pumping system. The chamber has two main 30-cm-diam arms, which intersect at 90 deg. in a center section 61 cm in diam. Each arm has approximately 90 cm of usable length with internal bulkheads along each beam axis for differential pumping. In the present work only one arm of the chamber and the center section were used for total cross section measurements.

The beam source is of the supersonic nozzle/skimmer type with a remotely adjustable skimmer. During operation, the beam intensity is peaked by adjusting the source pressure and the nozzle/skimmer separation distance until the maximum is reached. The pressure is then reduced to assure that operation is on the linear portion of the intensity curve where minimum clustering occurs.[6] It is assumed that the dimer fraction in the beam is less than 1%. The chamber containing the source gas is of high-purity copper and is connected to a Cryo-tip refrigerator for cooling. Precise temperature control of the source chamber to better than 0.03°K is obtained by overcooling and then proportionally driving heating coils to a null point, which is sensed by a platinum resistance thermometer. The range of temperature attainable in the source is approximately 30° to 400°K.

The target cell is placed downstream of the skimmer between two collimators. These collimators are externally adjustable during operation for proper alignment. The detector is a Granville-Phillips ionization gage, with the tube inlet covered by a 1.9-mm-diam orifice. Beam divergence at the detector is approximately 2 mrad. All components are of circular configuration and are aligned by means of a laser mounted behind the source.

The target cell and its associated calibration system deserve special mention because precise (and absolute) pressure measurement is required in the cell for the determination of absolute cross sections. Details of the system are shown in Fig. 2. The cell is mounted in the path of the beam by means of an adjustable bellows suspension system (h) which attaches to the chamber wall (i) via a vacuum feedthrough port. Calibration curves were first obtained for each gas at each cell temperature by measuring the flow as a function of the pressure in the calibrated volume(e) through the capil-

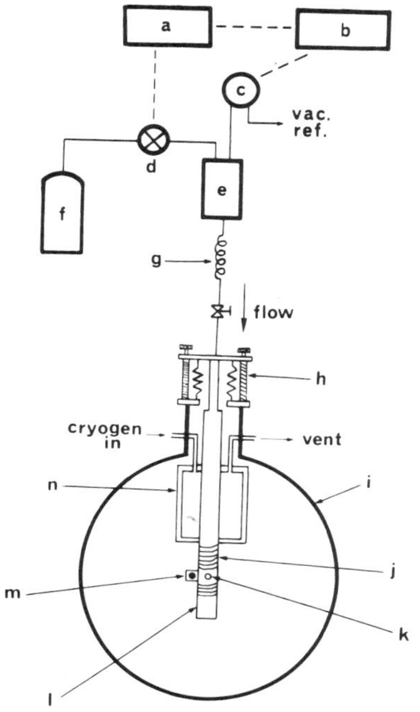

Fig. 2 Target cell and calibrated pressure control system. The components are described in the text.

lary leak(g) to the target cell(l). All of this flow must exit from the target cell orifices(k). The gas density in the cell is maintained in the free molecular flow regime therefore, the cell pressure can then be calculated from the effusion equation using the proper correction for the orifice wall effects. This procedure is repeated for many points throughout the desired cell pressure operating range, thus generating a calibration curve for each cell temperature used.

The experimental procedure is then to adjust the flow from the feed tank (f) and hence, the pressure in the calibrated volume to that value from the curve which corresponds to the desired target cell pressure. This pressure is maintained precisely by a servo valve/pressure controller system (d and a) which receives its signal from a capacitance manometer system (MKS Baratron - b and c). The factory calibrations of these manometers have been checked by Mound's secondary standards laboratory using the dead weight method and have been found to be accurate. The advantages of this method of cell pressure measurement include: 1) no effect on cell conditions by the measurement and 2) corrections due to thermal transpiration effects are unnecessary since the pressure measurement is made at relatively high pressures (2-30 torr). The temperature of the target cell is controlled in a manner similar to that used for the source. Overcooling is provided by a flow of cold nitrogen gas to a jacket (n) around the cell. Heating coils (j) in the region of the orifices are proportionally controlled by a platinum resistance thermometer (m) to maintain the selected temperature. Copper baffles were installed inside the cell to assure thermal equilibration of the gas with the wall.

The measurements are then made as follows: Temperatures T_s and T_t were established in the source and target cell, respectively. The unattenuated beam intensity was determined before introducing gas into the target cell, and then the beam attenuation was measured as a function of the cell pressure. In order to detect possible systematic errors in the system, the beam and target gases were inverted and conditions were set to yield the same relative center-of-mass velocities.

Experimental Results

Cross sections were obtained in the standard manner from Beer's Law by determining the slope, $-\ell_{eff}\, Q_{exp}$, in the equation

$$\ln {}^{I}/I_0 = -n\, \ell_{eff}\, Q_{exp}$$

where I and I_0 are the intensities of the attenuated and unattenuated beams, respectively, n is the target density and ℓ_{eff} is the effective path in the target cell. Q_{exp} is the desired "experimental" cross section. Following Beier,[7] the effective path length was ascertained to be 1.14 cm. An example of the raw experimental data is shown in Fig. 3 for an argon beam/krypton target in (a) and the inverse system in (b). Several points are worthy of note in this example. First, the corrected relative velocity cross sections $Q(v_r)$ should be equal for the same relative center-of-mass velocity upon inversion of the beam and target gases. In the example, which agrees to ~7%, the corrected $Q(v_r)$ were 417 and 388 Å^2, respectively. Secondly, the plots of $\ln {}^{I}/I_0$ vs. n should pass through the origin. We fitted the data by the method of

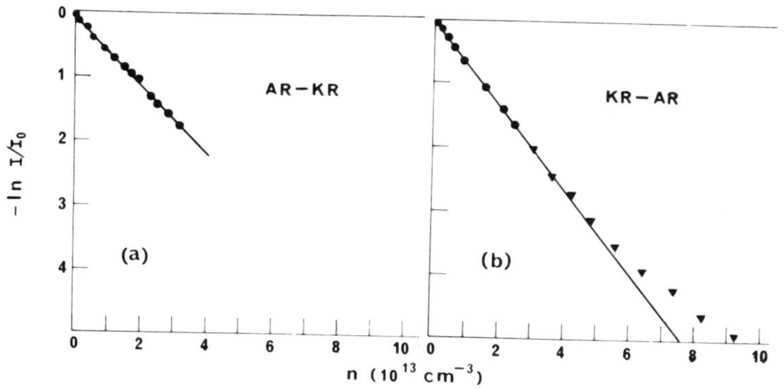

Fig. 3 a) Natural logarithm of the ratio of the attenuated beam intensity to the unattenuated beam intensity for an argon beam on krypton scattering gas as a function of the density of krypton in the target cell at $v_r = 5.29 \times 10^4$ cm sec^{-1}. The solid circles are measured intensity ratios down to an attenuation of $1/e^2$ which were used to obtain the solid straight line (2-parameter least square fit). The resultant cross sections were Q_{exp} = 466 Å^2 and $Q(v_r)$ = 425 Å^2. b) Legend same as preceding except that the experiment was inverted with a krypton beam on an argon scattering gas at $v_r = 5.30 \times 10^4$ cm sec^{-1}. The solid triangles are measured intensity ratios with attenuations below $1/e^2$ and show deviations from Beer's Law due to multiple collisions in the target cell. The resultant cross sections were Q_{exp} = 543 Å^2 and $Q(v_r)$ = 407 Å^2.

least squares to both a one-parameter function forcing the line through the origin and also to a two-parameter equation, allowing an intercept. The intercept was very close to the origin in all cases. However, a slight difference always was encountered in Q_{exp}. The results of the two methods of data reduction are shown with the experimental parameters in Table 1 as Q_{exp}. Thirdly, the density range of the data must be considered carefully. At high pressures, multiple collisions occur, and the data do not follow Beer's Law. This can be seen vividly by the "tailing-off" of the data in Fig. 3b. Careful consideration of the composite data revealed that multiple collisions were not significant down to an attenuation of approximately $1/e^2$. This was chosen as the cutoff point. After least squaring, those points more than three

Table 1 Experimental total cross sections for the argon/krypton system

v_r 10^4cm sec^{-1}	T_s/T_t (°K)	v_b (10^4cm sec^{-1})	Q_{exp}[a] (A^2)	ΔQ (A^2)	$Q(v_r)$[a] (A^2)	Q_{exp}[b] (A^2)	ΔQ (A^2)	$Q(v_r)$[b] (A^2)
3.33	75/130	2.79	634	6	551	600	5	521
3.33	75/130	2.79	654	6	569	621	6	540
3.33	75/130	2.79	657	7	572	589	5	512
3.70	100/130	3.23	554	7	503	555	7	504
3.70	100/130	3.23	575	8	522	540	7	490
3.93	75/300	2.79	668	8	498	629	7	468
3.93	75/300	2.79	672	8	501	640	8	477
3.97	120/130	3.53	555	9	512	516	8	476
3.97	120/130	3.53	540	8	498	527	8	486
4.25	100/300	3.23	625	8	497	600	8	478
4.25	100/300	3.23	611	8	486	582	7	463
4.48	120/300	3.53	584	8	481	545	7	449
4.48	120/300	3.53	570	8	470	552	7	455
[c]4.68	120/300	2.44	833	31	465	826	30	461
[c]4.68	120/300	2.44	831	30	463	830	30	463
5.14	180/300	4.33	482	9	425	454	8	400
5.14	180/300	4.33	484	9	427	462	9	408
5.29	196/300	4.52	497	11	445	470	10	420
5.29	196/300	4.52	491	11	440	466	10	417
5.29	196/300	4.52	476	10	426	457	10	409
[c]5.30	244/300	3.48	564	31	404	542	29	388
5.74	244/300	5.04	464	9	425	432	7	394
5.74	244/300	5.04	450	8	411	438	8	400
6.83	375/300	6.25	396	11	379	366	10	350
6.83	375/300	6.25	379	10	362	361	9	344

a Obtained from 1-parameter fit.

b Obtained from 2-parameter fit.

c Krypton beam; argon target gas.

standard deviations from the fitted curve were discarded and a final curve was fitted to the remaining points. At low pressures in the target cell, errors can occur when the background pressure outside of the cell is nonnegligible, thereby causing a change in the effective path length. Furthermore, our calibration method was least accurate at very low pressures. Data were taken above these limiting low pressures.

The values of Q_{exp} are too small because of the finite angle subtended by the detector from the target. A correction ΔQ is obtained using the methods derived by von Busch[8] for cylindrical beam geometry. The small angle differential cross section used in this calculation assumes an attractive potential proportional to the inverse 6.7 power. This modification to the usual inverse 6 power attractive potential appears to more accurately represent the true interaction.[3] An effective cross section Q_{eff} was obtained where $Q_{eff} = Q_{exp} + \Delta Q$. The correction ΔQ ranged from approximately 1% to 5% of our effective cross sections.

The effect of velocity distributions in the scattered and scattering gases must be taken into account and applied as a correction to Q_{eff}. This was done by use of the functions tabulated by Berkling et al.[9] The target gas possessed a Maxwellian velocity distribution. Because of the relatively high Mach number of the beam (>10) and the resultant peaking of the velocity distribution we assumed the beam velocity to be monochromatic and that an infinite beam speed ratio existed for the calculation of the beam velocity. The Berkling $F_{ao}(6,X)$ function was used to obtain the average relative velocity cross section in the equation

$$Q(v_r) = [Q_{eff}/F_{ao}(6,X)](v_b/v_r)^{2/5}$$

where $v_r = (v_b^2 + v_t^2)^{\frac{1}{2}}$ and $X = v_b/v_{pt}$. The velocities are the most probable beam velocity, v_b, the average target velocity, v_t, and the most probable target velocity, v_{pt}. The individual experiments at each velocity were averaged, and these results are shown in Fig. 4.

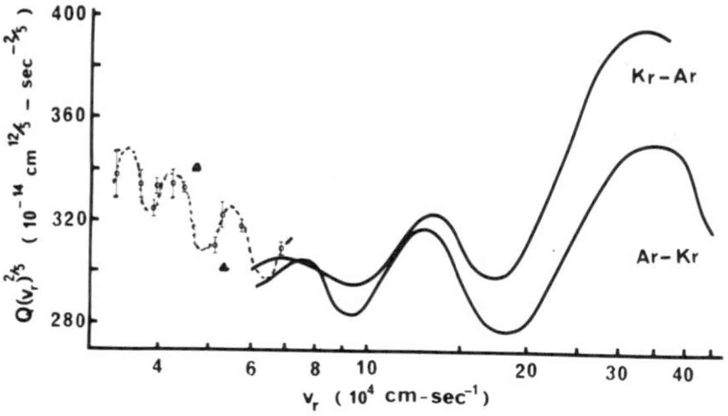

Fig. 4 Total cross sections of Ar-Kr with the primary velocity dependence removed vs. the relative velocity. The open circles are the averages of the individual experiments using an Ar beam on a Kr target at each velocity as given in Table 1. The solid triangle is the single experiment using a Kr beam on an Ar target for comparison. The dashed line is the estimated oscillatory behavior of the present work. The solid curves are the smoothed data for Ar-Kr and Kr-Ar given in Ref. 3.

The results of the 1- and 2-parameter fits were essentially the same within the experimental uncertainty. We have chosen to use the 2-parameter fit because of generally smaller average deviations. The slight shift of the line away from the origin is possibly because of a small systematic error, probably the lack of complete pressure equilibration between the individual points.

The measured cross section at each velocity was averaged and multiplied by $v_r^{2/5}$ to remove the primary velocity dependence. This quantity $Q(v_r)v_r^{2/5}$ is shown in Fig. 4 with the smoothed data of Bredewout.[3] The error bars represent the reproducibility of the data which ranges from less than 1% to 4.5%. The absolute cross sections are subject to an overall estimated uncertainty of $\pm 5\%$ in Q_{exp} based on consideration of possible systematic errors which include the absolute value of the target gas density, scattering path length, and signal detection uncertainties. The undulating dotted curve was drawn through the present Ar-Kr data to show five extrema in addition to those revealed by the previous work. The locations of the extrema were estimated from this curve and are given in Table 2 along with those given in Ref. 3.

The extrema locations may be used to obtain an estimate of the product $\varepsilon\sigma$ as mentioned previously. Specifically, the locations are given by[2]

$$N-3/8 = 0.3012\ (\varepsilon\sigma/hv_N)[1-0.354(\varepsilon/\mu)^{1/2}(1/v_N)]$$

where v_N is the velocity of the Nth extrema. Customarily, the maxima are represented by integers and the minima by ½ integers. The extrema positions are shown in Fig. 5 for both the data of Bredewout and the present work. A least square quadratic fit was made to the points, and the slope of this

Table 2 Relative velocity locations of experimental glory extrema for AR-KR[a]

N	1	1½	2	2½	3	3½	4	4½	5	5½
Present work	...	...	...	...	...	6.4	5.4	4.8	4.2	3.8
Ref. 3	37.4	18.6	13.05	9.67	7.55	...	...	...	...	...

[a] Velocities are in units of (10^4 cm-sec^{-1}).

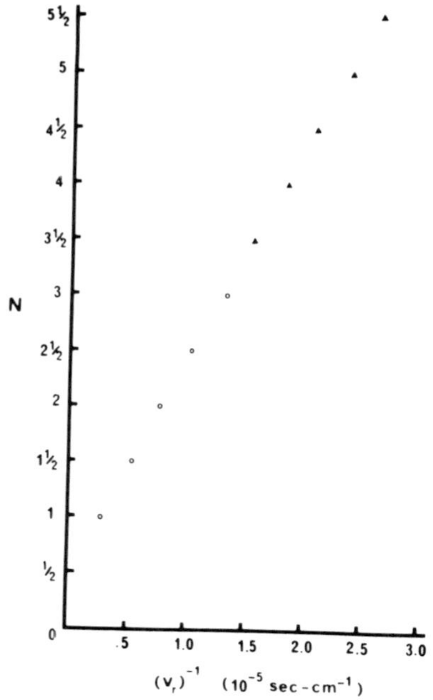

Fig. 5 Location of the glory extrema as a function of the inverse relative velocity. The triangles are the extrema as determined graphically from the dashed curve in Fig. 4. The circles are the extrema as given for Ar-Kr in Ref. 3.

curve was found at the intercept. Taking the derivative of the preceding equation

$$\partial N/\partial (1/v_N) = 0.3012\ (\varepsilon\sigma/\hbar)[1-0.708(\varepsilon/\mu)^{\frac{1}{2}}(1/v_N)]$$

and setting $1/v_N = 0$, we found the product $\varepsilon\sigma$ to be approximately 7.86×10^{-22} erg-cm.

Comparison with Theory

The theoretical cross sections were calculated by the equation

$$Q = [4\pi\hbar^2/(\mu v_r)^2] \sum_{\ell}(2\ell+1)\ (\sin \eta_\ell)^2$$

where μ is the reduced mass of the system and η_ℓ are the phase shifts. The phase shifts were evaluated by the WKB method as given by Marchi.[10] Phase shifts were calculated for several hundred ℓ values at each velocity until the magnitude dropped below 4 mrad.

Table 3 Selected intermolecular potentials for AR-KR

1. Barker[a] $V(r) = \varepsilon V(r)^*$

$$V(r)^* = \exp[a(1-X)]\sum_{i=0}^{5} A_i(x-1)^i - \sum_{j=0}^{2} c_{2j+6}/(\delta + x^{2j+6})$$

c_6	= 1.08756	A_0	= 0.25909
c_8	= 0.17017	A_1	= 4.62939
c_{10}	= 0.02394	A_2	= 8.7656
a	= 12.5	A_3	= 25.2696
δ	= 0.01	A_4	= 102.0195
		A_5	= 113.25

$\varepsilon/k = 167.04°K$; $r_m = 3.8859$ Å ($\sigma = 3.4712$);

2. Modified Buckingham (Exp-6)

$$V(r) = \frac{\varepsilon}{1-6/\alpha}\left[\frac{6}{\alpha}\exp[\alpha(1-r/r_m)] - (r/r_m)^6\right]$$

$\varepsilon/k = 166.92$; $r_m = 3.7955$ Å ($\sigma = 3.3846$); $\alpha = 14.46$

3. Lennard-Jones (12-6)

$$V(r) = 4\varepsilon\left[\left(\frac{\sigma}{r}\right)^{12} - \left(\frac{\sigma}{r}\right)^{6}\right]$$

$\varepsilon/k = 160.95$; $\sigma = 3.4635$ Å

a See Ref. 3.

The theoretical cross sections were evaluated with the above procedure using three interatomic potentials for Ar-Kr. The potentials and parameters used are given in Table 3. Parameters for the Modified Buckingham (exp-6) and Lennard-Jones (12-6) potentials were obtained from potentials given for the pure gases, which reproduced the thermal diffusion factors for argon and krypton within 10%.[11] It was reasoned that thermal diffusion is a property that is quite sensitive to the potential, and thus should yield a reasonably accurate value for the well depth. The potential parameters for Ar-Kr were then obtained from those for the pure gases by application of the usual combination rules. The parameter product $\varepsilon\sigma$ thus obtained from thermal diffusion for the Exp-6 and L-J (12-6) potentials were 7.80×10^{-22} and 7.70×10^{-22} erg-cm, respectively. The Barker potential was that form suggested to

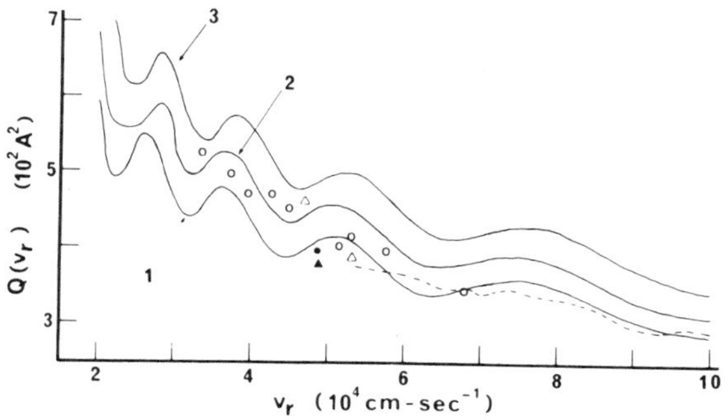

Fig. 6 Comparison of the experimental total cross section with theoretical cross sections calculated using three Ar-Kr potentials. The open circles are the present Ar-Kr results, the open triangle is the single Kr-Ar experiment in this work, the closed triangle is the work of Ref. 4, and the dashed curve is the smoothed Ar-Kr data of Ref. 3. The theoretical curves are the cross sections obtained using: 1) Barker potential; 2) Modified Buckingham (exp-6) potential; and 3) Lennard-Jones (12-6) potential.

Bredewout by Barker using his parameters for the Ar-Kr system. The $\varepsilon\sigma$ product in this potential was 8.00×10^{-22} erg-cm. The theoretical curves are shown with the experimental data in Fig. 6.

Conclusions

Experimental absolute total cross section measurements have been presented for the Ar-Kr scattering system at low relative velocities. The measured values agree with theoretical cross sections derived from potentials that were determined by experimentally independent methods. Locations of the glory extrema observed in the present work, along with those given by Bredewout, have determined the experimental value for $\varepsilon\sigma$ to be 7.86×10^{-22} erg-cm. This is in close agreement with the value of 7.80×10^{-22} erg-cm for the Exp-6 potential found experimentally by thermal diffusion measurements.

The results of the present work overlap the data of Rothe, et al. and Bredewout and are in good agreement in the higher relative velocity range, both with the previous data and with the conclusion that the Barker potential accurately describes the experimental data in this region. However, at relative velocities below approximately 4.5 x 10^4 cm/sec our results indicate larger cross sections for argon-krypton than predicted by the Barker potential and are better represented by the Exp-6 potential with the cited parameters.

In conclusion, we can state that the present work, which is consistent with both theory and earlier experiments, establishes a more accurate understanding of the Ar-Kr interaction.

References

[1]Bernstein, R. B. and Kramer, K. H., "Comparison and Appraisal of Approximation Formulas for Total Elastic Molecular Scattering Cross Sections," Journal of Chemical Physics, Vol. 38, May 1963, pp. 2507-2511.

[2]Rothe, E. W., Rol, P. K., and Bernstein, R. B., "Interaction Potentials From the Velocity Dependence of Total Atom-Atom Scattering Cross Sections," Physical Review, Vol. 130, June 1963, pp. 2333-2338.

[3]Bredewout, J. W., "Glory Undulations in the Total Cross Section for Scattering of Heavy Rare Gas Atoms," Thesis, Rijksuniversiteit Leiden (Netherlands), Sept. 1973. (Extensive references to related work are given here.)

[4]Rothe, E. W. and Neynaber, R. H., "Measurements of Absolute Total Cross Sections for Rare-Gas Scattering," Journal of Chemical Physics, Vol. 43, Dec. 1965, pp. 4177-4179.

[5]Barker, J. A., Fisher, R. A., Watts, R. O., "Liquid Argon: Monte Carlo and Molecular Dynamics Calculations," Molecular Physics, Vol. 21, 1971, p. 657.

[6]Golomb, D., Good, R. E., and Brown, R. F., "Dimers and Clusters in Free Jets of Argon and Nitric Oxide," Journal of Chemical Physics, Vol. 52, Feb. 1970, pp. 1545-51.

[7]Beier, H. J., "Scattering From Helium Atomic Beams in Rare Gases," Zeitschrift für Physik, Vol. 196, 1966, pp. 185-202.

[8]von Busch, F., "Der Einfluss des Winkelauflösungsvermögens auf die Messung totaler atomarer Streuquerschnitte," Zeitschrift für Physik, Vol. 193, 1966, pp. 412-25.

[9]Berkling, K., Helbing, R., Kramer, K., Pauly, H., Schlier,C. and Toschek, P., "Effective Collision Cross Sections in Scattering Experiments," Zeitschrift für Physik, Vol. 166, 1962, pp. 406-28.

[10]Marchi, R. P., "Intermolecular Potentials and Their Relation to Molecular Beam Scattering Data," Thesis, Purdue University, Jan. 1963.

[11]Taylor, W. L., and Weissman, S., "Isotopic Thermal Diffusion Factors for Argon and Krypton," Journal of Chemical Physics, Vol. 59, Aug. 1973, pp. 1190-95.

[illegible] Sommerfeld [illegible]

[illegible] Schulz [illegible] Cross Sections [illegible] Experiments [illegible]

[illegible] and Their Relation to Molecular Beam Scattering [illegible]

[illegible] and Mason, E. [illegible] Argon and [illegible] Journal of Chemical Physics [illegible]

EXPERIMENTAL COMPONENTS OF A CROSSED MOLECULAR BEAM APPARATUS DESIGNED TO STUDY THE REACTION $O + H_2O = OH + OH$

P. Poulsen,* D. R. Miller,+ and M. A. Fineman≠

University of California at San Diego, La Jolla, Calif.

Abstract

The components of a crossed molecular beam apparatus to study the reaction $O + H_2O = OH + OH$ have been completed and tested. With ground state $O(^3P)$ the reaction is endoergic by ~ 0.75 eV, requiring accelerated beams of $O(^3P)$ and H_2O. A microwave discharge source utilizes helium-oxygen mixtures (95% He) at high pressures (50 Torr) to give good dissociation (70%) and fast atomic oxygen (~ 0.3 eV). The effects of microwave power and source composition on conversion are reported. The required water beam source is an electron beam heated iridium nozzle operated with a 95% He/5% H_2O mixture at 1750°K and ~ 2200 Torr to provide H_2O beam energies ~ 1.25 eV. Mass spectrometer detection of the OH radical is accomplished by replacing the conventional ionization source on a quadrupole mass spectrometer by a barium chemi-ionization source. The ionization reaction $OH + Ba \rightarrow BaOH^+ + e$ is selective at thermal energies, not occurring with O, O_2, or H_2O.

Introduction

The reaction between atomic oxygen and water vapor which yields the reactive hydroxyl radical is important in H_2/O_2

Presented as Paper 54a at the 10th International Symposium on Rarefied Gas Dynamics, July 19-23, 1976, Aspen, Colo. Research supported by NSF Grant ENG71-02434A05.

*Ph.D. Candidate, Department of Applied Mechanics and Engineering Sciences.

+Associate Professor, Department of Applied Mechanics and Engineering Sciences.

≠Professor of Physics, Department of Astronomy and Physics, Lycoming College, Williamsport, Pa.

combustion kinetics, in atmospheric physics, and in jet or rocket exhaust plume interactions with the atmosphere.

The rates for the ground state (O^3P) and first excited state (O^1D) have been reported[1,2] to be $k(O^3P) \sim 6.8 \times 10^{13}$ exp $\{-9240/T\}$ cm^3 $mole^{-1}$ sec^{-1} and $k(O^1D) \sim 1.8 \times 10^{13}$ cm^3 $mole^{-1}$ sec^{-1}. The reaction with ground state (O^3P) is endoergic by ~ 0.75 eV, whereas the reaction with the first excited state is exoergic by ~ 1.25 eV. The pre-exponential values are about the same, and the cross sections are estimated to be about 2Å^2. We have initiated a crossed molecular beam program to study these reactions, and this paper describes the components that we have completed and tested to study the $O(^3P)$ reaction. In the sections that follow, we describe the major components, the oxygen source that provides atomic oxygen beams with energies up to 0.3 eV, the water beam source that accelerates H_2O up to 1.25 eV, and the special OH detector that provides the necessary signal to noise features to detect OH in this reaction.

Oxygen Atom Source

A microwave discharge is used to dissociate oxygen at high pressures, up to 60 Torr, in a quartz nozzle. Seeding the source gas with helium, for acceleration of the oxygen in the freejet expansion from the quartz nozzle, provides atomic oxygen with energies up to 0.3 eV. The microwave power is supplied by a Raytheon CMD-10 diatherm unit, operating at 2450 MHz. The power is coupled to the nozzle by a tunable Evanston cavity, commercially available from The Opthos Instrument Company. The cavity must be tuned in the vacuum, and small dc motors (Micro-Mo Electronics, Inc.) are used to adjust the tuning rods. The quartz nozzle that runs through the cavity is 0.8 cm o.d., with an exit orifice of 0.037 cm. The nozzle exit is located 0.8 cm outside the cavity. We have noted this latter distance, because apparently it is quite critical for maximizing the degree of dissociation. The beam properties of the discharge source are not defined completely, since several excited states of O and O_2 are produced. However, calculations of collisional quenching[3] in the freejet source indicate that, except at low source pressures, only $O_2(^1\Delta g)$ should survive in appreciable quantity. The detection of the excited species is discussed further in the following. Standard time-of-flight measurements were used to monitor the beam energies.

The atomic oxygen concentration in the beam was determined by monitoring the 16 and 32 atomic mass peaks with an Extranuclear quadrupole mass spectrometer, with and without the discharge on, in the manner described previously[3]; a numerical error in the original treatment has been corrected.[4] The reproducibility of these dissociation data was about 10%.

Figure 1 shows the dissociation of pure O_2 as a function of power (watts) and source pressure (Torr). Unlike our previous results with radio frequency discharges,[3] the dissociation falls off rapidly with pressure. It is desirable to operate at high pressures in order to have the advantages of a freejet expansion, but this requires more power, and perhaps a loss of atom intensity.

Figure 2 shows the dissociation as a function of composition, helium seeding, and pressure. The extremely high degree of dissociation, 70% is very encouraging, since it is achieved at high enough source pressures to have reasonable freejet expansion effects, especially the acceleration of the oxygen by the helium. As noted previously,[3] the effect of the helium is not only to increase the dissociation rate, but also to decrease the recombination rate by lowering the partial pressure of oxygen. The effects of adding of a few

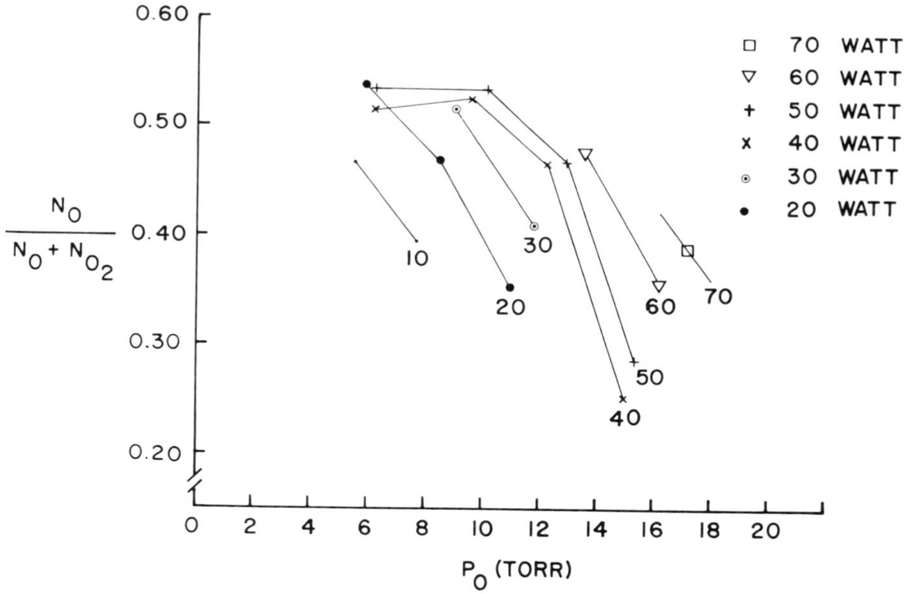

Fig. 1 The concentration of atomic oxygen vs microwave power (watts) and source pressure (Torr) for pure oxygen.

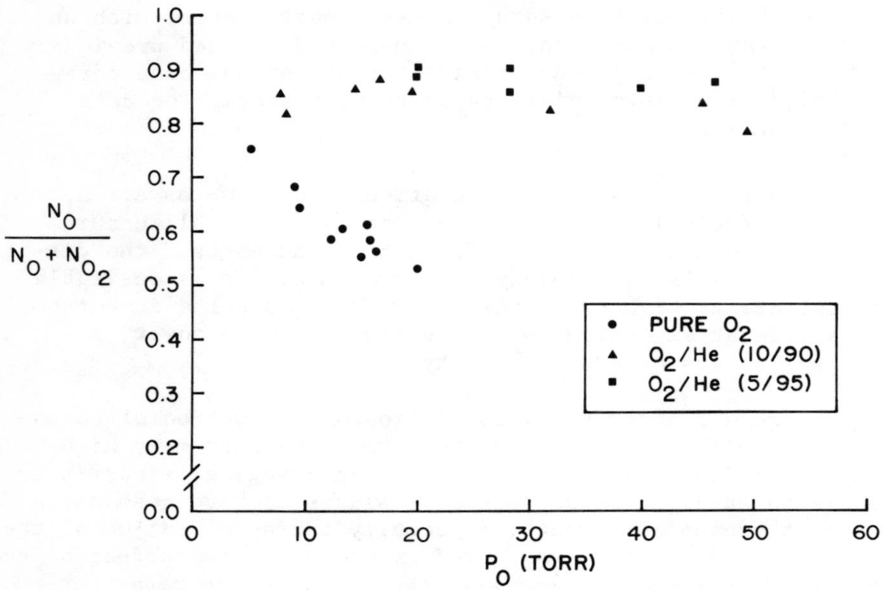

Fig. 2 The concentration of atomic oxygen vs source pressure (Torr) and molar ratios of He/O_2 mixtures.

percent H_2, and also H_2O, to the oxygen were examined, but there was negligible change in the dissociation at these pressures.

H_2O Source

The activation energy for the $O(^3P)$ reaction is believed to be about 0.75 eV. With our 0.3-eV oxygen source, this still requires an H_2O beam with an energy above 1 eV. Our calculations suggest that this may be accomplished by a freejet expansion of a 10% H_2O in helium mixture, heated to 1700°K. At temperatures much above this value, thermal dissociation of H_2O becomes a problem in our crossed-beam study.

We have developed the source shown in Fig. 3 to heat water-helium mixtures at high pressures (2500 Torr). The nozzle is made from a capped iridium tube (Engelhard Industries), 0.3 cm o.d., with a 0.025-cm wall, and a 0.008-cm orifice in the end. Although resistive heating is straightforward and desirable, the tube wall thickness caused a requirement of more than 150 A to reach 1700°K. Therefore, electron beam heating is used at 1000 eV and

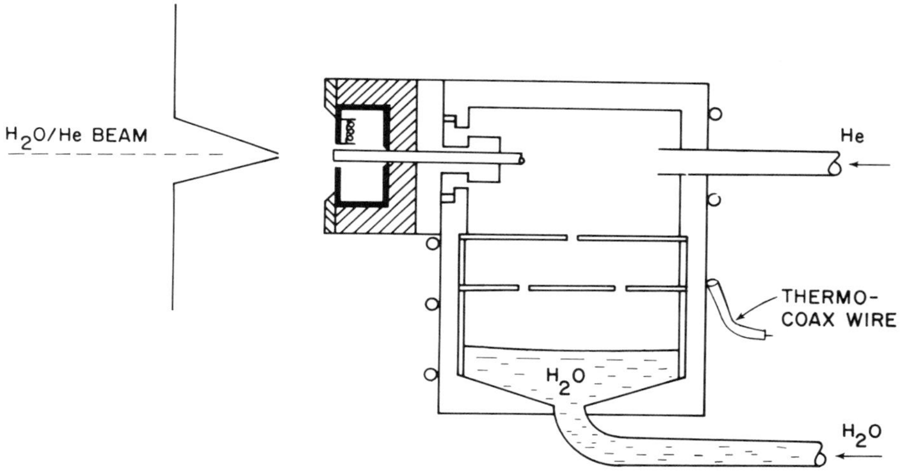

Fig. 3 The H_2O free-jet beam source.

40 mA to achieve the source temperature. The electron beam section is shielded carefully to trap as much of the electron current as possible. We have had few breakdown problems with the helium-seeded mixtures. The water partial pressure (up to 200 Torr) is maintained by heating a water reservoir mounted next to the source in the vacuum. Nozzle temperatures are read by an optical pyrometer and monitored during a run by a chromel-alumel thermocouple located $\sim$ 1.5 cm from the heated end of the nozzle. The iridium sources have held up extremely well, except for occasional blockage of the orifice due to particulate matter, the source of which confounds us.

Detector for OH

Detection of OH by a mass spectrometric technique in a vacuum system with a significant H_2O background pressure is difficult because of fragmentation of H_2O into OH^+ by electron bombardment. In this case, we have an H_2O beam with which to contend. Although it is possible to operate the ionizer of the mass spectrometer below the threshold of the fragmentation reaction ($\lesssim$ 18 eV), the loss in sensitivity is very great. We have developed a barium chemi-ionization source which can be used in conjunction with our quadrupole mass filter which is selective to OH.

Cohen[5] and his colleagues have published a series of papers on the chemi-ionization processes that occur when the group IIA metals react with O, OH, H_2O, and O_2. It appears the reaction $Ba + OH \rightarrow BaOH^+ + e$ is very selective, not occurring with O, O_2, or H_2O at thermal energies, and that it has a large cross-section, of order 5Å^2. We have since observed that excited O(^{1}D) also chemi-ionizes with Ba (see the following). We have utilized the results of Cohen to design a barium ionization source that mounts directly onto our Extranuclear quadrupole. An SSR model 1110 photon counter is used to count the arrival of ions at the multiplier.

Figure 4 shows the design that is employed. It consists of a heated oven (~ 900^{0}C), an inner radiation shield, and an outer radiation shield, which also serves to collect charged particles and to collimate the beam. Our original design was unsatisfactory because we did not carefully consider the problem of charged particles, both positive and negative, which come from the barium oven, and which contribute to background noise. During a typical run, the ionization region of the mass spectrometer is set at ~ 40 V, while $V_1 = V_2 = 0$ V, and $V_3 = -5$ V (see Fig. 4). We have tested the detector with beams of O(^{3}P), O_2, H_2O, and inert gases, and we substantiate the earlier results of Cohen, which state that chemi-ionization does not occur. We did produce OH by discharging H_2O, and observed the expected $BaOH^+$ signal. With the barium oven in operation and the quadrupole at full power, our background noise count is less than 0.05 counts/sec. With this low noise capability and selective sensitivity to OH, this is an excellent molecular beam detector for OH. Once we understand all of the chemistry involved with excited states, so that we can be confident about the conditions under which only $BaOH^+$ is produced, we intend to eliminate the quadrupole and simply

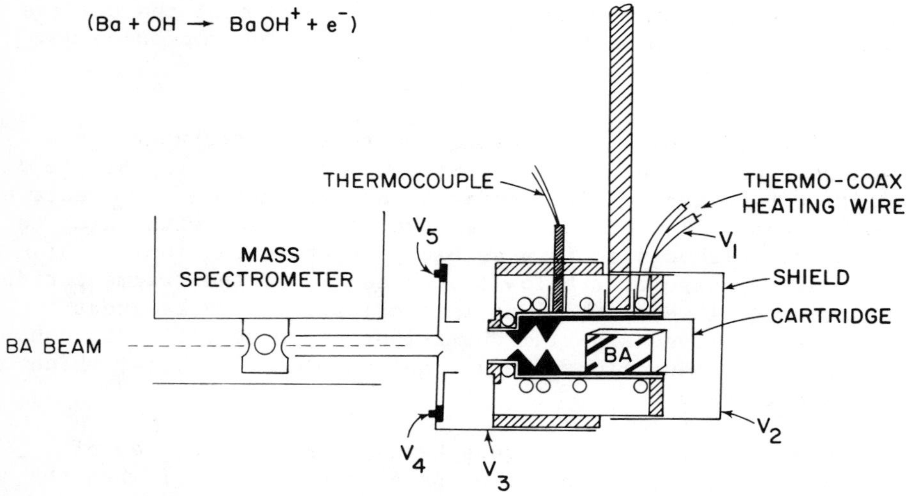

Fig. 4 The Ba atom chemi-ionization source.

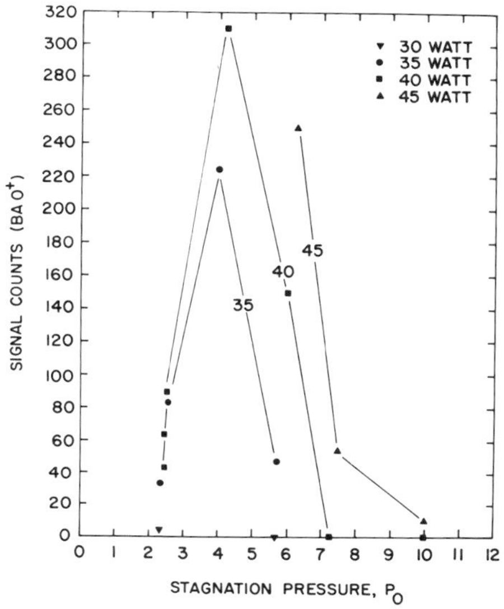

Fig. 5 Relative amounts of excited species in a beam formed from the discharge of pure O_2 vs microwave power and source stagnation pressure.

count with a multiplier. The geometry then can be optimized and the sensitivity greatly enhanced.

As mentioned previously, we have observed that excited states of atomic oxygen from the discharge, most likely $O(^1D)$, also chemi-ionize with barium to give BaO^+. Figure 5 shows the BaO^+ signal which we take as a measure of the relative amount of excited species in the beam as a function of microwave source pressure and power. Note that the amounts are in counts per second so that the actual concentrations are quite low. These results are in qualitative agreement with previous calculations on amounts of $O(^1D)$ remaining in such beams.[3] We have not been able to calibrate the detector because we do not yet have standard sources of either $O(^1D)$ or OH. As Fig. 5 suggests, we will operate the atomic oxygen source above pressures of 10 Torr in order to reduce the presence of excited states.

Crossed Molecular Beam System

The crossed molecular beam system is shown in Fig. 6. There are two source chambers individually pumped by a 4- and 32-in. diffusion pumps. The beams cross in a third chamber pumped by a 20-in. diffusion pump. The main features of the vacuum system have been described previously.[6]

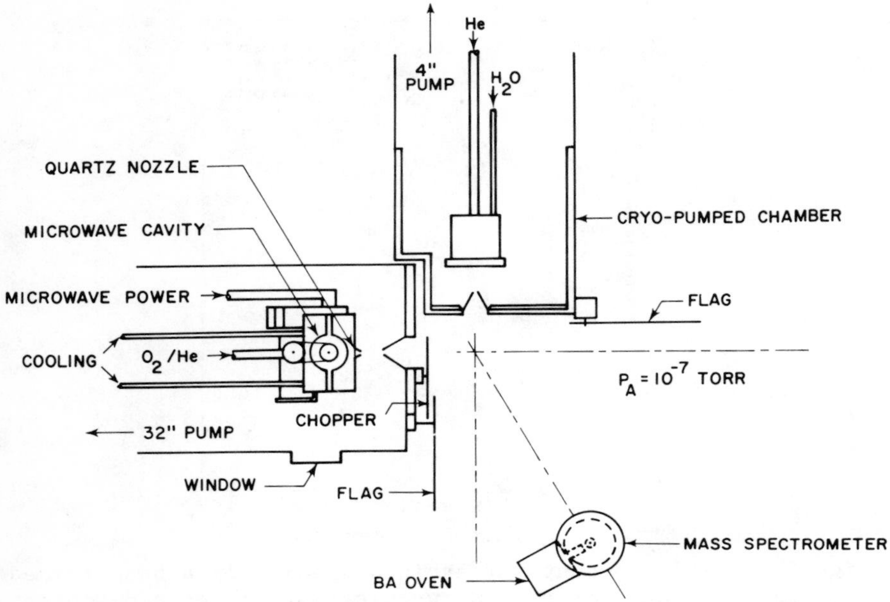

Fig. 6 The crossed molecular beam system.

Portions of the water beam source chamber are capable of being cooled by liquid nitrogen for additional cryogenic pumping of water. We have chosen initially to modulate the atomic oxygen beam in order to reduce the problem of detecting any OH coming directly from the H_2O source, either due to thermal dissociation or interaction with the electron beam heater. As mentioned previously, the detector, other than the barium chemi-ionization source, is a commercial Extranuclear unit, operated in an ion counting mode. The background noise, with all systems in operation, typically is 0.1 cps.

References

[1]Albers, E., Hoyermann, K., Wagner, H., and Wolfrum, J., "Absolute Measurements of Rate Coefficients for the Reactions of H and O Atoms with H_2O_2 and H_2O," Thirteenth Symposium (International) on Combustion," 1971, pp. 81-88.

[2]Biedenkaap, D., Hartshorn, L., and Bair, E., "The $O(^1D) + H_2O$ Reaction," Chemical Physics Letters, Vol. 5, May 1970, pp. 379-380.

[3]Miller, D. and Patch, D., "Design and Analysis of a High Intensity Fast Oxygen Atom Source," The Review of Scientific Instruments, Vol. 40, December 1969, pp. 1566-1569; also Patch, D., "Application of Free Jet Sources to Reactive Crossed Molecular Beam Experiments," Ph.D. Thesis, University of California, San Diego, 1973.

[4]Poulsen, P., "Development and Characterization of Free Jet Sources and an OH Detector, with Application to the $O + H_2O \rightarrow OH + OH$ Reaction," Ph.D. Thesis, University of California, San Diego, 1976.

[5]Cohen, R., Majeres, P., and Raloff, J., "Chemi-Ionization Reactions of Ba, Sr, Ca with OH," Chemical Physics Letters, Vol. 31, February 1975, pp. 176-180.

[6]Miller, D., "Development of a Supersonic Nozzle Molecular Beam Facility," Institute for Pure and Applied Physical Sciences, University of California, San Diego, Rept. No. 67/68-212, April 1968.

CHAPTER XI—CONDENSATION IN EXPANSIONS

THE SCALING LAW FOR HOMOGENEOUS CONDENSATION IN CO_2 FREE JETS

N. V. Gaisky, Yu. S. Kusner, A. K. Rebrov,
B. Ye. Semyachkin, P. A. Skovorodko,
and A. A. Vostrikov

Institute of Thermophysics, Siberian Department
of the USSR Academy of Sciences, Novosibirsk

Abstract

Experimental studies of CO_2 condensation in a free jet behind a sonic nozzle were performed with the molecular beam system of the Institute of Thermophysics. It is shown that the dependence of the minimum of the molecular beam intensity on stagnation pressure is essentially determined by the process of homogeneous condensation in the jet, and its location is correlated by

$$P_o T_o^{-4.45}\Big|_{d_*} = \text{const} \qquad \text{and} \qquad P_o d_*^{0.6}\Big|_{T_o} = \text{const}$$

The gasdynamics equations and those describing the process of homogeneous condensation in a jet according to classical nucleation theory were solved for comparison to the experimental results. The molecular beam at high pressures (behind the intensity minimum) is shown to consist of clusters throughout. Computed parameters of the flow were used for calculating molecular beam intensity behind the skimmer, and there is qualitative agreement between the experimental and theoretical intensity curves.

Introduction

It is known that physical processes in the free jet are completely determined by three parameters: stagnation pressure

---

Presented as Paper 143 at the 10th International Symposium on Rarefied Gas Dynamics, July 19-23, 1976, Aspen, Colo.

P_o, stagnation temperature T_o, and sonic nozzle diameter d_*. In this paper the scaling law for condensation in a free jet of CO_2 downstream of a sonic nozzle at constant T_o is considered.

In a number of papers devoted to the investigation of the homogeneous condensation in jets, it is shown experimentally that

$$P_o d_*^{\beta}\big|_{T_o} = \text{const} \ (0.5 < \beta < 1)$$

expresses the conditions necessary for the obtaining of dimer ion signal maximum[1], the same dimer mole fraction[2], and the same cluster mean size[3].

In classical nucleation theory, the limiting process is the critical nucleus formation by means of binary collisions. Physical processes in a jet determined by the binary collisions, such as translational-translational, translational-rotational, and translational-vibrational energy exchange, are known to be generalized by the relation

$$P_o d_*\big|_{T_o} = \text{const}$$

This corresponds with the conservation of the total number of binary collisions in a jet. Because the total number of termolecular collisions in a jet is proportional to $P_o d_*^{0.5}\big|_{T_o}$, the authors of Refs. 1-3 supposed that the correlation of experimental data by

$$P_o d_*^{\beta}\big|_{T_o} = \text{const} \ (\beta < 1)$$

is connected with the limiting condensation stage determined by the termolecular dimer formation (apparently by analogy with the above-mentioned processes determined by the binary collisions). This process is not considered in the classical nucleation theory.

The goal of this work is to show that the generalization of the experimental data by such relations can be explained in the context of classical nucleation theory without considering termolecular collisions.

Experiments

Experimental study of condensation in a CO_2 free jet downstream of a sonic nozzle was performed with a molecular beam generator having three sections with cryogenic pumping. The intensity and composition of the skimmed molecular beam were measured. The description of the generator and discussion on questions connected with the molecular beam formation are contained in Ref. 4.

A closed ionization gauge with an inlet orifice diameter of 3 mm was used to determine intensity, and a monopole mass spectrometer with an inlet diaphragm diameter of 2.5 mm was used to determine density. Measuring procedure and probe calibration are described in Ref. 5. The carbon dioxide used was vacuum cleaned, and the impurities were less than 0.01 percent.

Figure 1 shows the typical dependency of total intensity $\mathcal{J}$ and density probe ion signal on P_o. The intensity and density graphs are superposed at low P_o when there is no condensation. The experimental conditions were as follows: $d_* = 5.0$ mm; $T_o = 234$ K; nozzle-skimmer distance, $\bar{x} = x/d_* = 50$; skimmer diameter, $d_s = 1.9$ mm; collimator distance, 3 mm; skimmer-detector distance, 890 mm.

With decreasing T_o and increasing d_*, the minima of intensity and monomer density, as well as n-mer (n = 2,3,4) maxima, displace toward lower P_o.

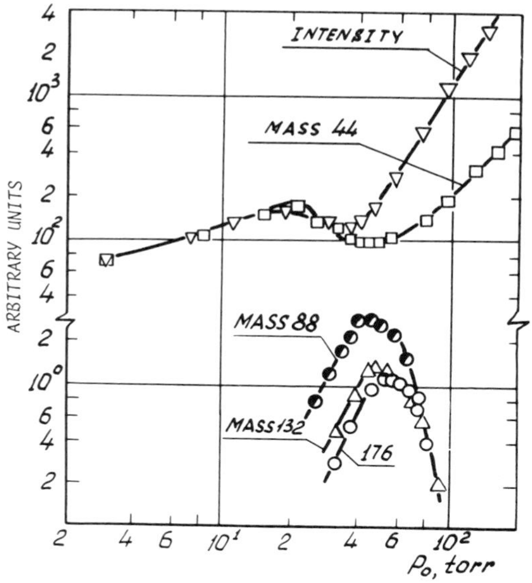

Fig. 1 Experimental intensity and ion signal of density probe as a function of stagnation pressure P_o.

The results were not changed qualitatively with changing of T_o, d_*, or d_s. The experiments showed that the position of minimum $\mathfrak{J} = \mathfrak{J}(P_o)$ curves depends weakly on nozzle-skimmer distance even in the region of considerable jet-skimmer interaction. It is noted that the conditions corresponding to the intensity minima are correlated by

$$P_o T_o^{-4.45}\Big|_{d_*} = \text{const} \quad \text{and} \quad P_o d_*^{0.6}\Big|_{T_o} = \text{const}$$

with T_o ranging from 234 to 300 K and d_* from 1.9 to 5.0 mm.

The experimental results suggest that the position of the intensity minimum is essentially determined by condensation.

The Calculation of Flow with Condensation

The method of computation of the free-jet parameters in the presence of homogeneous condensation is described in Ref. 6. It is based on classical nucleation theory[7] and equations of inviscid, nonconductive gas dynamics. The calculations were based on the three lowest moments of the drop-size distribution function. The condensate is assumed to be liquid. The basic system is comprised of the equations of

$$\text{continuity,} \quad \rho u a = \text{const} \tag{1}$$

$$\text{motion,} \quad \rho u du + dp = 0 \tag{2}$$

$$\text{energy,} \quad \frac{\gamma}{\gamma-1} RT + u^2/2 - Lq = \frac{\gamma}{\gamma-1} RT_o \tag{3}$$

$$\text{state,} \quad p = \rho RT\,(1-q) \tag{4}$$

The nucleation rate is

$$I = \left(\frac{p}{kT}\right)^2 \frac{1}{\rho_\ell} \left(\frac{2\sigma\mu}{\pi Na}\right)^{1/2} \exp\left(-\frac{4\pi r_*^2\sigma}{3kT}\right) \tag{5}$$

The critical nucleus radius is

$$r_* = \frac{2\sigma}{\rho_\ell\, RT\, \ell n\, [p/p_\infty\,(T)]} \tag{6}$$

The droplet growth date is

$$\frac{dr}{dt} = \frac{\alpha p}{\rho_\ell} \left(\frac{1}{\sqrt{2\pi RT}} - \frac{1}{\sqrt{2\pi RT_\ell}}\right) \tag{7}$$

Here ρu = density and velocity of gas-condensate mixture, p = pressure of gas, T = gas temperature, q = condensate mass fraction, a = current tube area, L = latent heat, γ = specific heat ratio for the gas, σ = surface tension, Na = Avogadro number, μ = molecular weight, ρ_ℓ = liquid density, p_∞ = saturated vapor pressure over flat surface, α = condensation coefficient, and r = droplet radius.

It was assumed that the droplet temperature T_ℓ is equal to saturation temperature corresponding to the gas pressure. Surface tension was determined by the Tolman relation

$$\sigma(r) = \sigma_\infty/(1 + 2\delta/r) \qquad (8)$$

The following relations for CO_2 thermodynamic properties were used[6,8]

$$\sigma_\infty[\text{dyn/cm}] = 54.96 - 0.184\ T$$

$$p_\infty[\text{atm}] = \exp\ (11.445 - 2126.9/T)$$

$$\rho_\ell[\text{g/cm}^3] = \begin{cases} -\ 1.0333 \times 10^{-3}\ T + 1.4387, & T < 173 \\ -\ 1.6250\times10^{-5}T^2+4.45\times10^{-3}T+0.97688, & T > 173 \end{cases}$$

with temperature in K. It was assumed that $\gamma = 1.4$, $\alpha = 1$, $L = 3.473 \times 10^9$ erg/g, and $2\delta = 3.89 \times 10^{-8}$ cm. The current tube area a was taken without considering condensation, just for an expansion with $\gamma = 1.4$. The two-dimensional calculation shows that the error in determining condensate parameters introduced by the last assumption is insignificant.

Molecular Beam Intensity

The computations show that the dispersion of the particle size distribution function $(\overline{r^2} - \overline{r}^2)^{1/2}$ is less than 10 percent from the value of $\overline{r}$. That is why it was assumed for the computation of molecular beam intensity that all of the clusters had the same mean radius $\overline{r}$ or the same mean number of molecules per cluster $\overline{N}$. The velocity distributions of both monomers and clusters in the skimmer plane were assumed to be Maxwellian functions at the same temperature. This means that the speed ratio for clusters is $\sqrt{\overline{N}}$ times more than that for monomers. The exact expressions for the contributions of monomer, $\mathfrak{J}_1$, and cluster, $\mathfrak{J}_c$, intensities in the total beam intensity, $\mathfrak{J}$, based on the assumption of a collisionless beam are given in Ref. 6.

With the approximations $S \gg 1$ and $\overline{N}S^2\phi^2 \ll 1$, where $S = u(m/2kT)^{1/2}$ is the speed ratio for monomers in the skimmer plane, and ϕ is the half-angle at which the skimmer is visible from the detector, the relations for intensities have a simple form

$$\mathfrak{J}_1 = (\rho u)_s \, \phi^2 \, (1 - q) \, S^2 \tag{9}$$

$$\mathfrak{J}_c = (\rho u)_s \, \phi^2 \, q \, \overline{N} \, S^2 \tag{10}$$

$$\mathfrak{J} = \mathfrak{J}_1 + \mathfrak{J}_c = (\rho u)_s \, \phi^2 \, S^2 \, (1 - q + q \, \overline{N}) \tag{11}$$

where $(\rho u)_s$ is the ρu value in the skimmer plane. As seen from Eqs. (9-11), when $q\overline{N} \gg 1$ the beam consists mainly of clusters. It should be noted that at high enough values of $\overline{N}$, Eq. (10) overestimates $\mathfrak{J}_c$ in comparison with the exact value. Equations (9-11) are convenient for analysis; however, results of the calculation of intensity given below are obtained by exact formulas.

Discussion of Calculations and Experiments

The parameters of the cluster beam are determined by ρ, u, $\overline{N}$, q, and S. Thus, it is important to compare the results of calculations of these quantities with the measurements. Such comparison allows verification of the condensation model used in the calculations.

The result of the calculations of the flow velocity at large distances from the nozzle as a function of P_o is shown in Fig. 2. The result of the monomer velocity measurements[1] is also displayed there. For low P_o the velocity is constant and 615 m/sec, which corresponds to the limiting velocity for $\gamma = 1.4$. This evidences "freezing" of vibrational degrees of freedom of CO_2 at these experimental conditions. The cause of the observed increase of flow velocity is the beginning of the condensation process and transformation of the liberated heat to the kinetic energy of the flow by expansion.

Figure 3 represents the calculated and experimental[3] values of P_o versus nozzle diameter, necessary for obtaining clusters with $\overline{N} = 500$ in the far flowfield. Comparison of the curves displayed in Figs. 2 and 3 enables one to say that there is qualitative agreement between the calculations and experiments, but the calculated values of P_o corresponding to the beginning of the velocity increase, or necessary for obtaining clusters of the same mean size, are systematically less than those from the experiments. Analysis of the condensation model

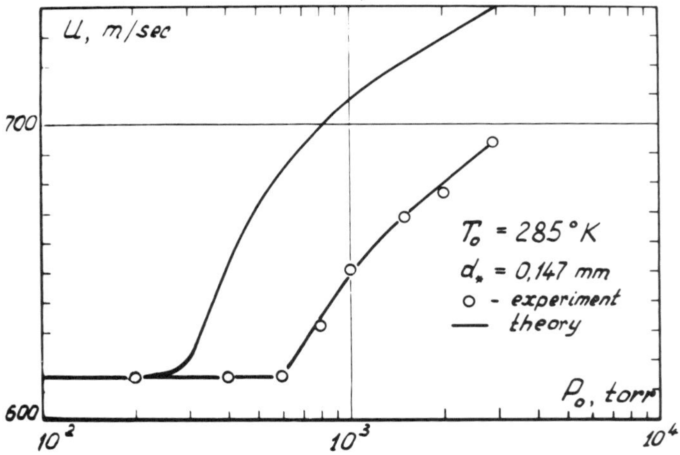

Fig. 2 Calculated flow velocity and measured monomer velocity versus stagnation pressure P_o.

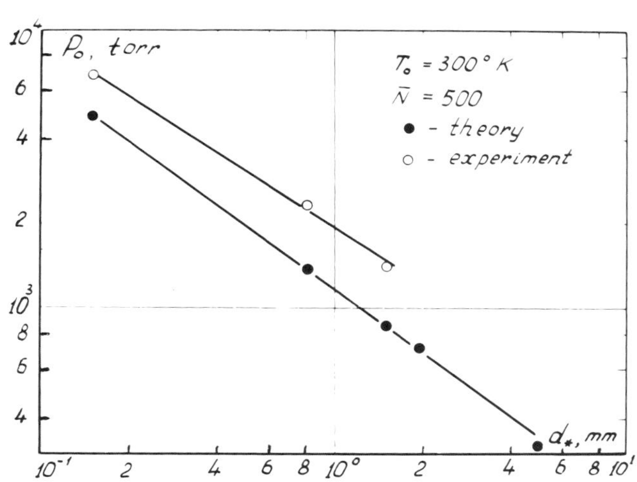

Fig. 3 Stagnation pressure versus orifice diameter for mean cluster size $\overline{N}$ = 500.

used in the calculations shows that the most probable cause of this discrepancy is the calculation of the liquid temperature T_ℓ on the basis of the liquid-vapor equilibrium curve for a plane interface surface, without considering the curvature of the drop. That leads to increase of drop growth rate, larger drop mean size, and condensate mass fraction. It is very complicated to make calculations taking into account the dependence of the drop growth rate on the drop radius. The abovementioned discrepancy is evident in all comparisons of calculations and experiments.

The results of calculation of the condensate mass fraction, speed ratio, and cluster mean size for typical conditions of our experiments are shown in Fig. 4. The condensate fraction has a sharp increase for $P_0 \simeq 100$ torr, and the condensation heat liberation leads to the significant decrease of the speed ratio. For further increasing P_0, the values of q and S change slowly, and the cluster mean size grows proportional to $P_0^{1.5}$. Such $\bar{N}(P_0)$ dependence is in agreement with experiments[3] for large P_0. The causes of the nonmonotonic behavior of $\bar{N}(P_0)$ are not yet clear.

The values q, S, $\bar{N}$ have been used for the calculations of monomer and total beam intensities as functions of P_0. Corresponding curves are given in Fig. 5. The monomer intensity has a deep minimum because of the decrease of the speed ratio. The minimum of the total beam intensity is less pronounced because the decrease of S is compensated by growing condensate fraction. The sharp increase of $\mathfrak{J}(P_0)$ behind the minimum is a consequence of the fact that under these conditions the beam

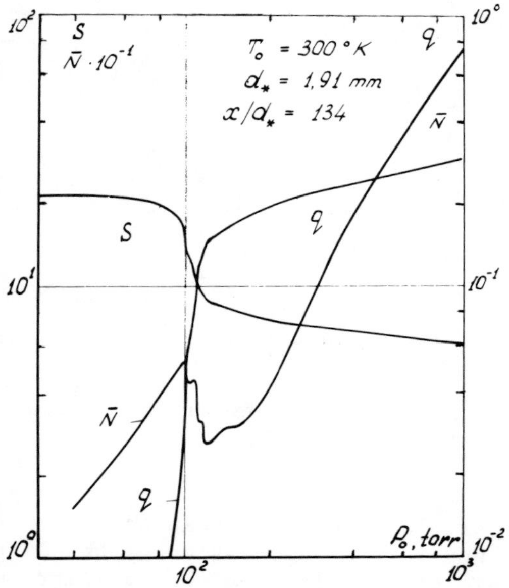

Fig. 4 Calculated speed ratio S, condensate mass fraction q and mean cluster size $\bar{N}$ versus stagnation pressure P_0.

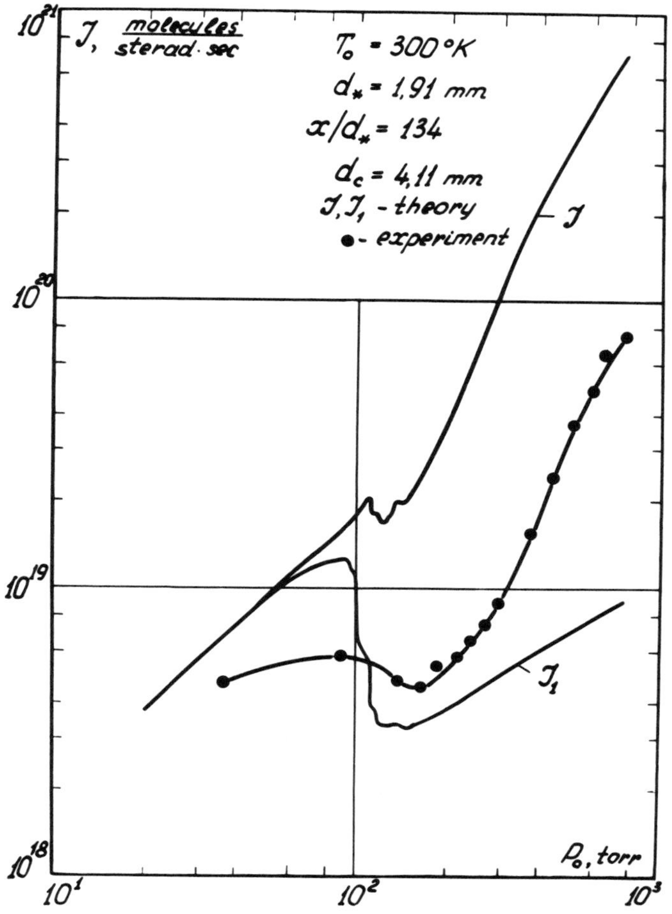

Fig. 5 Calculated cluster beam intensity, monomer beam intensity $\mathfrak{J}_1$ and experimental beam intensity versus stagnation pressure P_o.

consists mainly of clusters. In this region of P_o the values of q and S change weakly, $\overline{N} \sim P_o^{1.5}$, $(\rho u)_s \sim P_o$, and according to Eq. (11) we get $\mathfrak{J} \sim (\rho u)_s\, q\overline{N} \sim P_o^{2.5}$, in good agreement with experiments. Because of the qualitative agreement of the theoretical and experimental curves, we can state that the variation of the beam intensity in our experiments, as well as in experiments of other authors[1], is connected with the condensation process in the jet.

Figure 6 represents the calculated dependence of $q(P_o)$ on d_* and also shows the equilibrium value of condensate fraction ($d_* = \infty$, curve 4) computed on the assumption that all the states of the expanding gas, after it achieves saturation, are on the phase equilibrium curve. The shape of the curve $q(P_o)$ is the same for all d_*, although at some P_o there is a sharp change of slope. It is seen from Figs. 5 and 6 that the same P_o corresponds to the intensity minimum on the calculated curves of $\mathfrak{J}(P_o)$, with the condensate fraction being approximately the same and equal to 12-15 percent. For the larger P_o the condensate fraction asymptotically tends to its equilibrium value. This result was unexpected and means that, contrary to widespread notion, the condensation in a free jet behaves in a quasi-equilibrium way at large enough P_o.

The calculated curves $P_o(d_*)$ corresponding to constant values of the condensate fraction in the far flowfield are shown in Fig. 7. The calculated and experimental dependencies $P_o(d_*)$ corresponding to the minima of the intensity curves are also given. One can see that the calculation predicts very well the generalization of conditions corresponding to the intensity minimum by the relation

$$P_o d_*^{\beta}\Big|_{T_o} = \text{const} \quad (\beta = 0.6)$$

It follows from Figs. 3 and 7 that the conditions necessary for obtaining the same cluster mean size (Fig. 3), as well as the same condensate fraction (Fig. 7), are correlated by analogous relations with $\beta = 0.6 - 0.7$. Generalization of calculated results in this way can be qualitatively explained by approximate analysis. It is seen that the condensation process described by Eqs. (1-7) is completely determined by the total number of binary collisions Z after the point x_1, where the gas achieves its saturation state, i.e.

$$Z = \int_{x_1}^{\infty} \nu \, dx/u \sim d_* \int_{\bar{x}_1}^{\infty} \nu \, d\bar{x}/u \tag{12}$$

where ν is the frequency of binary collisions at the point x and $\bar{x}_1 = x_1/d_*$. Because of the way $\bar{x}_1$ depends on P_o, we can write $Z \sim P_o^{\gamma} d_*$, where $\gamma > 1$ and depends weakly on P_o, T_o, d_*. The approximate relation $P_o d_*^{1/\gamma} = \text{const}$ follows from the condition Z = const. Thus, this form of scaling is an approximation, valid only in the defined range of the source conditions. As seen from Fig. 7, conditions necessary for obtaining the same condensate fraction are generalized by expressions with different exponents β, which depend weakly on the condensate fraction.

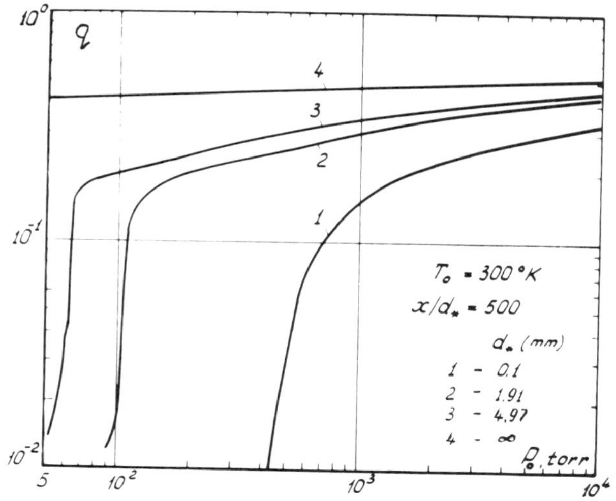

Fig. 6 Calculated condensate mass fraction q versus stagnation pressure P_o for different sonic nozzle diameters.

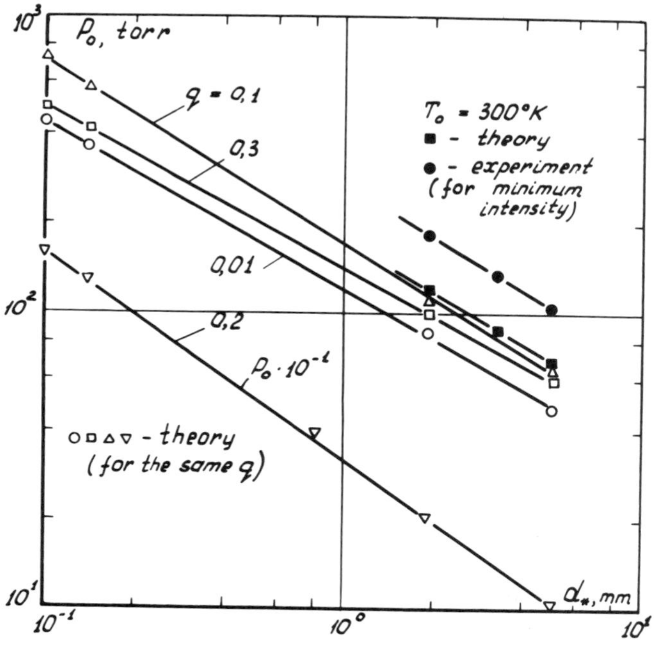

Fig. 7 Calculated condensate mass fraction q and condition for minimum molecular beam intensity in P_o and d_* coordinates

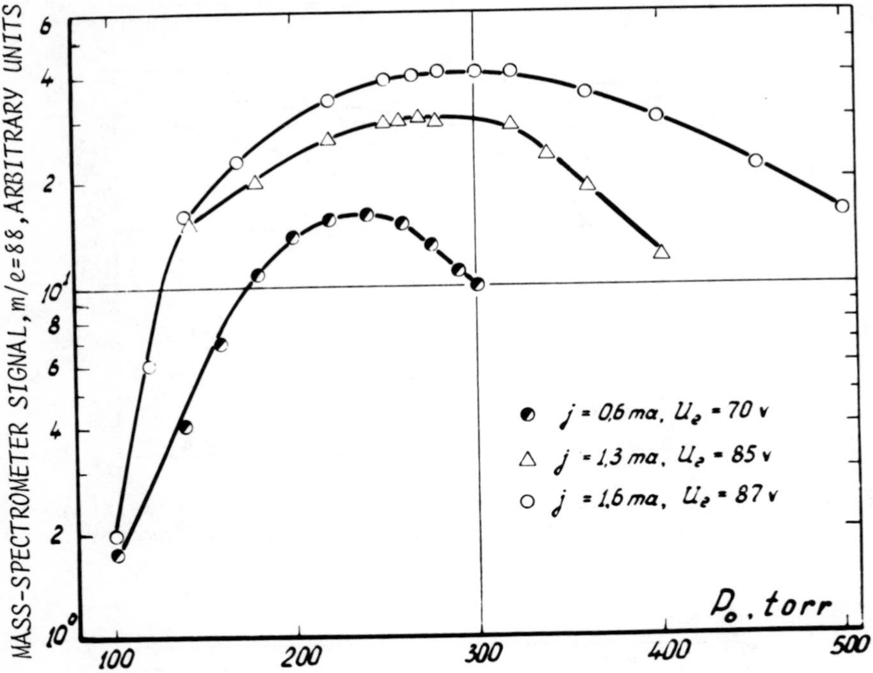

Fig. 8 Mass-spectrometric signal (m/e = 88) versus stagnation pressure P_o for various emission currents.

From the present results, one can state that the classical nucleation theory provides a qualitatively correct description of the homogeneous condensation process for rapid expansion of gas. Therefore, it is not evident that dimer formation resulting from termolecular collisions is the limiting stage for homogeneous condensation in a rapid gas expansion.

A number of experimental observations of dimers in free jets agree qualitatively[1,2,4] (Fig. 1). Yet, in our view, these data can be significantly distorted by the interaction of the molecular beam with the electron beam of the mass spectrometer ion source. Curves for mass/charge ratio = 88 as a function of P_o at T_o = 310°K, d_* = 1.91 mm, d_s = 1.89 mm, and $\overline{x}$ = 130, corresponding to three different emission currents, are shown in Fig. 8. The dependence of the location of the ion signal peak on the emission current (electron concentration in ion source) means that ion signal m/e = 88 is not proportional to the dimer concentration in the molecular beam and hence can not simply correspond to the number of dimers in the beam. Possibly the appearance of at least part of the ions for m/e = 88 is connected with the reactions in the ion source.

On the other hand, estimates show that at conditions when the condensation in the jet occurs, the number of dimers in the flow exceeds the number of critical nuclei obtained from calculations. One can only suppose that the process of dimer formation as a result of termolecular collisions is not the limiting stage for the condensation of rapidly expanding gas.

References

[1] Golomb, D., Good, R. E., Bailey, A. B., Busby, M. R., and Dawbarn, R., "Dimers, Clusters, and Condensation in Free Jets. II," Journal of Chemical Physics, Vol. 57, No. 9, November 1972, pp. 3844-3852.

[2] Milne, T. A., Vandergrift, A. E., and Greene, F. T., "Mass-Spectrometric Observations of Argon Clusters in Nozzle Beams. II. The Kinetics of Dimer Growth," Journal of Chemical Physics, Vol. 52, No. 3, February 1970, pp. 1552-1560.

[3] Hagena, O. F. and Obert, W., "Cluster Formation in Expanding Supersonic Jets," Journal of Chemical Physics, Vol. 56, No. 5, March 1972, pp. 1793-1802.

[4] Vostrikov, A. A., Kusner, Yu. S., Rebrov, A. K., and Semyachkin, B. Ye., "The Formation of CO_2 Molecular Beam by Gasdynamic Methods" (in Russian), Zhurnal Prikladnoi Mechaniki i Tecknicheskoi Physiki, No. 2, 1975, pp. 134-141.

[5] Vostrikov, A. A., Kusner, Yu. S., Rebrov, A. K., and Semyachkin, B. Ye., "The Measurements of Intensity and Density of Molecular Beam" (in Russian), Pribori i Tekhnika Experimenta, No. 1, 1975, pp. 177-179.

[6] Skovorodko, P. A., "The Influence of Homogeneous Condensation in Free Jet on the Molecular Beam Intensity" (in Russian), in collection Some Problems of Hydrodynamics and Heat Transfer, Institute of Thermophysics, Novosibirsk, 1976, pp. 106-114.

[7] Frenkel, J., Kinetic Theory of Liquids, Dover, New York, 1955, Chapter 7.

[8] Duff, K. M. and Hill, P. G., "Condensation of Carbon Dioxide in Supersonic Nozzles," Proceedings of the 1966 Heat Transfer and Fluid Mechanics Institute, Stanford University Press, 1966, pp. 268-294.

ARGON CLUSTERS IN A SUPERSONIC BEAM : SIZE TEMPERATURE AND MASS FRACTION ON CONDENSATE, IN THE RANGE 40 TO 1000 ATOMS PER CLUSTER

J. Farges,* M.F. de Feraudy,* B. Raoult,* and G. Torchet *

Université de Paris Sud, Orsay, France

Abstract

The use of both models with electron diffraction data allow the determination of the mean number $\bar{N}$ of atom of the cluster. We report several values of $\bar{N}$ obtained with a nozzle diameter of .2 mm, a source temperature of 300°K and different source pressures p_o . For decreasing p_o we find successively $\bar{N}$ = 1000 (p_o = 7600 Torr), $\bar{N}$ = 310 (p_o = 3800 Torr), $\bar{N}$ = 150 (p_o = 2500 Torr), and $\bar{N}$ = 40 (p_o = 1300 Torr). We find that $\bar{N}$ shows a $p_o^{1.7}$ dependence for $\bar{N} > 100$. This result is in good agreement with the one given by Hagena and Obert. The temperature is found to be 27 K ± 3°K and 34 K ± 3°K for $\bar{N}$ = 40 and $\bar{N}$ = 150 respectively. The mass fraction of condensate varies between 0,15 and 0,9 when p_o increases from 900 Torr to 3000 Torr.

Method

Our present investigations still concern the interpretation of diffraction patterns obtained by crossing an Argon cluster beam by an electron beam of 50 keV incident energy.[1] The main goal of this work is to determine precisely the structure of clusters as a function of their size. In our experiment, we make the cluster size vary by changing the source pressure p_o, while the source temperature is fixed at T_o = 300°K. The nozzle diameter D is 0.2 mm and the skimmer is located at x/D = 40.

Presented as Paper 132 at the 10th International Symposium on Rarefied Gas Dynamics, Aspen, Colo., July 19-23, 1976.

*Dr., Groupe des Agrégats Moléculaires, Laboratoire de Diffraction Electronique.

The method of the determination of the cluster structure dependence on size is as follows : The diffraction intensities are calculated for different cluster structures and sizes for several cluster temperatures. These computer-calculated intensities are then compared with the experimental results. When a perfect agreement is obtained, we say that the corresponding model provides a good description of the clusters in the cluster beam, for the considered source pressure. As an example of such agreement, Ref. 2 reports a comparison of experimental (p_o = 2500 Torr) and calculated intensities corresponding to an icosahedral model with 147 atoms at three different temperatures.

The method has been improved recently by taking into account, in intensity calculations, the free atoms present in the cluster beam, together with the clustered atoms. The relative proportion of each kind of atoms has been determined and, thus, so has the condensed mass fraction.

Results

1. Size

In Fig.1, giving the mean size $\overline{N}$ vs source pressure p_o, we report several points that have been determined with precision. These values correspond to remarkable changes in the structure of Argon clusters : for decreasing source pressures down from p_o = 7600 Torr, we find successively $\overline{N}$ = 1000 atoms, corresponding to clusters with crystalline face-centered cubic structure ; $\overline{N}$ = 310 (p_o = 3800 Torr) and 150 atoms (p_o = 2500 Torr), corresponding to icosahedral clusters with 4 and 3 complete shells, respectively ; and finally $\overline{N}$ = 40 atoms (p_o = 1300 Torr), corresponding to clusters with amorphous structure. For amorphous structure, size determination is more difficult than that for other structures, because the interference function is roughly independant of the cluster size. Fortunately, the 55-atom icosahedron scattered intensity, which is considerably different, appears and becomes rapidly predominent when p_o exceeds 1300 Torr. This indicates obviously that the given value $\overline{N}$ = 40 atoms is, in any case, an upper limit of the mean cluster size at p_o = 1300 Torr.

One can see in Fig.1 that, for mean sizes $\overline{N}$ > 100, experimental points lie on a straight line, and $\overline{N}$ shows a $p_o^{1.7}$ dependence. It is interesting to compare the present results to those obtained by Hagena and Obert with mass spectroscopic method and larger clusters[3] - their size range (N/Z)* > 100 corresponds to $\overline{N}$ > 300.[4] In order to make the comparison

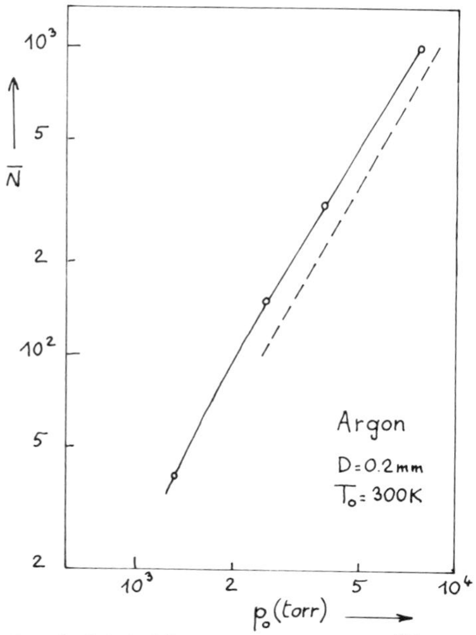

Fig. 1 Solid line : Mean size $\overline{N}$ as a function of source pressure p_o. With our method, the experimental uncertainties corresponding to the reported $\overline{N}$ values are estimated to be 1000 ± 100 at., 310 ± 50 at., 150 ± 30 at., 40 ± 15 at.. Dashed line : scaled data from Ref. 3.

easier by taking into account the difference in sonic nozzle diameter, it is possible to use the scaling rules reported by these authors. Thus, one obtains the expression $N \simeq 8 \times 10^{-5}\ p_o^{1.8}$, which is represented by the dashed line in Fig.1. Considering the approximations required to deduce the latter expression, it is seen that our N results are quite similar to those of Hagena and Obert, although both the method and the size range are different. The 40-atom value is located below our experimental line, and this probably means that, for source pressures below p_o = 2000 Torr, the variation in $\overline{N}$ is more rapid than a $p_o^{1.7}$ law. Such a sudden increase in size with increasing p_o at the very beginning of the condensation has already been observed by Lewis et al. for CO_2 clusters.[5]

2. Temperature

By use of the method presented previously,[2] the temperature of 40-atom clusters has been determined and a value

T = 27°K ± 3°K has been found. The temperature corresponding to 150-atom clusters is T = 34°K ± 3°K, as reported before. (There was a typing mistake in the value T = 32°K reported in Ref.1. The correct value T = 34 ± 3°K corresponds to the graphic determination shown in this Reference). This temperature represents the mean kinetic energy of clustered atoms. It is not surprising to observe a lower temperature for 40-atom clusters than for 150-atom clusters. In fact, after the growth period in the freejet, Argon clusters travel in vacuum during 10^{-4} sec before crossing the electron beam. While travelling they evaporate and their temperature is lowered down to some value T for which one more atom is not likely to escape during a time interval of about 10^{-4} sec.. This value T corresponds to our temperature measurement. Then, considering how evaporation phenomena depend on cluster size, we can expect that surface atoms are less strongly bounded for a small cluster ($N < 50$ atoms) than for a large one ($N > 100$ atoms), because the number of bulk atoms is smaller, and chiefly because the surface structure is unfavorable. Consequently, the smaller the cluster, the higher the evaporation rate and the lower the final temperature reached at the end of the process. It seems noteworthy to say that our recent results give evidence for clusters with less than 100 atoms to be in a solid phase at the diffraction point. More precisely, this means that the self-diffusion coefficient is less than 2×10^{-7} cm^2/sec,i.e., one-tenth of the bulk value at the triple point.

3. Mass fraction of condensate

As a consequence of the improvement cited previously, we get an estimate of the mass fraction of condensate g in the cluster beam. The g variation v p_o is shown if Fig.2. We must point out that this determination of g is relative to the diffraction point, far downstream from the freejet, after skimming and collimating the jet. This explains the rather high g values reported here, as compared to those provided by freejet experiments.[6]

Note

In our g estimate, we consider that each clustered atom contributes identically to the total scattered intensity (interference function + background). This is roughly true in the size range 10 to 50 atoms. For $N < 10$ atoms, the contribution becomes less and less important with decreasing size : as an example, one atom clustered in a dimer contributes to the interference function ten times less than if it were clustered in a large polymer ($10 < N < 50$). As a consequence, in our g determination we take into account only a small fraction

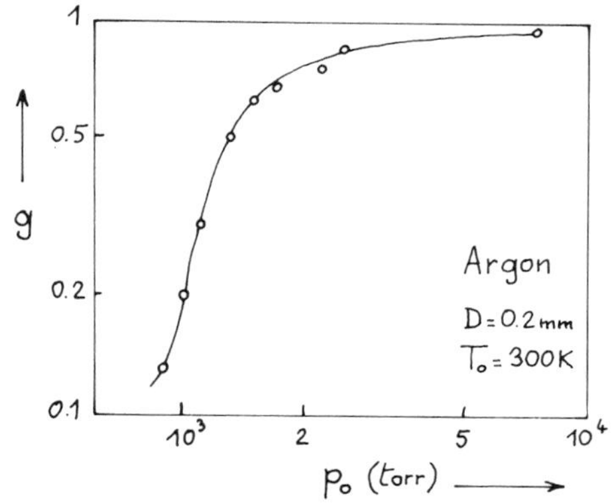

Fig. 2 Mass fraction of condensate g as a function of source pressure p_o. The relative uncertainty on g values is estimated to be 5%.

of dimers, most dimers beeing counted together with free atoms. In this way, we can estimate that g represents roughly the ratio of atoms clustered in clusters containing less than 5 or 10 atoms, to the total number of atoms present in the beam.

References

[1]Raoult, B., and Farges, J., "Electron Diffraction Unit with Supersonic Molecular Beam and Cluster Beam," The Review Scientific Instruments, Vol. 44, No 4, April 1973, pp.430-434.

[2]Farges, J., de Feraudy, M.F., Raoult, B., and Torchet, G., "Electron Diffraction Measurement of the Temperature of the Very Minute Argon Clusters in a Supersonic Beam," Proceedings of the Ninth International Rarefied Gas Dynamics Symposium, Vol. 2, DFVLR-Press, Porz-Wahn, Germany, 1974, F.8.

[3]Hagena, O.F., and Obert, W.,"Cluster Formation in Expanding Supersonic Jets : Effect of Pressure, Temperature, Nozzle Size, and Test Gas." The Journal of Chemical Physics,Vol. 56, No 5, March 1972, pp. 1793-1802.

[4] Falter, H., Hagena, O.F., Henkes, W., and Wedel, H.V., "Eingluss der Elektronenenergie auf das Massenspektrum Von Clustern in Kondensierten Molekularstrahlen," International Journal of Mass Spectroscopy and Ion Physics, Vol. 4, 1970, pp. 145-163.

[5] Lewis, J.W.L., Williams, W.D., and Powell, H.M., "Laser Diagnostics of a Condensing Binary Mixture Expansion Flow Field," Proceedings of the Ninth International Rarefied Gas Dynamics Symposium, Vol. 2, DFVLR-Press, Porz-Wahn, Germany, 1974, F.7.

[6] Lewis, J.W.L., and Williams, W.D., "Argon Condensation in Free-Jet Expansions," Arnold Engineering Developement Center, AEDC-TR - 74-32, July 1974.

FORMATION AND DETECTION OF HIGH-ENERGY CLUSTER BEAMS

O. F. Hagena* and W. Henkes†

Institut für Kernverfahrenstechnik
Kernforschungszentrum Karlsruhe, Karlsruhe, West Germany

and

U. Pfeiffer†

Institut für Aerobiologie
Fraunhofergesellschaft, West Germany

Abstract

Energetic particle beams with kinetic energies up to 10 keV per atom/molecule can be produced by ionizing and accelerating cluster beams obtained from partly condensed nozzle flows. The general requirements for the components of such a beam source are discussed, with close reference to an apparatus used for production of 1-MeV H_2 cluster beams. It consists of a cryopumped cluster beam source and an electron-impact ionizer to produce a cluster-ion beam, with both components being located in the high-voltage terminal of a 1-MV accelerator, and of a target chamber at ground with a gas cell to study the scattering and neutralization of the cluster beam. In addition to a description of the apparatus and its operational performance, data are given for the intensity and the scattering of

---

Presented as Paper 118 at the 10th International Symposium on Rarefied Gas Dynamics, Aspen, Colo., July 19-23, 1976. We are indebted to E. W. Becker and R. Klingelhöfer of the IKVT, Karlsruhe, and to H. Oldiges and W. Stöber of the IAe, Grafschaft, for continuous support. We are grateful to W. Althaus, L. Enderlein and S. Dürr, who assisted in setting up and operating the experiments.

*Dr.-Ing., wissenschaftlicher Mitarbeiter.

†Dr. phil., wissenschaftlicher Mitarbeiter.

the cluster beam. Finally, the problems of measuring the properties of energetic particle beams are discussed.

I. Introduction

High-energy cluster beams are produced by ionizing and electrostatically accelerating thermal-energy cluster beams,[1] which in turn are obtained from the core portion of partly condensed supersonic nozzle flows. The energy of the resulting cluster-ion beam is proportional to the accelerating voltage and to the specific charge of the cluster ion. For example, for 100 kV and singly-charged clusters of 100 atoms/molecules, the beam energy would be 1 keV per beam atom/molecule. Since cluster beams can be obtained from pure vapors or gas-vapor mixtures by a suitable choice of source pressure, source temperature, and nozzle geometry,[2] energetic cluster beams offer not only the high mass flux densities typical for cluster beams, but provide a rather unique possibility to produce beam kinetic energies in the range 0.1 up to some keV per molecule for molecular species for which the more conventional methods of beam production (such as high-temperature oven/nozzle beams, shock tubes, sputtering, or ionization-neutralization) do not work. This energy range is of interest for the study of chemical reactions, and of elastic or inelastic gas-surface interactions. Aside from these general applications for high-energy cluster beams, the Karlsruhe research project with energetic cluster beams is aimed at using energetic cluster beams of hydrogen for neutral injection heating of the plasma in a toroidal fusion experiment.[3] Compared to the neutral injection method, which uses molecular ion beams with subsequent neutralization by charge exchange, the use of cluster-ion beams promises to overcome the space-charge problems, which limit the available mass-flux intensities.[4] An alternative route to obtain the required high-temperature plasma is the gradual radial build up of a hot plasma by capturing an energetic particle beam in the confining magnetic field.[5] This requires very high mass-flux densities, which are most likely realized with energetic cluster beams only.[6]

The following paper discusses the general aspects for the production of energetic cluster beams and presents, as an example, the design and performance data for an apparatus to produce energetic hydrogen cluster beams with energies up to 1 MeV per cluster.

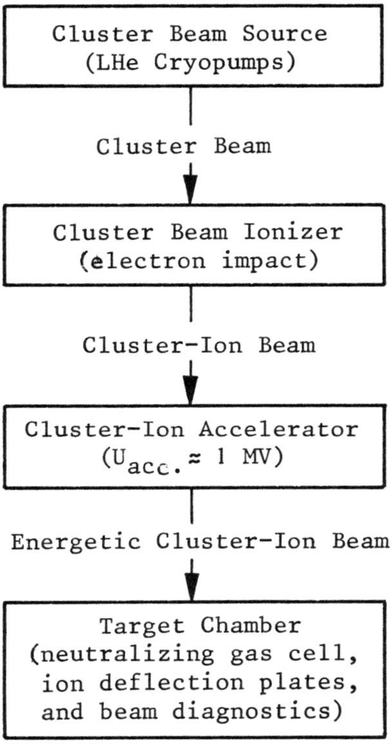

Fig. 1 Components of an apparatus for the production of energetic cluster beams. The data given in parenthesis refer to the 1 MeV hydrogen cluster source described in this paper.

II. Beam Source for High-Energy Clusters

1. General Design Considerations

The various components of a high-energy cluster beam source are shown schematically in Fig. 1. The cluster beam source delivers a cluster beam characterized by mean cluster size $\bar{N}$, mass flux, beam solid angle, and flow velocity into the cluster beam ionizer, where it is transformed into a beam of cluster ions characterized by the mean mass-to-charge ratio $(\overline{N/Z})$. The ions are extracted and accelerated by the cluster ion accelerator, where the accelerating voltage U_{acc}, together with the mass-to-charge ratio, determines the final kinetic energy per cluster molecule. The target chamber contains the respective beam target (gas, solid, or liquid target), whose interaction with the beam is to be studied. Notwithstanding the various constraints imposed by the specific experiment, a few general design characteristics can be given as follows.

Cluster beam source: Design and performance data for cluster beam sources have been reviewed in Ref. 2. Assuming no limitations by the capacity of the vacuum pumps that handle the nozzle mass flow, the characteristics of the cluster beam depend on the beam defining apertures and on the properties of the nozzle flowfield, which in turn is a unique function of nozzle geometry, pressure p_o, and temperature T_o of the gas prior to its expansion through the nozzle. Although the flow velocity (which becomes important for the efficiency of the ionizer) depends primarily on source temperature, the mean cluster size and the mass flux are sensitive to all three parameters: nozzle, p_o, and T_o. The use of an inert carrier gas[7] - the advantages of which have been discussed at the preceeding RGD symposium[8] - will be mandatory to obtain clusters for polyatomic gases with low vapor pressures and low values of the ratio of specific heats.

Cluster beam ionizer: It is essential that, during ionization, the cluster does not disintegrate into its component molecules. However, fragmentation of big clusters into smaller ones, due to multiple ionization with subsequent breakup of the charged cluster,[9,10] is a useful process. It allows to use the intensity increase, which can be realized by using bigger clusters, while still obtaining the small mass-to-charge ratios necessary to get high energies per molecule for a given upper limit of accelerating voltage. For polyatomic substances, e.g., organic molecules, ionization by electron impact will result in some fragment ions. The chemical nature, however, of most of the cluster molecules will be unaffected.

Cluster ion accelerator: Present technology limits the upper voltage of the necessary dc power supplies to 1 - 2 MV, with pressurized gas insulation minimizing the space requirements for the high-voltage components. For cluster ions with 100 molecules/charge this gives upper energies of 10 - 20 keV/molecule. Obviously it is much easier to operate at lower voltages, for which it may suffice to have only the ionizer at high potential, while cluster beam source and target chamber remain at ground potential. For high-current cluster-ion beams the design of the ion extraction and accelerating system must consider the differences in space charge compared to conventional ion or electron accelerators with the same accelerating voltage.[3]

2. 1-MeV Hydrogen Cluster Source

The first apparatus for producing hydrogen cluster beams with energies up to 1 MeV/cluster was built and operated during the last 4 years using the 1-MV open-air accelerator of the Institut für Aerobiologie. The length of the accelerating tube is 4 m, and it is not designed for high currents. Compared to an earlier cluster accelerator at Fontenay-aux-Roses, with a maximum energy of 600 keV,[11] the present apparatus uses a cluster beam source with cryopumps instead of diffusion pumps to handle the nozzle mass flow. With reference to Fig. 1, cluster beam source, ionizer, and their respective control-supply systems were sufficiently compact to fit into the high-voltage terminal of the accelerator, and a special target chamber was built to accomodate the various components for the experiments with the cluster beam. The operational experience gained with this apparatus was used for the construction of a second apparatus to test the high-gradient acceleration of medium-size cluster currents at the 1-MV accelerator of the Institute de Physique Nucléaire of the University of Lyon[12] and for the design of a high-power cluster injector, which now is being constructed at Karlsruhe.[3]

Figure 2 gives a schematic cross-sectional view of the vacuum chamber, which contains the cluster beam source and the ionizer. The nozzle source system (1) has a converging-diverging hypersonic nozzle, a throat diameter of 0.16 mm, with

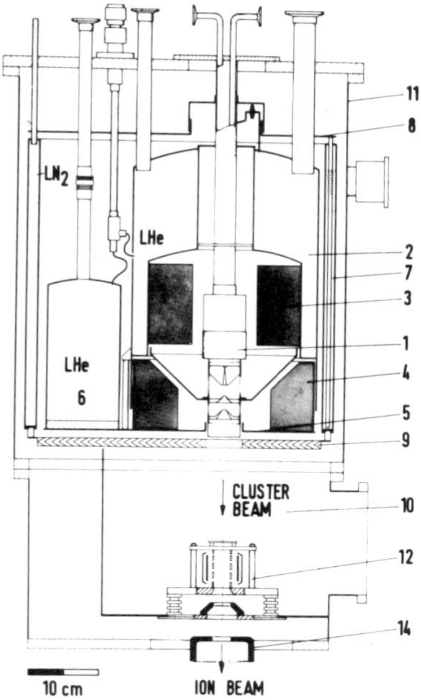

Fig. 2 Vacuum chamber with cryopumped cluster beam source and ionizer, schematic.

skimmer and collimator as beam-defining elements. The nozzle source is surrounded by the LHe-filled container (2), which cools a series of radial copper plates (3), which serve as cryopumps for the nozzle mass flow. The skimmer-collimator region is pumped by copper plates (4) and base plate (5) in contact with a second LHe bath (6), which can be kept at reduced pressure (0.3 bar), and thus at reduced temperature. The cryopumps are surrounded by LN_2-cooled radiation shields (7), (8). The chevron baffle (9) allows the base plate (5) to act as cryopump for the ionizer chamber (10). The top flange of the vacuum vessel (11) contains the vacuum feedthroughs for the cryogenic fill and vent lines, and for the nozzle supplies. Source pressure p_o is pneumatically regulated to a preset value, and source temperature T_o is given by the vapor pressure of a neon bath cooling the nozzle, with the necessary refrigeration power supplied to the neon bath by part of the helium boil-off gas. By using N_2 or H_2 instead of Ne, higher or lower nozzle temperatures are possible. The cryopumps contain 8 and 2 l helium, and under typical operating conditions, the boil-off rates are 1 and 0.2 l LHe/hr. The cluster beam passes through the ionizer (12) inside of a slotted anode cage of quadratic cross section. Four indirectly heated cathodes in front of the four sides of the anode produce an electron current up to 600 mA, which intersects the cluster beam and produces cluster ions. The electron energy is typically 200 eV. The ions are extracted in the direction of the cluster beam by a voltage of 6 - 20 kV between anode and extracting electrode, and are further accelerated into the accelerating tube by the first of five accelerator electrodes (14). More details on the construction and operation of the ionizer are presented in Ref. 13. The lower end of the 4-m ceramic accelerator tube is attached to a 1-m-long base section with the connections for the oil diffusion pump. In order to reduce the partial pressure of condensable gases within the accelerator tube, a LN_2 cooled cylindrical cryopump was fitted into the vacuum line leading to the diffusion pump. Typical pressures measured with an ionization gage connected to the base section are 1×10^{-6} Torr. Following in the direction of the cluster ion beam is the target chamber, as shown in Fig. 3. Along the beam axis one recognizes an upper pair of deflection plates, a scattering gas cell 0.4 m long, a lower pair of deflection plates, and in the lower compartment the beam detectors. They are mounted on a rotating detector support so as to bring any of the detectors in line with the beam axis. In addition, the support can be moved so that it is perpendicular to the beam axis for beam profile measurements.

The gas cell has been operated with water vapor and with hydrogen as scattering gas. For H_2O, a LN_2-filled trap next to the gas cell serves as cryopump, whereas for H_2, a water-cooled titanium sublimation pump was installed with an effective pum-

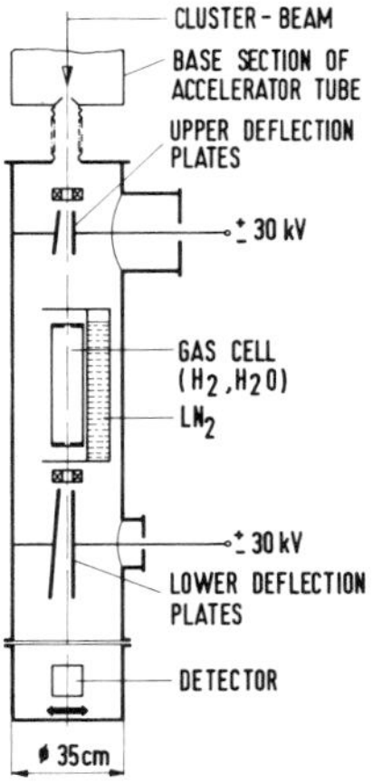

Fig. 3 Target chamber with ion-deflection plates, gas cell for scattering of the cluster beam, and beam diagnostics, schematic.

ping speed for H_2 between 4000 and 12000 l/sec, depending on the titanium evaporation rate and the H_2 flow rate out of the gas cell. Base pressure within the target chamber maintained by a diffusion pump with water-cooled baffle was 1 - $2 \cdot 10^7$ Torr.

Not shown in the schematic of Fig. 3 are a number of collimating apertures to limit the geometric cross section of the beam. The alignment of the beam with respect to the target chamber axis can be checked by observing the visible light emitted by those parts of the apertures exposed to the energetic beam.

III. Discussion of Experimental Results

1. Thermal Energy Cluster Beam

The nozzle source system with cryopumps allows steady-state as well as pulsed beam operation. The intensities for the present hypersonic nozzle showed the same dependence on pressure p_0 and temperature T_0 as that observed in studies with pulsed cluster beams.[2,14] For p_0 = 350 Torr, T_0 = 28 K, the intensity of $2 \cdot 10^{21}$ H_2 molecules/sr/sec was sufficiently high to to be measured with a simple ionization gage located in the target chamber, 7 m downstream of the nozzle. This intensity corresponds to a collimating efficiency[2,8] of CE = 3.25/sr, and is higher by over a factor 5 than the maximum intensity one could expect from a sonic nozzle with the same mass flow. This points to the importance of proper nozzle design.[2,14]

An important parameter in designing a beam apparatus is the background pressure, which can be tolerated without marked loss in beam intensity. Cluster beams generally are much less sensitive to such scattering losses because of their high mass and the resulting forward scattering.[7] It has been pointed out, however, that in addition to depending upon the mass, the attenuation depends upon the collision cross-section of the cluster (which increases as $N^{2/3}$), and upon the effects of reactive collisions (which reduce the cluster mass).[2] An example of these effects is shown in the attenuation curves of Fig. 4, which gives the beam intensity (logarithmic scale) as function of scattering pressure p_s in the gas cell of the target chamber. The dashed curve indicates the slope to be expected for a beam with an attenuation cross-section equal to the gaskinetic cross-section of hydrogen molecules, $\sigma = 2.4 \cdot 10^{-15}$ cm^2. The cluster beam data for p_0 = 128 Torr, with $\overline{N}$ estimated from Ref. 14 to be between 10^3 and 10^4 molecules/cluster, show a marked variation in the attenuation cross-section, which decreases from a high of $\sigma = 13 \cdot 10^{-15}$ cm^2 down to $5.6 \cdot 10^{-15}$ cm^2 in the range investigated. This is still higher than the expected loss for a molecular H_2-beam. For p_0 = 350 Torr, where $\overline{N}$ is estimated to be in the range of 10^5 molecules/cluster, the attenuation is now slightly less, as compared to molecules. Note that for these clusters the geometric cross section, obtained from the density of solid H_2, amounts to about $3 \cdot 10^{-12}$ cm^2. A third factor, which must be considered when determining the scattering loss, is the angular resolution of the respective experiment. This is demonstrated by the beam profiles of Fig. 5 for

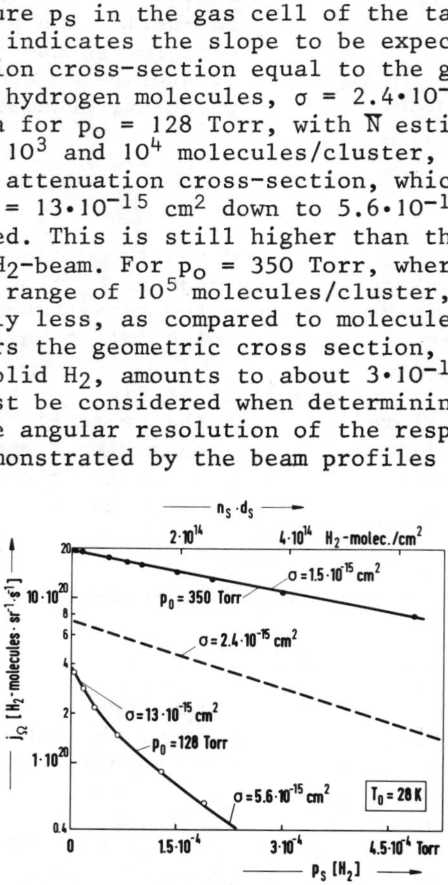

Fig. 4 The intensity of thermal-energy cluster beams as function of scattering pressure in the gas cell, p_s, for two values of the cluster beam source pressure p_0. The dashed curve represents the expected slope of the intensity of a molecular beam with attenuation cross-section $\sigma = 2.4 \cdot 10^{-15}$ cm^2. The upper abscissa scale gives the target thickness $n_s d_s$, where n_s is the particle number density, and d_s=40 cm is the length of the gas cell.

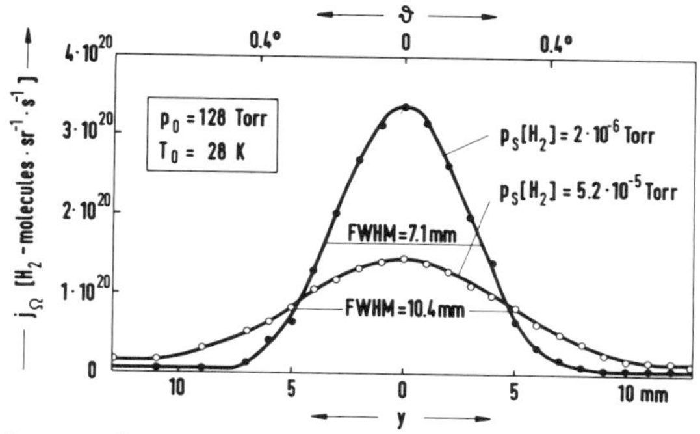

Fig. 5 The intensity profile of a thermal-energy cluster beam with and without attenuation due to scattering gas in the gas cell. y is the perpendicular distance of the detector position from the beam axis, and θ is the angle under which the detector is seen from the center of the gas cell. Distance gas cell center - detector entrance is 94 cm.

the cluster beam with p_0 = 128 Torr, with two different scattering pressures p_s. As in the experiment shown in Fig. 4, the detector gage had an entrance diameter of 6 mm, which when seen from the center of the scattering gas cell corresponds to an angle of 0.4°. The beam profiles show that the loss in intensity on the axis is accompanied by a marked widening of the beam. Thus, at least part of the axial intensity loss is due to small-angle scattering, and will not be observed in experiments with lower angular resolution. Similar scattering results were obtained using H_2O instead of H_2 as scattering gas. Because of the increase in mass, the possible energy and momentum transfer is higher, which explains why, for H_2O, the same loss in intensity occurred at about 1/3 the pressure p_s that was necessary with hydrogen.

2. High-Energy Cluster Beam

The main objective of the present apparatus was the study of the interaction of cluster ions with a gas target to produce a beam of energetic neutral particles. Typical results of a neutralization experiment are shown in Fig. 6. It gives the beam signal I_{se} as function of the gas cell scattering pressure p_s. I_{se} is the secondary electron current emitted from a Ni target hit by the beam. The detector entrance diameter was 14 mm, whereas the beam cross section was about rectangular, 6 by 11 mm^2. Three curves are shown: A represents the total beam signal, i.e., no electric field is applied to the upper and lower ion deflection plates (see Fig. 3). This signal A re-

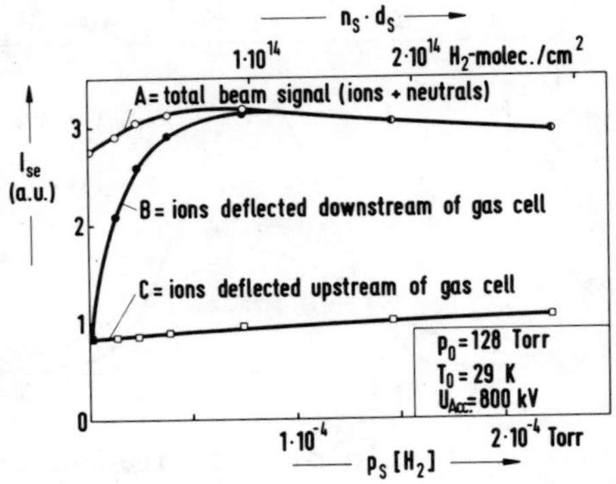

Fig. 6 Neutralization of high-energy cluster beam. Cluster beam and range of p_S are the same as in Fig. 4. Beam signal I_{se} is the secondary electron current from a Ni target.

mains essentially constant over the range of p_s investigated, with a slight maximum at about $0.5 \cdot 10^{-4}$ Torr. For signal C, which is about 1/3 of signal A, all ions have been deflected out of the beam upstream of the gas cell by the upper deflection plates. Thus, this signal is produced by energetic neutral particles formed from the cluster ions upstream of the deflection plates, i.e., primarily in the accelerator tube itself, with its integrated target thickness of about 10^{13} molecules/ cm^2. The modest increase of C with increasing p_S may be explained by the fact that the pressure outside the gas cell increases by about 1 - 2 % of p_S also, and results in additional neutralizing scattering processes. Finally, signal B is obtained when only the lower deflection plates are removing ions from the beam, and one observes that the difference A - B due to ions not neutralized in the gas cell decreases rapidly with increasing p_S.

For a target thickness $n_S d_S = 5 \cdot 10^{13}$ H_2 molecules/cm^2, the neutralized signal B - C is about 90 % of the initial ion signal A - C. Above $1.2 \cdot 10^{14}$ H_2 molecules/cm^2 there is no detectable charged component in the beam downstream of the gas cell, i.e., A = B. This result may be compared with the neutralization of protons. Here the charge-exchange thickness is smallest for the energy range of 5 - 10 keV/proton, with an e-folding value for that energy of $1.1 \cdot 10^{15}$ H_2 molecules/cm^2. For the 90% point, the target thickness needs to be $2.5 \cdot 10^{15}$ H_2 molecules/cm^2, which exceeds the value for cluster ions by a factor of about 50. This example demonstrates the advantages of cluster beams when producing energetic neutrals from an ion beam.[3] More details on the neutralization of hydrogen cluster ions will be presented elsewhere.[15]

Another aspect in discussing the data of Fig. 6 is the problem of quantitative measurements of high-energy cluster beams, which is complicated not only by the high particle energy, which may give the reason for a variety of secondary processes, but by the range of cluster sizes within the beam also. Considering that the secondary electron current I_{se} is produced by a whole spectrum of charged and neutral particles, and that the SE-coefficient will depend not only on cluster energy, but also on charge and mass of the cluster, and considering further that the mass spectrum will be changed by the interaction with the scattering gas and that the SE-coefficient may depend on the background pressure (which varies for the experiment shown from $1 \cdot 10^{-7}$ up to about $3 \cdot 10^{-6}$ Torr), it is no longer surprising that signal A exhibits a maximum for an intermediate value of p_s. This maximum does not stand for an increase of beam intensity or beam energy, and only the relative changes in signal with and without ion deflection can be interpreted without ambiguity. For similar reasons, the ionization gage used in Figs. 4 and 5 for the intensity measurements of the thermal cluster beam does not give signals for the accelerated beam, which can be equated simply with beam intensity. However, by repeating the experiment of Fig. 6 with this detector, one confirms the result that, at target densities above $1.2 \cdot 10^{14}$ H_2 molecules/cm^2, there is no longer a detectable charged component downstream of the gas cell. Consequently, these simple detectors can be used to measure the neutralization of cluster ions and its dependence on the parameters of the cluster-ion beam, whereas for the absolute determination of the properties of the beam calorimetric detectors for energy, microbalances for momentum, and specially shielded Faraday cups for ion current will be used.

References

[1]Becker, E. W., Bier, K., and Henkes, W., "Strahlen aus kondensierten Atomen und Molekeln im Hochvakuum," Zeitschrift für Physik, Vol. 146, 1956, pp. 333-338.

[2]Hagena, O. F., "Cluster Beams from Nozzle Sources," Molecular Beams and Low Density Gasdynamics, Marcel Dekker, New York, 1974, pp. 93-181.

[3]Becker, E. W., Falter, H. D., Hagena, O. F., Henkes, W., Klingelhöfer, R., Körtin, K., Mikosch, F., Moser, H., Obert, W., and Wüst, J., "Development and Construction of an Injector Using Hydrogen Cluster Ions for Nuclear Fusion Devices," report KFK 2016, July 1974, Gesellschaft für Kernforschung, Karlsruhe.

[4]Henkes, W., "Production of Plasma by Injection of Charged Hydrogen Clusters," Physics Letters, Vol. 12, October 1964, pp. 322-323.

[5]Falter, H. D. and Henkes, W., "Plasma Buildup in Toroidal Reactors by Neutral Injection," Proceedings of the 6th European Conference on Controlled Fusion and Plasma Physics, Lomonosov Moscow State University, 1973, pp. 385-387.

[6]Falter, H.D. and Henkes, W., "The Radial Buildup of Plasma by Neutral Injection into Toroidal Configurations," Proceedings of the 7th European Conference on Controlled Fusion and Plasma Physics, Centre de Recherches en Physique des Plasmas, Lausanne, 1975, Vol. I, p. 26.

[7]Becker, E. W., Klingelhöfer, R., and Lohse, P., "Strahlen aus kondensiertem Wasserstoff, kondensiertem Helium und kondensiertem Stickstoff im Hochvakuum," Zeitschrift für Naturforschung, Vol. 17a, 1962, pp. 432-438.

[8]Hagena, O. F. and v.Wedel, H., "Cluster Beams from Gas Mixtures: Effects of Carrier Gas on Cluster Size and Beam Intensity," Proceedings of the 9th International Symposium on Rarefied Gas Dynamics, DFVLR Press, Porz-Wahn, 1974, Vol. II, pp. F.10-1 -F.10-10.

[9]Falter, H. D., Hagena, O.F., Henkes, W., and v.Wedel, H., "Einfluß der Elektronenenergie auf das Massenspektrum von Clustern in kondensierten Molekularstrahlen," International Journal of Mass Spectrometry and Ion Physics, Vol. 4, 1970, pp. 145-163.

[10]Isenberg, G., "Untersuchung der Verteilung von Massen- und Ladungszahl bei Stickstoff-Agglomerat-Ionen nach einem Laufzeitverfahren," Doctoral thesis, Universität Karlsruhe, 1969.

[11]Coutant, J., Fois, M., and LeBihan, A. M., "INGRA: Resultats des Essais en Azote de la Source de Granions," Euratom-CEA Fontenay aux Roses, DPh-PFC/Note Interne 1051, 1969.

[12]Moser, H. O., Martin, J., and Salin, R., "High-Gradient Acceleration of Hydrogen Cluster Beams," Proceedings of the Symposium on Experimental and Theoretical Aspects of Heating of Toroidal Plasmas, Grenoble, 1976 (to be published).

[13]Mikosch, F., "Untersuchung einer Instabilität bei der Extraktion von Clusterionen aus einem Elektronenstoßionisator," Doctoral thesis, Universität Karlsruhe, 1975.

[14]Obert, W., "Production of Intense Condensed Molecular Beams (Cluster Beams)," Proceedings of the 6th Int. Cryogenic Eng. Conf., Grenoble, 1976 (to be published by IPC Science and Technology Press Ltd., Guildford).

[15]Hagena, O. F., Henkes, W., and Pfeiffer, U., "Neutralization of 1 MeV Hydrogen Cluster Beam," Proceedings of the 9th Symposium on Fusion Technology, Garmisch-Partenkirchen, 1976 (to be published by the Commision of the European Communities (CID), Luxembourg).

EXPERIMENTAL STUDY OF THE RESERVOIR TEMPERATURE SCALING OF CONDENSATION IN A CONICAL NOZZLE FLOWFIELD

W. D. Williams* and J. W. L. Lewis+
ARO, Inc., Arnold Air Force Station, Tenn.

Abstract

The onset of condensation and subsequent cluster growth were studied for conical nozzle expansions of the individual gases N_2, O_2, CO, and Ar. The Rayleigh scattering technique was used to investigate the variation of the spatial location of condensation onset and growth with change in the reservoir temperature, and scaling laws for the intensity of light scattering and axial location of condensation onset were determined. The reservoir temperature was varied over the range of 290 to 500K, and the reservoir pressure ranged from 0.8 atm for Ar to 7.9 atm for N_2. The use of intermolecular parameters for predicting the laws of condensation onset in the pressure-temperature plane is presented.

Nomenclature

b = a function of reservoir pressure and temperature and of nozzle throat diameter and expansion half angle
c = a factor which depends on reservoir pressure and nozzle throat diameter and expansion half angle
D_t = conical nozzle throat diameter

Presented as Paper 110 at the 10th International Symposium on Rarefied Gas Dynamics, Aspen, Colo., July 19-23, 1976. The research reported herein was conducted by the Arnold Engineering Development Center, Air Force Systems Command, and was sponsored by the Air Force Rocket Propulsion Laboratory, Air Force Systems Command. Research results were obtained by personnel of ARO, Inc., contract operator at AEDC.

*Physicist, Aerospace Projects Branch, von Kármán Facility.
+Senior Scientist, Aerospace Projects Branch, von Kármán Gas Dynamics Facility.

f	=	Rayleigh scattering function defined by Eq. (4)
$I(\|\|)$	=	relative Rayleigh scattering intensity defined by Eq. (1)
$I'(\|\|)$	=	relative Rayleigh scattered intensity normalized to the relative Rayleigh scattering intensity of a gas sample of number density N_o and with polarization parallel to the plane of polarization of the incident laser beam
K	=	constant in Eq. (1)
k	=	Boltzmann's constant
MOCS	=	method of characteristics solution
N	=	number density of gas species
N_i, N_o	=	i-mer and reservoir number density, respectively
N_1^o, N_T	=	isentropic monomer number density and total local number density, respectively
n_1, n_2	=	scaling constants
P_o	=	reservoir pressure
P_s, P_θ	=	saturation pressure and pressure at condensation onset, respectively
P_θ^*	=	P_θ normalized by the intermolecular potential parameter ε/σ^3
r	=	radial distance from flowfield centerline
$(s_\theta)^o$	=	isentropic supersaturation ratio
$(s_\theta')^o$	=	isentropic degrees of supercooling
T_o	=	reservoir temperature
T_s, T_θ	=	saturation temperature and temperature at condensation onset, respectively
T_θ^*	=	T_θ normalized by the intermolecular potential parameter ε/k
X_i	=	i-mer mole fraction
$\hat{x}$	=	axial position in flowfield from nozzle throat normalized by the nozzle throat diameter
$\hat{x}_s$, $\hat{x}_\theta$	=	axial location of saturation and condensation onset, respectively
α, α_i	=	polarizability and i-mer polarizability, respectively
$\varepsilon/k, \varepsilon/\sigma^3$	=	intermolecular potential parameters
λ	=	wavelength of scattered radiation

Introduction

Although early studies of condensation processes in expansion flowfields were primarily for the purpose of determining reservoir parameters required for producing uncondensed high-speed flowfields in ground-based simulation facilities, more recent investigations have broadened the area of applicability of such data to plume visibility calculations,[1] and clustering mechanisms.[2,3] The primary interest of the experiments reported herein is determining the spatial location of condensation onset in conical nozzle expansion flowfields and the variation of the onset location with the reservoir parameters P_0 and T_0 for a variety of gases. Additionally, it is desired to know the geometrical locus of onset throughout the flowfield, and the condensate rate of growth, and far-field asymptotic characteristics, all as manifested by the scattering of injected laser radiation. Finally, it is desired to verify suggested scaling laws for condensation processes using intermolecular potential constants,[4] thereby enabling the prediction of condensation phenomena for each class of molecular species.

Because of its spatial resolution and sensitivity capabilities optical scattering is most useful for determining the location of condensation onset as well as the spatial distribution of condensate within the flowfield. The use of laser Rayleigh scattering for such a characterization of condensing flowfields has been demonstrated by Wegener[5,6,7] and Daum[8] and their co-workers, Beylich,[9,10] and the present authors.[11-13] The existing experimental evidence amply demonstrates the size range of condensate to be typically 10-100 Å and of number density sufficiently low to allow the use of single-scattering Rayleigh-Gans theory. Further, the present authors have demonstrated[14] the application of laser Raman scattering to condensing flowfields for the measurement of monomer density and molecular rotational temperature. As was shown in Ref. 13, the combined use of Rayleigh and Raman scattering in conjunction with a specific cluster model enables the estimation of cluster sizes within the cluster growth region. The results reported in this work, however, will be restricted to Rayleigh scattering data.

Experimental Apparatus

The experimental apparatus has been described in detail previously.[13-15] Vacuum chamber pumping is provided by 78K liquid N_2 and 20K gaseous He cryopanels. Chamber pressure during the experiments was between 10^{-7} and 10^{-3} Torr, depending on the reservoir pressure, and underexpanded flows were insured. A 10.5^o half-angle conical nozzle source of 1.0 mm throat diameter was mounted on a motor-driven mechanism to provide movement in the x-y-z directions. The nozzle reservoir was instrumented with calibrated pressure and temperature gages. The gases were supplied from high pressure bottles and were passed through 250 Å particulate filters to minimize heterogeneous condensation processes. The stated gas purities were as follows: N_2, 99.998%; O_2, 99.5%; CO, 99.5%; and Ar, 99.99%.

The laser source was an argon ion laser operating at 1.0 W power at 514.5 nm wavelength. The directions of the incident radiation, flowfield axis, and scattered radiation collection system were all perpendicular to each other. The incident radiation was focused into a cylindrical volume of 50 to 100 μm diameter. The demagnification of the lens system used for collection of scattered radiation and the entrance aperture of the spectrometer used for wavelength selection provided observation of a 1.5 mm length of the scattering volume. Detection was accomplished using a cooled RCA C31034A photomultiplier tube. Following processing by a photon counting system, data registration was achieved by either digital display or strip chart recording. Rayleigh scattering in situ calibrations were performed routinely during the course of this study.

Results and Analysis

It will be recalled that the scattered intensity $I(\|)$ of a laser beam of intensity I_o and wavelength λ which is incident on a gas sample of number density N is given by

$$I(\|) = K\, N\alpha^2/\lambda^4 \quad (1)$$

where $I(\|)$, the scattered intensity with polarization parallel to that of the incident laser beam, has been normalized by I_o. The polarizability of the single molecular species present is α, and K includes all geometrical, transmission, and calibration factors. Assuming the flowfield to be composed of an ideal gas mixture of monomers and clusters, or i-mers, where i is the number of molecules per cluster, the Rayleigh intensity can be written as

$$I'(\|) = \sum_{i=1} (N_i/N_o)(\alpha_i/\alpha_o)^2 \quad (2)$$

where N_o is the reservoir number density and $I'(\|)$ includes further normalization provided by the scattered intensity from a collection of monomers of number density N_o. For an uncondensed, isentropic expansion

$$I'(\|) = (N_1/N_o)^o \equiv I^o(\|) \quad (3)$$

where super- and subscripts designate isentropic-uncondensed and reservoir parameters, respectively. Predicted values of $I^o(\|)$ are provided by the method of characteristics solutions (MOCS); deviation of $I'(\|)$ from $I^o(\|)$ is indicative of condensation.

The scattering function f is defined as

$$f = [I'(\|)/I^{o}(\|)] - 1 \simeq \sum_{i=2} X_i(\alpha_i/\alpha_1)^2 \tag{4}$$

where X_i is the i-mer mole fraction. The approximation in Eq. (4) results from the assumption that $N_1 \simeq N_1^o \simeq N_T$, where N_T is the total, local number density. Support for this approximation is provided by Refs. 13 and 14. Therefore, the initial deviation of the measured value of f from zero, or $I'(\|)/I^o(\|)$ from unity, serves to locate the position in the flow of the condensation onset; this position is denoted by $\hat{x}_\theta = x_\theta/D_t$, where D_t is the nozzle throat diameter. The value of $\hat{x}_\theta$ in conjunction with the MOCS then enables the determination of the static pressure (P_θ) and temperature (T_θ) at onset. If a supersaturation ratio $(s_\theta)^o$ is defined as

$$(s_\theta)^o = P_s/P_\theta \tag{5}$$

where P_s is the saturation pressure, the combined use of P_θ and the MOCS, as well as the species vapor pressure, then provides an experimental value of $(s_\theta)^o$. Similarly, the degrees of supercooling $(s'_\theta)^o$ achieved by the expansion prior to onset is defined as

$$(s'_\theta)^o = T_s - T_\theta \tag{6}$$

and $(s'_\theta)^o$ can be determined.

The scattered intensity function $I'(\|)$ was measured for each gas species over a range of axial and radial distances for each value of reservoir temperature (T_o), which was varied over the range of 290 to 500K. A single reservoir pressure (P_o) was used for the T_o variation. The axial variations of $I'(\|)$ for N_2 and Ar are shown in Figs. 1 and 2 for a range of T_o values studied. Shown in Fig. 3 is the axial variation of f for N_2 expansions with various values of T_o. The agreement of $I'(\|)$ with the isentropic MOCS prior to condensation onset is obvious. Further, it is seen that $\hat{x}_\theta$ increases with T_o and the magnitude of scattering decreases as T_o increases. Figures 4 and 5 show the measured radial profiles of $I'(\|)$ at various axial positions for N_2 for the T_o values 288 and 355K. Off-axis condensation is observed and rapidly disappears with increasing T_o.

From the axial variation of $I'(\|)$ for each species the variation of the scattering function f with $\hat{x}$ was found to be

$$\hat{x} = \hat{x}_\theta \exp(bf) \tag{7}$$

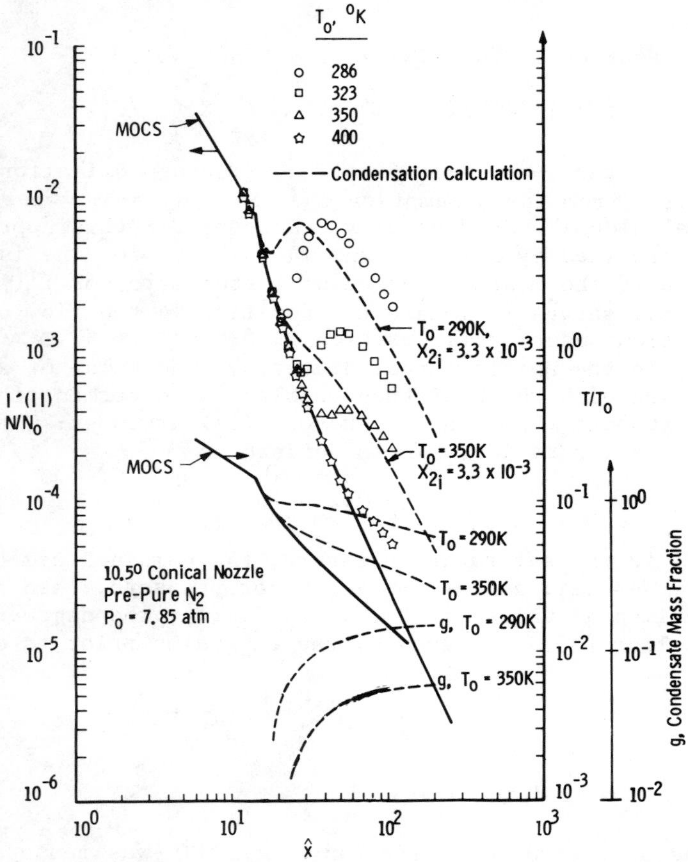

Fig. 1 Axial variation of I'(‖) for all N_2 reservoir temperatures investigated.

where the parameter b was found to depend upon T_0 according to

$$b = T_0^{-n_2} \tag{8}$$

The values of n_2 for each species are listed in Table 1.

It was found that $\hat{x}_\theta$ varied with T_0 as

$$\hat{x}_\theta = c\, T_0^{n_1} \tag{9}$$

where c is a factor which depends on the reservoir pressure and nozzle throat diameter and half angle,[15] all of which are constant for this work. The values of n_1 are listed in Table 1 for each species. Using the MOCS results and the vapor pressure data of Hilsenrath, et al.,[16] values of $(s_\theta)^0$ and $(s_\theta')^0$

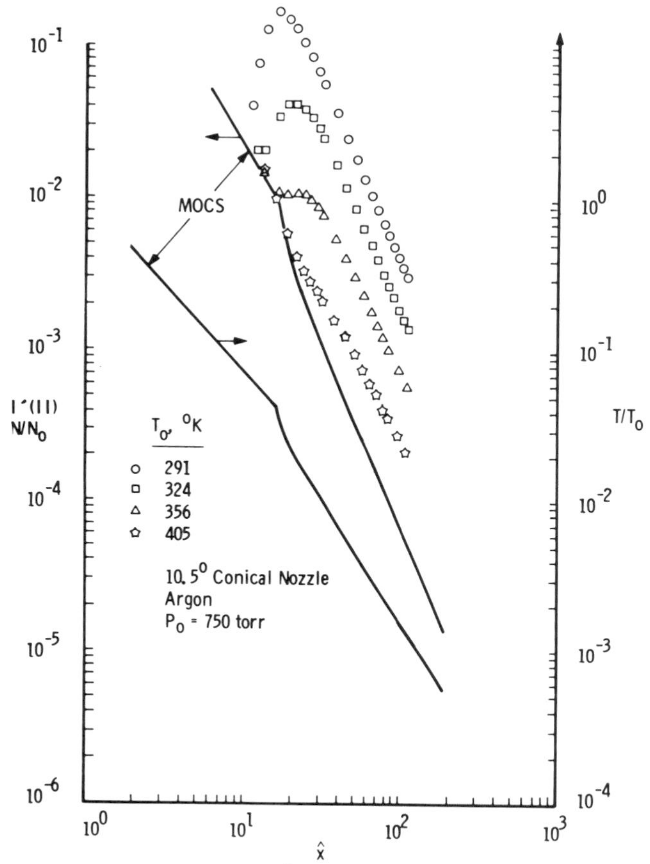

Fig. 2 Axial variation of I'(||) for all Ar reservoir temperatures investigated.

were found. Table 2 lists these supersaturation ratios and degrees of supercooling for each species as a function of T_o.

By using the well-depth and range parameters, ε and σ, respectively, of the Lennard-Jones 12-6 intermolecular potential function, reduced onset pressures (P^*_θ) and temperatures (T^*_θ) have been determined using

$$P^*_\theta = P_\theta/(\varepsilon/\sigma^3) \tag{10}$$

and

$$T^*_\theta = T_\theta/(\varepsilon/k) \tag{11}$$

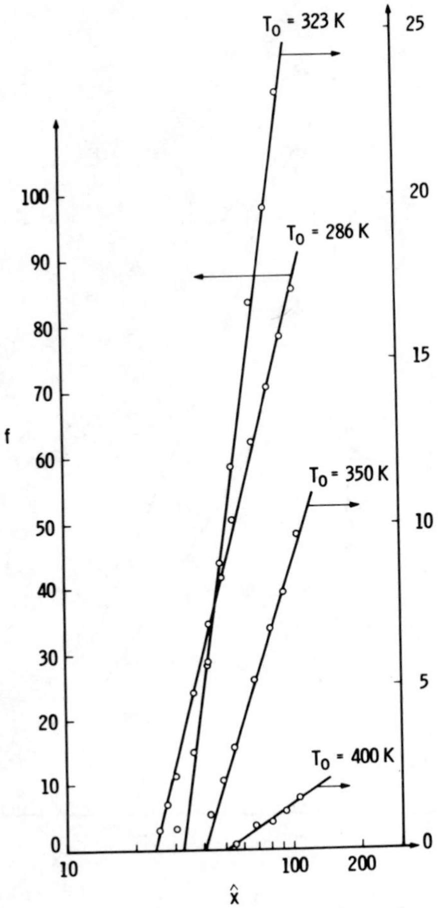

Fig. 3 Axial variation of scattering function, f, for N_2 expansions.

where k is Boltzmann's constant. With these values for each species, the locus of condensation onset in the pressure-temperature plane is shown in Figs. 6 and 7. The values for CO also are shown in Fig. 7 even though it is recognized that the use of the Stockmayer potential is ultimately necessary; however, at present the authors have insufficient data for other heteronuclear polar diatomic molecules to allow meaningful comparisons. The common locus for the homonuclear diatomic molecules N_2 and O_2 is to be noted. Additionally, the difference between the Ar results and those of N_2 and O_2 is to be noted.

A liquid-drop, monodisperse distribution condensation model was developed by M. Kinslow of ARO, Inc. for use by the authors to acquire information regarding condensate mass frac-

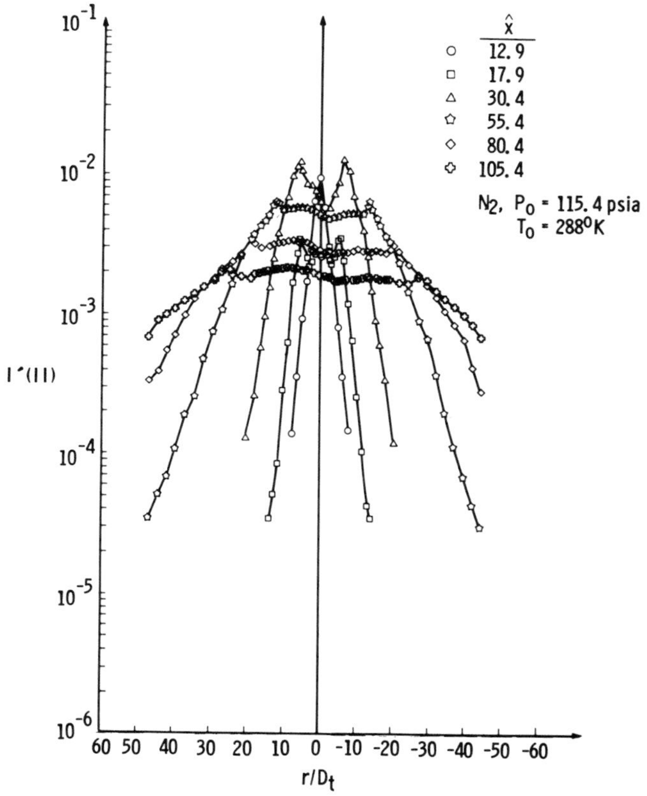

Fig. 4 Radial variation of I'(||) for a N_2 reservoir temperature of 288K at six axial positions.

tion and cluster size along the centerline of the expansion flowfield. This calculation assumes the condensing flowfield to be inviscid, adiabatic, and one-dimensional with no mass transfer across the stream tube boundary. The gas and condensate are assumed to obey the perfect gas relation. The condensed phase is assumed to be of the form of monodisperse spherical drops or particles which are characterized by bulk properties and to be in the free molecular flow regime relative to the uncondensed phase. Condensate-gas velocity slip effects are ignored, and the condensate growth rate is determined by gas condensate interaction only. The mass accommodation coefficient is assumed to be unity. Initial size and number density of spontaneous nucleation sites are adjustable parameters and may be selected so as to reproduce as closely as possible the experimental scattering results.

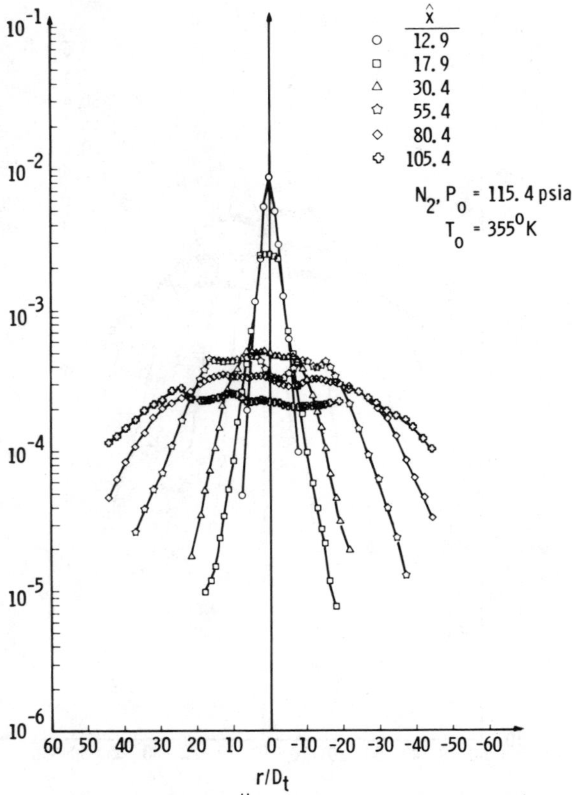

Fig. 5 Radial variation of I'(||) for a N_2 reservoir temperature of 355K at six axial positions.

For N_2 the condensation calculations were made for reservoir temperatures of 290 and 350K, and the flowfield conditions at the saturation point, $\hat{x}_s$, were used as the starting data. The results of the calculations are shown in Fig. 1. A dimer mole fraction on the order of 10^{-3} was used to obtain the fair

Table 1
SCALING CONSTANTS FOR GASES STUDIED

Gas	n_1	n_2
O_2	2.03	9.33
Ar	1.6	11.0
CO	1.97	10.5
N_2	2.12	9.59

Table 2

SATURATION AND CONDENSATION ONSET PARAMETERS FOR GASES STUDIED

Gas	T_0 (K)	P_0 (atm)	$\hat{X}_s$	P_s (torr)	T_s (K)	$\hat{X}_\theta$	P_θ (torr)	T_θ (K)	P_θ^*	T_θ^*	$(s_\theta)^0$	$(s'_\theta)^0$ (K)
O_2	294	6.12	7.1	25.7	66.2	22.6	0.338	19.4	1.27×10^{-6}	0.165	76.0	46.8
	325		8.4	16.8	64.6	25.4	0.193	18.2	7.27×10^{-7}	0.155	87.0	46.4
	353		10.0	11.1	62.6	30.3	0.101	16.6	3.81×10^{-7}	0.141	109.9	46.0
	401		12.6	6.33	60.5	40.8	0.0343	13.4	1.29×10^{-7}	0.114	184.5	47.1
Ar	291	0.987	0.37	17.3	63.5	11.0	0.816	18.6	2.6×10^{-6}	0.155	21.2	44.9
	324		0.41	12.2	62.3	13.8	0.415	15.7	1.32×10^{-6}	0.131	29.4	46.6
	356		0.45	9.36	61.1	15.9	0.270	14.2	8.6×10^{-7}	0.119	34.7	46.9
	405		0.48	6.34	59.9	19.0	0.0656	9.32	2.09×10^{-7}	0.0778	96.6	50.6
CO	285	4.76	7.9	14.9	59.3	20.5	0.402	21.1	2.07×10^{-6}	0.211	37.1	38.2
	324		10.0	8.69	57.4	27.8	0.107	16.5	5.52×10^{-7}	0.165	81.2	40.9
	353		11.7	5.70	56.1	33.6	0.0525	14.8	2.71×10^{-7}	0.148	108.6	41.3
	402		14.4	3.24	54.1	40.0	0.0279	13.9	1.44×10^{-7}	0.139	116.1	40.2
N_2	286	7.85	8.5	20.8	56.7	24.4	0.309	17.02	1.59×10^{-6}	0.179	67.3	39.7
	323		10.5	12.0	54.7	32.4	0.0990	13.89	5.10×10^{-7}	0.146	121.2	40.8
	350		12.6	8.34	53.2	40	0.0461	12.08	2.38×10^{-7}	0.127	180.9	41.1
	400		14.7	4.60	51.3	50	0.0211	11.04	1.10×10^{-7}	0.116	218	40.3

agreement between the calculations and Rayleigh scattering measurements. Also shown are the predicted variation of temperatures and condensate mass fraction with axial distance. Due to condensation there is a noticeable increase in temperature past the onset location. As shown in Ref. 15 this predicted temperature increase for a 14.5° nozzle agrees well with the experimentally determined temperatures obtained using laser Raman scattering.

Conclusions

The Rayleigh scattering measurements of condensing conical nozzle expansions of N_2, O_2, CO, and Ar have yielded scaling laws for the scattered light intensity and the axial condensation onset location as a function of reservoir temperature. A strong inverse dependence of the scattering function on the reservoir temperature ($f \propto T_0^{-10.4}$) was found which was both expected and consistent with mass spectrometric sampling studies of cluster beams.[2] The axial location of the onset of condensate growth was found to vary directly as T_0^2. Further, the use of intermolecular potential parameters for determining the condensation onset locus in the pressure-temperature plane was investigated. The onset of condensation is controlled

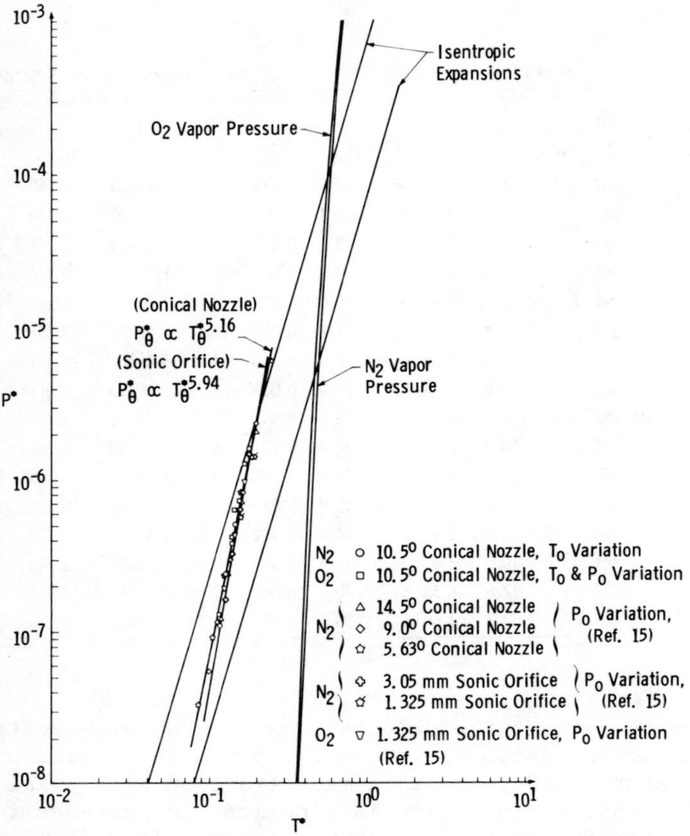

Fig. 6 N_2 and O_2 condensation onset locus using reduced values.

primarily by the formation of molecular dimers, the process of which depends upon both the equilibrium number density of dimers and the gas dynamic expansion rate. Since the former is a function of the intermolecular potential parameters and the latter depends upon the specific heat ratio., it is not unreasonable to find differences in these results between N_2 and O_2, on the one hand, and Ar, on the other, even though both classes of molecules are well described by the 12-6 Lennard-Jones potential. The understandably distinct behavior of CO in this regard was noted, and additional heteronuclear polar molecule data are required for comparison and formulation of scaling laws. Finally, a comparison of the experimental re-

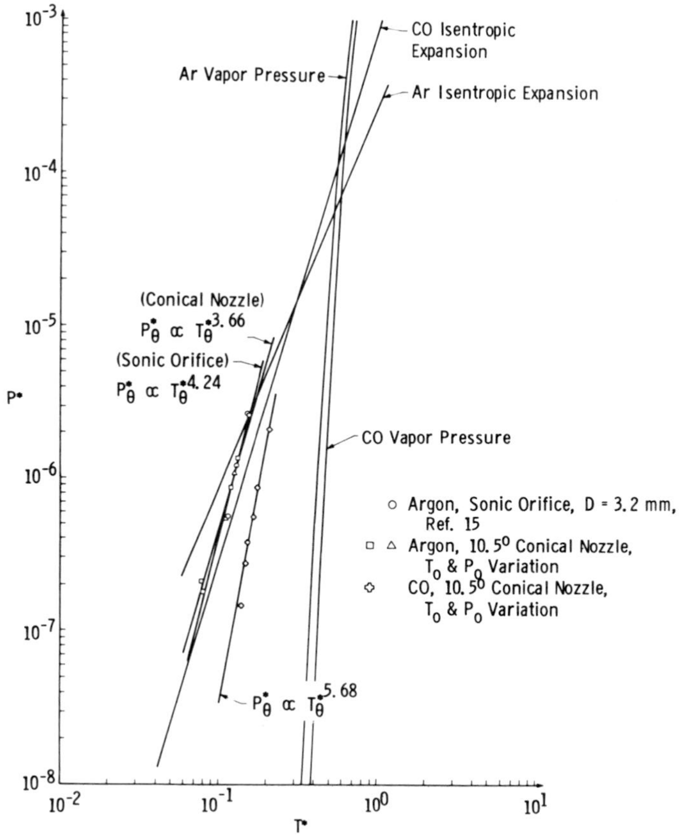

Fig. 7 Ar and CO condensation onset locus using reduced values.

sults and the predictions based upon a classical, liquid drop condensation model was presented, and the discrepancies were noted.

References

[1]Kung, R. T. V., Cianciolo, L., and Myer, J. A.,"Solar Scattering from Condensation in Apollo Translunar Injection Plume." AIAA Journal, Vol. 13, April 1975, pp. 432-437.

[2]Hagena, O. F. and Obert, W., "Cluster Formation in Expanding Supersonic Jets: Effect of Pressure, Temperature, Nozzle Size and Test Gas," The Journal of Chemical Physics, Vol. 56, March 1972, pp. 1793-1802.

[3]Yealland, R. M., Deckers, J. M., Scott, I. D., and Touri, C. T., "Dimer Concentrations in Supersonic Free Jets," Canadian Journal of Physics, Vol. 50, Oct. 1972, pp. 2464-2470.

[4]Hagena, O. F., "Condensation in Supersonic Free Jets," Rarefied Gas Dynamics, edited by L. Trilling and H. Y. Wachman, Vol. II, Academic Press, New York, 1969, p. 1465.

[5]Wegener, P. P. and Stein, G. D., "Light Scattering Experiments and Theory of Homogeneous Nucleation in Condensing Supersonic Flow," International Symposium on Combustion, Combustion Institute, University of Poiters, Poiters, France, pp. 1183-1190.

[6]Stein, G. D., "Angular and Wavelength Dependence from a Cloud of Particles Formed by Homogeneous Nucleation," Dept. of Engineering and Applied Sciences, Yale University, New Haven, Conn., Dec. 1968.

[7]Clumpner, J. A., "Light Scattering from Ethyl Alcohol Droplets Formed by Homogeneous Nucleation," The Journal of Chemical Physics, Vol. 55, Nov. 1971, pp. 5042-5045.

[8]Daum, F. L. and Farrell, C. A., "Light Scattering Instrumentation for Detecting Air Condensation in a Supersonic Wind Tunnel," Fourth International Congress on Instrumentation in Aerospace Simulation Facilities, June 1971, Brussels, Belgium, pp. 209-215.

[9]Beylich, A. E., "Condensation in Carbon Dioxide Jet Plumes," AIAA Journal, Vol. 8, May 1970, pp. 965-967.

[10]Beylich, A. E., "Experimental Investigation of Carbon Dioxide Jet Plumes," The Physics of Fluids, Vol. 14, May 1971, pp. 898-905.

[11]Lewis, J. W. L. and Williams, W. D., "Argon Condensation in Free Jet Expansions," Arnold Engineering Development Center, Arnold Air Force Station, Tenn., AEDC-TR-74-32, July 1974.

[12]Lewis, J. W. L., Williams, W. D., Price, L. L., and Powell, H. M., "Nitrogen Condensation in a Sonic Orifice Expansion Flow," Arnold Engineering Development Center, Arnold Air Force Station, Tenn., AEDC-TR-74-36, July 1974.

[13]Lewis, J. W. L. and Williams, W. D., "Profile of an Anisentropic Nitrogen Nozzle Expansion," Arnold Engineering Development Center, Arnold Air Force Station, Tenn., AEDC-TR-74-114, Feb. 1975.

[14]Lewis, J. W. L. and Williams, W. D., "Measurement of Temperature and Number Density in Hypersonic Flow Fields Using Laser Raman Spectroscopy," AIAA Paper 75-175; also, AIAA Journal, Vol. 13, Oct. 1975, pp. 1269-1270.

[15]Williams, W. D. and Lewis, J. W. L., "Experimental Study of Condensation Scaling Laws for Reservoir and Nozzle Parameters and Gas Species," AIAA Paper 76-53, Jan. 1976, Washington, D.C.

[16]Hilsenrath, J., Beckett, C. W., Benedict, W. S., Faro, L., Hoge, H. J., Masi, J. F., Nuttall, R. L., Touloukian, L. S., and Wooley, H. W., Tables of Thermal Properties of Gases, National Bureau of Standards, Washington, D.C., Circular 564, Nov. 1955.

ISOTOPE EFFECT ON THE FORMATION OF HYDROGEN CLUSTER BEAMS

W. Obert*

Institut für Kernverfahrenstechnik
Kernforschungszentrum Karlsruhe, Karlsruhe, West Germany

Abstract

Experiments with cluster beams formed by the hydrogen isotopes H_2 and D_2 show considerable differences in cluster size and intensity. It is found that, in spite of the marked differences in the thermodynamic properties of both isotopes, under identical stagnation conditions, the cluster size is nearly equal for both isotopes at 30°K nozzle temperature. At 77°K nozzle temperature, the H_2 clusters are about twice, and at 110°K about 2.5 times as large as that of D_2; however, the vapor pressure of D_2 is smaller than that of H_2, and therefore larger D_2 clusters would be expected. On the basis of the well-known marked differences in the thermodynamic properties of the two isotopes that control the clustering process, the differences in cluster size and intensity are discussed. It is shown that the pronounced isotope effect in vapor pressure will be nearly compensated for by the effect in molar volume and condensation energy. The differences resulting from the variation of the nozzle temperature are explained by the differences caused by quantum mechanics in the number of excited rotational energy terms of the two isotopes. The number of these terms determines the specific heat, and thus the curve for the adiabatic expansion of both gases. Accordingly, it is possible to discuss cluster beam formation in the case of the ortho- and paramodification of hydrogen, just as in the case of tritium.

1. Introduction

Condensed molecular beams formed by rapid expansion of gas into a vacuum[1,2] are of growing interest both for the study of

---

Presented as Paper 101 at the 10th International Symposium on Rarefied Gas Dynamics, Aspen, Colo., July 19-23, 1976
*Dr.-Ing., Dipl.-Phys.

the basis condensation process[3] itself, and with regard to their application to heating and fueling fusion devices.[4] It is particularly interesting, in this context, to compare the two hydrogen isotopes H_2 and D_2 with respect to the formation of cluster beams. Unlike most of the other gases, hydrogen and its isotopes show marked differences in their thermodynamic properties, which are decisive for the condensation process. In addition, the formation of heavy hydrogen cluster beams is of exceptional interest for fusion experiments.

2. Experimental Setup

The experimental setup for cluster beam formation and detection is similar to that discussed at earlier symposia.[5,6] The experiments were carried out with a hypersonic nozzle at temperatures of 30°, 77°, and 110°K. The geometry of the nozzle used is given in the inset of Fig. 1. It is a trumpet-shaped nozzle, which proved to be particularly appropriate to form intense cluster beams.[7] The gas was discharged without any skimmer or collimator into a high vacuum chamber with a base pressure of 10^{-6} Torr. To minimize the pumping effort, the nozzle flow was pulsed for periods up to 50 msec, and by observation of the time-dependence of the beam signal, it was noted that the pressure rise in the discharge chamber due to the gas pulse did not change the flow pattern of the freely expanding jet.

3. Experiments

The cluster size and beam intensity were measured for both hydrogen isotopes as a function of the stagnation pressure p_0

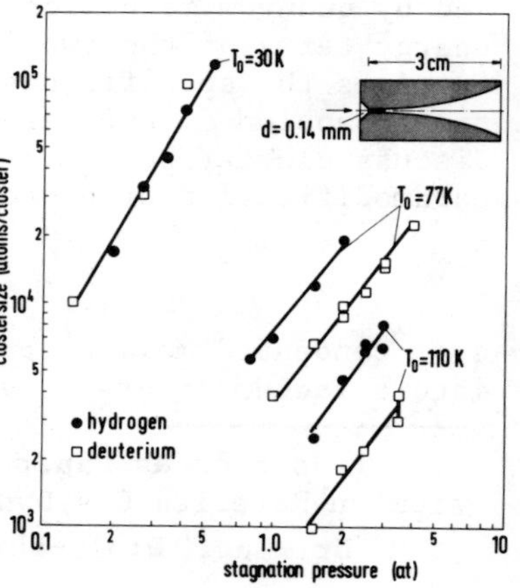

Fig. 1 Cluster size as a function of stagnation pressure for three different nozzle temperatures in the case of H_2 and D_2. The used nozzle is also plotted in this figure.

and the nozzle temperature T_o, using a retarding field detector and an ionization gage, respectively. The different nozzle temperatures were chosen to divide the effect of the specific heat and the other thermodynamic properties (see Sec. 4).

3.1 Cluster Size

Figure 1 shows, for both isotopes, the cluster size on the beam axis as a function of the stagnation pressure p_o, for different nozzle temperatures T_o. Figure 1 illustrates 1) the known increase of cluster size with rising stagnation pressure p_o and decreasing nozzle temperature T_o; 2) comparing the two isotopes, one notes, for given stagnation conditions, marked differences in the cluster size, which depends on the nozzle temperature: at 30°K nozzle temperature, nearly the same cluster size results for H_2 and D_2; at 77°K nozzle temperature, the H_2 clusters are about twice the size and at 110°K they are about 2.5 times as large as that of D_2.

Regarding these results, two facts seem relevant. The first is the marked difference in cluster size under a given stagnation condition. On the other hand, one notes that these differences are strongly dependent on the nozzle temperature. In order to discuss this result, the thermodynamic properties of both hydrogen isotopes must be compared with one-another.

3.2 Intensity

Figure 2 shows the intensity on the beam axis as a function of the stagnation pressure p_o for different nozzle temperatures T_o. Figure 2 illustrates 1) for both gases the well-known increase in intensity with rising stagnation pressure p_o and decreasing nozzle temperature T_o. This increase is paralleled by the increase in cluster size; 2) comparing both gases, a given stagnation condition p_o/T_o in any case yields more in-

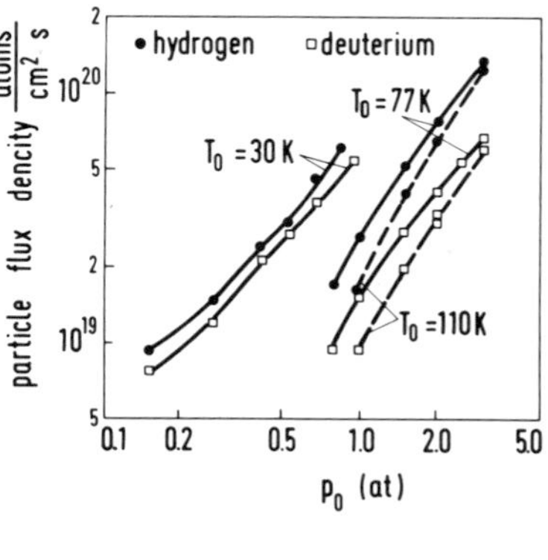

Fig. 2 Intensity as a function of stagnation pressure for three different nozzle temperatures in the case of H_2 and D_2.

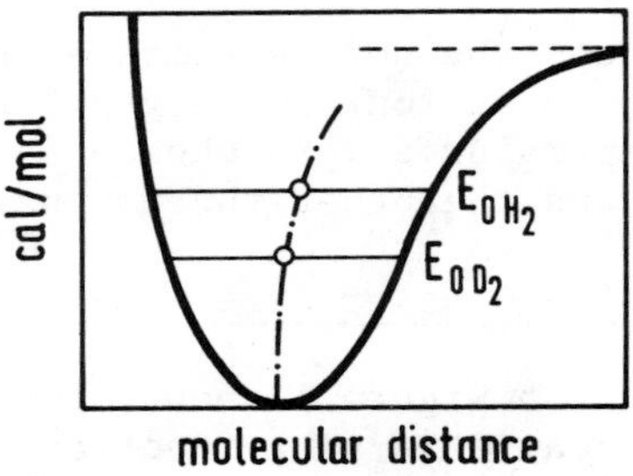

Fig. 3 Van der Waals potential curve for H_2 and D_2.

tensity for H_2 than for D_2; 3) the difference in intensity at given temperature is nearly independent of the pressure p_0, but increases with the nozzle temperature (it amounts to a factor of 1.2 and 2 at 30° and 77°K, respectively).

4. Thermodynamic Properties of Hydrogen Isotopes

As opposed to other gases (except He), the hydrogen isotopes differ markedly in their thermodynamic properties because of the considerable mass differences and the different quantum-mechanical symmetry. The differences in the thermodynamic properties can be understood when observing the potential curve of the van der Waals binding energy and the ground energy term, as shown in Fig. 3. In spite of the same potential curve, the energy term for H_2 is higher than that for D_2, corresponding to the smaller mass of H_2. Therefore, the condensation energy gets smaller and the vapor pressure gets higher in the case of H_2. Because of the anharmonicity of the potential curve, the mean distance for the van der Waals molecule is greater in the case of H_2. This corresponds to a higher molar volume, and also to a larger cluster surface for a given cluster size (atoms/cluster) in the case of H_2.

It is known that there also are effects due to the quantummechanical modification of the gas. Although they generally yield only small effects, the specific heat at low temperatures depends strongly on the given modification. In the experiments discussed in this paper, we can assume that the n-modification is valid, because the gas is at room temperature, except for the short time when it passes through the cooled nozzle.

Details of the thermal properties of hydrogen and its isotopes[8,9] are listed in Fig. 1 for the different modifications. Figure 4 shows the rotational energy portion of the specific heat as function of the temperature.

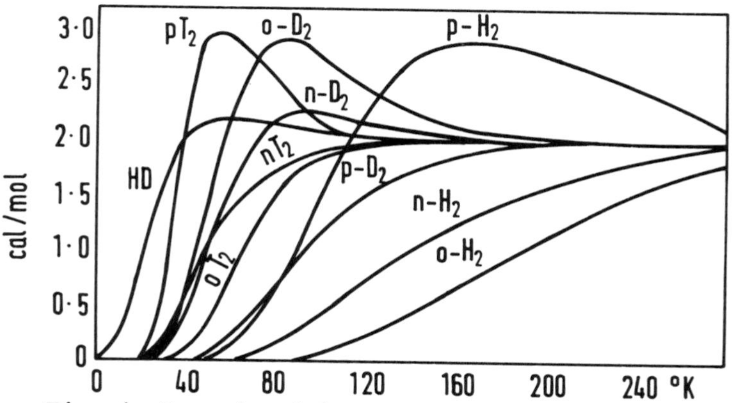

Fig. 4 Rotational heat curves for H_2 and D_2.

5. Discussion

5.1 Cluster Size

The cluster size measurements at different temperatures show how the different thermodynamic properties control the clustering process. This becomes obvious from Fig. 5, which shows the curve for the vapor pressure and the curve for adiabatic expansion for H_2 and D_2 starting from two significant stagnation conditions. The expansion curve is given by the i-

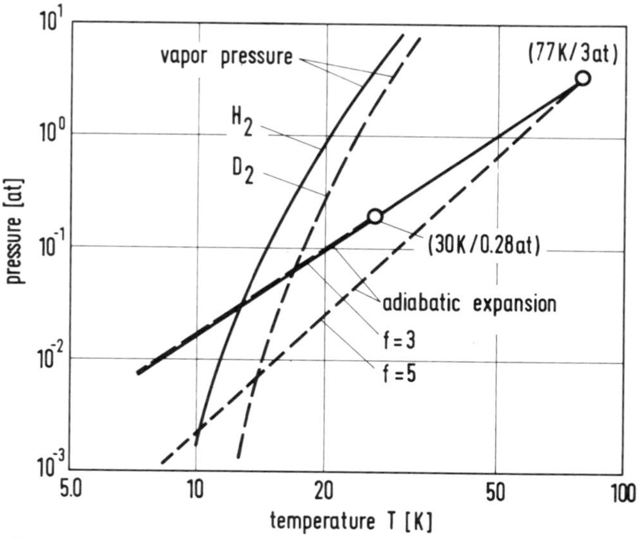

Fig. 5 Curves for vapor pressure and adiabatic expansion starting from two significant stagnation conditions.

sentrope $p \sim T^{(f+2)/2}$ (where p, T, and f are the pressure, temperature, and the number of excited degrees of freedom, respectively). Therefore, it is determined by the specific heat. Accordingly, quite different expansion curves will result for the hydrogen isotopes at the different temperatures, corresponding to the different specific heats shown in Fig. 4.

A simple clustering model is assumed in which the clustering process is determined by the removal of the condensation energy from the cluster. The cluster growth, and thus the resulting cluster size, will be higher when the number of collisions that take place on the cluster is higher. This means that the cluster size will be higher when the density of the surroun

the cluster size will be higher when the density of the surrounding gas is higher the cluster surface is larger, and the expansion rate is slower. On the other hand, when the growth is higher, the amount of condensation energy that must be removed is smaller.

Because of the differences in the vapor pressure and expansion curve, very different conditions are obtained for clustering. This will be discussed now for particular stagnation conditions.

<u>Condition 1</u> (200 Torr/30°K): In this case, both H_2 and D_2 give the same curve for the adiabatic expansion, which corresponds to a monoatomic gas (both gases have not yet excited rotational terms). Therefore, D_2 will reach saturation at higher densities, and thus larger D_2 clusters can be expected. However, the experiments show the same cluster size as for H_2. Taking into account the clustering model mentioned previously, it can be shown that in case of D_2 the higher density, which is due to the lower vapor pressure, is nearly compensated for by the isotope effect of larger condensation energy and smaller molar volume. The resulting equal cluster sizes thus can be explained.

<u>Condition 2 (3 at/77°K):</u> In the case of H_2, this condition yields the same cluster size as condition 1 because both of them lie on a common adiabatic expansion curve.[3] In the case of D_2, the expansion curve shows a stronger decline because of the excited rotational energy (see Fig. 4). The saturation is, therefore, reached at lower densities, and thus smaller clusters result than in the case of condition 1.

Compared with H_2, the effect of larger condensation energy and smaller molar volume of D_2 is added to the effect of smaller density at saturation. This explains the smaller cluster size of D_2, in spite of the smaller vapor pressure.

5.2 Intensity

The intensity, i.e., the particle flux density of the H_2 and D_2 cluster beams differ roughly with the nozzle particle flux, which decreases with rising particle mass and increasing specific heat. The higher mass of the D_2 molecules explains the smaller intensity of the D_2 cluster beams. The additional effect of higher specific heat, which occurs at 77° and 110°K in the case of D_2 (see Fig. 4), explains the higher difference in the H_2 intensities at these temperatures.

5.3 Principle of Corresponding States

A similar result is derived when the principle of corresponding states, used in an earlier RGD paper,[5] is applied to compare clustering of different gases. The principle predicts similar reduced thermodynamic properties for different gases, and thus the same cluster size for identical stagnation conditions, assuming that the expansion curves are the same. It should be noted, however, that, strictly speaking, this principle is no longer valid for quantummechanical effects, as in the case of hydrogen.

Table 1 Thermal properties for the hydrogen

Hydrogen modification	nH_2	pH_2	nD_2	oD_2	T_2
Critical point					
-temperature, T_0 (K)	33.19		38.34		40.44
-pressure, p_c (Torr)	9865		12488		13878
Heat of vaporization at boiling point, W (J/mole)	899.1	898.3	1210.	1229.	1393.3
Reduced value, W/T_c	27.09		31.58		34.4
Molar volume at triple point, v (cm^3 $mole^{-1}$)	23.25	23.30	20.48	20.58	18.62
Reduced value, $v/(T_c/p_c)$	6910		6670		6389
Vapor pressure at 10 K p_s (Torr)	1.73	1.92	0.05	0.05	0.005[a]

[a]Estimated

The reduced properties are given in Table 1 for the n-modification of H_2 and D_2. If the stagnation condition 1 (300 Torr/30°K) is reduced in the same manner, two reduced stagnation conditions result: one for H_2 and one for D_2. Both are close to one-another, on a common adiabatic curve; therefore, similar cluster sizes can be expected when this model is used if the specific heats are equivalent, as under condition 1.

5.4 Conditions for Equal Cluster Sizes

As known from earlier experiments, stagnation conditions p_o/T_o, which will yield equal cluster sizes, lie close to a common adiabatic expansion curve (see, also, Sec. 5.1). On the other hand, it may be possible to use the curve for the adiabatic expansion to determine the actual effect of specific heat. This will be done by observing these stagnation conditions, which yield equal cluster sizes, as shown in Fig. 6.

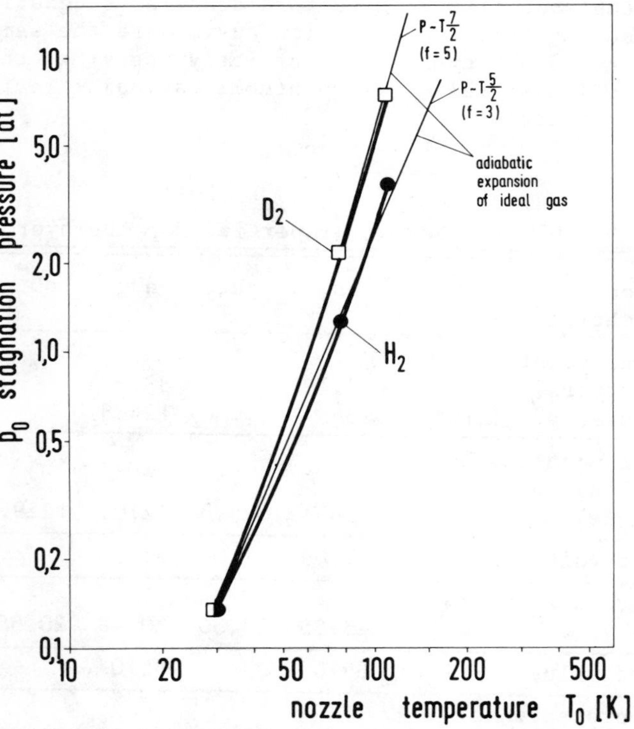

Fig. 6 Stagnation pressure p_o vs. nozzle temperature T_o for a given cluster size in the case of D_2 and H_2.

One notes that, in fact, almost up to 77°K, hydrogen shows a slope corresponding to a monoatomic gas; the same applies for D_2 around 30°K. Between 30° and 110°K, as expected, both gases differ as to the decline of the curve according to the specific heat values given in Fig. 4.

6. Conclusion

These results show that the specific heat, and thus the adiabatic expansion curve, respectively, are the dominant effect in hydrogen cluster beam formation. In addition, the marked differences in the thermodynamic properties of the hydrogen isotopes nearly compensate themselves. Accordingly, the formation of cluster beams can be discussed in the case of the ortho- and paramodifications in the same manner as in the case of tritium; the resulting adiabatic expansion curve can be used according the specific heats, which are given in Fig. 4

References

[1] Becker, E. W., Bier, K., and Henkes, W., "Strahlen aus kondensierten Atomen und Molekeln im Hochvakuum", Zeitschrift für Physik, Vol. 146, 1956, pp. 333-338

[2] Becker, E. W., Klingelhöfer, R., and Lohse, P., "Strahlen aus kondensiertem Wasserstoff, kondensiertem Helium und kondensiertem Stickstoff im Hochvakuum", Zeitschrift für Naturforschung, Band 17a, Heft 5, 1962, pp. 432-438

[3] Hagena, O. F., and Obert, W., "Cluster Formation in Expanding Supersonic Jets: Effect of Pressure, Temperature, Nozzle Size, and Test Gas", The Journal of Chemical Physics, Vol. 56, No. 5, March 1972, pp. 1793-1802

[4] Henkes, W., "Production of Plasma by Injection of Charged Hydrogen Clusters", Physics Letters, Vol. 12, October 1964, pp. 322-323

[5] Hagena, O. F., and Obert, W., "Condensation in Supersonic Free Jets: Experiments with Different Gases", Proceedings of the 7th International Symposium on Rarefied Gas Dynamics, Pisa, 1970, Vol. I, pp. 585-591

[6] Hagena, O. F., and v. Wedel, H., "Cluster Beams from Gas Mixtures: Effects of Carrier Gas on Cluster Size and Beam Intensity", Proceedings of the 9th International Symposium on Rarefied Gas Dynamics, DFVLR Press, Porz-Wahn, 1974, Vol. II, pp. F.10-1 - F.10-10

[7] Obert, W., "Production of Intense Condensed Molecular Beams (Cluster Beams)", Proceedings of the 6th Int. Cryogenic Eng. Conf., Grenoble, 1976 (to be published by IPC Science and Technology Press Ltd., Guildford).

[8] Farkas, A., "Light and Heavy Hydrogen", The Cambridge Series of Physical Chemistry, Cambridge University Press, Cambridge, 1935

[9] Roder, H. M., et al., "Survey of the Properties of the Hydrogen Isotopes below the Critical Temperatures", National Bureau of Standards, TN-641, 1973, 113 pp.

NOZZLE-INLET DESIGN FOR AEROSOL BEAM INSTRUMENTS

B. Dahneke* and D. Padliya†

University of Rochester, Rochester, N.Y.

Abstract

This paper describes an investigation of a nozzle-inlet design for aerosol beam instruments which, it is hoped, will extend the range of particle sizes that can be counted and measured in aerosol beam instruments. The use of aerosol beams for measuring the atmospheric aerosol is discussed. As an example, a time-of-flight aerosol beam spectrometer (TOFABS) is described, together with preliminary analytical and experimental calibration curves for this instrument. Growth of the beam particles by condensation to extend the particle size range and performance of aerosol beam instruments is discussed. Numerical results show that condensational growth of small particles in a converging nozzle is not adequate because the particle residence time is not long enough in the supersaturated region of the nozzle flow. Calculations of particle growth in a pre-nozzle growth chamber show that this type of inlet-nozzle system would provide sufficient particle growth to extend the operating range of many aerosol beam instruments substantially.

Introduction

To determine the toxicity of airborne particles, it is essential to measure both particle size and composition, since the size fixes the probability of the particle penetrating into the working regions of the lung, whereas the composition fixes the particle's influence on the lung tissue and other

---

Presented as Paper 38 at the 10th International Symposium on Rarefied Gas Dynamics, Aspen, Colo., July 19-23, 1976.

*Asst. Professor, Radiation Biology and Biophysics.
†Postdoctoral Research Fellow, Radiation Biology and Biophysics.

body materials. The respirable particle size range extends from a few hundredths to a few micrometers in diameter. Smaller particles are removed by diffusion and larger particles by sedimentation and inertial impaction before they reach the working regions of the lung.

Because so many chemical species are involved, together with variations in time of day, humidity, location, wind speed and direction, etc., a vast amount of data must be obtained and processed to discover and monitor the public health danger of the atmospheric aerosol and to determine its origins. For this reason, the composition analysis of respirable airborne particulates needs to be obtained automatically, in situ, and in real time at a number of sampling locations.

Aerosol beam spectrometry[1] provides techniques for the size sorting of airborne particulates wherein the size-sorted samples are available in an environment suitable for composition analysis of the particulate material. Moreover, such instruments are simple and relatively inexpensive and seem, at this point in their development, well suited for the task of measuring atmospheric particulates.

We discuss in this paper certain aspects of the nozzle design for various aerosol beam instruments. In particular, nozzle designs for time-of-flight aerosol beam spectrometers (TOFABS) are investigated in view of their use in counting the total number concentration of airborne particulates, in measuring particulate size distributions, and in providing size-separated samples for composition analysis.

Time-of-Flight Aerosol Beam Spectrometry (TOFABS)

A schematic diagram of a TOFABS is shown in Fig. 1. The aerosol sample is drawn downward into the vertical sampling tube by the action of a small pump, which maintains fresh sample continuously throughout the sampling tube. A small fraction of the aerosol flow in the sampling tube is drawn through the metering orifice and entrained in the center of a clean air flow. This total flow is expanded through the conveying nozzle into the vacuum chamber. By maintaining a small sample-to-total-flow ratio, the aerosol sample is confined to the central core of the expanding air jet, resulting in a focusing of the aerosol beam in the vacuum chamber.[2] This focusing of the beam occurs in a twofold sense: 1) the solid angle of the aerosol beam in the vacuum chamber is reduced, since the radial expansion of the gas jet in the vacuum chamber imparts a smaller radial velocity component to the beam particles when

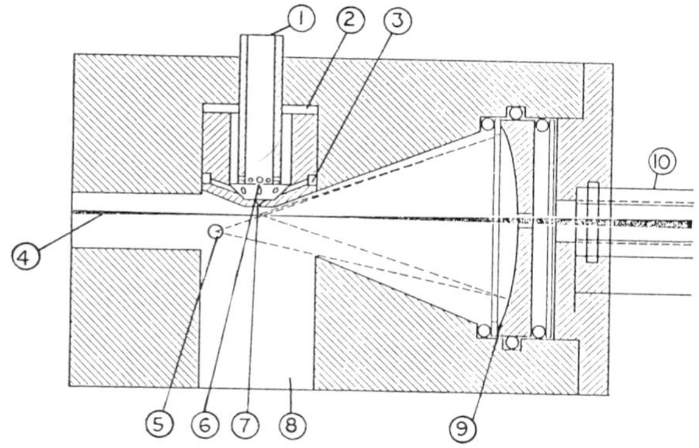

Fig. 1 Schematic diagram of TOFABS. 1) sampling tube; 2) slight vacuum; 3) clean fresh air duct; 4) orifice; 5) nozzle; 6) laser beam; 7) deflecting mirror; 8) vacuum chamber; 9) spherical mirror; 10) light trap.

they are confined to the axial region of the jet; and 2) particles of the same aerodynamic size obtain the same terminal velocity in the vacuum chamber, since the axial gas flow does not vary with radius near the jet axis.

Of course, particles of different aerodynamic size obtain different terminal velocities in the vacuum chamber, the smaller particles obtaining the higher velocities. Because of this property of aerosol beams, the measurement of particle TOF over a fixed path length in the vacuum chamber infers the particle's aerodynamic size.

Because the TOFABS of Fig. 1 uses a scattered light signal to detect the particles and because the amplitude of the scattered light signal may depend on particle size, orientation, and on index of refraction in addition to particle location in the beam, electronic circuits were designed to "normalize" the signal from the photomultiplier tube (PMT) generated when a particle passed through the light beam, scattering light into the spherical mirror and, eventually, onto the PMT. The electronic circuitry is shown schematically in Fig. 2. An example PMT signal (upper trace) and processed logic pulse signal (lower trace) are shown in Fig. 3 for a beam of 2.02 μm latex spheres. The intensity distribution in the light beam is gaussian causing the time variation of the scattered light (PMT) signal as the particle traverses the beam to be gaussian. The constant amplitude logic pulse of Fig. 3 has the particle TOF as its width. This pulse turns on when the instantaneous PMT signal reaches 40% of its maximum value and turns off when

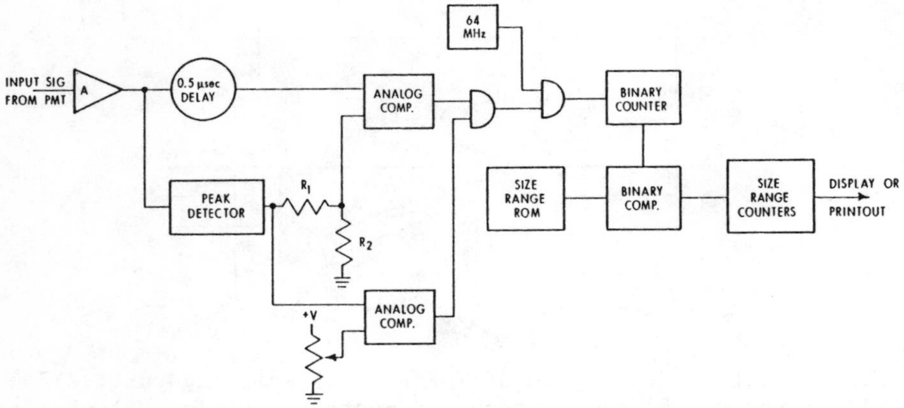

Fig. 2 Schematic diagram of electronic circuitry.

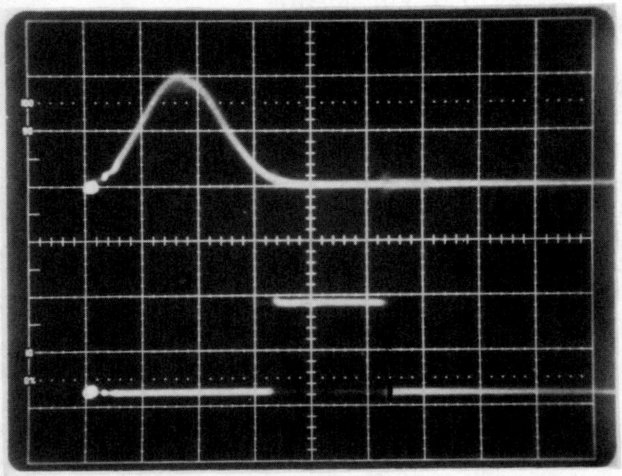

Fig. 3 Example PMT signal (upper trace) for a 2.02 μm latex particle. The logic pulse (lower trace) has width equal to the particle's TOF through the light beam. Horizontal scale is 200 nsec/cm.

the signal falls below 40% of its maximum. This procedure insures that the particle TOF is always measured over a fixed path length in the vacuum chamber.

Analytical Calibration and Nozzle Design

The motion of a particle in a fluid stream is described by

$$m_p \, (dv_p/dt) = - f(v_p - v_f) \tag{1}$$

where m_p is the particle mass, v_p the particle velocity, t the time, f the particle-fluid friction coefficient, and v_f the local fluid velocity (neglecting the influence of the particle). For spherical particles

$$\alpha(t) \equiv f/m_p = 9\mu\kappa/2a^2\rho_p C_s \qquad (2)$$

where μ is the local gas viscosity, a the particle radius, ρ_p the particle mass density, C_s the slip correction factor, and κ the dynamic shape factor.

The solution of (1) is

$$v_p(t) = v_f(o)\ \exp\left(-\int_o^t \alpha d\nu\right) + \exp\left(-\int_o^t \alpha d\nu\right) \int_o^t \exp\left(\int_o^\theta \alpha d\nu\right) v_f(\theta)\alpha(\theta)d\theta \qquad (3)$$

where t = 0 corresponds to the instant the particle passes the nozzle entrance plane x = 0. Particle location in the nozzle-free jet system is given by

$$x(t) = \int_o^t v_p(t)dt \qquad (4)$$

Particle velocity $v_p(x)$ is obtained by first obtaining the gas flow field $v_f(x)$ in the nozzle[3] and free jet[4] system and then by numerical evaluation of (3) and (4). The terminal velocities in the vacuum chamber so obtained for several particle sizes provide the calibration curves like those shown in Fig. 4 for three different nozzle geometries. The particle TOF has been obtained by dividing the calculated terminal velocity of the particle into the path length (72 μm). Measured data for latex spheres accelerated through nozzle 1 are compared with theory in Fig. 5.

A TOFABS nozzle design is regarded as satisfactory when it provides (a) sufficient acceleration of the expanding gas to cause sufficient variation in particle terminal velocity for adequate resolution, and (b) not too much variation in particle terminal velocity (or TOF) so that the full particle size range of interest can be measured and size selected by the detection and separation systems. Most nozzle designs we have investigated analytically and experimentally have satisfied these two requirements. Some additional requirements of a special nature have not been satisfied, however. In the fol-

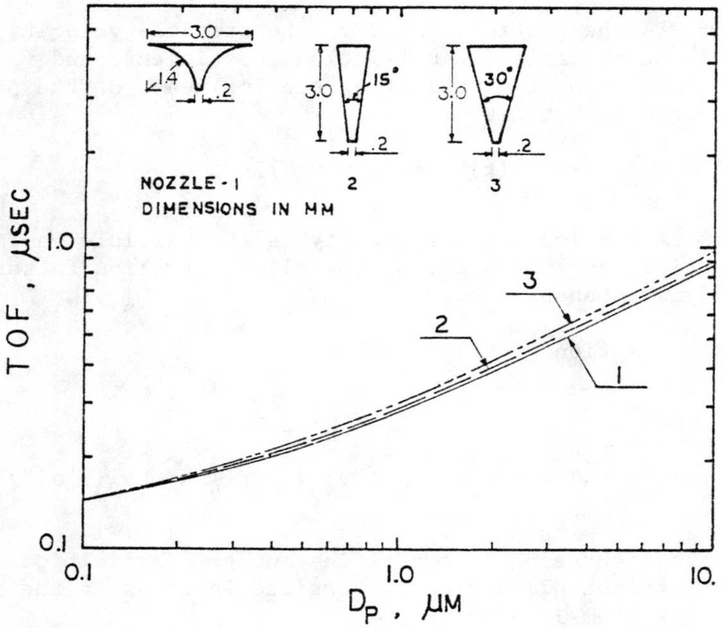

Fig. 4 Analytical calibration curves for different nozzle geometries.

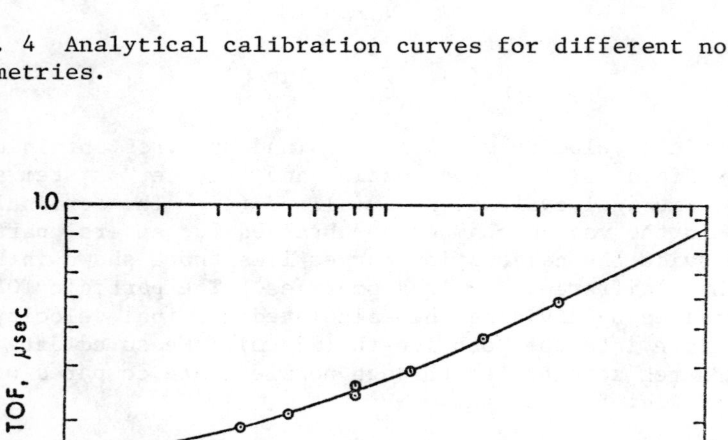

Fig. 5 Analytical and measured calibration data for unit density spheres. The analytical curve was fitted to the measured data by selecting the flight path length of 72 μm as the light beam "thickness".

lowing sections we will describe these additional requirements and the nozzle inlet system we propose to meet them.

Particle Growth by Condensation

One problem of the TOFABS herein described with the scattered light-PMT system for detecting the particles is its limitation in detectable particle size. The lower particle size limit of the present instrument is 0.3 μm particle diameter. Improved scattered light detection systems should be able to extend this lower limit to below 0.1 μm, but a significant range of respirable particle sizes remains which cannot be detected by the present instrument.

To count and measure smaller particles we have investigated growing the particles by condensation of a supersaturated vapor onto them, a technique long used in condensation nuclei counters. The larger particles thus obtained provide several advantages. Beams of larger aerosol particles are better focused and are, therefore, more efficiently measured. Also, if the growth is sufficient the particles can be detected by scattered light or by other methods such as fluorescence of the condensed material or surface ionization of the condensed material on a suitable heated surface with subsequent measurement of the ions by mass spectrometry.

Particle motion in a jet with condensation or evaporation is described by

$$m_p \; (dv_p/dt) = - \; [f + (dm_p/dt)] \, (v_p - v_f) \qquad (5)$$

where $m_p = m_p(t)$ is the particle (or droplet) mass. If we now define α by

$$\alpha(t) = [f + (dm_p/dt)]/m_p \qquad (6)$$

instead of (2), we obtain the same solution of (5) as before, namely (3) and (4), with the previous definition of α replaced by (6). However, condensation or evaporation can alter the temperature of the jet so that an energy equation coupled with (5) must also be solved. Our calculations have neglected the energy expression. That is, we have assumed the particle and gas temperatures are equal. This point will be discussed briefly later.

Because condensation or evaporation in the jet spans the full range of Knudsen numbers Kn from continuum to free molecule flow, the following expressions were used to approximate the condensation rate

$$I_f = \pi(n-n_o)(\bar{c} + |v_f - v_p|)a^2 \tag{7}$$

$$I_c = 2.4893\pi(n-n_o)D^{2/3}|v_f-v_p|^{1/3}a^{4/3} \tag{8}$$

$$I = I_c I_f/(I_c + I_f) \tag{9}$$

where I_f is the free molecule rate of condensation or evaporation[5], for $K_n \to \infty$, I_c the continuum rate of condensation or evaporation[6] for $K_n \to 0$, I the rate of condensation or evaporation for intermediate Kn[7], a the droplet radius, n the local number concentration of vapor molecules in the gas, n_o the equilibrium vapor number concentration over a surface with radius of curvature a, D the diffusion coefficient of the vapor molecules, $\bar{c}$ the average thermal velocity of the vapor molecules, Kn = ℓ/a, and ℓ the average mean free path of the gas molecules. These expressions with the assumption that the droplet temperature remained equal to the local gas temperature allowed calculation of the terminal particle velocity and size.

Table 1 shows example results of the final particle diameters calculated for four vapor species and for two initial diameters when the vapor pressure at the nozzle inlet was assumed to be the equilibrium vapor pressure at 293°K, i.e., 100% vapor saturation. The results of Table 1 indicate that growth of particles by condensation in the jet is not significant. The particle growth is limited because the particles pass through the supersaturated region so quickly there is simply not sufficient time for significant growth to occur.

Table 1. Final Particle Diameters for Condensational Growth in a Converging Nozzle

Condensible vapor	Vapor pressure at 293°K	Final diameters for initial particle diameter of:	
		0.5 μm	0.1 μm
	Torr	μm	μm
Water	17.404	0.5004	0.1004
Acetone	177.810	0.5084	0.1080
Benzene	74.662	0.5007	0.1007
Ethanol	42.947	0.5021	0.1021

We also found that particle growth in nozzles contoured to give particle residence times in the supersaturated region fifty times longer still were not sufficient to cause adequate growth of small particles.

Actual particle growth will be less than the calculated values of Table 1 because of the assumption that the particle temperature remains equal to the gas temperature in the jet. This assumption is conservative because (a) the particle will cool more slowly than the adiabatically expanding gas, and (b) the latent heat of condensation will further retard cooling of the particle. Thus, as indicated by the results of Table 1, condensational growth of particles in adiabatically expanding jets does not appear to be significant.

To obtain significant particle growth in the supersaturated vapor it therefore appears necessary to expand a vapor-aerosol mixture into a growth chamber preceding the nozzle expansion to cause particle growth as in a condensation nuclei counter. The particle growth in the supersaturated vapor is greater because of the relatively low velocity (long residence time) in the growth chamber. Moreover, particle growth can be controlled by adjusting the growth chamber length or velocity.

Example calculations of the growth obtained by expanding a small aerosol flow and a much larger flow of water saturated air from 730 to 540 torr in the growth chamber are shown in Fig. 6. This expansion gives a supersaturation of 400%. Thus, particles above 17.8 Å diameter will serve as condensation nuclei upon which droplets will grow. In a period of 1.3 msec these droplets will grow to a size of 1.49 μm, easily detectable by scattered light methods.

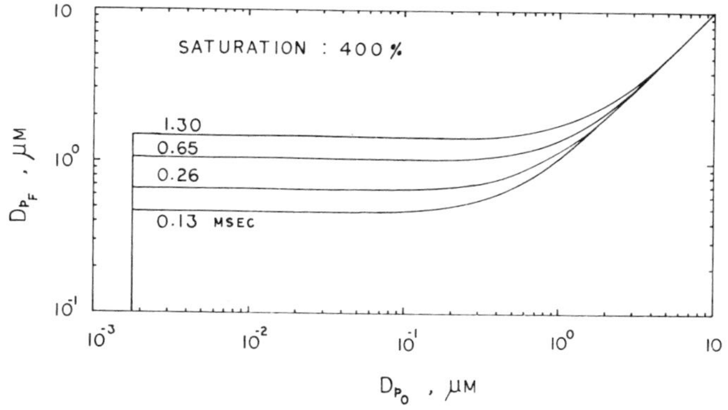

Fig. 6 Calculated final particle diameter vs. initial particle diameter for various particle residence times in the growth chamber containing water vapor at 400% supersaturation.

Table 2. Minimum Particle Size for Condensational Growth at Various Supersaturations of Water Vapor[a]

Saturation in chamber %	Critical diameter Å	Final minimum diam. for a residence time of 1.3 msec in growth chamber μm
400	17.8	1.491
300	22.4	1.217
200	35.5	0.859
150	60.7	0.604
110	258.	0.240
102	1235.	0.199

[a] Growth chamber (total) pressure, 540 Torr; growth chamber temperature, 268.7°K.

Finally, we point out in Table 2 that it seems possible to measure not only the total number of nuclei present in the aerosol sample, but also their size distribution. For this measurement the supersaturation in the growth chamber is decreased for each of several measurements. Because of the Kelvin effect, the minimum particle size which can serve as a condensation nucleus will increase as the supersaturation decreases, as shown in Table 2. Differencing the respective measured concentrations may provide the condensation nuclei size distribution.

References

[1] Dahneke, B., "Aerosol Beam Spectrometry," Nature Physical Science, Vol. 244, July 23, 1973, pp. 54-55.

[2] Dahneke, B. and Flachsbart, H., "An Aerosol Beam Spectrometer," Aerosol Science, Vol. 3, September, 1972, pp. 345-349.

[3] Shapiro, A. H., The Dynamics and Thermodynamics of Compressible Fluid Flow, Ronald Press Co., New York, 1953.

[4] French, J. B., "Continuum-Source Molecular Beams," AIAA Journal, Vol. 3, June 1965, pp. 993-1000.

[5] Dahneke, B., "Slip Correction Factors for Nonspherical Bodies II Free Molecular Flow," Aerosol Science, Vol. 4, March 1973, pp. 147-161.

[6] Levich, V. G., Physicochemical Hydrodynamics, Prentice-Hall, Inc., Englewood Cliffs, N.J., 1962.

[7] Fuchs, N. A. and Sutugin, A. G., Highly Dispersed Aerosols, Ann Arbor Science Publishers, Ann Arbor, Mich., 1970.

CHAPTER XII—CONDENSATION AND EVAPORATION

NONLINEAR NUMERICAL SOLUTIONS FOR AN EVAPORATION-EFFUSION PROBLEM

Shee-Mang Yen* and Terrence J. Akai†

University of Illinois at Urbana-Champaign

Abstract

An evaporation problem in which pure vapor flows between the interphase boundary and a downstream flat sink at which uniform equilibrium flow is assumed to occur and an equivalent problem of effusion from a perforated wall are considered. The problem is simulated by one similar to an evaporation-condensation problem of vapor flow between two walls, and the Boltzmann equation for elastic spheres and Maxwellian molecules as well as the Krook equation are solved. The results for the transport and flow properties as functions of other properties obtained from these solutions and a solution from the Boltzmann transport equations are in good agreement. It also appears that the effect of intermolecular collision law on most of these functions is small.

Introduction

Nonlinear evaporation-condensation problems require accurate kinetic theory treatment. For example, the Boltzmann equation (for two intermolecular collision laws) and the Krook equation were solved for an evaporation-condensation

---

Presented as Paper 19 at the 10th International Symposium on Rarefied Gas Dynamics, Aspen, Colorado, July 19-23, 1976. This work was supported by the Joint Services Electronics Program (U.S. Army, U.S. Navy, and U.S. Air Force) under Contract DAAB-07-72-C-0259. We are indebted to Dr. T. Ytrehus who suggested the evaporation-effusion problem to us and to Dr. F. G. Cheremisin who first solved problems of nonequilibrium gas flows between emitting and absorbing walls and stimulated our interest in these problems. We are appreciative of a NATO grant (No. 1075) that has provided travel funds needed in this cooperative research effort with Dr. Ytrehus and Professor Wendt.

*Professor.

†Research Assistant.

problem to study the mass transports and the related flow characteristics under two different nonequilibrium conditions.[1,2] In this paper we shall present the results of such a study for a similar problem which we chose to simulate realistically certain evaporation-effusion problems. Such problems have a wide range of applications, e.g., in the design of spacecraft experiments, certain meteorology tests, and nuclear reactors. Ytrehus[3,4] formulated these problems using a Mott-Smith type Ansatz, solved the Boltzmann transport equations for Maxwellian molecules, and compared his calculations with available experimental and other theoretical results. He indicated that further study of these problems using more accurate kinetic theory treatment is necessary to study the effect of different boundary conditions and collision laws.

Simulation of the Evaporation-Effusion Problem

The evaporation and effusion problems and our simulation of them are shown schematically in Fig. 1. In the evaporation problem, the flow of pure vapor occurs between the interphase boundary and the downstream flat sink at which

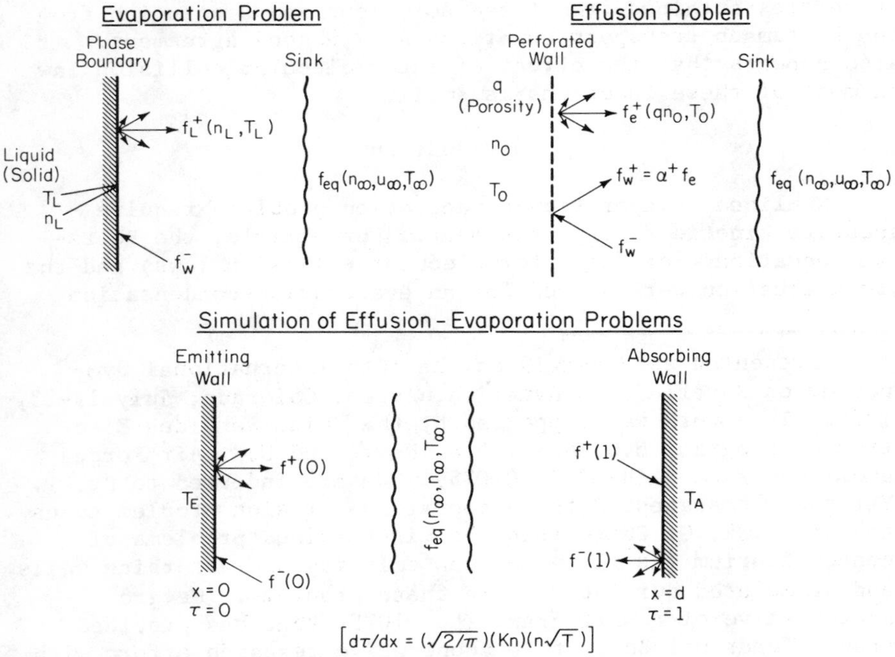

Fig. 1 Schematic of the evaporation, effusion and simulation problems.

uniform equilibrium flow is assumed to occur. At the interphase, T_L = temperature of the liquid (solid) and P_L = saturation vapor pressure. At the downstream boundary, $f = f_{eq}(n_\infty, u_\infty, T_\infty)$. It is assumed that the evaporated molecules have a half-Maxwellian distribution function and all impinging molecules are condensed and re-emitted through evaporation. The problem of effusion from a thin, perforated wall[3-5] toward a flat sink is equivalent to the evaporation problem, and the boundary parameters of the two problems are related as follows: $n_L = (1 + \alpha^+)(qn_0)$, $T_L = T_0$, $P_L=(1 + \alpha^+)P_e$ (α^+ = a reflection parameter). In the simulation problem, the vapor flows between an evaporating surface at temperature T_E and an absorbing wall at temperature T_A with the sink located at some intermediate position where equilibrium conditions exist. Basically, the simulation problem is the same as the emitting and absorbing problem considered first by Cheremisin[6] and later by Yen[2]; however, the nonequilibrium conditions to be used here should be compatible to the two problems under consideration. We shall consider the case of low Knudsen numbers such that equilibrium flow occurs away from the emitting wall.

There are three parameters in this problem: the length parameter Kn, the temperature ratio T_A/T_E, and the mass transport parameter β. The parameter

$$\beta = \dot{m}^+(d)/\dot{m}^-(d) \qquad (1)$$

in which $\dot{m}^+(d)$ = mass flux reaching the absorbing wall and $\dot{m}^-(d)$ = mass flux emitting from the absorbing wall.

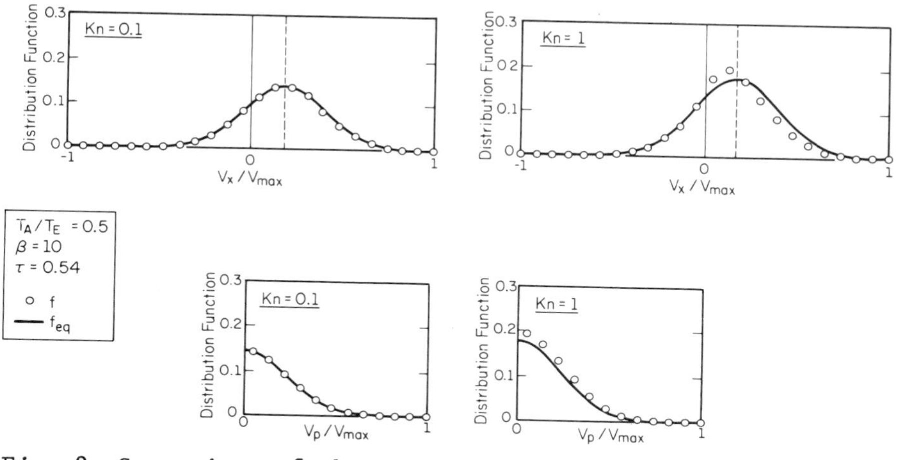

Fig. 2 Comparison of the actual and equilibrium distribution functions at τ = 0.54 for Kn = 0.1 and 1, T_A/T_E = 0.5, and β = 10.

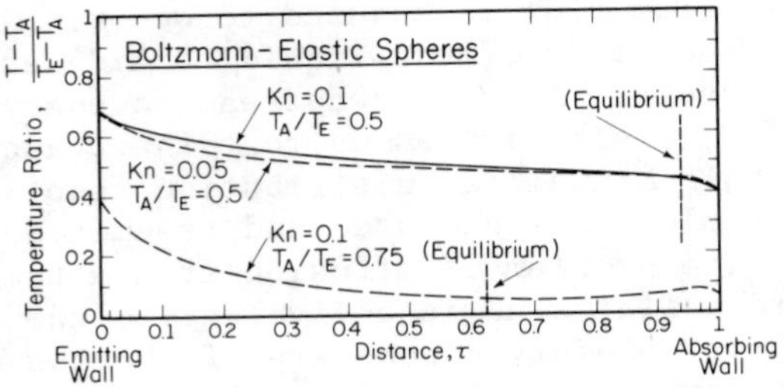

Fig. 3 Temperature distribution between emitting and absorbing walls for three cases. $\beta = 10$.

Results

We have obtained the Boltzmann solutions for elastic spheres and Maxwellian molecules and the Krook solutions for elastic spheres for the following set of parameters: $T_A/T_E = 0.5$ and 0.75; Kn = 1, 0.1, and 0.05; $\beta = 2.5$ and 10. For each solution, we first verify by examining the distribution function that equilibrium or near-equilibrium conditions do exist downstream from the emitting wall. As shown in Fig. 2, the distribution function at $\tau = 0.54$ is very close to f_{eq} for Kn = 0.1 ($T_A/T_E = 0.5$, $\beta = 10$); however, there is a significant departure for Kn = 1. As illustrated in Fig. 3 in plots of temperature vs τ, the location at which such an equilibrium state occurs depends on the temperature ratio T_A/T_E. These plots also show that Kn of 0.1 is low enough to obtain simulation solutions for the case of elastic spheres.

The transport and flow properties as functions of other properties for the case of elastic spheres are in good agreement with Ytrehus' results.[3] Also, the Krook results are in excellent agreement with the corresponding Boltzmann values. For Maxwellian molecules, the agreement is not as good in some cases; however, these solutions do not match the downstream equilibrium state as well as those for elastic spheres. Nevertheless, it seems that the effect of intermolecular law is small on most of the properties studied. We feel that a larger effect is produced by the presence of the absorbing wall which is peculiar to the simulation problem but not to the actual problem. Figures 4-6 show, respectively, the results for the fractional evaporation rate $\dot{m}_\infty/\dot{m}_L$ vs the downstream speed ratio S_∞, the backscattering flux $\dot{m}^-(0)/\dot{m}^+(0)$ vs the pressure ratio P_L/P_∞, and the downstream speed ratio vs the pressure ratio. Figures 7 and 8 show, respectively, the temperature jump $T(0)/T_L$ vs the downstream speed ratio and the

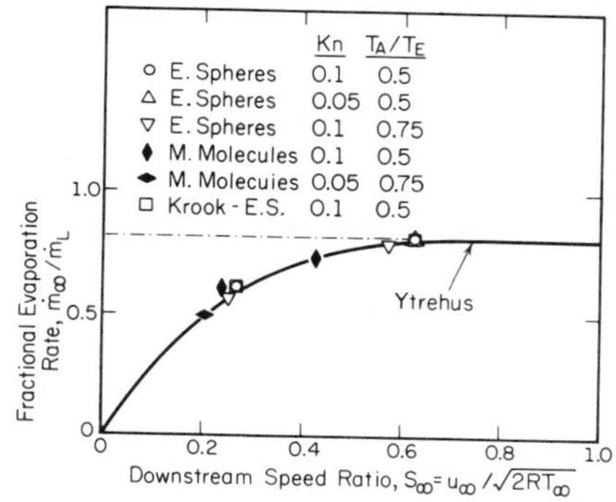

Fig. 4 Fractional evaporation rate vs downstream speed ratio.

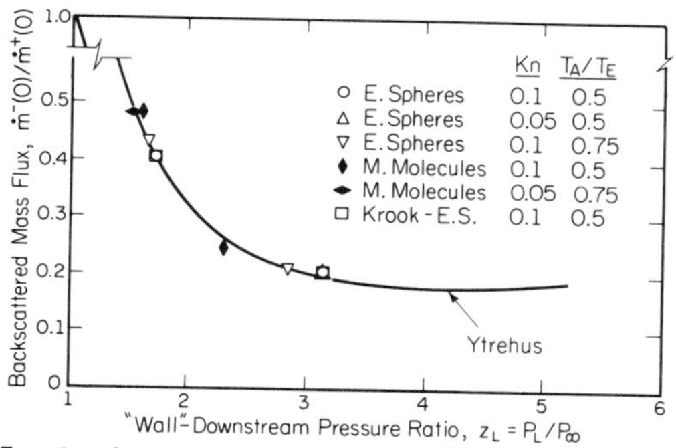

Fig. 5 Backscattering mass flux vs pressure ratio.

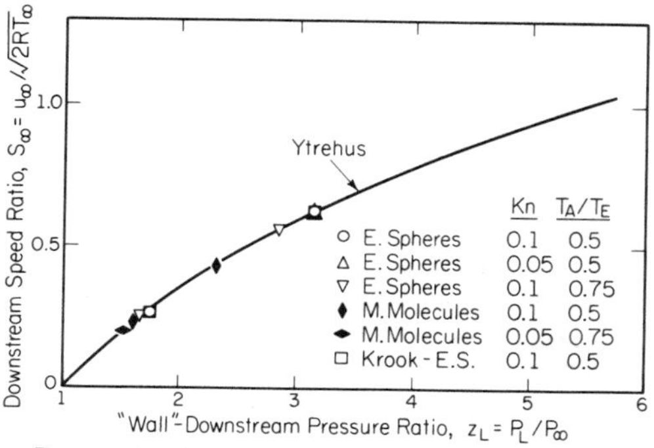

Fig. 6 Downstream speed ratio vs pressure ratio.

downstream mach number M_∞ vs the number flux rate $n(0)u(0)/(n_L\sqrt{2RT_L})$. These results, together with the parameters α^+ and $(\lambda_L/n(0))(dn(0)/dx)$ (λ_L = mean free path at the phase boundary), are summarized in Table 1.

It should be emphasized that the agreement among results obtained from solutions of different kinetic equations for

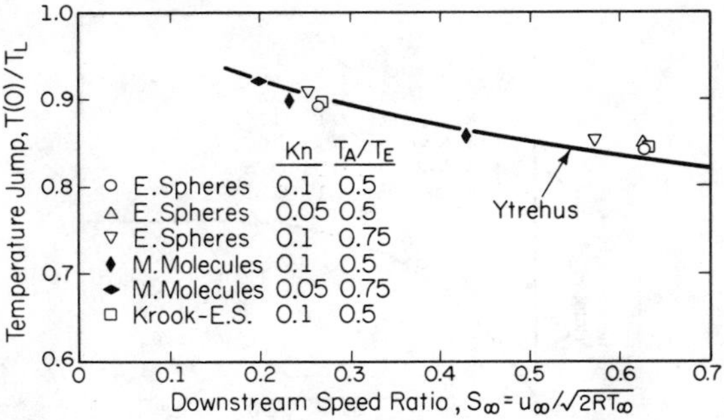

Fig. 7 Temperature jump at the emitting wall vs downstream speed ratio.

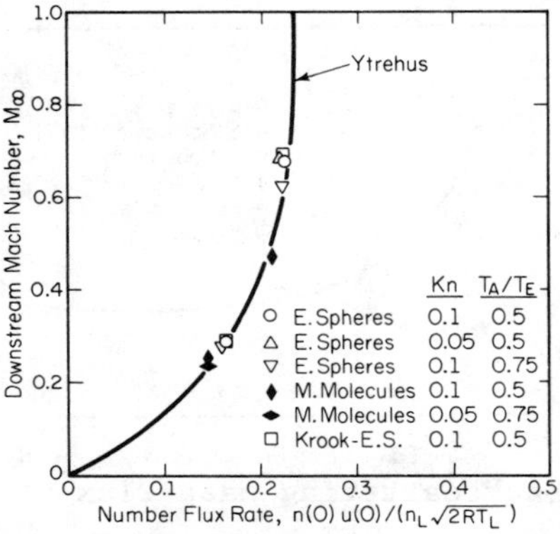

Fig. 8 Downstream Mach number vs number flux rate.

different intermolecular collision laws concerns properties as functions of other properties. Our findings provide further verification that the moments of the distribution function in a nonequilibrium flow are strongly coupled and that this coupling depends weakly on the intermolecular collision law and the kinetic equation used. Such an observation was made for the problem of a shock wave.[7] The principal difference among solutions of different kinetic equations and for different intermolecular collision laws is in the relaxation rate, i.e., the variation of properties with respect to distance. Figure 9 gives the difference in density distribution in the Knudsen layer between elastic spheres and Maxwellian molecules for the case of Kn = 0.1, $T_A/T_E = 0.5$, and $\beta = 10$.

Conclusions

The evaporation and effusion problems with a downstream equilibrium boundary condition can be simulated by a problem

Table 1 Summary of Results

Kn	β	$\frac{T_A}{T_E}$	S_∞	Z_L	$\frac{n(0)u(0)}{n_L\sqrt{2RT_L}}$	$\frac{\dot{m}_\infty}{\dot{m}_L}$	$\frac{\dot{m}^-(0)}{\dot{m}^+(0)}$	α^+	M_∞	$\frac{n(0)}{n_L}$	$\frac{T(0)}{T_L}$	$\frac{\lambda_L}{n(0)}\left(\frac{dn}{dx}\right)_0$
Boltzmann - Elastic Spheres												
0.1	10.0	0.5	0.6248	3.151	0.2243	0.8265	0.2064	0.2601	0.6845	0.664	0.8465	0.4704
0.1	2.5	0.5	0.2655	1.730	0.1655	0.6075	0.4144	0.7078	0.2908	0.767	0.8933	0.2846
0.05	10.0	0.5	0.6258	3.149	0.2216	0.8285	0.2143	0.2728	0.6855	0.666	0.8507	0.4739
0.1	10.0	0.75	0.5712	2.873	0.2233	0.8033	0.2119	0.2688	0.6257	0.664	0.8538	0.4669
0.05	2.5	0.75	0.2542	1.665	0.1606	0.5754	0.4338	0.7662	0.2785	0.770	0.9119	0.2358
0.1	2.5	0.75	0.2536	1.669	0.1603	0.5756	0.4341	0.7671	0.2778	0.772	0.9092	0.2792
Krook - Elastic Spheres												
0.1	10.0	0.5	0.6318	3.185	0.2254	0.8273	0.2025	0.2539	0.6921	0.656	0.8435	0.5313
0.1	2.5	0.5	0.2691	1.727	0.1658	0.6162	0.4137	0.7055	0.2948	0.761	0.8963	0.3189
0.1	10.0	0.75	0.6131	3.075	0.2246	0.8081	0.2070	0.2611	0.6716	0.658	0.8489	0.5189
0.1	2.5	0.75	0.2542	1.665	0.1607	0.5769	0.4327	0.7628	0.2785	0.769	0.9067	0.3068
Boltzmann - Maxwell Molecules												
0.1	10.0	0.5	0.4289	2.315	0.2128	0.7465	0.2473	0.3285	0.4699	0.688	0.8582	0.8211
0.1	2.5	0.5	0.2338	1.619	0.1460	0.6170	0.4835	0.9362	0.2561	0.803	0.8996	0.4329
0.1	2.5	0.75	0.2493	1.652	0.1470	0.5797	0.4810	0.9269	0.2731	0.797	0.9162	0.3908
0.05	2.5	0.75	0.2014	1.524	0.1480	0.4972	0.4782	0.9163	0.2206	0.794	0.9214	0.4088
0.05	2.5	0.5	0.2446	1.622	0.1424	0.6612	0.4963	0.9853	0.2679	0.805	0.9105	0.3950

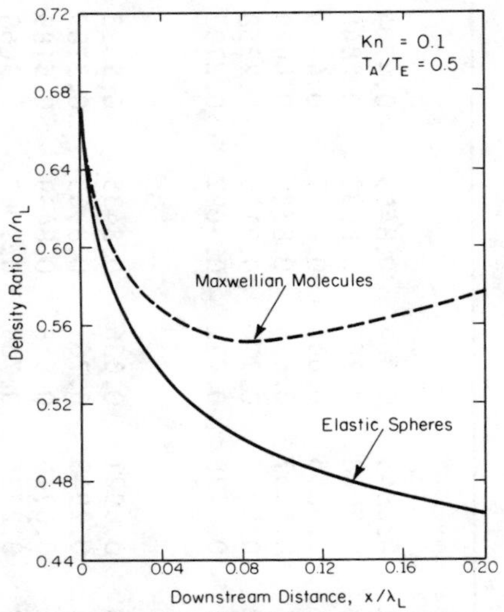

Fig. 9 Comparison of density distribution in the Knudsen layer for Kn = 0.1, T_A/T_E = 0.5, and β = 10.

of nonequilibrium gas flow between an emitting and an absorbing wall. The flow conditions may be obtained by varying three flow parameters - namely, the Knudsen number, the temperature ratio, and the mass transport parameter. The macroscopic and transport properties as functions of other properties obtained from solutions of two kinetic equations and two intermolecular collision laws are in agreement not only with each other but with those obtained by Ytrehus[3] with a moment method. This finding indicates that the coupling of moments of the distribution function depends only weakly on the kinetic equation and collision law and that the study of evaporation-effusion problems for properties such as the evaporation rate, backscattering flux, and downstream speed ratio can be made by using a simplified method such as that suggested by Ytrehus.

References

[1]Yen, S. M., "Numerical Solutions of Non-Linear Kinetic Equations for a One-Dimensional Evaporation-Condensation Problem," International Journal of Computers and Fluids, Vol. 1, 1973, pp. 367-377.

[2]Yen, S. M., "Solutions of Kinetic Equations for the Non-Equilibrium Gas Flow Between Emitting and Absorbing Surfaces," Proceedings of the 9th International Symposium on Rarefied Gas Dynamics, Vol. 1, 1974, pp. A.15-1 - A.15-10.

[3]Ytrehus, T., "Kinetic Theory Description and Experimental Results for Vapor Motion in Arbitrary Strong Evaporation," von Kármán Institute for Fluid Dynamics, Rhode Saint Genese, Belgium, June 1975.

[4]Ytrehus, T., "One-Dimensional Effusive Flow by the Method of Moments," Proceedings of the 9th International Symposium on Rarefied Gas Dynamics, Vol. 1, 1974, pp. B.4-1 - B.4-11.

[5]Ytrehus, T., "A Study of the Large-Scale Rarefied Flow Field that can be Obtained Through Molecular Effusion from a Perforated Boundary," von Kármán Institute for Fluid Dynamics, Rhode Saint Genese, Belgium, February 1973.

[6]Cheremisin, F. G., "Rarefied Gas Flows Between Infinite Parallel Emitting and Absorbing Surfaces," Izvestiya Akademie Nauk SSSR, Makhamika Zhidkosti i gaza, Vol. 2, 1972, pp. 176-178.

[7]Yen, S. M., and Ng, W., "Shock Wave Structure and Intermolecular Collision Laws," Journal of Fluid Mechanics, Vol. 65, Pt. 1, 1974, pp. 127-144.

[illegible] Kinetic Equations for the [illegible] Surfaces [illegible]

[illegible] Propagation and [illegible]

[illegible] of the method of [illegible]

[illegible] Large-Scale Three-Dimensional Flow [illegible]

[illegible]

[illegible]

QUASISTEADY ONE-DIMENSIONAL EVAPORATION PROBLEM USING ENTROPY-BALANCE RELATION

T. Soga*

Stosswellenlabor, Rheinisch-Westfaelische Technische, Hochschule Aachen, Aachen, West Germany

Abstract

The approximate solution for the half-space steady-state evaporation problem is obtained from the two-surface problem using entropy-balance relation. Steady-stateevaporation flow is connected with the evaporation wave, making use of shock-tube relation. Then, the evaporation rate in the quasisteady evaporation is specified with the boundary conditions that are given before the evaporation takes place. When the evaporation is weak, evaporation rate is dependent only upon the pressure fifference between saturated vapor and gas phase. On the contrary, it is strongly dependent upon the humidity of the gas phase, as well as the pressure difference when the evaporation is strong. Present results show a remarkable agreement with the Monte Carlo simulation of the Boltzmann equation.

I. Introduction

The transient behavior of the one-dimensional evaporation problem is studied by Shankar and Marble[1] for weak evaporation and by Murakami and Oshima[2] for arbitrary evaporation. Both results give the asymptotic evaporation rate at $t = \infty$. Especially, the results of Monte Carlo simulation show the formation of the steady flow near the interphase surface. In

---

Presented as Paper 62 at the 10th International Symposium on Rarefied Gas Dynamics, Aspen, Colo., July 19-23, 1976. This work was done in the term of the fellowship of the Alexander von Humboldt-Stiftung in the West Germany.

* Research associate; presently research associate of the department of the Aeronautical Engineering, Nagoya University, Nagoya, Japan.

the steady flow, temperature, number density, and flow velocity seem to be the unique functions of the evaporation rate, which is specified from the boundary conditions that are given before the evaporation takes place. The purpose of this paper is to propose an approximate method by which the aforementioned unique functions in steady evaporation are obtained, and the evaporation rate in quasisteady evapotation will be obtained. This paper is concerned only with quasisteady evaporation problem, which corresponds to the asymptotic behavior of the aforementioned treatment.[1, 2]

The assumptions that are used in the analysis are the following:

1) Quasisteady flowfields are composed of two parts: one is the steady-state flowfield near the interphase surface, and the other is the wave part propagating into the undisturbed region (see, Fig. 1).
2) Two parts of the flowfield are combined making use of the shock-tube relation.
3) All particles emitting from the surface have a Maxwellian distribution corresponding to the saturated vapor at the temperature of the condensed phase (evaporation coefficient = 1).
4) All molecules impinging into the interphase surface are captured completely by the condensed phase (absorption coefficient = 1).
5) To supply the emitting particles satisfying condition 3, saturated vapor is provided instantaneously by the condensed phase, and the specific entropy of the condensed phase is negligible compared with that of the gas phase.
6) Absorbed particles are thermalized through the very thin layer inside of the interphase surface.

Assumptions 1 and 2 seem resonable from the results of Monte Carlo simulation and from the analogy of the evaporation problem to the shock-tube problem. Assumptions 3 and 4 are the same as the assumptions employed in Refs. 1 and 2. The fifth and sixth assumptions are not postulated directly, but these conditions seem to be necessary so long as we employ the saturated vapor condition at the interphase surface.

II. Steady-state evaporation

Instead of considering the half-space evaporation problem, we treat the two-surface problem,[3] where the vapor is evaporating from one interphase surface located at $x = L$ and condensing onto the other interphase surface located at $x = 0$. This means that we consider the counterpart of the half-space evaporation problem, i.e., the half-space condensation problem.

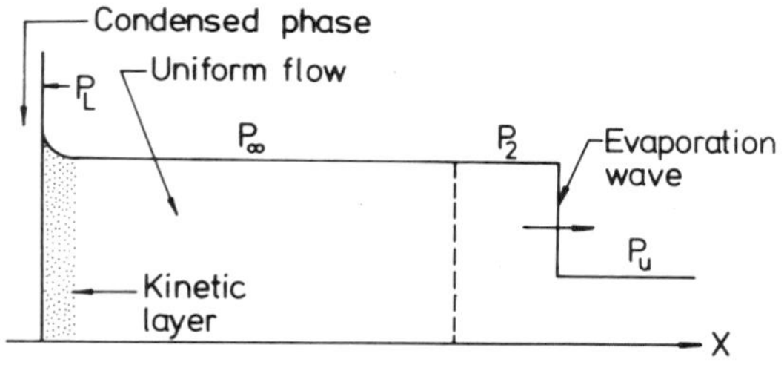

Fig. 1 Schematic drawing of the quasi-steady evaporation problem.

Condensed phase in $x < 0$ is kept at a constant temperature T_0 and the other condensed phase in $x > L$ at temperature T_L, respectively.

We assume, as mentioned before, that at the interphase surfaces the condensed phases have the saturated vapor pressure corresponding to their temperature. Then, there may exist steady mass flux, momentum flux, and energy flux from $x = L$ to $x = 0$ (we assume $T_L > T_0$). In addition, we assume that the flowfield far away from the interphase surfaces is in thermal equilibrium at temperature T_∞, number density n_∞, and flow velocity u_∞. Suffixes L, 0, and ∞ denote the conditions of the saturated vapor at the interphase surface L, interphase surface 0, and the condition of the thermal equilibrium state far away from the interphase surfaces, respectively. Mass flux $\dot{m}$, momentum flux $\dot{P}$, and energy flux $\dot{E}$ may be specified with the flow properties T_∞, n_∞, and u_∞ as follows

$$\dot{m} = mn_\infty u_\infty \tag{1a}$$

$$\dot{P} = P_\infty + mn_\infty u_\infty^2 \tag{1b}$$

$$\dot{E} = mn_\infty u_\infty \{c_p T_\infty + (1/2)u_\infty^2\} \tag{1c}$$

where P is the pressure, c_p is the specific heat at constant pressure, and m is the mass of the molecule. The state equation is

$$P_\infty = mn_\infty RT_\infty$$

where R is the gas constant.

The net flux of W, which is the moment of the velocity distribution function f, is determined from the difference of its flux at the interphase surface. For steady-state flow, $\dot{W}_- = W_L^+ + W_L^- = W_0^+ + W_0^-$. Here $\dot{W}$ is the net flux of W, and W^+ and W^- are the flux of W through the interphase surface in the directions of x > 0 and x < 0, respectively. W_L^- and W_0^+ are written as follows

$$W_L^- = \iiint_{V_x<0} Wf_L d\vec{V} \ ; \quad W_0^+ = \iiint_{V_x>0} Wf_0 d\vec{V}$$

where f_L and f_0 are the Maxwellian distribution functions pertinent to the saturated vapor conditions at the interphase surfaces L and 0, respectively

$$f_L = n_L(2\pi RT_L)^{-3/2}\exp(-\vec{V}^2/2RT_L)$$

$$f_0 = n_0(2\pi RT_0)^{-3/2}\exp(-\vec{V}^2/2RT_0)$$

Here $\vec{V}^2 = V_x^2 + V_y^2 + V_z^2$, $d\vec{V} = dV_x dV_y dV_z$, and V_x, V_y, and V_z are the velocity components of the gas particles in the x, y, and z directions, respectively. Flux components W_L^+ and W_0^- are not known a priori. However, they may be expressed by making use of retardation coefficient Γ_w in the following form: $W_L^+ + W_L^- = (W_0^+ + W_L^-)(1-\Gamma_w)$. The retardation coefficient must be some complicated function of the T_L/T_0 and n_L/n_0, and another definition also is possible. But we use this coefficient in the sense of dimensional analysis. Thus, three conservation equations are rewritten in terms of retardation coefficients as follows

$$n_\infty u_\infty = \frac{n_L c_{mL}}{2\sqrt{\pi}}\{-1+\frac{n_0}{n_L}(\frac{T_0}{T_L})^{1/2}\}(1-\Gamma_1) \tag{2a}$$

$$\frac{P_\infty}{m} + n_\infty u_\infty^2 = \frac{P_L}{2m}\{1+\frac{n_0 T_0}{n_L T_L}\}(1-\Gamma_2) \tag{2b}$$

$$n_\infty u_\infty(\frac{\gamma R}{\gamma-1}T_\infty + \frac{1}{2}u_\infty^2) = \frac{\gamma+1}{4(\gamma-1)(2\sqrt{\pi})} n_L c_{mL}^3\{-1+\frac{n_0}{n_L}(\frac{T_0}{T_L})^{3/2}\}(1-\Gamma_3) \tag{2c}$$

where γ is the specific heats ratio, $c_{mL} = (2RT_L)^{1/2}$, and Γ_1, Γ_2, Γ_3 are the retardation coefficients of mass flux, momentum flux, and energy flux, respectively.

Three conservation equations are obtained as the lowest-order moment equations of the Boltzmann equation, and it is obvious that higher-order moment equations are necessary in order to connect the saturated vapor condition L with the thermal equilibrium state ∞ , because two states are combined with each other through the nonequilibrium kinetic layer (Knudsen layer). Now we recall the Rankine-Hugoniot relation in a shock wave, where two thermal equilibrium states are related through one paramrter, i.e., flow Mach number. As mentioned in Sec. I, the saturated vapor just inside of the interphase surface is assumed to be in thermal equilibrium, and nonequilibrium phenomena take place just at the surface and in the kinetic layer adjacent to the interphase surface. Then, two equilibrium states L and ∞ may be related to each other with the nonisentropic relation as follows

$$T_\infty/T_L = (n_\infty/n_L)^{\gamma-1}\exp\{(S_\infty-S_L)/c_v\} \tag{3}$$

where S is the specific entropy of unit mass defined by[4]

$$S = R\ln\{T^{1/(\gamma-1)}/n\} + \text{const.}$$

and c_v is the specific heat at constant volume. Relation (3) may be the result of the integration of the entropy production rate equation.[5] Entropy difference $\Delta S = S_\infty - S_L$ is produced during the expansion process through the kinetic layer. But this layer is so thin compared with the distance L between surface L and surface 0 that the average entropy of the fluid in this interval may be considered as S_∞.

In the steady state, the total number of the molecules between the two interphase surfaces is constant, and the entropy production of the whole system of this evaporation and condensation problem must be equal to $\dot{m}(S_0-S_L)$ in unit time, where S_0 and S_L are the entropies of saturated vapor of the condensed phases 0 and L, respectively. Balance of the entropy production can be written formally as follows

$$\dot{m}(S_0-S_L) = m\dot{N}S_\infty - (mn_Lc_{mL}S_L+mn_0c_{m0}S_0)/2\sqrt{\pi} \tag{4}$$

The first term of the right-hand side means the entropy flux outward through two interphase surfaces, where $\dot{N}$ is the total number flux of the molecules emitting into the gas phase from the two interphase surfaces, and $\dot{N} = (n_Lc_{mL}+n_0c_{m0})/2\sqrt{\pi}$. The second term of the right-hand side is the entropy flux, which is carried with the molecules emitting inward from the interphase surfaces. Rearranging Eq. (4), we obtain

$$\begin{aligned} S_\infty &= (n_0c_{m0}S_L+n_Lc_{mL}S_0)/(n_Lc_{mL}+n_0c_{m0}) \\ &= (\alpha S_L+S_0)/(1+\alpha) \end{aligned} \tag{5}$$

where $\alpha = n_0c_{m0}/n_Lc_{mL} = (1+2\sqrt{\pi}\dot{m}/mn_Lc_{mL})/(1-\Gamma_1)$. When we substitute S_∞ into Eq. (5), Eq. (3) may be rewritten as follows

$$T_\infty/T_L = (n_\infty/n_L)^{\gamma-1}\{(T_0/T_L)/(n_0/n_L)^{\gamma-1}\}^{1/(1+\alpha)} \tag{6}$$

The following fact must be emphasized: for the given boundary conditions, n_L, T_L, n_0, and T_0, innumerable thermal equilibrium states with n_∞, T_∞, and u_∞, which satisfy the conservation equations, are corresponding, but, making use of the entropy relation, only one state will be predicted which is relevant to the present problem.

In order to obtain the closed form of the equations, the retardation coefficients must be evaluated. From the results of Siewert and Thomas,[6] however, Γ_i are evaluated for weak evaporation as follows: $\Gamma_1 \simeq 0.08$, $\Gamma_2 \simeq -0.07$, $\Gamma_3 \simeq 0.06$. For the strong evaporation, Γ_i are estimated from the result of Anisimov[7] and Murakami and Oshima[2], and Γ_1, Γ_2, $\Gamma_3 \lesssim 0.03$. From these estimates of Γ_i, we assume $\Gamma_i \cong 0$ for the first approximation. But this assumption is only the quantitative one, and it does not mean that we employ the free molecular condition.

As Eqs. (2a-2c and 6) include five unknowns T_∞/T_L, n_∞/n_L, u_∞/c_{mL}, T_0/T_L, and n_0/n_L, each thermodynamic property can be expressed with one parameter. Thus, the flow properties T_∞/T_L, n_∞/n_L, and u_∞/c_{mL} will be obtained as the unique functions of the evaporation rate $\dot{m}/mn_L c_{mL}$ if we choose it as the parameter.

Linear Solution

For the first, we consider the linear solution. Let us assume that $\Delta n_0 = (n_L-n_0)/n_L \ll 1$, $\Delta T_0 = (T_L-T_0)/T_L \ll 1$, $\Delta n_\infty = (n_L-n_\infty)/n_L \ll 1$, $\Delta T_\infty = (T_L-T_\infty)/T_L \ll 1$, and then $\hat{u}_\infty = u_\infty/c_{mL} \ll 1$. Retaining the first order of Δn_∞, ΔT_∞, $\hat{u}_\infty$, ΔT_0, and Δn_0, we obtain the analytical form of the solution of Eqs. (2a-2c and 6) as follows

$$\Delta n_\infty = \{(5\gamma-6)/4(\gamma-1)\}\sqrt{\pi}(-\hat{u}_\infty) \tag{7}$$

$$\Delta T_\infty = \{(2-\gamma)/2(\gamma-1)\}\sqrt{\pi}(-\hat{u}_\infty) \tag{8}$$

$$\Delta n_0 = \{(\gamma+3)/2(\gamma-1)\}\Delta T_0 \tag{9}$$

Here, relation (9) may be the necessary boundary condition that is required for the surface 0, and (if we set $\gamma = 5/3$) this relation is quite the same as that which is obtained by Pao[3] for zero temperature gradient in his two-surface problem. But in this paper surface 0 can be considered to be a sort of artificial surface, which is introduced so as to match the two-surface problem to the half-space problem. We can, therefore, say that the solution obtained is not dependent upon the substance. It is interesting to see that present results (7) and (8) coincide with Pao's approximate relation for the half-space problem when we set $\gamma = 5/3$. It is obvious that Schrage's method[8], which Pao used for the estimation of mass flux and energy flux, gives the trivial result $\Delta T_\infty = \Delta n_\infty = 0$ when we take into account momentum flux, too.

Nonlinear Solution

Eliminating n_0 and T_0 from Eqs. (2a-2c and 6), we obtain the algebraic equations for n_∞ and T_∞, and these equations are solved numerically for the fixed value of $\dot{m}$. The solution of the equations becomes discontinuous beyond the critical value of the evaporation rate $\dot{m}$ at which the local flow Mach number of the uniform flow reaches unity. From the analogy of the problem to the shock-tube problem, the maximum evaporation rate may be obtained for $M_\infty = u_\infty/(\gamma R T_\infty)^{1/2} = 1$, and the solution for $M_\infty > 1$ may be physically insignificant. Nonlinear solutions are shown in Figs. 2 and 3. If we set $M_\infty = 1$, n_∞ and T_∞ are expressed explicitly as a function of the evaporation rate as follows

$$T_\infty/T_L = \sqrt{\pi}\{(2/\gamma)^{1/2}(\gamma+1)-\omega f_1^{1/2}\}^2$$

$$n_\infty/n_L = \{(\gamma+1)/\gamma\pi\}(1-f_1^2)\{1-(2/\gamma)^{-1/2}(\gamma+1)^{-1}\omega f_1^{1/2}\}$$

where $\omega = \{(2/\gamma)(\gamma+1)^2-\gamma\pi\}^{1/2}$, and $f_1 = 1-2\pi^{1/2}\dot{m}/mn_L c_{mL}$. The equation that determines the maximum evaporation rate is

$$\dot{m}/mn_L c_{mL} = (\gamma/2)^{1/2} f_1^{(3\gamma-1)/\{2(\gamma-1)(1+f_1)\}}$$

$$\cdot\{(1-\gamma T_\infty/T_L+\gamma f_1 T_\infty/T_L)^{1/(1+f_1)}/(T_\infty/T_L)\}^{(\gamma+1)/2}$$

The maximum evaporation rates obtained are 0.236693 for $\gamma = 5/3$ and 0.229961 for $\gamma = 7/5$.

III. Quasisteady Evaporation

An evaporation wave is driven by the pressure difference between the saturated vapor and undisturbed region. The formation of the wave front is seen in the results of Shankar et al.[1] and Monte Carlo simulation.[2] So, the evaporation wave may be a sort of the shock wave, and, as a consequence, the steady flow outside of the kinetic layer may be corresponding to the flow behind the contact surface in the shock-tube problem. Under these assumptions, shock-tube relation may be applicable to connect the uniform flow outside of the kinetic layer with the state in the undisturbed region in front of the evaporation wave.

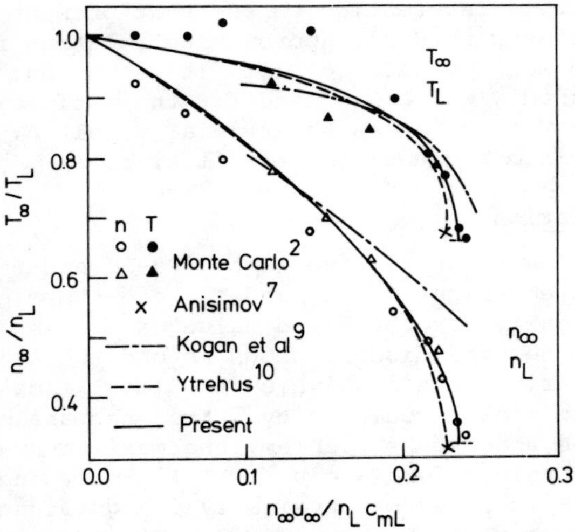

Fig. 2 Number density and temperature in the hydrodynamic region vs evaporation rate.

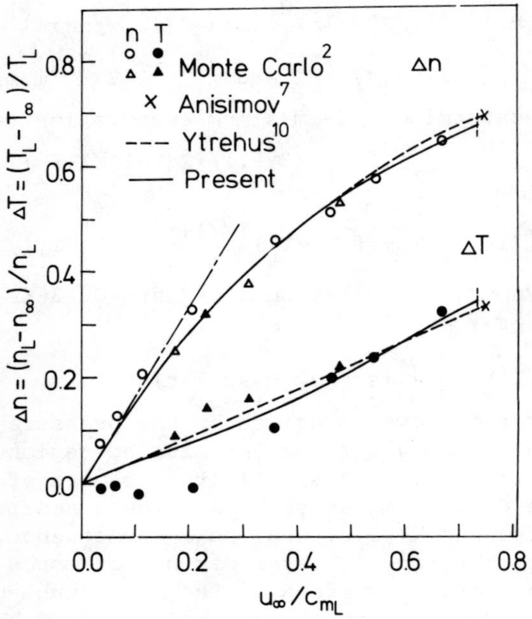

Fig. 3 Macroscopic jumps of number density and temperature vs flow velocity in the hydrodynamic region.

Let M_s be the shock Mach number of the evaporation wave and P_2 be the pressure behind the evaporation wave. From the well-known shock-tube relation

$$P_2/P_u = P_\infty/P_u = \{2\gamma M_s^2 - (\gamma-1)\}/(\gamma+1) \tag{10}$$

where suffix u denotes the state in the undisturbed region. Here we introduce the parameter β, defined by $\beta = (n_L/n_u-1)/(T_L/T_u-1)$. Then, $P_L/P_u = F(T_L/T_u, \beta) = (T_L/T_u)\{1+\beta(T_L/T_u-1)\}$. Relation (10) can be rewritten as follows

$$\{2\gamma/(\gamma+1)\}(P_\infty/P_L)F(T_L/T_u,\beta)/\{M_s^2-(\gamma-1)/(2\gamma)\} = 1 \tag{11}$$

where M_s can be obtained from the following shock-tube relation

$$u_\infty/c_{mL} = \{2/(\gamma+1)\}(a_u/c_{mL})(M_s-1/M_s)$$

where a is the sound speed, defined by $a = (\gamma RT)^{1/2}$.

As mentioned in Sec. II, P_∞/P_L, T_∞/T_L, and u_∞/c_{mL} are the unuque functions of the evaporation rate $\dot{m}/mn_Lc_{mL}$. Thus, Eq. (11) is the relation with which we can obtain the evaporation rate from the given boundary conditions n_L, T_L, n_u, and T_u. For the case $\dot{m}/mn_Lc_{mL} \ll 1$, we can obtain the analytical form of the solution of Eq. (11) as follows

$$\dot{m}/mn_Lc_{mL} = \{(9\sqrt{\pi}/8)+(2\gamma)\}^{-1}\{(P_L-P_u)/P_L\}$$

If we set $\gamma = 5/3$, this result coincides with the asymptotic solution of Shankar et al.[1] with four-moment equations, although they could not obtain the quasisteady solution for the evaporation problem. Nonlinear solutions of Eq. (11) are shown in Fig. 4 for $\beta = 1$ and for $\beta = 20$, and the present results show a remarkable agreement with the asymptotic solution of Monte Carlo simulation. In Fig. 5, the isoevaporation rate curves obtained from the solution of Eq. (11) are shown schematically as a function of the boundary conditions.

Boundary conditions that give the maximum evaporation rate may be obtained after some algebraic calculation as follows

$$T_L/T_u = (1/2\beta)[k_1+\{k_1^2+4\beta[(3\gamma+1)/(\gamma+1)]/(P_\infty/P_L)_{M_\infty=1}\}^{1/2}]$$

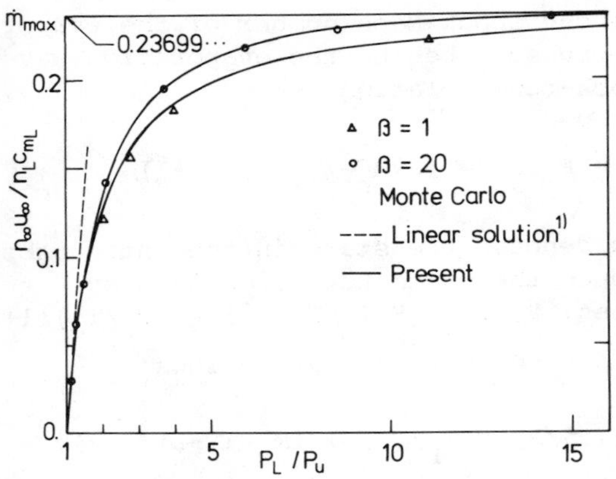

Fig. 4 Evaporation rate vs pressure ratio P_L/P_u.

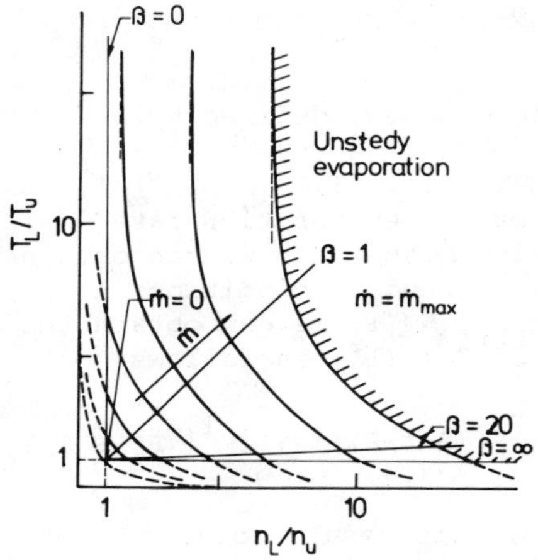

Fig. 5 Isoevaporation curves.

where $k_1 = \beta-1+\gamma(\gamma+1)/(2n_\infty/n_L)_{M_\infty=1}$. At $\beta \to 0$ (which may correspond to a supersaturated gas), $T_L/T_u \to \infty$ and $n_L/n_u \simeq \gamma(\gamma+1)/(2n_\infty/n_L)_{M_\infty=1}$, and, at $\beta \to \infty$ (which may mean dry gas), $T_L/T_u = 1$ and $n_L/n_u \simeq \gamma(\gamma+1)/(4n_\infty/n_L)_{M_\infty=1}+(3\gamma+1)/\{(\gamma+1)(P_\infty/P_L)_{M_\infty=1}\}-1/2$. Figures 4 and 5 show that when the evaporation is weak evaporation rate is proportional to the pressure difference, and it does not depend upon the parameter β (or humidity), and, when the evaporation is strong, on the contrary, evaporation rate is strongly dependent upon the parameter β (or humidity).

IV. Discussion

In spite of the crude assumptions, present results show a remarkable agreement with those of the Monte Carlo simulation to the Boltzmann equation.[2] Agreement with other results[1, 3, 6, 7, 9, 10] is also reasonable. The small disagreement between present results and those of Siewert et al. may be caused partly by the assumption on the retardation coefficient. But it is questionable whether or not these difference are substantial, because the results of Siewert et al. are based upon the BGK kinetic model equation, and present results, on the contrary, seem to be corresponding to the Boltzmann equation.

The entropy balance relation (5) shows that the entropy in the uniform flow is specified from the "lever principle." This fact seems physically resonable, and without this relation we cannot predict the equilibrium state of the uniform flow from the conservation equations. "Lever principle" of the entropy may be the condition for the minimum entropy production rate in this evaporation and condensation system. Even if we take into account the entropy of the condensed phase, the entropy balance relation seems to be valid so long as the saturated vapor is supplied at the surface of the condensed phase, because the decrease of the entropy due to the phase change may be corresponding to the decrease of the heat flux inward of the condensed phase.

References

[1] Shankar, P. N. and Marble, F. E., "Kinetic Theory of Transient Condensation and Evaporation at a Plane Surface," Physics of Fluids, Vol. 14, March 1971, pp. 510-516.

[2] Murakami, M. and Oshima, K., "Kinetic Approach to the Transient Evaporation and Condensation Problem," Rarefied Gas Dynamics, Vol. 2, edited by M. Becker and M. Fiebig, DFVLR Press, Porz-Wahn, West Germany, 1974, F. 6.

[3] Pao, Y. P., "Application of Kinetic Theory to the Problem of Evaporation and Condensation," Physics of Fluids, Vol. 14, February 1971, pp. 306-312.

[4] Chapman, S. and Cowling, T. G., The Mathematical Theory of Nonuniform Gas, Cambridge University Press, London, 1964, pp. 81.

[5]Hirschfelder, J. O., Curtiss, C. F., and Bird, R. B., Molecular Theory of Gas and Liquids, Wiley, New York, 1964, Chap. 11.

[6]Siewert, C. E. and Thomas, J. R., Jr., "Half-space Problem in Kinetic Theory of Gases," Physics of Fluids, Vol. 16, September 1973, pp. 1557-1559.

[7]Anisimov, S. I., "Vaporization of Metal Absorbing Laser Radiation," Soviet Physics JETP, Vol. 27, July 1968, pp. 182-183.

[8]Shrage, R. W., A Theoretical Study of Interphase Mass Transfer , Columbia University Press, New York, 1953, Chap. 2.

[9]Kogan, M. N. and Makashev, N. K., "Role of the Kunudsen Layer in the Theory of Heterogeneous Reactions and in Flows with Surface Reactions," Fluids Dynamics, Vol. 6, June 1974, pp. 913-920.

[10]Ytrehus, T., "Kinetic Theory Description and Experimental Results for Vapor Motion in Arbitrary Strong Evaporation" TN 112, June 1975, Von Kármán Institute, Rode Saint-Genese, Belgium.

THEORY AND EXPERIMENTS ON GAS KINETICS IN EVAPORATION

Tor Ytrehus*

The Norwegian Institute of Technology, Trondheim, Norway

Abstract

A nonlinear solution for the quasisteady half-space evaporation problem has been obtained using a moment method of the Mott-Smith, Liu-Lees type. The problem contains one single independent driving parameter, such as the pressure ratio p_L/p_∞, and downstream flow conditions, mass and heat flux from the interphase boundary, are obtained as unique functions of this parameter in the range $0 < p_L/p_\infty \lesssim 4.85$. Sonic downstream flow conditions occur at the limiting value $p_L/p_\infty \approx 4.85$, beyond which no solution is found that approaches a Maxwellian state far downstream. The limiting mass and heat fluxes are $\dot{m}/\dot{m}_L = 0.815$ and $\dot{q}_w/\dot{q}_L = 0.910$, respectively. Experimental values for the mass flux at the interphase boundary, and for the speed ratio in the downstream flow, are obtained from measurements in an equivalent effusive flow from a perforated wall. The theoretical results are in substantial agreement with the experimental findings, and with numerical solutions of the nonlinear evaporation problem as recently obtained by other authors.

Introduction

Research activity on the kinetics of vapors close to interphase boundaries has increased significantly over the

---

Presented as Paper 28 at the 10th International Symposium on Rarefied Gas Dynamics, Aspen, Colo., July 19-23, 1976.

The cooperation with Professors S.M. Yen and J.F. Wendt promoted by NATO Research Grant 1075 is acknowledged gratefully.

*Senior Lecturer, Institute of Applied Mechanics.

last few years, thus reflecting the large number of problems in science and technology for which evaporation and condensation phenomena are of practical importance. Problems of this type are encountered in such diversified areas as upper atmosphere meteorology, the sodium cooling of nuclear reactors, design of spacecraft experiments, petrochemical engineering, vacuum technology, and the interaction of high-power laser radiation with metal surfaces. In addition, the vapor kinetic problem has intrinsic fundamental interest, because it invokes a general Knudsen layer problem in which noncontinuum boundary conditions are essential, and where all the gasdynamic variables: density, velocity, and temperature undergo significant changes.

Although the problem is strongly nonlinear in the general case, previous theoretical efforts have mostly been devoted to the linearized version, which is of relevance to cases of weak evaporation and condensation, only. A semiempirical formula for the mass flux in the nonlinear case has been devised by Shrage[1], and some theoretical results have been derived for the limiting case of strong evaporation into vacuum.[2,3] No theoretical solution is available that describes the problem satisfactorily over a wide range of flow conditions. Numerical solutions, however, have appeared recently in which the kinetic equations have been solved for a wide range of nonequilibrium conditions, [4,5,6] and for different molecular interaction laws.[6]

The present paper is intended to demonstrate that a simple nonlinear moment solution, combining features of the Mott-Smith[7] and Liu-Lees[8] approaches, can predict accurately most of the quantities that are of practical interest in cases of arbitrary strong evaporation. The validity of the solution is assessed by comparisons with accurate numerical results; in particular those of Yen and Akai,[6] and with experimental values. The method is identical to the one used by the author in previous treatments of effusive flow from a perforated wall,[9,10] and further details from the solution may be found in Ref. 11. Here we present the main points, only, such as, 1) the importance of the conservation equations in determining the downstream flow conditions, mass flux, and related quantities, 2) the restriction to subsonic flow implied by the ξ_x^2 moment equation, and 3) the equivalence between the effusion and evaporation, which gives access to new experimental information on vapor kinetics in evaporation. The method is applicable to cases with net condensation as well, but only the case of evaporation has been treated in detail so far.

Formulation of the Problem

We consider the steady-state limit of the following one-dimensional, timedependent problem: a liquid (or solid) is initially in equilibrium with its pure vapor which occupies the half-space $x \geq 0$ at uniform temperature and pressure T_L and p_L, respectively. At time $t = 0$ the pressure level in the vapor changes discontinuously to the value p_∞ and is kept constant at this value throughout the procedure. Then, evaporation or condensation begins through the phase boundary according to whether the pressure level p_∞ is below or above the saturation pressure p_L.

Let us further assume that, far downstream of the phase boundary, there is an idealized, flat sink or source for the vapor, that can match instantaneously the mass flow created. Then it is reasonable to assume that, after a time sufficiently long for transients to have died out or to have propagated through the system, a steady state will be accomplished in which the flow far from the phase boundary is a uniform equilibrium flow with constant parameters n_∞, u_∞, and T_∞ (Fig. 1). A kinetic boundary layer then will form between the phase boundary and the downstream equilibrium region, it which nonequilibrium effects may influence significantly the motion of the vapor. We take the Boltzmann equation[12] for the oneparticle distribution function to describe the behaviour of the vapor in this layer.

The emission from the interphase boundary is described in the usual way by a Maxwellian distribution in the velocity half-space

$$f_L^+ = [n_L/(2\pi RT_L)^{3/2}]\exp(-\xi^2/2RT_L), \quad \xi_x > 0 \qquad (1)$$

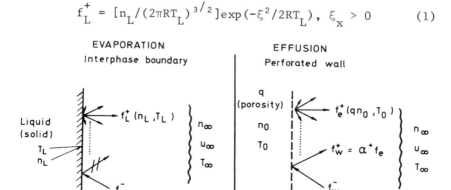

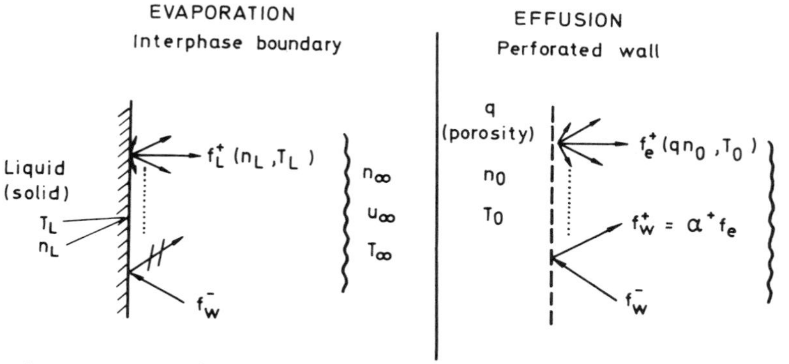

Fig. 1 Physical models for half-space evaporation and effusion problems.

where n_L is the saturation density corresponding to the temperature T_L of the dense phase. It is assumed further that all impinging molecules are condensed into the phase boundary, and that re-emission occurs through evaporation, only.

In the downstream equilibrium region the vapor is Maxwellian through the entire velocity space, and is described by the distribution function

$$f_\infty = [n_\infty/(2\pi RT_\infty)^{3/2}]\exp\{-[(\xi_x - u_\infty)^2 + \xi_y^2 + \xi_z^2]/2RT_\infty\} \qquad (2)$$

where n_∞, u_∞, T_∞ are *a priori* unknown parameters.

Because there are no analytical solutions available for the detailed nonlinear Boltzmann equation for this type of problem, we base our analysis on a set of Maxwell moment equations[13]

$$(\partial/\partial x)\int_{\vec{\xi}} \xi_x Q_\mu(\vec{\xi}) f \underline{d\xi} = \int_{\vec{\xi}} Q_\mu(\vec{\xi})(\partial f/\partial t)_{coll} \underline{d\xi} \equiv \Delta Q_\mu \qquad (3)$$

obtained by integrating the Boltzmann equation in velocity space, after multiplying with functions Q_μ of molecular velocity and making a Mott-Smith,[7] Liu-Lees[8] Ansatz for the distribution function. We consider the particular set $Q_\mu = 1,\ \xi_x,\ \frac{1}{2}\xi^2,\ \xi_x^2$ for $\mu = 1, 2, 3, 4$, corresponding to the three collisional invariants in one-dimensional flow, plus the nonconserved quantity ξ_x^2. In this way, the Boltzmann equation is satisfied in an average sense, only.

Gas Kinetic Connection Problem

Most of the information usually required in an evaporation problem can be obtained at the level of the conservation equations, and in fairly general terms. The general, one-dimensional conservation equations are obtained from Eq. (3) when the three collisional invariants are inserted for Q_μ. Then the collisional contributions ΔQ_μ vanish identically, and we have the general result

$$\int_{\vec{\xi}} \xi_x Q_\mu(\vec{\xi}) f \underline{d\xi} = C \quad , \ \mu = 1,2,3 \qquad (4)$$

where the C_μ's are the constant fluxes of mass, momentum, and energy in the system. In particular, Eq. (4) evaluated at $x = 0$ and at $x = \infty$ yields the relations

$$[\int_{\vec{\xi}} \xi_x Q_\mu(\vec{\xi}) f \underline{d\xi}]_{x=0} = [\int_{\vec{\xi}} \xi_x Q_\mu(\vec{\xi}) f \underline{d\xi}]_{x=\infty}, \quad \mu = 1,2,3 \qquad (5)$$

which connect the downstream flow conditions to the state at the interphase boundary. Let the distribution function at that boundary be represented as

$$f(0,\vec{\xi}) = \Big\} \qquad (6)$$

where f_L^+ is the Maxwellian specified for the emission by Eq. (1), and where f_w^- denotes the as yet unknown distribution function for molecules arriving at the boundary. By performing the indicated integrations in velocity-space, the system of Eqs. (5) may be written as

$$n_L u_L - \dot{m}_w^- = n_\infty u_\infty \qquad (7a)$$

$$(n_L RT_L/2) + (1/m)\dot{\sigma}_w^- = n_\infty u_\infty^2 + n_\infty RT_\infty \qquad (7b)$$

$$2n_L u_L RT_L - (1/m)\dot{\varepsilon}_w^- = n_\infty u_\infty[(u_\infty^2/2) + (5RT_\infty/2)] \qquad (7c)$$

with $u_L = \sqrt{RT_L/2\pi}$, and with the quantities $\dot{m}_w^-$, $\dot{\sigma}_w^-$, and $\dot{\varepsilon}_w^-$ representing the number flux, momentum flux, and energy flux in the stream impinging upon the wall, expressible as moments of the distribution function f_w^- in the following way

$$\dot{m}_w^- = \int_{\xi_x<0} |\xi_x| f_w^- \underline{d\xi} \qquad (8a)$$

$$(1/m)\dot{\sigma}_w^- = \int_{\xi_x<0} \xi_x^2 f_w^- \underline{d\xi} \qquad (8b)$$

$$(1/m)\dot{\varepsilon}_w^- = 1/2 \int_{\xi_x<0} |\xi_x| \xi^2 f_w^- \underline{d\xi} \qquad (8c)$$

Equations (8) are the present problem's analog to the Rankine-Hugoniot equations in shock-wave problems, relating the state of the flow at the downstream side of some narrow layer, containing large gradients, to the conditions that are given at the upstream side of this layer. An important difference lies, however, in the fact that the present system's left-hand side does not represent an equilibrium state and therefore is not expressible in terms of simple equilibrium parameters, only, such as those occurring on the right, and in the Rankine-Hugoniot relations.

Equations (7), if interpreted as conservation statements on the continuum level, constitute three equations for the six unknown quantities n_∞, u_∞, T_∞, $\dot{m}_w^-$, $\dot{\sigma}_w^-$, and $\dot{\varepsilon}_w^-$. A kinetic theory interpretation is, however, essential for the terms on the left, because the corresponding state is described by purely kinetic boundary conditions, Eq. (6). The definitions of $\dot{m}_w^-$, $\dot{\sigma}_w^-$, and $\dot{\varepsilon}_w^-$ as moments of the distribution function f_w^-, as given i Eqs. (8), therefore must be taken into account, and, since the function f_w^- is the same in all three expressions, this adds two independent functional relations between three of the unknowns: $\dot{m}_w^-$, $\dot{\sigma}_w^-$, and $\dot{\varepsilon}_w^-$ in Eqs. (7). One degree of freedon is therefore left in Eqs. (7), in agreement with the idea that only one of the downstream parameters, the pressure $p_\infty = n_\infty k T_\infty$, say, may be chosen freely in the analysis.

To obtain the explicit solution, the form of the unknown distribution function f_w^- at the wall must be assumed, as it must in any moment method. An obvious choice will be

$$f_w^- = \beta^-[n_\infty/(2\pi RT_\infty)^{3/2}]\exp\{-[(\xi_x-u_\infty)^2+\xi_y^2+\xi_z^2]/2RT_\infty\} = \beta^- f_\infty^- \quad (9)$$

which combines features of the Mott-Smith bimodal[7] and the Liu-Lees half-range[8] representation. The amplitude β^- is unknown and must be obtained in the solution. It represents essentially nonequilibrium effects caused by collisions in the Knudsen layer. Any value of β^- larger than unity would indicate that the group of backscattered molecules is enriched, relative to the downstream equilibrium value f_∞^-, principally as a result of products from collisions between molecules in the evaporated stream.

When this expression for f^- is inserted in the definitions (8) for the fluxes at the wall, these quantities become

$$\dot{m}_w^- = \beta^- n_\infty \sqrt{RT_\infty/2\pi}\,\tilde{F}^- \quad (10a)$$

$$(1/m)\dot{\sigma}_w^- = \beta^- (1/2) n_\infty RT_\infty \tilde{G}^- \quad (10b)$$

$$(1/m)\dot{\varepsilon}_w^- = \beta^- 2 n_\infty RT_\infty \sqrt{RT_\infty/2\pi}\,\tilde{H}^- \quad (10c)$$

where $\tilde{F}^-$, $\tilde{G}^-$, and $\tilde{H}^-$ are functions of the downstream speed ratio $S_\infty = u_\infty/\sqrt{2RT_\infty}$, resulting from the half-range integrations in the velocity-space, and being given as follows

$$\tilde{F}^- = -\sqrt{\pi}S_\infty \operatorname{erfc} S_\infty + e^{-S_\infty^2} \tag{11a}$$

$$\tilde{G}^- = (2S_\infty^2+1)\operatorname{erfc} S_\infty - (2S_\infty/\sqrt{\pi})e^{-S_\infty^2} \tag{11b}$$

$$\tilde{H}^- = -(\sqrt{\pi}S_\infty/2)(S_\infty^2+5/2)\operatorname{erfc} S_\infty + (1/2)(S_\infty^2+2)e^{-S_\infty^2} \tag{11c}$$

with $\operatorname{erfc} S_\infty$ denoting the complementary error function, i.e.,

$$\operatorname{erfc} S_\infty = 1 - \operatorname{erf} S_\infty = 1 - (2/\sqrt{\pi}) \int_0^{S_\infty} e^{-t^2} dt \tag{12}$$

Equations (7) then can be recast into a convenient nondimensional form

$$z_L\sqrt{T_\infty/T_L} - \beta^-\tilde{F}^- = 2\sqrt{\pi}S_\infty \tag{13a}$$

$$z_L + \beta^-\tilde{G}^- = 4S_\infty^2 + 2 \tag{13b}$$

$$z_L - \beta^-\sqrt{T_\infty/T_L}\tilde{H}^- = \sqrt{T_\infty/T_L}\sqrt{\pi}S_\infty(S_\infty^2 + 5/2) \tag{13c}$$

and solved for β^-, $\sqrt{T_\infty/T_L}$ and S_∞ in terms of the pressure-parameter

$$z_L = p_L/p_\infty \tag{14}$$

Related quantities, such as the backscattered flux at the wall, $\dot{m}_w^-$, and the net evaporated flux then are readily obtained from the first of Eqs. (10) and (7). The solution is summarized in Table 1, and we note that the amplitude β^- is always larger than unity, in accordance with the physical interpretation just given. Typical results are shown in Figs. 2 - 5.

The same calculations have been repeated with more general expressions for the distribution function f_w^-, to see the influence upon the results from the arbitrary elements in the analysis. In particular the Ansatz

$$f_w^- = \alpha^- f_*^- + \beta^- f_\infty^- \tag{15}$$

was studied, where f_*^- represents first collision products, from $f_L^+ - f_L^+$ collisions and where an upper bound for the amp-

Table 1 Summary of theoretical results

S_∞	z_L	n_∞/n_L	T_∞/T_L	$\dot{m}/\dot{m}_L$ [a]	$\dot{q}_w/\dot{q}_L$ [b]	β^-	λ_L/ℓ
0.00	1.0000	1.0000	1.0000	0.0000	0.0000	1.0000	1.3333
0.05	1.1107	0.9205	0.9781	0.1614	0.1975	1.0081	1.2236
0.10	1.2307	0.8494	0.9567	0.2945	0.3536	1.0198	1.1187
0.15	1.3604	0.7856	0.9357	0.4041	0.4769	1.0365	1.0189
0.20	1.5002	0.7283	0.9152	0.4940	0.5742	1.0597	0.9240
0.25	1.6506	0.6768	0.8952	0.5675	0.6509	1.0916	0.8339
0.30	1.8120	0.6303	0.8756	0.6272	0.7112	1.1349	0.7485
0.35	1.9849	0.5882	0.8565	0.6754	0.7585	1.1932	0.6676
0.40	2.1698	0.5501	0.8378	0.7140	0.7955	1.2711	0.5909
0.45	2.3670	0.5155	0.8195	0.7445	0.8244	1.3746	0.5183
0.50	2.5770	0.4841	0.8016	0.7682	0.8467	1.5115	0.4494
0.55	2.8002	0.4554	0.7842	0.7863	0.8640	1.6917	0.3841
0.60	3.0369	0.4292	0.7671	0.7996	0.8772	1.9284	0.3221
0.65	3.2876	0.4053	0.7505	0.8090	0.8872	2.2384	0.2631
0.70	3.5527	0.3834	0.7342	0.8151	0.8947	2.6437	0.2069
0.75	3.8325	0.3632	0.7183	0.8185	0.9003	3.1727	0.1534
0.80	4.1273	0.3447	0.7028	0.8196	0.9044	3.8625	0.1023
0.85	4.4376	0.3277	0.6877	0.8188	0.9073	4.7612	0.0535
0.90	4.7635	0.3120	0.6729	0.8165	0.9092	5.9316	0.0067
0.95	5.1055	0.2975	0.6584	0.8129	0.9106	7.4557	-0.0381

[a] $\dot{m}_L = n_L u_L$, $u_L = \sqrt{RT_L/2\pi}$

[b] $\dot{q}_L = 2n_L u_L RT_L$

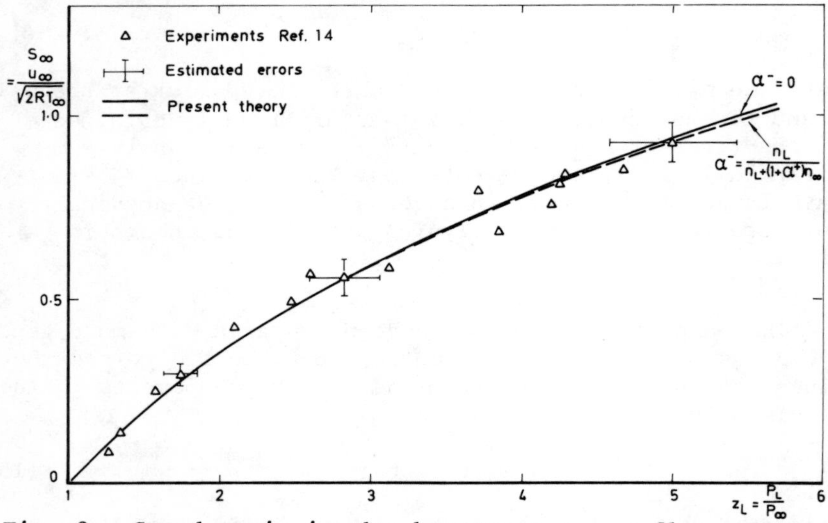

Fig. 2 Speed ratio in the downstream vapor flow compared with experimental values.

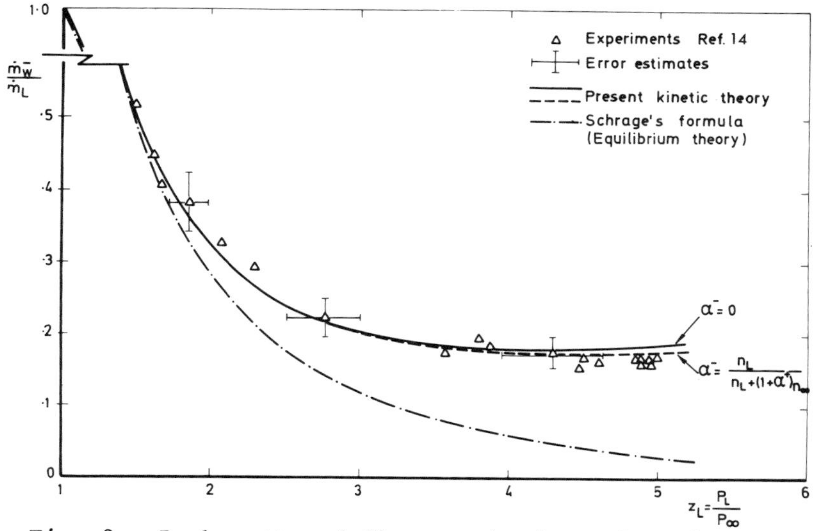

Fig. 3 Backscattered flux to the interphase boundary compared with experimental values.

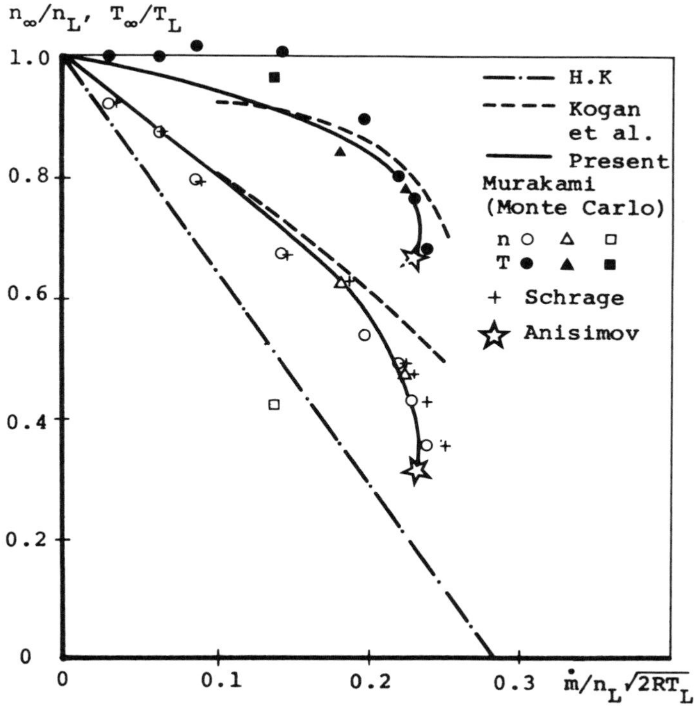

Fig. 4 Number density and temperature in the downstream vapor flow compared with numerical results of Kogan and Murakami.

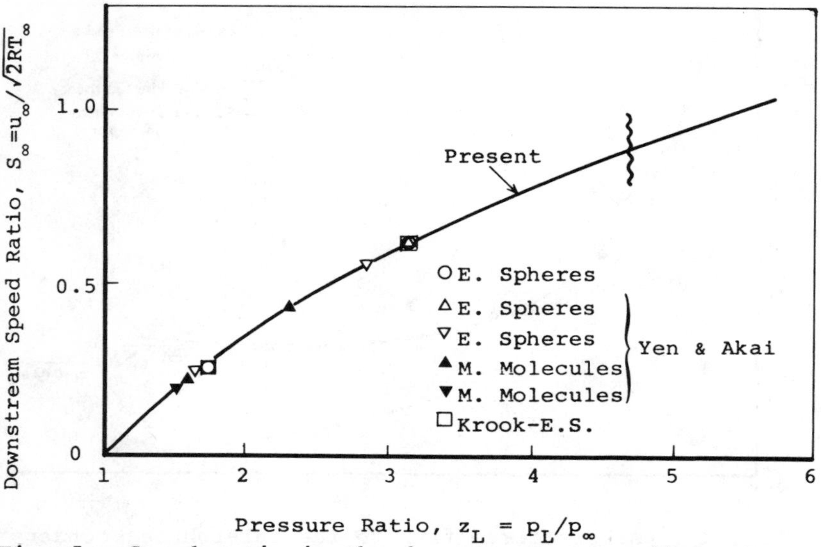

Fig. 5 Speed ratio in the downstream vapor flow compared with numerical results of Yen and Akai.

litude α was established from further collisional considerations[14]. The influence was found to be very small (Figs. 2,3) under flow conditions such that the Boltzmann H-theorem for the system was valid[14]. It is to be noted that the solution to the present gas kinetic connection problem is independent of the properties of the gas, such as the intermolecular collision law.

Structure of the Vapor Knudsen Layer

The transition in local quantities from the state at the phase boundary through the Knudsen layer to the downstream equilibrium state is computed from the nonconserved moment equation obtained with $Q_\mu = \xi_x^2$ for $\mu = 4$ in Eq. (3), using classical results for Maxwell molecules[15] to evaluate the collisional contribution ΔQ_4. The form of the distribution function must be compatible with Eqs. (1), (2), and (9), and is taken as the linear combination of half-range Maxwellians

$$f(x,\vec{\xi}) = a_1^+(x)f_L^+ + a_3^+(x)f_\infty^+ + a_3^-(x)f_\infty^- \qquad (16)$$

where the individual amplitude functions must satisfy the boundary conditions

$$x = 0:\quad a_1^+ = 1,\quad a_3^+ = 0,\quad a_3^- = \beta^- \qquad x = \infty:\quad a_1^+ = 0,\quad a_3^+ = 1,\quad a_3^- = 1 \tag{17}$$

with β^- being the value obtained from the gas kinetic connection problem (Table 1). This is exactly the Mott-Smith Ansatz that Anisimov[2] used to compute the limiting case of evaporation at sonic conditions, but he evaluated the collisional contribution based upon the BGK collision term and lost some interesting features in the solution.

Since there are three amplitude functions $a_i^\pm(x)$, and four moment equations for $\mu = 1, 2, 3, 4$, a degeneracy exists so that only two of the local conservation equations are linearly independent, and two of the amplitude functions thereby can be expressed as linear functions of the third one[11], e.g.

$$a_1^+(x) = [a_3^-(x) - 1]/(\beta^- - 1) \tag{18a}$$

$$a_3^+(x) = [\beta^- - a_3^-(x)]/(\beta^- - 1) \tag{18b}$$

The nonconserved moment equation for $Q_4 = \xi_x^2$ can therefore be written in terms of the function $a_3^-(x)$ alone, and the result is the following nonlinear differential equation

$$(da_3^-/dx) = -(A/\lambda_L)(a_3^- - 1)(a_3^- - r) \tag{19}$$

where A and r are parameters that depend upon the flow conditions through S_∞ or through z_L[11]. The solution to this equation, subject to the boundary condition at x = 0: $a_3^-(0) = \beta^-$, where $\beta^- > 1$, is given by

$$[a_3^-(x)-1]/[a_3^-(x)-r] = [(\beta^- -1)/(\beta^- -r)]\cdot\exp(-x/\ell) \tag{20}$$

where the scaling length ℓ has been defined as

$$\ell = \lambda_L/[A(1-r)] \tag{21}$$

and is contained in the inverse quantity λ_L/ℓ in Table 1 (λ_L is the mean free path for Maxwellian molecules[15] at saturation conditions).

The most striking property of the solution is that the scaling length, and therefore the thickness of the Knudsen layer, strongly increases and becomes infinite as the downstream flow conditions approach the sonic point. Beyond this point; i.e. for $M_\infty > 1$ or $S_\infty \gtrsim 0.91$, corresponding to $z_L=4.85$, there is no solution of Eq. (19) that approaches the correct downstream equilibrium value, $a_3 = 1$, because the parameter r then exceeds unity. This result i s found to be virtually independent of the particular form chosen to represent the distribution function.[14]

Effusion - Evaporation Analogy

Experimental and additional theoretical information on the kinetics in evaporation may be obtained from previous work on effusion from a perforated wall,[9,10,14] by observing the simple analogy that exists between the two problems. The molecular effusion from a thin, uniformly perforated wall may, under suitable conditions, be represented by the distribution function

$$f_e^+ = [n_e/(2\pi RT_o)^{3/2}]\exp(-\xi^2/2RT_o), \quad \xi_x > 0 \tag{22}$$

where the effusion density $n_e = qn_o$ is the product of the wall porosity q and the stagnation density n_o, and T_o is the stagnation temperature (Fig. 1). Diffuse reflection, after complete accomodation to the wall temperature $T_w = T_o$, is assumed for the molecules impinging on the wall, and capturing of these molecules by the wall orifices is neglected. The condition on the distribution function at the boundary is then

$$x = 0: \quad f = (1+\alpha^+)f_e^+ \quad , \quad \xi_x > 0 \tag{23}$$

where α^+ is the ratio between the density in the reflected and the effusing stream; $\alpha^+ = n_r/n_e$. For evaporation the corresponding condition is given by Eq.(1) and reads

$$x = 0: \quad f = f_L^+ \quad , \quad \xi_x > 0 \tag{24}$$

The two problems are similar from the kinetic point of view if the boundary conditions, Eqs. (23) and (24), are equal, assuming that the Boltzmann equation is satisfied to the same

extent in the two cases. The condition for similarity therefore becomes

$$f_L^+ = (1+\alpha^+) f_e^+ \tag{25}$$

which gives the explicit transformation rules

$$n_L = (1+\alpha^+) n_e \quad , \quad T_L = T_o \tag{26a}$$

$$p_L = (1+\alpha^+) q p_o \quad , \quad z_L = (1+\alpha^+) z \tag{26b}$$

where $z = q p_o / p_\infty$ is the pressure-porosity parameter in
effusive flow.[9,10,14]

     The transform function α^+ can be determined experimen-
tally simply by measuring the flux scattered back on the
perforated wall in effusion[9,14], and the transformation of
experimental results from effusion to evaporation can be
performed in a rigorous way, independent of limitations laid
down in the theoretical treatment of the two problems. Typical results are shown in Figs. 2 and 3, in which downstream
speed ratio and backscattered flux, obtained by free molecular orifice probe measurements in effusive flow[9,14], have
been transformed to evaporation and plotted versus
$z_L = (1+\alpha^+) z$.

                    Discussion and Conclusions.

     From the results shown in Figs. 2 and 3 we may conclude
that 1) the theoretical solution is practically insensitive
to changes in the arbitrary element α^- in the assumed distribution function f_w^-, and 2) there is substantial agreement
between the theoretical predictions and the experimental values for the downstream speed ratio and the backscattered mass
flux throughout the whole regime of quasisteady evaporation,
i.e. for parameter values in the range $0 < p_L/p_\infty \lesssim 4.85$,
$0 < M_\infty \lesssim 1$. Outside of this range the theoretical solution
becomes invalid, because the postulated downstream Maxwellian
state is not attained. The evaporation and effusion Knudsen
layer problem in one dimension appears therefore to be of
entirely subsonic nature. Figure 3 furthermore shows that
the Shrage formula[1] underestimates the backscattered mass
flux $\dot{m}_w$, in particular at high evaporation rates, due to its
neglect of collisional effects in the Knudsen layer. This
leads to a slightly overestimated net mass flux $\dot{m} = \dot{m}_L - \dot{m}_w$
at extreme conditions, as may be seen from Fig. 4. In that
figure the numerical BGK solution of Kogan and Makasev[4] and
the Monte Carlo results of Murakami and Oshima[5] are also

included, and they show good agreement with the present theory in their respective regimes of high accuracy - e.g. in the low and moderately strong evaporation regime for the BGK solution, and in the strong and moderately strong evaporation regime for the Monte Carlo results. The classical Hertz-Knudsen formula[16] is seen to underestimate the mass flux by a factor of approximately two, even in the case of weak evaporation (H.K. in Fig. 4).

Critical conditions, for which the downstream state becomes sonic, occur at the parameter values

$$z = p_L/p_\infty \simeq 4.85, \ \dot{m}_{cr} \simeq n_L\sqrt{RT_L/2\pi}\cdot 0.815 = \dot{m}_L\cdot 0.815$$

$$(\dot{q}_w)_{cr} \simeq 2n_L u_L RT_L\cdot 0.910 = \dot{q}_L\cdot 0.910$$

in agreement with the result of Anisimov[2]. Further comparisons are given by 1), Murakami and Oshima:[5] $\dot{m}_{cr} = 0.85\dot{m}_L$, 2), Edwards and Collins' calculation of evaporation from a spherical source into vacuum:[3] $\dot{m}_{cr} = 0.81\ \dot{m}_L$, 3), Golubtsov's experimental value:[17] $\dot{m}_{cr} = 0.87\dot{m}_L$, and 4) by the effusion experiments in Refs. 9 and 14: $\dot{m}_{cr} = (0.83 \pm 0.03)\dot{m}_L$. Also in the Monte Carlo simulations[5], no equilibrium state seemed to be attained at or beyond the critical conditions, in agreement with the results from the present ξ_x^2 moment equation.

In Fig. 5 a typical comparison between the present moment solution and the numerical results of Yen and Akai[6] is shown. The agreement is remarkable, and shows in particular that the present treatment of the gas kinetic connection problem as being of one-parameter nature, independent of the collisional dynamics, is essentially correct. This means that the relations between the macroscopic variables in evaporation do not depend upon the particular molecular interaction law and collision model considered, but will be conditioned by the general conservation equations for the problem.

The present study has thus provided new theoretical and experimental information on the nonlinear vapor kinetic problem in the intermediate and strong evaporation regime. The study has shown that a simplified moment method *a la* Mott-Smith, with the Liu-Lees half-range character imbedded, yields analytical results that are in substantial agreement with extensive computational solutions, and with experiments, throughout the whole regime of quasisteady evaporation.

References.

[1]Shrage, R.W., *A Theoretical Study of Interphase Mass Transfer*, Columbia University Press, New York, 1953.

[2]Anisimov, S.I., "Vaporization of Metal absorbing Laser Radiation", *Soviet Physics JETP*, Vol. 27, No. 1, 1968, pp. 182-183.

[3]Edwards, R.H. and Collins, R.L., "Evaporation from a Spherical Source into a Vacuum", *Rarefied Gas Dynamics*, Academic Press, New York and London, 1969, pp. 1489-1496.

[4]Kogan, M.N. and Makashev, N.K., "Role of the Knudsen Layer in the Theory of Heterogeneous Reactions and in Flows with Surface Reactions", *Fluid Dynamics*, Vol. 6, No. 6, 1974, pp. 913-920.

[5]Murakami, M. and Oshima, K., "Kinetic Approach to the Transient Evaporation and Condensation Problem", *Rarefied Gas Dynamics*, DFVLR Press, Porz-Wahn, 1974, Paper F6.

[6]Yen, S.M. and Akai, T.I., "Numerical Solutions of the Nonlinear Boltzmann and Krook Equations for an Evaporation - Effusion Problem", To be published in *Proceedings of Tenth International Symposium on Rarefied Gas Dynamics*, 1976.

[7]Mott-Smith, H.M., "The Solution of the Boltzmann Equation for a Shock Wave", *Physical Review*, Vol. 82, 1951, pp. 885-892.

[8]Liu, C.H. and Lees, L., "Kinetic Theory Description of Plane Compressible Couette Flow", *Rarefied Gas Dynamics*, Academic Press, New York and London, 1961, pp. 391-428.

[9]Ytrehus, T., "A study of the Large Scale Rarefied Flow Fields that Can Be Obtained Through Molecular Effusion from a Perforated Boundary", *von Karman Institute Tech. Note* 79, February, 1973.

[10]Ytrehus, T., "One-dimensional Effusive Flow by the Method of Moments", *Rarefied Gas Dynamics*, DFVLR Press, Porz-Wahn, 1974, Paper B4.

[11]Ytrehus, T., "Kinetic Theory Description and Experimental Results for Vapor Motion in Arbitrary Strong Evaporation", *von Karman Institute Tech. Note* 112, June 1975.

[12]Boltzmann, L., *Lectures on Gas Theory*, University of California Press, Berkely, 1964.

[13]Maxwell, T.C., *Collected Works*, Vol. II, Cambridge University Press, London 1890.

[14]Ytrehus, T., "Theoretical and Experimental Study of a Kinetic Boundary Layer Produced by Effusion from a Perforated Wall", Doctoral Thesis, 1975, Vrije Universiteit Brussel.

[15]Vincenti, W.G. and Kruger, C.H., *Introduction to Physical Gas Dynamics*, Wiley, New York, 1965, p. 364.

[16]Knudsen, M., "Die maximale Verdampfungsgeschwindigkeit des Quicksilbers", *Annalen der Physik*, Vol. 47, 1915, pp. 697-708.

[17]Golubtsov, I.V., "Investigation of the Evaporation of Tantalium into Vacua", *Heat Transfer Soviet Res*, Vol. 5, 1973, pp. 18-21.

EVAPORATION AND CONDENSATION IN A VAPOR-GAS MIXTURE

Tadashi Matsushita *

National Space Development Agency of Japan, Tokyo, Japan

Abstract

The two-surface problem of evaporation and condensation in the presence of noncondensable gas is considered using the reduced, linearized Gross and Krook equation as the governing equation and the finite-element method as a tool. The effects of the noncondensable gas on the flowfield are studied. It is shown that the presence of a small amount of noncondensable gas in the bulk of the vapor can cause a large buildup of the noncondensable gas near the lower-temperature interphase surface and can have a decisive effect on the vapor mass transfer from the higher-temperature surface to the lower-one.

Introduction

The mechanism and rate of transfer of vapor to and from an interphase surface differ, depending upon whether the medium is a chemically pure substance or contains a noncondensable gas. In the latter case, only a small mass fraction of the noncondensable gas can have a decisive effect in retarding the vapor mass-transfer rate because the vapor molecules approaching the interphase surface must diffuse through the noncondensable gaseous component. Although the practical importance of noncondensable gases has been established clearly by experiment, the analytical prediction of these effects from the kinetic theory viewpoint has proven to be a difficult problem. This problem has been studied by Pao[1] by the use of the modified linearized BGK equations to estimate the effects of the noncondensable gas on the mass flux rate and the heat flux rate. In a previous paper,[2] it was demonstrated that the tem-

---

Presented as Paper 52 at the 10th International Symposium on Rarefied Gas Dynamics, Aspen, Colo., July 19-23, 1976.

*Senior Engineer, Satellite Design Group.

perature gradient in the pure vapor field between two interphase surfaces can be opposite in sign to the externally maintained temperature under some particular conditions.

The purpose of the present paper is to investigate analytically the effects of the noncondensable gas on the negative temperature gradient as well as on the mass flow rate between two interphase surfaces. In this work, we also intend to present the solution of the preceding problem on the basis of the kinetic model equations that were proposed originally by Gross and Krook[3] (henceforth to be called GK equations). In order to solve these equations, we employ the finite-element method.

Theoretical Formulation

We consider the one-dimensional two-surface problem of evaporation and condensation of the vapor gas in the presence a noncondensable gas, as illustrated in Fig. 1. The two interphase surfaces have slightly different temperatures $T_s - \Delta T/2$ and $T_s + \Delta T/2$ and are placed at a distance L apart, and n_{1s} and n_{2s} are the number densities of saturated vapor and noncondensable gas at the temperature T_s, respectively. The ratio n_{2s}/n_{1s} is the measure of the contamination of the vapor by the noncondensable gas. If $\Delta T > 0$, then there will be generally steady mass flow from $X = L$ to $X = 0$.

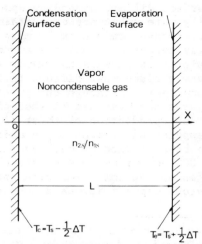

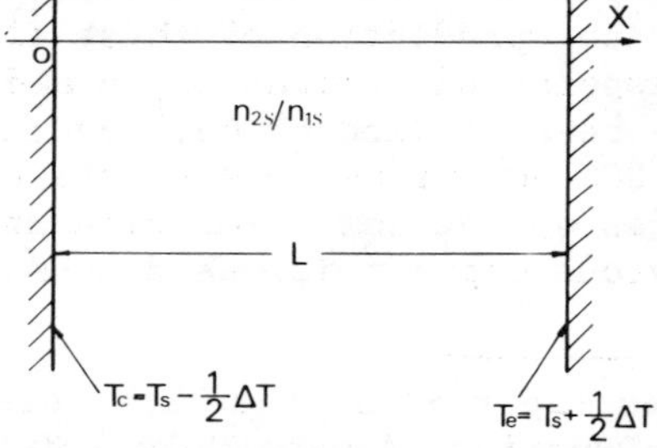

Fig. 1 Schematic representation of two-surface problem of evaporation and condensation in presence of noncondensable gas.

For the one-dimensional, steady-state case of a binary mixture in the absence of external forces, the GK equation for species i may be written as

$$\xi_i (\partial f_i/\partial X) = \nu_{ii}(M_{ii} - f_i) + \nu_{ij}(M_{ij} - f_i) \quad (1)$$

where

$$\left.\begin{array}{l} i = 1,\ 2 \\ j = 1,\ 2 \end{array}\right\} i \neq j$$

The subscripts 1 and 2 relate to the vapor and the noncondensable gas, respectively, f_i represents the distribution function for species i, and the collisional parameters ν_{ii} and ν_{ij} are proportional to the collision frequencies for the encounters, respectively, between species i and j. M_{ii} is the local Maxwellian distribution associated with the number density n_i, the velocity U_i, and the temperature T_i as follows

$$M_{ii} = n_i (m_i/2\pi k T_i)^{3/2} \exp\left[-(m_i/2kT_i)\{(\xi_{ix} - U_i)^2 + \xi_{iy}^2 + \xi_{iz}^2\}\right] \quad (2)$$

M_{ij} is the local Maxwellian distribution function associated with the velocity U_{ij} and the temperature T_{ij} as follows

$$M_{ij} = n_i (m_i/2\pi k T_{ij})^{3/2} \exp\left[-(m_i/2kT_{ij})\{(\xi_{ix} - U_{ij})^2 + \xi_{iy}^2 + \xi_{iz}^2\}\right] \quad (3)$$

As shown in Refs. 4 and 5, the velocity U_{ij} and temperature T_{ij} are given by

$$U_{ij} = (1 - \delta_i)U_i + \delta_i U_j \quad (4a)$$

$$T_{ij} = T_i + \frac{2m_i}{m_i + m_j}\delta_i (T_j - T_i) + \frac{m_i}{3k}\delta_i\left(\frac{2m_j}{m_i + m_j} - \delta_i\right)(U_j - U_i)^2 \quad (4b)$$

Here the parameter δ_i has the relation with the collisional parameter ν_{ij} for the cross-collision frequency ν_{ij}^* given by

$$\nu_{ij}^* = \left(\frac{8kT_i}{\pi m_i} + \frac{8kT_j}{\pi m_j}\right)^{1/2} \pi \left(\frac{d_i + d_j}{2}\right)^2 n_j \quad (5)$$

as follows

$$\delta_i = (4/3)[m_j/(m_i + m_j)](\nu_{ij}^*/\nu_{ij}) \quad (6)$$

where the value of δ_i is chosen to be equal to 5/6 from the semiempirical formula that has been proposed by Fay[6] for the viscosity of one component in the mixture.

In order to make numerical computation by the finite-element method more manageable, the following reduced distribution functions are defined

$$g_i(X,\xi_{ix}) = \int_{-\infty}^{\infty}\int_{-\infty}^{\infty} f_i(X,\xi_{ix},\xi_{iy},\xi_{iz})\,d\xi_{iy}\,d\xi_{iz} \tag{7}$$

$$h_i(X,\xi_{ix}) = \int_{-\infty}^{\infty}\int_{-\infty}^{\infty} (\xi_{iy}^2+\xi_{iz}^2) f_i(X,\xi_{ix},\xi_{iy},\xi_{iz})\,d\xi_{iy}\,d\xi_{iz} \tag{8}$$

Similarly, reduced local equilibrium distribution functions are defined as

$$G_{ii}(X,\xi_{ix}) = \int_{-\infty}^{\infty}\int_{-\infty}^{\infty} M_{ii}\,d\xi_{iy}\,d\xi_{iz} = n_i\left(\frac{m_i}{2\pi k T_i}\right)^{1/2}\exp\left\{-\frac{m_i}{2kT_i}(\xi_{ix}-U_i)^2\right\} \tag{9}$$

$$G_{ij}(X,\xi_{ix}) = \int_{-\infty}^{\infty}\int_{-\infty}^{\infty} M_{ij}\,d\xi_{iy}\,d\xi_{iz} = n_i\left(\frac{m_i}{2\pi k T_{ij}}\right)^{1/2}\exp\left\{-\frac{m_i}{2kT_{ij}}(\xi_{ix}-U_{ij})^2\right\} \tag{10}$$

$$H_{ii}(X,\xi_{ix}) = \int_{-\infty}^{\infty}\int_{-\infty}^{\infty} (\xi_{iy}^2+\xi_{iz}^2) M_{ii}\,d\xi_{iy}\,d\xi_{iz} = n_i\left(\frac{2kT_i}{\pi m_i}\right)^{1/2}\exp\left\{-\frac{m_i}{2kT_i}(\xi_{ix}-U_i^2)\right\} \tag{11}$$

$$H_{ij}(X,\xi_{ix}) = \int_{-\infty}^{\infty}\int_{-\infty}^{\infty} (\xi_{iy}^2+\xi_{iz}^2) M_{ij}\,d\xi_{iy}\,d\xi_{iz} = n_i\left(\frac{2kT_{ij}}{\pi m_i}\right)^{1/2}\exp\left\{-\frac{m_i}{2kT_{ij}}(\xi_{ix}-U_{ij}^2)\right\} \tag{12}$$

where n_i, U_i, and T_i are defined by

$$n_i = \int_{-\infty}^{\infty} g_i(X,\xi_{ix})\,d\xi_{ix} \tag{13a}$$

$$U_i = \frac{1}{n_i}\int_{-\infty}^{\infty} \xi_{ix}\, g_i(X,\xi_{ix})\,d\xi_{ix} \tag{13b}$$

$$\frac{3}{2}\frac{k}{m_i}T_i = n_i\left\{\int_{-\infty}^{\infty}\frac{1}{2}(\xi_{ix}-U_i)^2 g_i(X,\xi_{ix})\,d\xi_{ix} + \int_{-\infty}^{\infty}\frac{1}{2}h_i(X,\xi_{ix})\,d\xi_{ix}\right\} \tag{13c}$$

If Eq. (1) is multiplied by unit and integrated with respect to ξ_{iy} and ξ_{iz}, the following equation is obtained

$$\xi_{ix}(\partial g_i/\partial X) = \nu_{ii}(G_{ii}-g_i) + \nu_{ij}(G_{ij}-g_i) \tag{14}$$

If the multiplying factor is $\xi_{iy}^2+\xi_{iz}^2$, the corresponding equation yields

$$\xi_{ix}(\partial h_i/\partial X) = \nu_{ii}(H_{ii}-h_i) + \nu_{ij}(H_{ij}-h_i) \tag{15}$$

In the system whose deviation from an equilibrium state is presumed to be small, the following linearizations are possible

$$n_i = n_{is}(1+\zeta_i) \tag{16a}$$

$$T_i = T_s(1+\tau_i) \tag{16b}$$

$$T_{ij} = T_s\left\{1+\tau_i + \frac{2m_i}{m_i+m_j}\delta_i(\tau_j-\tau_i)\right\} \tag{16c}$$

$$G_{ii} = g_{is}\left\{1+\zeta_i+2u_ic_i+\tau_i\left(c_i-\frac{1}{2}\right)\right\} \tag{16d}$$

$$G_{ij} = g_{is}\left[1+\zeta_i+2\left\{(1-\delta_i)u_i+\delta_i u_j\left(\frac{m_i}{m_j}\right)^{1/2}\right\}c_i + \left\{\tau_i+\frac{2m_i}{m_i+m_j}\delta_i(\tau_j-\tau_i)\right\}\left(c_i^2-\frac{1}{2}\right)\right] \tag{16e}$$

$$H_{ii} = h_{is}\left\{1+\zeta_i+2u_ic_i+\tau_i\left(c_i^2+\frac{1}{2}\right)\right\} \tag{16f}$$

$$H_{ij} = h_{is}\left[1+\zeta_i+2\left\{(1-\delta_i)u_i+\delta_i u_j\left(\frac{m_i}{m_j}\right)^{1/2}\right\}c_i + \left\{\tau_i+\frac{2m_i}{m_i+m_j}\delta_i(\tau_j-\tau_i)\right\}\left(c_i^2+\frac{1}{2}\right)\right] \tag{16g}$$

$$g_i = g_{is}(1+\varphi_{gi}) \tag{16h}$$

$$h_i = h_{is}(1+\varphi_{hi}) \tag{16i}$$

where

$$\zeta_i = \pi^{-1/2}\int_{-\infty}^{\infty}\exp(-c_{ix}^2)\varphi_{gi}\,dc_{ix} \tag{17a}$$

$$u_i = \pi^{-1/2}\int_{-\infty}^{\infty}c_{ix}\exp(-c_{ix}^2)\varphi_{gi}\,dc_{ix} \tag{17b}$$

$$\tau_i = \frac{2}{3}\pi^{-1/2}\left\{\int_{-\infty}^{\infty}\left(c_{ix}^2-\frac{3}{2}\right)\exp(-c_{ix}^2)\varphi_{gi}\,dc_{ix} + \int_{-\infty}^{\infty}\exp(-c_{ix}^2)\varphi_{hi}\,dc_{ix}\right\} \tag{17c}$$

and g_{is} and h_{is} are defined by

$$g_{is} = n_{is}(m_i/2\pi kT_s)^{1/2}\exp\left[-(m_i/2kT_s)\xi_{ix}^2\right] \tag{18a}$$

$$h_{is} = n_{is}(2kT_s/\pi m_i)^{1/2}\exp\left[-(m_i/2kT_s)\xi_{ix}^2\right] \tag{18b}$$

and c_i and u_i are the normalized molecular velocity and the macroscopic velocity, which are defined by

$$c_{ix} = \frac{\xi_{ix}}{\sqrt{2(k/m_i)T_s}}, \qquad u_i = \frac{U_i}{\sqrt{2(k/m_i)T_s}}$$

Thus, the fundamental equations are

$$c_1\frac{\partial\varphi_{g1}}{\partial x} + (\alpha_{11}+\alpha_{12})\varphi_{g1} = (\alpha_{11}+\alpha_{12})\left\{\zeta_1+2A_1c_1u_1+A_2\left(c_1^2-\frac{1}{2}\right)\tau_1 + 2A_3c_1u_2+A_4\left(c_1^2-\frac{1}{2}\right)\tau_2\right\} \tag{19}$$

$$C_1\frac{\partial\varphi_{A1}}{\partial x}+(\alpha_{11}+\alpha_{12})\varphi_{A1}=(\alpha_{11}+\alpha_{12})\Big\{\zeta_1+2A_1C_1u_1+A_2\big(C_1^2+\tfrac{1}{2}\big)\tau_1+2A_3C_1u_2+A_4\big(C_1^2+\tfrac{1}{2}\big)\tau_2\Big\} \tag{20}$$

$$C_2\frac{\partial\varphi_{B2}}{\partial x}+(\alpha_{22}+\alpha_{21})\varphi_{B2}=(\alpha_{22}+\alpha_{21})\Big\{\zeta_2+2B_1C_1u_2+B_2\big(C_2^2-\tfrac{1}{2}\big)\tau_2+2B_3C_2u_1+B_4\big(C_2^2-\tfrac{1}{2}\big)\tau_1\Big\} \tag{21}$$

$$C_2\frac{\partial\varphi_{A2}}{\partial x}+(\alpha_{22}+\alpha_{21})\varphi_{A2}=(\alpha_{22}+\alpha_{21})\Big\{\zeta_2+2B_1C_1u_2+B_2\big(C_2^2+\tfrac{1}{2}\big)\tau_2+2B_3C_2u_1+B_4\big(C_2^2+\tfrac{1}{2}\big)\tau_1\Big\} \tag{22}$$

where the spatial coordinate has been nondimensionalized by the surface separation L, and the subscript x has been omitted for brevity. Besides

$$A_1=1-\frac{\alpha_{12}}{\alpha_{11}+\alpha_{12}}\delta_1, \qquad A_2=1-\frac{\alpha_{12}}{\alpha_{11}+\alpha_{12}}\frac{2m_1}{m_1+m_2}\delta_1$$

$$A_3=\frac{\alpha_{12}}{\alpha_{11}+\alpha_{12}}\delta_1\left(\frac{m_1}{m_2}\right)^{1/2}, \qquad A_4=\frac{\alpha_{12}}{\alpha_{11}+\alpha_{12}}\frac{2m_1}{m_1+m_2}\delta_1$$

$$B_1=1-\frac{\alpha_{21}}{\alpha_{22}+\alpha_{21}}\delta_2, \qquad B_2=1-\frac{\alpha_{21}}{\alpha_{22}+\alpha_{21}}\frac{2m_2}{m_1+m_2}\delta_2$$

$$B_3=\frac{\alpha_{21}}{\alpha_{22}+\alpha_{21}}\delta_2\left(\frac{m_2}{m_1}\right)^{1/2}; \qquad B_4=\frac{\alpha_{21}}{\alpha_{22}+\alpha_{21}}\frac{2m_2}{m_1+m_2}\delta_2$$

and α_{ii} and α_{ij} are the inverse Knudsen numbers related to the collisional parameters ν_{ii} and ν_{ij}, which are given by

$$\alpha_{ii}=\frac{\nu_{ii}L}{(2kT_s/m_i)^{1/2}}, \qquad \alpha_{ij}=\frac{\nu_{ij}L}{(2kT_s/m_i)^{1/2}} \tag{23}$$

The collisional parameter ν_{ij} is related to ν_{ii} by the equation

$$\frac{\nu_{ij}}{\nu_{ii}}=\left(\frac{m_j}{m_i+m_j}\right)^{1/2}\left(\frac{d_i+d_j}{2d_i}\right)^2\frac{n_j}{n_i}\frac{1}{\delta_i} \tag{24}$$

with d_i the equivalent hard-sphere diameter of i species. From Eqs. (5), (6), (23), and (24), the following relations are reduced

$$\alpha_{12}=\left(\frac{m_2}{m_1+m_2}\right)^{1/2}\left(\frac{d_1+d_2}{2d_1}\right)^2\frac{n_2}{n_1}\alpha_{11} \tag{25}$$

$$\alpha_{22}=\left(\frac{d_2}{d_1}\right)^2\frac{n_2}{n_1}\frac{\delta_2}{\delta_1}\alpha_{11} \tag{26}$$

$$\alpha_{21}=\left(\frac{m_1}{m_1+m_2}\right)^{1/2}\left(\frac{d_1+d_2}{2d_1}\right)^2\frac{1}{\delta_1}\alpha_{11} \tag{27}$$

so that, once only the measure of the contamination of the vapor gas by the noncondensable gas and the inverse Knudsen number α_{11} for the pure vapor are given for an arbitrary mixture, the other inverse Knudsen numbers α_{12}, α_{22}, and α_{21} can be determined.

Furthermore, by applying the half-range method in which φ is divided into two parts and introducing the following transformations of the dependent variables

$$\phi_{gi}^{\pm} = \varphi_{gi}^{\pm} - \psi_{gi}^{\pm} \quad , \qquad \phi_{Ri}^{\pm} = \varphi_{Ri}^{\pm} - \psi_{Ri}^{\pm} \tag{28}$$

to the fundamental equations in order to bring in the homogeneous form of the boundary conditions, we get the final form of the governing equations. For vapor gas

$$\begin{Bmatrix} F_{g1}(\phi_{g1}^{\pm}, \phi_{R1}^{\pm}, \phi_{g2}^{\pm}, \phi_{R2}^{\pm}, x, c_1, c_2) \\ F_{R1}(\phi_{g1}^{\pm}, \phi_{R1}^{\pm}, \phi_{g2}^{\pm}, \phi_{R2}^{\pm}, x, c_1, c_2) \end{Bmatrix} = c_1 \begin{Bmatrix} \partial\phi_{g1}^{\pm}/\partial x \\ \partial\phi_{R1}^{\pm}/\partial x \end{Bmatrix} + (\alpha_{11}+\alpha_{12}) \begin{Bmatrix} \phi_{g1}^{\pm} \\ \phi_{R1}^{\pm} \end{Bmatrix}$$

$$-(\alpha_{11}+\alpha_{12})\pi^{-1/2} \Bigg[\Bigg[\begin{bmatrix} \{1 - A_2(c_1^2 - 1/2)\} \\ \{1 - A_2(c_1^2 + 1/2)\} \end{bmatrix} \int_{-\infty}^{\infty} exp(-c_1^2)(\phi_{g1}^{+} + \phi_{g1}^{-})dc_1$$

$$+ 2A_1 c_1 \int_{-\infty}^{\infty} c_1 \, exp(-c_1^2)(\phi_{g1}^{+} + \phi_{g1}^{-})dc_1 + 2A_3 c_1 \int_{-\infty}^{\infty} c_2 \, exp(-c_2^2)(\phi_{g2}^{+} + \phi_{g2}^{-})dc_2$$

$$+ \frac{2}{3} A_2 \begin{Bmatrix} (c_1^2 - 1/2) \\ (c_1^2 + 1/2) \end{Bmatrix} \left\{ \int_{-\infty}^{\infty} c_1^2 exp(-c_1^2)(\phi_{g1}^{+} + \phi_{g1}^{-})dc_1 + \int_{-\infty}^{\infty} exp(-c_1^2)(\phi_{R1}^{+} + \phi_{R1}^{-})dc_1 \right\}$$

$$+ \frac{2}{3} B_4 \begin{Bmatrix} (c_1^2 - 1/2) \\ (c_1^2 + 1/2) \end{Bmatrix} \left\{ \int_{-\infty}^{\infty} c_2^2 exp(-c_2^2)(\phi_{g2}^{+} + \phi_{g2}^{-})dc_2 + \int_{-\infty}^{\infty} exp(-c_2^2)(\phi_{R2}^{+} + \phi_{R2}^{-})dc_2 \right.$$

$$\left. - A_4 \begin{Bmatrix} (c_1^2 - 1/2) \\ (c_1^2 + 1/2) \end{Bmatrix} \int_{-\infty}^{\infty} exp(-c_2^2)(\phi_{g2}^{+} + \phi_{g2}^{-})dc_2 \Bigg] + \begin{Bmatrix} \Phi_{g1}^{\pm} \\ \Phi_{R1}^{\pm} \end{Bmatrix} = 0 \tag{29}$$

where

$$\begin{Bmatrix} \Phi_{g1}^{\pm} \\ \Phi_{R1}^{\pm} \end{Bmatrix} = (\alpha_{11}+\alpha_{12}) \begin{Bmatrix} \psi_{g1}^{\pm} \\ \psi_{R1}^{\pm} \end{Bmatrix} - (\alpha_{11}+\alpha_{12})\pi^{-1/2} \Bigg[\Bigg[\begin{bmatrix} \{1 - A_2(c_1^2 - 1/2)\} \\ \{1 - A_2(c_1^2 + 1/2)\} \end{bmatrix}$$

$$\times \int_{-\infty}^{\infty} exp(-c_1^2)(\psi_{g1}^{+} + \psi_{g1}^{-})dc_1 + 2A_1 c_1 \int_{-\infty}^{\infty} c_1 \, exp(-c_1^2)(\psi_{g1}^{+} + \psi_{g1}^{-})dc_1$$

$$+ 2A_3 c_1 \int_{-\infty}^{\infty} c_2 \, exp(-c_2^2)(\psi_{g2}^{+} + \psi_{g2}^{-})dc_2 + \frac{2}{3} A_2 \begin{Bmatrix} (c_1^2 - 1/2) \\ (c_1^2 + 1/2) \end{Bmatrix}$$

$$\times \left\{ \int_{-\infty}^{\infty} c_1^2 exp(-c_1^2)(\psi_{g1}^{+} + \psi_{g1}^{-})dc_1 + \int_{-\infty}^{\infty} exp(-c_1^2)(\psi_{R1}^{+} + \psi_{R1}^{-})dc_1 \right\}$$

$$- A_4 \begin{Bmatrix} (c_1^2 - 1/2) \\ (c_1^2 + 1/2) \end{Bmatrix} \int_{-\infty}^{\infty} exp(-c_2^2)(\psi_{g2}^{+} + \psi_{g2}^{-})dc_2 + \frac{2}{3} A_4 \begin{Bmatrix} (c_1^2 - 1/2) \\ (c_1^2 + 1/2) \end{Bmatrix}$$

$$\times \left\{ \int_{-\infty}^{\infty} c_2^2 exp(-c_2^2)(\psi_{g2}^{+} + \psi_{g2}^{-})dc_2 + \int_{-\infty}^{\infty} exp(-c_2^2)(\psi_{R2}^{+} + \psi_{R2}^{-})dc_2 \right\} \Bigg]\Bigg]$$

For noncondensable gas

$$\begin{Bmatrix} F_{g2}(\phi_{g1}^{\pm}, \phi_{R1}^{\pm}, \phi_{g2}^{\pm}, \phi_{R2}^{\pm}, x, C_1, C_2) \\ F_{R2}(\phi_{g1}^{\pm}, \phi_{R1}^{\pm}, \phi_{g2}^{\pm}, \phi_{R2}^{\pm}, x, C_1, C_2) \end{Bmatrix} = C_2 \begin{Bmatrix} \partial\phi_{g2}^{\pm}/\partial x \\ \partial\phi_{R2}^{\pm}/\partial x \end{Bmatrix} + (\alpha_{22}+\alpha_{21}) \begin{Bmatrix} \phi_{g2}^{\pm} \\ \phi_{R2}^{\pm} \end{Bmatrix}$$

$$-(\alpha_{22}+\alpha_{21})\pi^{-1/2} \Bigg[\Bigg[\begin{bmatrix} \{1-B_2(C_2^2-1/2)\} \\ \{1-B_2(C_2^2+1/2)\} \end{bmatrix} \int_{-\infty}^{\infty} \exp(-C_2^2)(\phi_{g2}^+ + \phi_{g2}^-)\,dC_2$$

$$+ 2B_1C_2 \int_{-\infty}^{\infty} C_2 \exp(-C_2^2)(\phi_{g2}^+ + \phi_{g2}^-)\,dC_2 + 2B_3C_2 \int_{-\infty}^{\infty} C_1 \exp(-C_1^2)(\phi_{g1}^+ + \phi_{g1}^-)\,dC_1$$

$$+ \frac{2}{3} B_2 \begin{Bmatrix} (C_2^2-1/2) \\ (C_2^2+1/2) \end{Bmatrix} \left\{ \int_{-\infty}^{\infty} C_2^2 \exp(-C_2^2)(\phi_{g2}^+ + \phi_{g2}^-)\,dC_2 + \int_{-\infty}^{\infty} \exp(-C_2^2)(\phi_{R2}^+ + \phi_{R2}^-)\,dC_2 \right\}$$

$$+ \frac{2}{3} B_4 \begin{Bmatrix} (C_2^2-1/2) \\ (C_2^2+1/2) \end{Bmatrix} \left\{ \int_{-\infty}^{\infty} C_1^2 \exp(-C_1^2)(\phi_{g1}^+ + \phi_{g1}^-)\,dC_1 + \int_{-\infty}^{\infty} \exp(-C_1^2)(\phi_{R1}^+ + \phi_{R1}^-)\,dC_1 \right\}$$

$$- B_4 \begin{Bmatrix} (C_2^2-1/2) \\ (C_2^2+1/2) \end{Bmatrix} \int_{-\infty}^{\infty} \exp(-C_1^2)(\phi_{g1}^+ + \phi_{g1}^-)\,dC_1 \Bigg]\Bigg] \tag{30}$$

where

$$\begin{Bmatrix} \Phi_{g2}^{\pm} \\ \Phi_{R2}^{\pm} \end{Bmatrix} = (\alpha_{22}+\alpha_{21}) \begin{Bmatrix} \psi_{g2}^{\pm} \\ \psi_{R2}^{\pm} \end{Bmatrix} - (\alpha_{22}+\alpha_{21})\pi^{-1/2} \Bigg[\Bigg[\begin{bmatrix} \{1-B_2(C_2^2-1/2)\} \\ \{1-B_2(C_2^2+1/2)\} \end{bmatrix}$$

$$\times \int_{-\infty}^{\infty} \exp(-C_2^2)(\psi_{g2}^+ + \psi_{g2}^-)\,dC_2 + 2B_1C_2 \int_{-\infty}^{\infty} C_2 \exp(-C_2^2)(\psi_{g2}^+ + \psi_{g2}^-)\,dC_2$$

$$+ 2B_3C_2 \int_{-\infty}^{\infty} C_2 \exp(-C_2^2)(\psi_{g2}^+ + \psi_{g2}^-)\,dC_2 + \frac{2}{3} B_2 \begin{Bmatrix} (C_2^2-1/2) \\ (C_2^2+1/2) \end{Bmatrix}$$

$$\times \left\{ \int_{-\infty}^{\infty} C_2^2 \exp(-C_2^2)(\psi_{g2}^+ + \psi_{g2}^-)\,dC_2 + \int_{-\infty}^{\infty} \exp(-C_2^2)(\psi_{g2}^+ + \psi_{g2}^-)\,dC_2 \right\}$$

$$- B_4 \begin{Bmatrix} (C_2^2-1/2) \\ (C_2^2+1/2) \end{Bmatrix} \int_{-\infty}^{\infty} \exp(-C_1^2)(\psi_{g1}^+ + \psi_{g1}^-)\,dC_1 + \frac{2}{3} \begin{Bmatrix} (C_2^2-1/2) \\ (C_2^2+1/2) \end{Bmatrix}$$

$$\times \left\{ \int_{-\infty}^{\infty} C_1^2 \exp(-C_1^2)(\psi_{g1}^+ + \psi_{g1}^-)\,dC_1 + \int_{-\infty}^{\infty} \exp(-C_1^2)(\psi_{R1}^+ + \psi_{R1}^-)\,dC_1 \right\} \Bigg]\Bigg]$$

and

$$\phi_{gi}^{\pm}(x = {}^{0}_{1},\ C_x \gtrless 0) = 0, \qquad \phi_{Ri}^{\pm}(x = {}^{0}_{1},\ C_x \gtrless 0) = 0$$

where $\psi_i^{\pm}$ denote the boundary conditions on $\varphi_i^{\pm}$, respectively, and are obtained by assuming that, for vapor gas, molecules emitted from the interphase surface have a Maxwellian distribution corresponding to a saturated vapor at the temperature of the interphase, and, for noncondensable gas, molecules reflected from the surface also have a Maxwellian distribution characteristic of the surface temperature, and with these assumptions the boundary conditions become the following. For vapor gas,

$$\left.\begin{aligned}\psi^{+}_{g1}(0,C_1) &= -\frac{1}{2}\frac{\Delta n_1}{n_{1s}} - \frac{1}{2}\frac{\Delta T}{T_s}\left(C_1^2 - \frac{1}{2}\right)\\ \psi^{+}_{h1}(0,C_1) &= -\frac{1}{2}\frac{\Delta n_1}{n_{1s}} - \frac{1}{2}\frac{\Delta T}{T_s}\left(C_1^2 + \frac{1}{2}\right)\end{aligned}\right\} \quad C_1 > 0 \tag{31a}$$

$$\left.\begin{aligned}\psi^{-}_{g1}(1,C_1) &= \frac{1}{2}\frac{\Delta n_1}{n_{1s}} + \frac{1}{2}\frac{\Delta T}{T_s}\left(C_1^2 - \frac{1}{2}\right)\\ \psi^{-}_{h1}(1,C_1) &= \frac{1}{2}\frac{\Delta n_1}{n_{1s}} + \frac{1}{2}\frac{\Delta T}{T_s}\left(C_1^2 + \frac{1}{2}\right)\end{aligned}\right\} \quad C_1 < 0 \tag{31b}$$

For noncondensable gas

$$\left.\begin{aligned}\psi^{+}_{g2}(0,C_2) &= -\frac{1}{2}\frac{\Delta n_2}{n_{2s}} - \frac{1}{2}\frac{\Delta T}{T_s}\left(C_2^2 - \frac{1}{2}\right)\\ \psi^{+}_{h2}(0,C_2) &= -\frac{1}{2}\frac{\Delta n_2}{n_{2s}} - \frac{1}{2}\frac{\Delta T}{T_s}\left(C_2^2 + \frac{1}{2}\right)\end{aligned}\right\} \quad C_2 > 0 \tag{32a}$$

$$\left.\begin{aligned}\psi^{-}_{g2}(1,C_2) &= \frac{1}{2}\frac{\Delta n_2}{n_{2s}} + \frac{1}{2}\frac{\Delta T}{T_s}\left(C_2^2 - \frac{1}{2}\right)\\ \psi^{-}_{h2}(1,C_2) &= \frac{1}{2}\frac{\Delta n_2}{n_{2s}} + \frac{1}{2}\frac{\Delta T}{T_s}\left(C_2^2 + \frac{1}{2}\right)\end{aligned}\right\} \quad C_2 < 0 \tag{32b}$$

There exists the following simple relation between $\Delta n/n_{1s}$ and $\Delta T/T_s$ in the boundary conditions (31)

$$\Delta n_1/n_{1s} = \beta\,(\Delta T/T_s)$$

where β is the slope of the saturated vapor density-temperature curve at the temperature T_s and relates to the latent heat h_{lg} of phase transition as follows

$$\beta = (h_{lg}/RT_s) - 1$$

According to the demand that the overall gas flux rate be zero, $\Delta n_2/n_{2s}$ may be written as

$$\frac{\Delta n_2}{n_{2s}} = -\frac{1}{2}\frac{\Delta T}{T_s} + 2\int_{-\infty}^{\infty} C_2\, \exp(-C_2^2)(\phi^{+}_{g2} + \phi^{-}_{g2})\,dC_2 \tag{33}$$

We can solve the preceding governing equations simultaneously using the similar finite-element method used in Ref. 2. Here the assumed shape functions of kth element are approximated by

$$\phi^{\pm}_{gik}(\eta,C_i) = \left(1-\frac{\eta}{\ell}\right)\sum_{s=1}^{S} C_i^{s-1}e^{-C_i^2}(g^{\pm}_{2s-1})_{gik} + \left(\frac{\eta}{\ell}\right)\sum_{s=1}^{S} C_i^{s-1}e^{-C_i^2}(g^{\pm}_{2s})_{gik} \tag{34a}$$

$$\phi^{\pm}_{hik}(\eta,C_i) = \left(1-\frac{\eta}{\ell}\right)\sum_{s=1}^{S} C_i^{s-1}e^{-C_i^2}(g^{\pm}_{2s-1})_{hik} + \left(\frac{\eta}{\ell}\right)\sum_{s=1}^{S} C_i^{s-1}e^{-C_i^2}(g^{\pm}_{2s})_{hik} \tag{34b}$$

where η is the local space coordinate of x for the kth element, and ℓ is the width of the subdomains in the x direction.

Once the constants $q^{\pm}_{rs}$ are known as a result of the computation by the finite-element method, the macroscopic flow quantities for i species within the finite-element subdomain are given in the following manner

$$\frac{n_i}{n_{is}} = 1 + \pi^{-1/2} \sum_{s=1}^{S} \Big[\left(1 - \frac{\eta}{\ell}\right) \left\{ I_s^+ (q^+_{2s-1})_{gi} + I_s^- (q^-_{2s-1})_{gi} \right\} + \left(\frac{\eta}{\ell}\right) \left\{ I_s^+ (q^+_{2s})_{gi} + I_s^- (q^-_{2s})_{gi} \right\} \Big] \tag{35a}$$

$$u_i = -\frac{1}{2} \pi^{-1/2} \left(\frac{\Delta n_i}{n_{is}} + \frac{1}{2} \frac{\Delta T}{T_s} \right) + \pi^{-1/2} \sum_{s=1}^{S} \Big[\left(1 - \frac{\eta}{\ell}\right) \left\{ I^+_{s+1} (q^+_{2s-1})_{gi} + I^-_{s+1} (q^-_{s+1})_{gi} \right\} + \left(\frac{\eta}{\ell}\right) \left\{ I^+_{s+1} (q^+_{2s})_{gi} + I^-_{s+1} (q^-_{2s})_{gi} \right\} \Big] \tag{35b}$$

$$\frac{T_i}{T_s} = 2 - \frac{n_i}{n_{is}} + \frac{2}{3} \pi^{-1/2} \sum_{s=1}^{S} \Big[\left(1 - \frac{\eta}{\ell}\right) \left\{ I^+_{s+2} (q^+_{2s-1})_{gi} + I^-_{s+2} (q^-_{2s-1})_{gi} \right\} + \left(\frac{\eta}{\ell}\right) \left\{ I^+_{s+2} (q^+_{2s})_{gi} + I^-_{s+2} (q^-_{2s})_{gi} \right\} \Big] + \frac{2}{3} \pi^{-1/2} \sum_{s=1}^{S} \Big[\left(1 - \frac{\eta}{\ell}\right) \left\{ I^+_s (q^+_{2s-1})_{Ai} + I^-_s (q^-_{2s-1})_{Ai} \right\} + \left(\frac{\eta}{\ell}\right) \left\{ I^+_s (q^+_{2s})_{Ai} + I^-_s (q^-_{2s}) \right\} \Big] \tag{35c}$$

$$T \simeq T_i + [n_j/(n_i + n_j)](T_j - T_i) \tag{35d}$$

where T is the temperature of the gas mixture, and

$$I_s^+ = \int_0^{\infty} C_{ix}^{s-1} \exp(-2C_{ix}^2) dC_{ix}, \quad I_s^- = \int_{-\infty}^{0} C_{ix}^{s-1} \exp(-2C_{ix}^2) dC_{ix}$$

Numerical Results and Discussion

The calculations were done for imaginary vapor-gas mixtures with arbitrary values of β, but the values of the ratio of molecular mass m_2/m_1 and that of molecular diameter d_2/d_1 were chosen to be close to the ones for the air-water vapor mixture. Figure 2 gives the temperature distributions in the bulk of gas mixture between two interphase surfaces for the case with β = 10.0. The negative temperature gradient in the bulk of the gas between two surfaces which was found in calculations for the pure vapor in the previous paper is observed also under cases of the rather small noncondensable gas con-

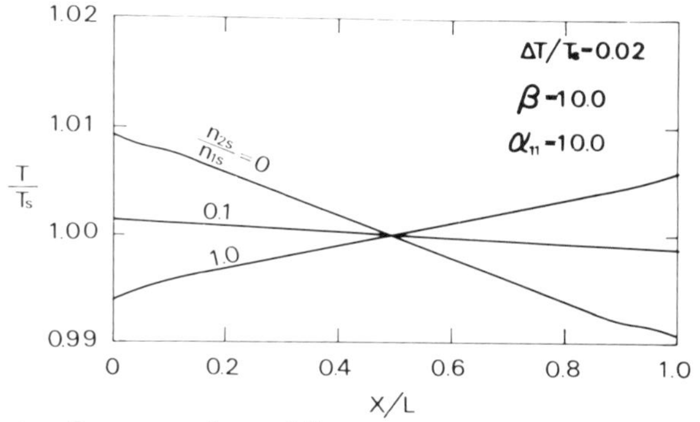

Fig. 2 Effect of noncondensable gas on temperature distribution.

centration. It is seen, however, that the increase of the noncondensable gas concentration makes the occurrence of this negative temperature difficult. Figure 3 shows the density distributions for the noncondensable gas. From this figure, it should be noted that molecules of noncondensable gas have been transferred considerably in the direction of the lower-temperature surface because of collisions with the vapor molecules. This result suggests that, in the experiments, the noncondensable gas tends to collect at, or be concentrated near, the lower-temperature surface. The dependence of the condensation rate on the noncondensable gas concentration is

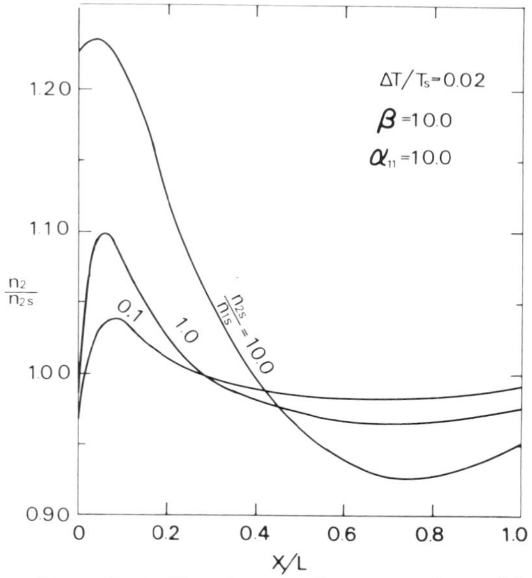

Fig. 3 Density distributions of noncondensable gas.

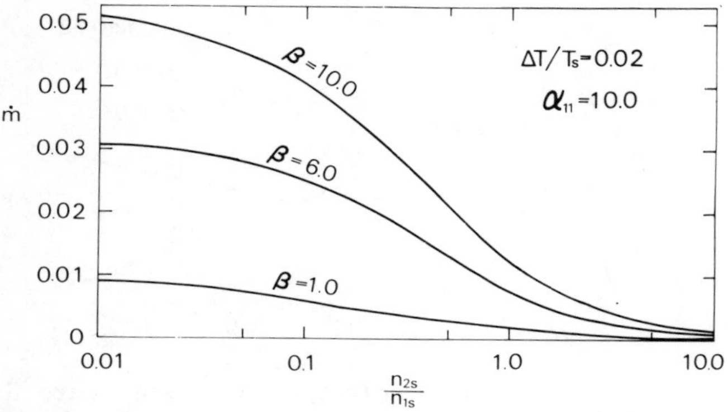

Fig. 4 Effect of noncondensable gas on condensation mass transfer.

illustrated for several values of β in Fig. 4. Inspection of this figure reveals that the increase of a noncondensable gas reduces the condensation rate, and its presence has a decisive effect in reducing the condensation mass transfer. The result allows us to state that the 10% mass fraction of noncondensable gas reduces the condensing flux to 77% of condensation rate for the pure vapor with β = 10.0. Such a strong influence of the noncondensable gas on the condensation rate may be explained by the resistance caused by the diffusion of vapor molecules through the layer of the noncondensable gas.

The dependence of the condensation rate on the noncondensable gas can be explained well by using Fig. 5, which shows the vapor mass flow rate of the two streams by means of the half-range method. As seen in this figure, the mass flux $\dot{m}^-$ of the vapor molecules moving in the direction of the lower-

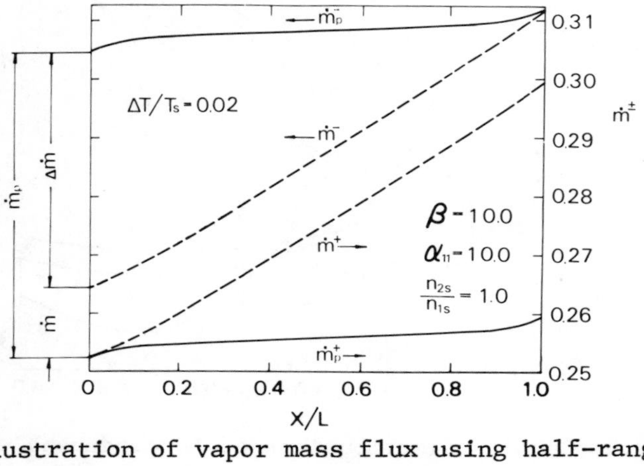

Fig. 5 Illustration of vapor mass flux using half-range method.

temperature starts at the same absolute Maxwellian mass flux rate corresponding to the surface condition as the value for the case of pure vapor from the higher-temperature surface. At the lower-temperature surface, this vapor mass flux becomes lower by $\Delta\dot{m}$ than the computed value for the case of pure vapor owing to the considerable backscattering of the vapor molecules in the direction of the higher-temperature surface as a result of molecular collisions with the molecules of noncondensable gas through the flowfield between two interphase surfaces. On the other hand, the mass flux $\dot{m}^+$ of the vapor molecules moving in the direction of the higher-temperature surface, at first, also has the same absolute Maxwellian mass flux rate corresponding to the surface condition as the value for the case of pure vapor at the lower-temperature surface. Hence, in this case, the net vapor mass flow rate $\dot{m}$ is lower by $\Delta\dot{m}$ than the one computed for the case of pure vapor.

References

[1] Pao, Y. P., "Evaporation in a Vapor-Gas Mixture," Journal of Chemical Physics, Vol. 59, Dec. 1973, pp. 6688-6689.

[2] Matsushita, T., "An Analytical Study of the Problem of Evaporation and Condensation Using the Finite Element Method," Proceedings of the Symposium on Rarefied Gas Dynamics, Vol. II, pp. F3-1 to F3-12.

[3] Gross, E. P. and Krook, M., "Model for Collision Processes in Gases: Small-Amplitude Oscillation of Charged Two-Component Systems," Physical Review, Vol. 102, May 1956, pp. 593-604.

[4] Oguchi, H., "A Kinetic Model for a Binary Mixture and its Application to a Shock Structure," Proceedings of the Symposium on Rarefied Gas Dynamics, Vol. I, 1967, pp. 745-757.

[5] Abe, K., "A Numerical Analysis of Plane Couette Flow of Rarefied Binary Gas Mixture," Report No. 440, July 1969, Institute of Space and Aeronautical Science, University of Tokyo.

[6] Fay, J. F., "Hypersonic Heat Transfer in the Air Laminar Boundary Layer," AMP 71, March 1962, AVCO Everett Labs.

TIME-EVOLUTION OF A MULTIDROPLET SYSTEM IN THE KINETIC THEORY OF CONDENSATION-VAPORIZATION

N. Bellomo,* R. Loiodice,+ and G. Pistone≠

Politecnico, Torino, Italy

Abstract

This paper deals with a stochastic problem in the kinetic theory of vaporization and condensation and studies the time evolution of a multidroplet system vaporizing in a vapor medium at given, fixed, thermodynamical conditions. In particular, the vaporization/condensation of a single droplet is studied with a pahse-transition model proposed in a previous paper, whereas a law of evolution of the distribution function of the diameters and temperatures of the droplets is deduced here on the basis of Liouville's theorem, and treated analytically and numerically, by means of a computing sequence proposed in this paper, in order to study, at given initial conditions, the time-evolution of the system. We belive that this paper proposes a rigorous and useful theory that is a necessary basis for the study of precisely defined physical problems and for accurate interpretation of experimental results in this field.

Presented as Paper 5 at the 10th International Symposium on Rarefied Gas Dynamics, Aspen, Colo., July 19-23, 1976. This work has been realized within the activities of the Italian Council of the Research, C.N.R.

*Prof. stab. of Meccanica Razionale, Istituto di Meccanica Razionale.

+Researcher of the Centro Studi sulla Dinamica dei Fluidi, C.N.R.

≠Ph.D. Fellow, Istituto Matematico.

Introduction

The problem of vaporization and condensation of spherical droplets in a vapor medium has been studied, by means of the equations both of the kinetic theory and of the continuum mechanics, by several authors.[1-5] These papers are, however, based mainly on uknown parameters and consequently cannot give qualitative exact results. On the other hand, in another paper,[6] a phase-transition model (for spherical droplets in rarefied gas-vapor conditions) has been proposed and compared successfully with experimental results of other authors. This model, which is based on the statistical methods of kinetic theory and of surface interaction,[7-9] has proved to be able to describe also evolutions in strong nonequilibrium conditions.[10]

In practice, however, the study of one single droplet is just a preliminary approach to the more realistic problem of the time-evolution of a system of several droplets. This paper deals with the analysis of a condensing/vaporizing multidroplet system in a vapor volume at constant thermodynamical conditions, and studies the time-evolution of the distribution function of the diameters D(t) and temperatures T(t) of the droplets at given initial distribution (on the initial diameters and temperatures) and at given thermodynamical conditions of the vapor volume. In particular, a general theory, which shows that the study of the evolution of the distribution function of D and T is equivalent to an initial-value problem, has been proposed.[11] In this paper, we shall apply this theory to the computing of the time-evolution of the aforementioned distribution and analyze the analytical problems involved.

Equations of Evolution of the Thermodynamical System

The evolution of one single droplet is described by the law of variation, in function of time, of the diameter and temperature of the droplet itself. The evolution of a multidroplet system is described by the evolution of the distribution $f(D,T;t)$ at given initial distribution $f_0(D_0,T_0;t=0)$. In another paper,[11] this evolution has shown (on the basis of an elaboration of a phase-transition model[6] and of the theory of stochastic differential equations[12,13]) to be equivalent to the study of an ordinary, nonlinear, differential vector equation with random initial conditions.

The basic hypotheses of the condensation-vaporization model are the ones of Ref.6, also reported in Ref.11; their mathematical elaboration, together with the theory of ordinary differential equations with random initial conditions, gives the following set of equations

$$dD/dt = F_1(T;\underline{\beta}) = [(2/\pi)(k/m)T]^{1/2}\,[(n_\infty/n_\ell)f^-(T) - f^+(T)] \tag{1}$$

$$dT/dt = F_2(D,T;\underline{\beta}) = [(2/\pi)(k/m)T]^{1/2}(1/D)\{4(n_\infty/n_\ell)(T_\infty/T)^{1/2}\cdot \cdot(T_\infty - T) + T[(n_\infty/n_\ell)f^-(T) - f^+(T)]\} \tag{2}$$

$$dJ/dt = F_3(D,T;\underline{\beta}) = -J(t;D_0,T_0)(\partial F_1/\partial D + \partial F_2/\partial T) \tag{3}$$

$$f[T(T_0,D_0),D(T_0,D_0);t] = f_0(T_0,D_0;t=0)J(t;T_0,D_0) \tag{4}$$

$$D_0 = D(t=0); \quad T_0 = T(t=0) \;; \quad J(t=0) = 1 \tag{5}$$

$$\underline{\beta} = (n_\infty, n_\ell, T_\infty, k/m, \sigma) \tag{6}$$

$$f^-(T) = \frac{\exp[y(T)] - 1 - y(T)}{\exp[y(T)]} \;; \quad f^+(T) = \frac{1 + x(T)}{\exp[x(T)]} \tag{7}$$

$$x(T) = [\sigma(T)/kT][\,8\pi(3/4\pi n_\ell)^{2/3} - 12/(Dn_\ell)] \tag{8}$$

$$f^+[x(T)] = (n_\infty/n_\ell)_{Eq.}\, f^-[y(T)] \tag{9}$$

$$x(T) = [V_e(T)/c(T)]^2 \;; \quad y(T) = [V_c(T)/c(T)]^2; \quad c = [2(k/m)T]^{1/2} \tag{10}$$

V_e is the vaporization velocity, V_c is the condensation velocity and $\sigma(T)$ is the surface tension.

We shall not deduce the set of Eqs.(1-10) which has been dealt with in previous papers, see Ref.11, but just shall comment that Eqs.(1) and (2) describe, respectively, the mass and energy balance (in particular, accomodation of the condensing/vaporizing molecules at the liquid-surface temperature is assumed[11]) whereas Eq.(9) states that at equilibrium conditions the vaporizing flux must equal the condensing one. Moreover, Eqs.(3) and (4), which are deduced directly by the theorem of Liouville in stochastic differential equations (see the next section), describe the evolution of the distribution function, and this is the main aim of our paper.

Proof of Eqs.(3) and (4): Liouville Theorem

Let us consider an ordinary vector differential equation of the type

$$\dot{\underline{X}} = \underline{F}(\underline{X};t) \;;\quad \underline{X}(t) \in R_+^n \;;\quad t \in [0, t_f[\tag{11}$$

with $\underline{F}$ function C^1 in $\underline{X}$ and with initial conditions $\underline{X}_o = \underline{X}(t=0)$ distributed according to a probability density $P(\underline{X}_o)$. Let us also consider the Jacobian $G(t;\underline{X}_o)$ such that

$$P[\underline{X}(t;\underline{X}_o)] = P(\underline{X}_o) \cdot J(t;\underline{X}_o) \;;\quad G = \partial \underline{X} / \partial \underline{X}_o \tag{12}$$

or the inverse mapping $J = G^{-1}$ such that

$$P[\underline{X}(t;\underline{X}_o)] = P(\underline{X}_o) \cdot J(t;\underline{X}_o) \tag{13}$$

We shall prove the following theorem:

Theorem (Liouville): Supposing that a solution of Eq.(11) exists, then the density $P(\underline{X};t)$ of this solution satisfies the Liouville equation

$$\partial P / \partial t + \nabla(\underline{F}\, P) = 0 \tag{14}$$

In fact, multiplying Eq.(14) by G and taking into account the known,[12-13] result of the theory of differential equations ($dG/dt = G \cdot \nabla F$), the following result can be obtained

$$G(\partial P/\partial t) + G \cdot \underline{F}(\partial P/\partial \underline{X}) + P(dG/dt) = 0 \qquad d(G \cdot P)/dt = 0 \tag{15}$$

thus,

$$P \cdot G = \text{cost} \Longrightarrow G(t=0) P(\underline{X}_o) = G(t) P(\underline{X}) \;;\quad G(t=0) = 1 \tag{16}$$

which proves the theorem.

Consequently, since the solution of Liouville's equation is known as follows

$$P(\underline{X};t) = P(\underline{X}_o;0) \exp\left\{ - \int_0^t \nabla \underline{F} \cdot dt \right\} \tag{17}$$

the comparison with Eq.(13) yields

$$\ln(J) = - \int_0^t \nabla \underline{F} \cdot dt \Longrightarrow \frac{dJ}{dt} = - J \cdot \nabla \underline{F} \tag{18}$$

and, if $\underline{X} = (D,T)$, then Eqs.(3) and (4) are proved by Eq.(18).

Initial Distribution Function

Before approaching the general, aforementioned, problem, some discussion about the initial condition on $f_o(D,T)$ must be made. For this purpose, let us assume that the initial maxima and minima values of T and D (respectively T_{max}, D_{max} and T_{min}, D_{min}) and also the mean values of T and D (respectively $\langle T_o \rangle$ and $\langle D_o \rangle$) are given at t=0. These limit and average values should be a consequence of the physical problem that is being considered.

The initial distribution $f_o(D_o,T_o;t=0)$ can be written as follows

$$f_o = P_{oD}(D;D_p,a_D,b_D,t=0)\cdot P_{oT}(T;T_p,a_T,b_T,t=0) \qquad (19)$$

where D_p and T_p are the most probable diameter and temperature, which, together with $a_{D,T}$ and $b_{D,T}$, depend on D_o and T_o as well as on the ranges B_{oT} and B_{oD} defined as follows

$$B_{oT} = \left[T_{min}/T_p, T_{max}/T_p\right] \subset \mathbb{R}_+ \; ; \; B_{oD} = \left[D_{min}/D_p, D_{max}/D_p\right] \subset \mathbb{R}_+ \qquad (20)$$

Of course the position of independent initial probabilities, stated by Eq.(19), is not strictly necessary.

T_p, D_p, $a_{D,T}$, and $b_{D,T}$ can be determined by means of the following set of equations that describe some known properties of statistical mechanics

$$1 = \int_{B_{oT}} P_{oT}\, d\left(\frac{T}{T_p}\right) = \int_{B_{oD}} P_{oD}\, d\left(\frac{D}{D_p}\right) \; ; \; T=T_p \Leftrightarrow \max(P_{oT}) ; \; D=D_p \Leftrightarrow \max(P_{oD})$$

$$T_p \int_{B_{oT}} P_{oT}\left(\frac{T}{T_p}\right) d\left(\frac{T}{T_p}\right) = \langle T_o \rangle \; ; \quad D_p \int_{B_{oD}} P_{oD}\left(\frac{D}{D_p}\right) d\left(\frac{D}{D_p}\right) = \langle D_o \rangle \qquad (21)$$

Computation of the Distribution Function

In this section, we shall consider the problem of finding, at given vapor conditions and initial distribution f_o, the distributions $f(D;T_f,t_f)$ and $f(T;D_f,t_f)$ where T_f and D_f are fixed temperatures and diameters at the final time t_f. This problem requires both a preliminary analysis of the evolution equation and the application of a computing sequence, which is proposed here.

Therefore, let us first point out some basic properties of the evolution of one droplet which also can be proved rigorously, see Ref.(14). We refer to problems with predominant vaporization, that is, according to the hypothesis of rarefied medium:

1) The rates of variation of D and T for small and hot drops are higher than the ones of large and cold drops

$$D_{o1} \gtrless D_{o2},\ T_{o1}=T_{o2} \Longrightarrow \left|\dot{D}(t;D_{o1},T_{o1})\right| \lesseqgtr \left|\dot{D}(t;D_{o2},T_{o2})\right| ,$$
$$\left|\dot{T}(t;D_{o1},T_{o1})\right| \lesseqgtr \left|\dot{T}(t;D_{o2},T_{o2})\right|$$

$$D_{o1}=D_{o2},\ T_{o1} \gtrless T_{o2} \Longrightarrow \left|\dot{D}(t;D_{o1},T_{o1})\right| \gtrless \left|\dot{D}(t;D_{o2},T_{o2})\right| ,$$
$$\left|\dot{T}(t;D_{o1},T_{o1})\right| \gtrless \left|\dot{T}(t;D_{o2},T_{o2})\right|$$

2) The droplet reaches asimptotically a temperature $T_{as} \leq T_\infty$, and afterwards the evolution is such that $\dot{T}=0$ and $\dot{D}$=cost.

A consequnce of property 1 is that the ranges of the minima and maxima temperatures and diameters of the multidroplet system can be defined as follows

$$B_T(t_f) = \left[T_{min}(t_f;T_{min},D_{max}),T_{max}(t_f;T_{max},D_{max})\right] \tag{22}$$

$$B_D(t_f) = \left[D_{min}(t_f;T_{max},D_{min}),D_{max}(t_f;T_{max},D_{max})\right] \tag{23}$$

Moreover, for simplicity of writing, let us compact Eqs.(1-4) as follows

$$\dot{\underline{X}}=\underline{F}(\underline{X},t)\ ;\ \underline{X}(D,T)\ ;\ \underline{F}(F_1,F_2)\ ;\ \underline{X}_o=(D_o,T_o) \tag{24}$$

and let us call ΔT and ΔD the steps by which the variables T and D can be discretized. Thus we shall apply the following process:

1) Fix T_f, $T_f \in B_T(t_f)$, t_f and $D_o=D_{min}$.

2) Solve the following problem with limits: find $\underline{X}_o=(D_o,T_o)$ such that $\dot{\underline{X}}=\underline{F}(\underline{X},t)$, $\underline{X}_o \Longrightarrow \underline{X}(t_f)=(D,T_f)$.

3) If problem 2 has not a solution, $D_o+\Delta D \rightarrow D_o$, start from 2. If problem 2 has a solution, apply Eq.(4) to compute $f(D;T_F,t_f)$

4) $D_o < D_{max}$, $D_o+\Delta D \rightarrow D_o$, start from 2. If $D_o \geq D_{max}$, $T_f+\Delta T \rightarrow T_f$.

5) If $T_f \in B_T$, start from 2. If $T_f \notin B_T$, stop.

Moreover, in order to compute $f(T;D_f,t_f)$, we shall use the same method in the following:

1) Fix D_f, $D_f \in B_D(t_f)$, t_f and $T_o = T_{min}$.

2) Solve the following problem with limits: find $\underline{X}_o(D_o,T_o)$ such that $\dot{\underline{X}} = \underline{F}(\underline{X},t)$, $\underline{X}_o \Rightarrow \underline{X}(t_f) = (D_f, T)$.

3) If problem 2 has not a solution, $T_o + \Delta T \rightarrow T_o$, start from 2. If problem 2 has a solution, apply Eq.(4) to compute $f(T;D_f t_f)$

4) If $T_o < T_{max}$, $T_o + \Delta T \rightarrow T_o$, start from 2. If $T_o \geq T_{max}$, $D_f + \Delta D$ D_f.

5) If $D_f \in B_D$, start from 2. If $D_f \notin B_D$, stop.

Numerical Calculations

Some numerical calculations have been made for water drops in vapor medium. In particular, Figs.1 and 2 show the evolution of one single droplet and visualize the effects described by properties 1 and 2. With regard to the evolution of the

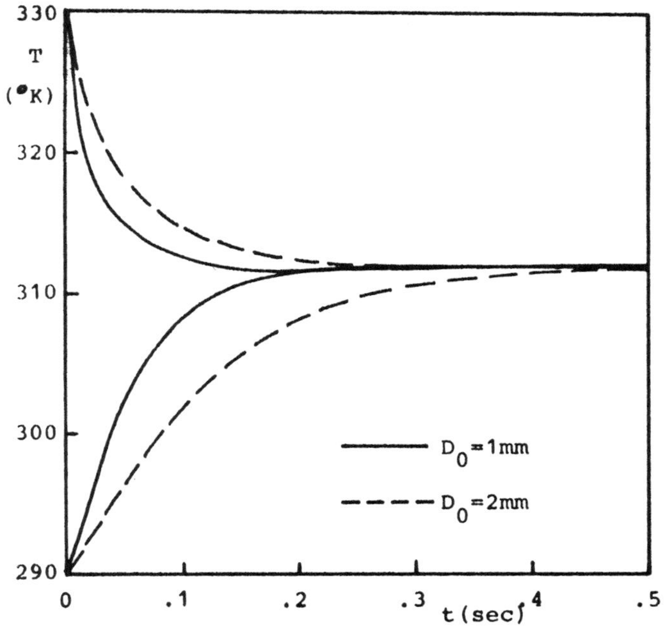

Fig. 1 T vs t: water droplet.

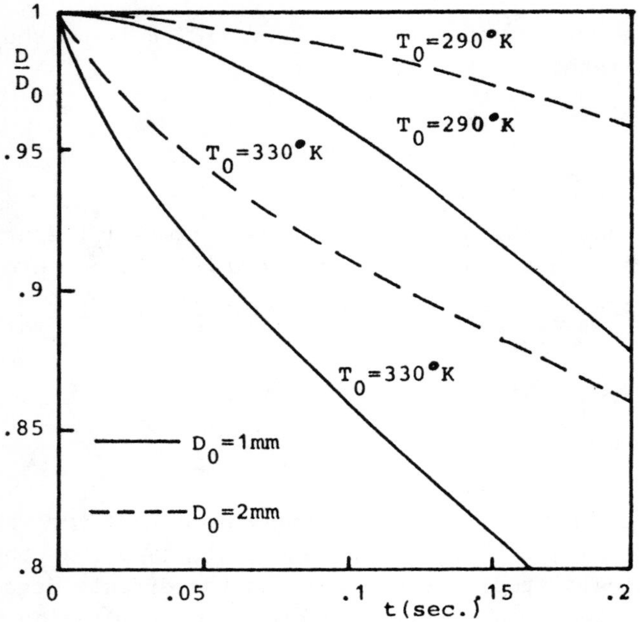

Fig 2 D/D_0 vs t: water droplet.

multidroplet system, the following initial conditions have been chosen

$$P_{oT}=a_T(T/T_p)\exp\left\{-b_T(T/T_p)^2\right\}$$

$$P_{oD}=a_D(D/D_p)\exp\left\{-b_D(D/D_p)^2\right\}$$

$$T_{min}=290^oK; \quad T_{max}=330^oK; \quad D_{min}=0.2\text{ mm}; \quad T_o=308.5^oK$$

$$D_o = 1.1\text{ mm}$$

According to Eq.(21), the following result is obtained

$$a_T=12.498; \quad a_D=1.308; \; b_D=b_T=0.5; \; T_p=302.62^oK; \; D_p=1.15\text{ mm}$$

The initial distribution is shown in Fig.3 and is concave with respect to both variables T and D, even if the concavity with respect to D is more evident. On the other hand, the sequence indicated in the preceding section give, at t_f=0.025 sec, the result shon in Figs.4 and 5, which point out that the distributions, for particular values of T_f and D_f, depending on the

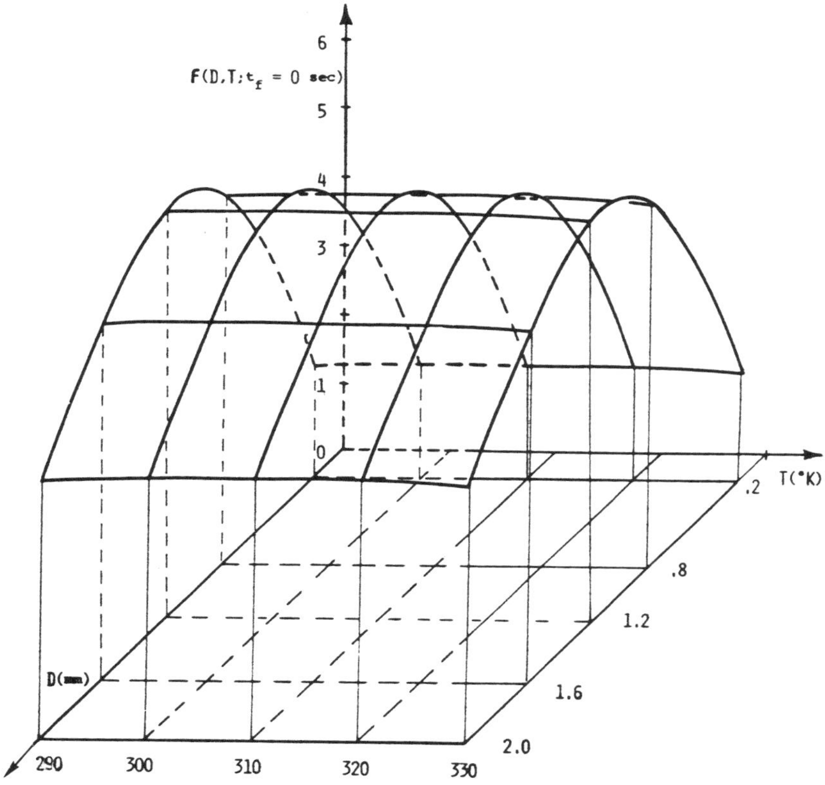

Fig 3 Space representation of $f(D,T;t=0)$.

initial conditions, can become convex. This effect is explained by Fig.6, which shows the space distribution of $J(D,T;t_f)$, and this is the operator transforming the initial distribution into the final one. Figure 7, in conclusion, shows $f(D,T;t_f)$ obtained by Figs.4 and 5. In particular, on the (D,T) plane, the variation of the range of D and T (which at t=0 is a rectangular one; see Fig.1) also is shown.

We shall not give any further comment on these numerical calculations; in fact, we have not taken into account any particular physical problem, but (according to the main aim of the paper) we have visualized the evolution of the system and tested, successfully, and for arbitrary f_o and $\underline{\beta}$, the theory described in the preceding section.

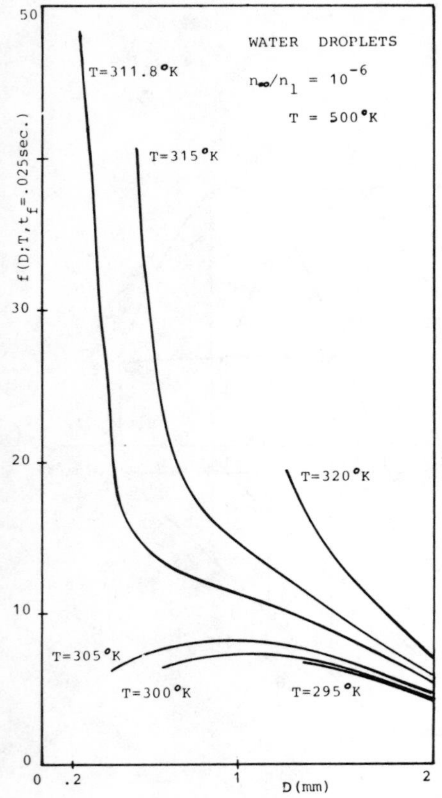

Fig 4 $f(D;T_f,t_f=0.025\ \text{sec})$ vs D.

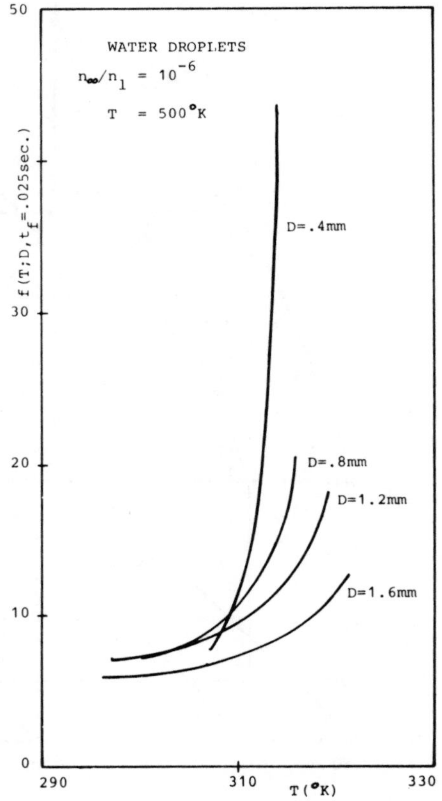

Fig 5 $f(T;D_f,t_f=0.025\ \text{sec})$ vs T.

Discussion

The subject dealt with in this paper has been proposed in a general form in order to supply a rigorous method, in statistical mechanics, for the study of the time-evolution of a system of a lerge number of objects (the state of each object being described by some variables, and the evolution of each variable being described by an ordinary differential equation) when the initial conditions of these objects are distributed on the said variables. In fact,in our paper, we deduce on the basis of Liouville's theorem, a law of evolution for the distribution function of the said variables and propose a computing sequence in order to find the time-evolution of the said distribution.

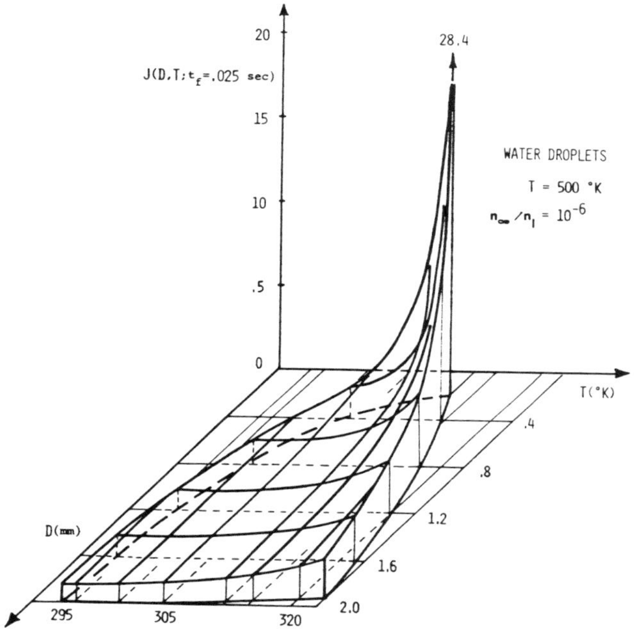

Fig 6 Space representation of $J(D,T;t_f=0.025$ sec).

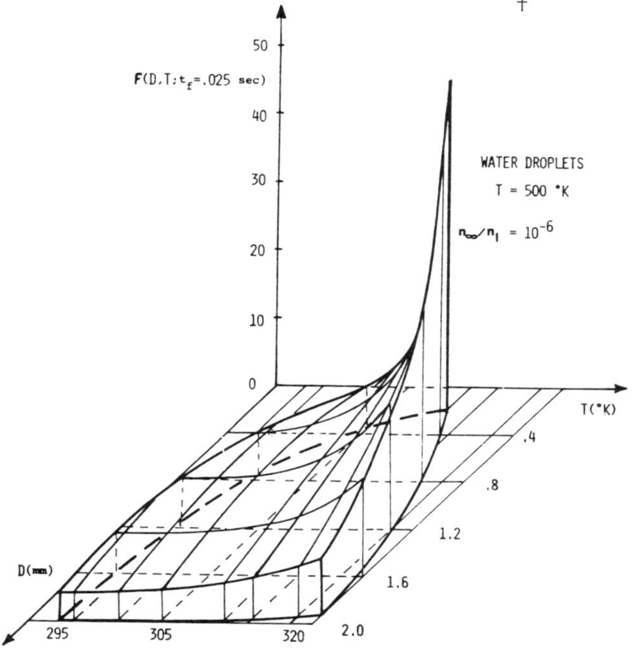

Fig 7 Space representation of $f(D,T;t_f=0.025$ sec).

This methodology can, of course, use other phase-transition models as long as the evolution still is described by an ordinary differential equation. In particular, a rigorous theory for liquid-vapor phase-transition is proposed in Ref. 15, without the limitation of rarefied gas conditions. On the other hand, the model that this paper deals with has given theoretical results very close to experimental ones, see Ref.6, when it is applied to small droplets in strong nonequilibrium vaporization conditions.

References

[1] Kurzius, S. and Raab, F., "Vaporization and Decomposition Kinetics of Candidate Re-entry Blackout Suppressant in Low-pressure Flames," CR 1330, 1969, NASA.

[2] Edward, R. and Collins, R., "Evaporation of a Spherical Source into Vacuum," Rarefied Gas Dynamics, Vol.2, edited by L. Trilling and H. Wachman, Academic Press, New York, 1969, pp.1489-1496.

[3] Nix, N. and Fukuta, N., "Nonsteady-state Theory of Droplet growth," Journal of Chemical Physics, Vol.58, February 1973, pp. 1735-1740.

[4] Vdovin, Y., Ermachenko, V. and Rubenzhny, Y., "Evaporation of a Drop under Irradiation," Soviet Physics, Vol.20, March 1976, pp. 391-395.

[5] Brock, J., "Evaporation and Condensation of Spherical Bodies in Noncontinuum Regime," Journal of Physical Chemistry, Vol. 68, October 1964, pp. 2857-2862.

[6] Bellomo, N., "Kinetic Theory of Vaporization and Condensation of Liquid droplets in Rarefied Gas Conditions," Rarefied Gas Dynamics, Vol.2, edited by M. Becker and M. Fiebig, DFVLR Press, Porz-Whan, West Germany, 1974, pp. F1.1-F1.10.

[7] Cercignani, C., Mathematical Methods in Kinetic Theory, Plenum Press, New York, 1969.

[8] Chapman, S. and Cowling, T., The Mathematical Theory of Non Uniform Gases, Cambridge University Press, London, 1970.

[9] Kogan, M., Rarefied Gas Dynamics, Plenum Press, New York, 1969.

[10] Bellomo, N. and Chiadò Piat, M., "Liquid Droplets Vaporization under Free-Molecule Gas Beam Conditions," Meccanica Journal of the Italian Association of Theoretical and Applied Mechanics, Vol.1, January 1975, pp.57-60.

[11] Bellomo, N., Loiodice, R. and Pistone G., "Mathematical Formulation of a Stochastic Problem in Statistical Fluid Dynamics," l'Aerotecnica Journal of the Italian Association of Aeronautics and Astronautics, 1976, (to be published).

[12] Soong, T., Random Differential Equations, Academic Press, New York, 1973.

[13] Saaty, T., Modern Non-Linear Equations, McGraw-Hill, New York, 1973.

[14] Bellomo, N., Loiodice, R., and Pistone, G., "Qualitative Analysis of a Non-Linear Differential Equation with Random Initial Conditions," Proceedings of the III AIMETA Congress, 1976 (to be published).

[15] Bellomo, N., Loiodice, R., and Rizzi, G., "A New Statistical Kinetic Theory of Phase-Transition in Liquid-Vapour Systems, Mechanics Research Communications, 1976 (to be published).

DEPOSITION COEFFICIENT MEASUREMENTS OF WATER VAPOR ONTO ICE OVER AN EXTENDED TEMPERATURE AND VAPOR FLUX RANGE

J. Armstrong* and N. Fukuta+
University of Denver, Denver, Colo.

Abstract

A new method has been developed to determine the deposition coefficient of water vapor onto ice. In this method, the transfer of latent heat by a condensable gas (water vapor) between the surfaces of a narrow crack made along a lattice plane of a single ice crystal from a glacier is considered. The experiment consists of applying a steady-state temperature gradient through a single ice crystal cylinder. The ice cylinder is cleaved along the lattice plane normal to its axis. After cracking, a new linear temperature profile is established with a step existing at the cleavage gap. By measuring the temperature difference across the crack, the deposition coefficient can be determined. The results of measurements made over an extended temperature and vapor flux range indicate that the coefficient is dependent upon the ice supersaturation.

Introduction

The deposition coefficient γ is the fraction of water molecules impinging upon an ice surface that is captured and incorporated into the ice lattice. The coefficient describes the efficiency of the heat and vapor exchange between the gaseous and solid phases, and thus plays an important role in ice crystal growth and evaporation kinetics.[1] Past measurements of γ for water vapor onto ice, however, vary as much as

---

Presented as Paper 66 at the 10th International Symposium on Rarefied Gas Dynamics, July 19-23, 1976, Aspen, Colo., this work was supported by the Atmospheric Sciences Division, National Science Foundation, under Grant GA 43900.

*Research Environmental Engineer, Denver Research Institute.
+Head, Cloud Physics Division, Denver Research Institute.

Table 1 Reported deposition coefficients of water vapor on ice.

Temperature range, °C	γ	Investigator
-6 to - 7	0.04	Vulfson and Levin[2]
-10 to -11	0.7	Vulfson and Levin[2]
-2 to -13	0.0144	Delaney, et al.
-40 to -60	1.0	Kramers and Stemerding[4]
-45	0.36	Davy and Somorjai[5]
-50 to -80	0.07	Isono and Iwai[6]
-60 to -80	0.67	Davis and Strickland-Constable[7]
-85	1.0	Davy and Somorjai[5]
-110	0.4	Isono and Iwai[6]
-128	0.38	Koros, et al.[8]

two orders of magnitude for the same temperature range, as shown in Table 1.[2-8] The previous measurements usually were based on a rapid sublimation technique in which water molecules are transported by free molecular flow from an ice surface, with the sublimation rate being given by Knudsen as

$$(dm/dt)_s = - (1/2\pi RT)^{\frac{1}{2}} \gamma(p_s - p_\Delta) \qquad (1)$$

where R is the specific gas constant of water vapor, T is the temperature of the vapor, p_s is the saturated vapor pressure of ice, and p_Δ is the pressure of arriving molecules at the ice surface. Difficulties inherent in experiments using this technique have been cited in detail previously.[9] They include the necessity of maintaining high vacuums, of accurately measuring the temperature of rapidly subliming ice surface, and of guarding against contamination of the ice surface. The method discussed here avoids these basic problems. Surface contamination is eliminated by utilizing narrow cracks created along the cleavage planes of pure single ice crystals obtained from the Mendenhall Glacier of Alaska. Free molecular flow of water vapor occurs only between the cracks, with the pressure difference between the cracked surfaces being typically about 0.06 mb. Since the mass flux across the crack therefore is gentle, the ice surface temperatures can be determined accurately. In addition, the measurements can be extended over a large temperature range and can be conducted on both the prism and basal planes of ice to determine if the coefficient is dependent on lattice orientation. Finally, the possibility of extending this method to other chemical systems appears feasible.

Theory

The deposition coefficient determination is based on an analysis that follows Knudsen's treatment of free-molecule heat conduction between parallel plates for a noncondensable gas.[10] The present theory extends this treatment to consider the total thermal energy transferred by a condensable gas (water vapor) across the surfaces of a crack made along the lattice plane of a single crystal of ice. The complete derivation has been reported previously.[9]

The total heat transferred by the net vapor flux across the crack per unit time is

$$Q_{total} = \frac{\alpha}{2-\alpha}\left(\frac{1}{2\pi RT'}\right)^{\frac{1}{2}} p'(c_v+\tfrac{1}{2}R)\Delta T + \frac{\gamma}{2-\gamma}\left(\frac{1}{2\pi RT'}\right)^{\frac{1}{2}} \frac{L^2 p_2 \Delta T}{R(T')^2} \qquad (2)$$

where the top ice surface temperature, T_1, is higher than that of the bottom, T_2. R is the specific gas constant for water vapor, α is the thermal accommodation coefficient, T' is the mean temperature of the vapor across the crack, p' is the mean vapor pressure across the crack, c_v is the specific heat at constant volume, γ is the deposition coefficient, L is the latent heat of sublimation, p_2 is the saturated vapor pressure of ice at temperature T_2, and $\Delta T = T_1 - T_2$.

The first term of the equation is the sensible heat contribution, whereas the second term is the latent heat contribution, which is based on a mass flux consideration. Both terms are seen to have the same form. With $\alpha = 1$, which is the maximum possible value, the sensible heat contribution has been calculated to be of the order of 0.5% as compared to that due to latent heat, based on our experimental observations; therefore, it can be ignored without affecting the measurement appreciably.

The heat transfer through the bulk of the ice can be written as

$$Q_{int} = K\left(\frac{dT}{dZ}\right)_{int} \qquad (3)$$

where Q_{int} is the heat flux per unit time, K is the thermal conductivity of ice, and $(dT/dZ)_{int}$ is the temperature gradient through the ice. Once the ice is cracked and a new steady-state condition is established

$$Q_{int} = Q_{total}$$

Now, after cracking

$$(dT/dZ)_{int} = \left[(T_B - T_E) - \Delta T\right] / (Z_B - Z_E) \qquad (4)$$

as shown in Fig. 1. Substitution of Eqs. (2) (disregarding the sensible heat term) and (4) into (3) and solving for γ yields

$$\gamma = 2K \left[\frac{(T_B - T_E) - \Delta T}{Z_B - Z_E}\right] \Bigg/ \left[\left(\frac{1}{2\pi RT'}\right)^{\frac{1}{2}} \frac{L^2 p_2}{R(T')^2} \Delta T + K \frac{(T_B - T_E) - \Delta T}{Z_B - Z_E}\right] \qquad (5)$$

With the latent heat transfer, there exists a net flux of water molecules, evaporating from the warm top ice surface and condensing on the cold lower surface. Thus, with time, the crack will migrate in the direction of the warmer zone. The migration velocity of the crack can be calculated

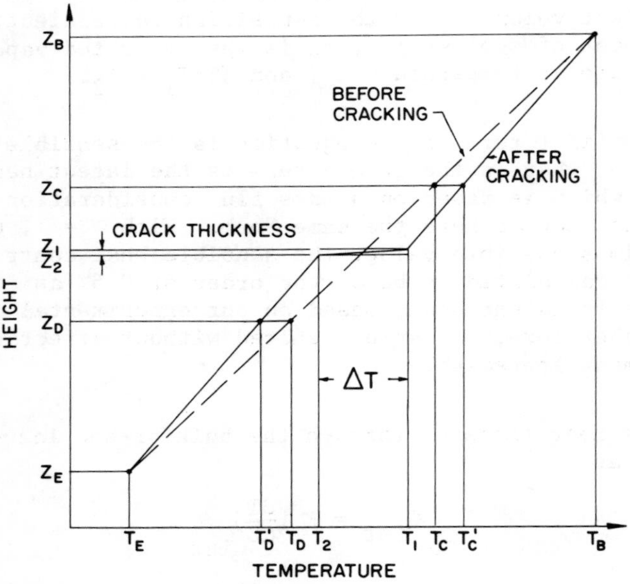

Fig. 1 Temperature profiles of ice crystal before and after cracking.

as a function of the bulk temperature gradient through the ice, which is a further check of the validity of the experiment. Equation (3) can be rewritten as

$$K\ (dT/dZ)_{int} = L\rho v_c \tag{6}$$

where ρ is the density of ice and $v_c = dZ/dt$ is the migration velocity. Substituting Eq. (4) into (6) and solving for v_c gives

$$v_c = (K/L\rho)\left[[(T_B - T_E) - \Delta T]/(Z_B - Z_E)\right] \tag{7}$$

The temperature difference across the crack, ΔT, also is a measure of the supersaturation, which influences the nucleation of water molecules impinging upon the lower ice surface. The ice supersaturation is defined as

$$SS_{ice} = (p_1 - p_2)/p_2 \tag{8}$$

where p_1 and p_2 are the saturation vapor pressures of ice at temperatures T_1 and T_2 of the top and bottom ice surfaces, respectively.

Experimental Apparatus and Procedure

The lattice orientation of a single ice crystal, typically 10 to 15 cm in diam, is determined first by examining the specimen under polarized light. The hexagonal ice crystal is optically uniaxial, i.e., the optical axes will, by symmetry, coincide with the c-axis. When the crystal is positioned so that the polarized light is travelling approximately along the optic axis, bright interference colors are observed. The exact prism and basal plane orientation of the crystal lattice then is identified by growing hoar crystals on that ice surface which is determined to be approximately normal to the c-axis. A cylinder whose axis is perpendicular to the lattice plane that is to be studied is cut from the ice crystal by use of a mechanism that consists of a heated copper ring attached to a plexiglas tube. After four thermocouple holes are drilled into the ice at predetermined intervals, the cylinder is ready to be placed in the ice chamber shown in Fig. 2. The chamber walls consist of two plexiglas cylinders that are joined together by a sliding O-ring seal. Plexiglas is used here, since it has a thermal conductivity of one-thirtieth of that of ice. The end plates of the ice chamber are O-ring sealed copper blocks. The temperatures of these blocks are controlled

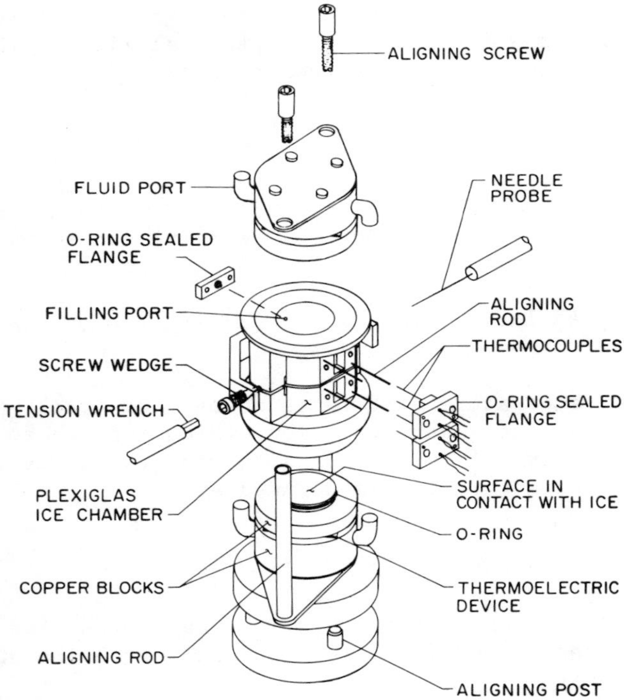

Fig. 2 Ice chamber apparatus for measuring the deposition coefficient of water vapor onto ice.

by use of thermoelectric devices. After regulating the top and bottom cooling blocks to $0^{\circ}C$, the ice is positioned in the chamber and four platinum sheathed ultrafine thermocouples are inserted into the previously drilled holes in the ice. The ceramic bases of the thermocouples are O-ring sealed to the ice chamber. The temperature of the bottom cooling block then is lowered by several degrees centigrade, so that the ice cylinder is frozen to the block surface. The small gap between the ice and chamber walls is filled from the bottom with clean degassed water. The water is allowed to overflow from the top of the chamber until the gap volume has been replaced several times. This is done in order to eliminate air that is either present as bubbles or is in solution in the water that has melted from the ice surface. The top cooling block assembly then is clamped into place and a vacuum is applied to the sealed ice chamber through the chamber port by means of an O-ring sealed syringe. This further eliminates any entrapped air and insures that the thermocouple seals are tight. The top cooling block assembly then is removed, the gap is again flushed several times with degassed water, and the top cooling assembly is clamped back into place. The bottom cooling block is next lowered to -20°C so that the added water freezes from the bottom. As the freezing progresses, the ice expands and forces excess water out of the filling port. Once the freezing is complete, the filling port is sealed and the system is left operating for at least 40 hr with a temperature gradient through the ice of 18° to 20°C. This is done as a final purification process to insure that any vapor figures originally in the glacier ice, or small entrapped air bubbles that have come out of solution when the gap water froze, will migrate through the ice to the top cooling surface. The water that fills the gap freezes with the same crystal orientation of the original ice cylinder as confirmed under polarized light.

The apparatus finally is placed in a vacuum chamber, which then is evacuated to the 10^{-4}-Torr range in order to eliminate external heat transfer to the ice chamber due to gaseous conduction and convection. Further, the vacuum chamber is located in a cold room, which is operated at -18°C to reduce radiation effects between the ice chamber and the surrounding environment.

Once the vacuum reaches the previously mentioned range, the cooling blocks are regulated to give the desired temperature gradient through the ice with the top being warmer than the bottom. After the system achieves steady state, as determined by the thermocouple readings (shown in Fig. 1 by the dashed line), tension is applied vertically to the ice using the mechanism shown in Fig. 2. This consists of two circular tapered wedges connected to a metal U. The wedge in which the cracking needle passes is fixed to the U, and the other is threaded to it and is located in line with the first. The wedges ride on notched blocks attached to the plexiglas cylinders. Uniform tension to the ice is applied by tightening the threaded screw wedge. This causes both wedges, simultaneously, to spread the plexiglas cylinder evenly. The needle is driven into the ice, initiating a crack that progresses completely across the crystal along the desired lattice plane. The needle then is extracted and an open-ended hollow tube is inserted into the cracking port of the ice chamber for several minutes. This allows the crack to be vacuum pumped as a further safeguard to eliminate any air that may have been trapped at the junction of the two plexiglas cylinders. The hollow tube then is extracted and the cracking needle is reinserted part way into the cracking port until it seals against an O-ring attached to the chamber port. Once the ice chamber is sealed, only water vapor is present between the crack.

The temperature distribution through the ice cylinder is linear and, after cracking, the linearity will be recovered as soon as the steady state is reestablished, although the slope will be different with a new step at the crack. The temperature difference, ΔT, across the crack is determined from the new steady-state profile (as seen in Fig. 1). In addition, after cracking, the mass flux across the crack can be varied for the same mean crack temperature, by changing the top and bottom copper block temperatures. As the experiment progresses, the crack migration velocity is determined by periodically illuminating the ice chamber and measuring the position of the crack front relative to a millimeter scale attached to the top plexiglas cylinder. At the end of an experiment, the final migration distance is determined accurately by use of a caliper.

Discussion of the Results

The results of eleven basal plane experiments, conducted over a wide range of mean crack temperatures and vapor fluxes, are presented in the following figures. A direct check of the validity of the experimental technique is the measurement of the migration velocity of the crack surfaces, since the technique is based on a mass flux consideration. Figure 3 shows the ratio of the measured vs calculated migration velocities, v_m/v_c, of ten experiments for various mean crack temperatures. For all of the experiments conducted at temperatures warmer than $-25^{\circ}C$, the ratio is seen to be close to unity, indicating that the technique is valid. For the first four experiments conducted at colder temperatures (cases 4, 5, 6, and 8), the measured velocities were considerably lower than the calculated velocities. The experimental program is based on the premise that heat transport between the crack is due to latent heat transfer. The results of cases 4, 5, 6, and 8 indicate that there was at least one heat-transfer mechanism occurring other than that due to vapor transport. (The sensible heat-transfer contribution of water vapor with α taken as 1 again has been calculated to be negligible for these cases.) These results strongly suggest that the cracked surfaces were not remaining separated, but that bridging occurred. The reason for this became apparent when the supersaturation with respect to ice was calculated for all of the experiments conducted up to that time. (The ice supersaturations also are noted in Fig. 3.) For cases 4, 5, 6, and 8, the supersaturation varied from 7.3 to 15.6%, whereas for the cases warmer than $-25^{\circ}C$, the ice supersaturation ranged from 0.7 to 2.6%. The nucleability and growth of water molecules on the cold lower sur-

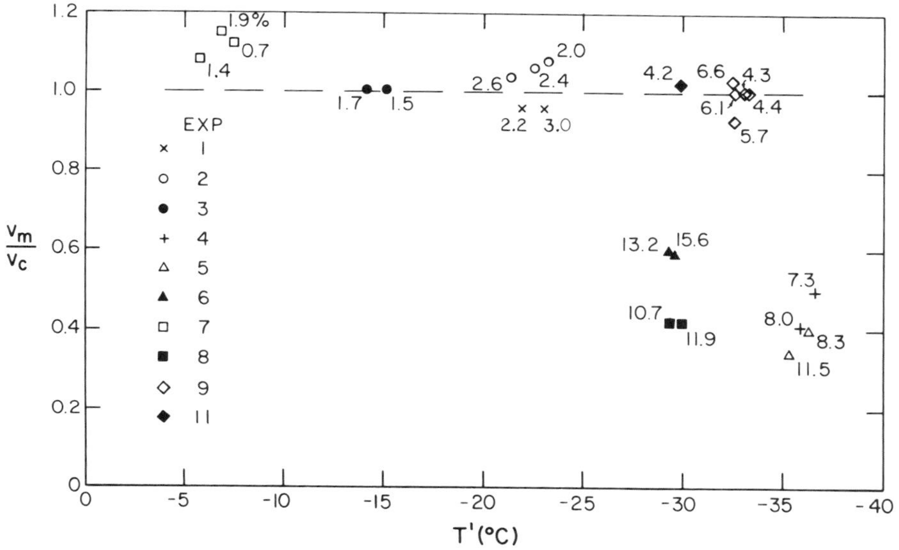

Fig. 3 Ratio of measured to calculated migration velocity vs mean crack temperature (basal plane).

face will be a function of the supersaturation. For such supersaturation values as calculated for the above cold cases, it is conceivable that steps and dislocations on the lower ice surface grew quickly in the vertical direction and ultimately contacted the upper surface. A large fraction of the heat transfer then would be due to conduction through the bridges. The ice crystals were examined at the end of all of the experiments. The crack planes of crystals from experiments 4, 5, 6, and 8 appeared more opaque than the planes of the warmer experiments. Subsequently, two experiments were performed with smaller bulk temperature gradients, T_B-T_E, for mean crack temperatures colder than $-25^{\circ}C$ (see cases 9 and 11, Fig. 3). With the lower ice supersaturations across the cracks the measured migration velocities approached the calculated values.

It is evident that the deposition coefficient is strongly dependent upon supersaturation as well as temperature. This is illustrated further in Fig. 4. Here the data from four experiments, in which $v_m/v_c \approx 1$, is plotted. In each case, as the supersaturation was increased, for approximately constant mean crack temperatures, the deposition coefficient decreased. At first this was surprising, but it must be kept in mind that the deposition coefficient is a measure of the apparent capture coefficient, and is the result of a complicated sequence of events, which includes molecule adsorption, surface migration, nucleation, and incorporation at steps and

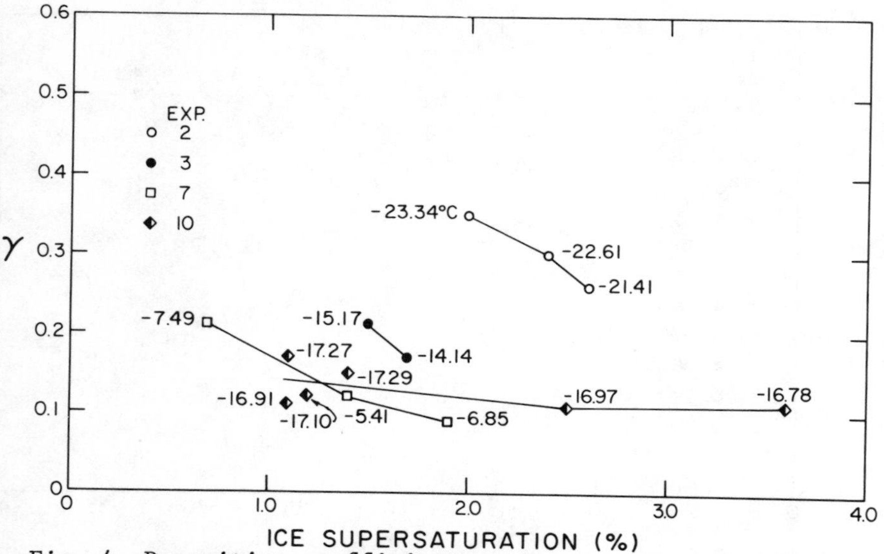

Fig. 4 Deposition coefficient vs ice supersaturation for various mean crack temperatures (basal plane).

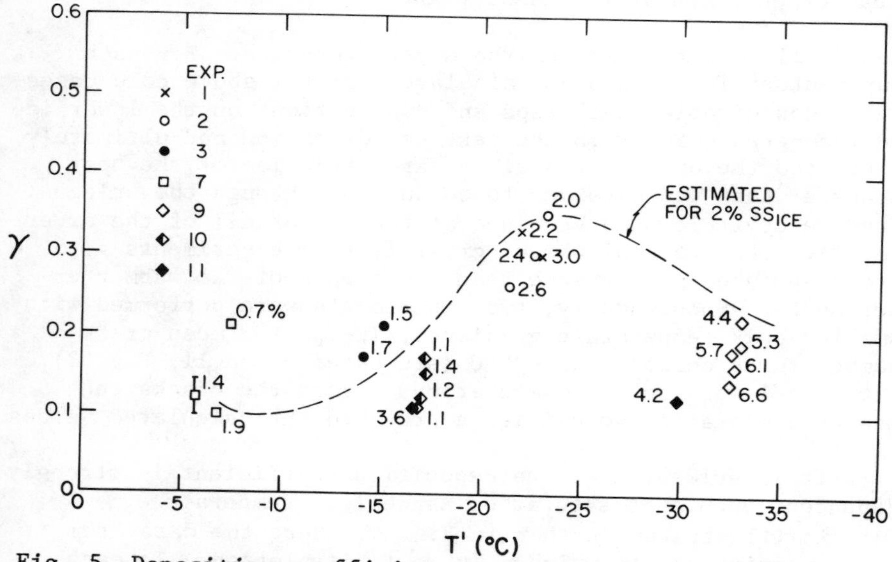

Fig. 5 Deposition coefficient vs mean crack temperature for various ice supersaturations (basal plane).

dislocations on the ice surface. For steady-state experimental conditions, the net flux of water molecules is constant, and because of this, whatever happens to a water molecule on the ice surface is indirectly revealed in the process of apparent water molecule incorporation into the ice lattice from the gaseous phase through the behavior of the measured deposition coefficient. Additional basal plane measurements over a wider supersaturation or molecular flux range will be conducted in the future to identify the dependency relationships further.

The deposition coefficient vs mean crack temperature for all the experiments performed so far on the basal plane, in which v_m/v_c was close to unity, is given in Fig. 5. The ice supersaturation for the various data points is also presented. The estimated variation of the deposition coefficient with respect to mean crack temperature is shown by the dashed line for 2% ice supersaturation. These measurements are in agreement with the well known ice crystal habit behavior[11] where the observed ice crystal growth near -6 and -23°C is predominantly columnar in the direction of the c-axis. The above deposition coefficient determination is quantative, within a small range of error, and provides basic information for ice crystal growth kinetics. In the future, the deposition coefficient measurement over an extended temperature and vapor flux range for the prism plane also will be made.

References

[1] Fukuta, N. and Walter, L. A., "Kinetics of Hydrometeor Growth from a Vapor-Spherical Model," Journal of Atmospheric Science, Vol. 27, November 1970, pp. 1160-1172.

[2] Vulfson, N. I. and Levin, L. M., "Studies of Clouds, Precipitation, and Thunderstorm Electricity," American Meteorological Society, Vol. 33, 1965.

[3] Delaney, L. J., Houston, R. W., and Eagleton, L. C., "The Rate of Vaporization of Water and Ice," Chemical Engineering Science, Vol. 19, January 1964, pp. 105-114.

[4] Kramers, H. and Stemerding, S., "The Sublimation of Ice in Vacuum," Applied Science Research, Vol. 3, January 1953, p. 73.

[5] Davy, J. G. and Somorjai, G. A., "Studies of the Vaporization Mechanism of Ice Single Crystals," The Journal of Chemical Physics, Vol. 55, October 1971, pp. 3624-3636.

[6] Isono, K. and Iwai, K., "Growth Rate and Habit of Ice Crystals in Air at Low Pressure," Journal of the Meteorological Society of Japan, Vol. 49, Special Issue, 1971, pp. 836-844.

[7] Davis, A. and Strickland-Constable, R. F., "Evaporation and Growth of Crystals from the Vapor Phase," International Symposium on Condensation and Evaporation of Solids, Dayton, Ohio, 1962, pp. 665-679.

[8] Koros, R. M., Deckers, J. M., and Bondart, M., "Sticking Probability of Water and Ice," International Symposium on Condensation and Evaporation of Solids, Dayton, Ohio, 1962, pp. 681-682.

[9] Fukuta, N. and Armstrong, J. A., "A New Method for Precision Measurement of the Deposition Coefficient of Water Vapor onto Ice," Rarefied Gas Dynamics-Proceedings of the Ninth International Symposium, Vol. 2, 1974, pp. C.5-1 - C.5-10.

[10] Kennard, E. H., Kinetic Theory of Gases, McGraw-Hill, New York, 1938, p. 312.

[11] Fukuta, N., "Experimental Studies on the Growth of Small Ice Crystals," Journal of Atmospheric Science, Vol. 26, May 1969, pp. 522-531.

CHAPTER XIII—INSTRUMENTATION

PLANETARY ATMOSPHERES WITH MASS SPECTROMETERS CARRIED ON HIGH-SPEED PROBES OR SATELLITES

Alfred O. Nier*
University of Minnesota, Minneapolis, Minn.

Abstract

The analysis of complex gas mixtures containing unknown constituents, some of which may be highly reactive chemically or very low in abundance, poses problems that are difficult to solve, even in the laboratory. For an Earth satellite or a high-speed probe carried to a distant planet, the restrictions on weight, power consumption, and time for analysis, along with other constraints, complicate the problem even further. Properly designed mass spectrometers appear to have the greatest promise for providing the scientific data sought. In the case of the Atmosphere Explorer Satellites C, D, and E, an open-source mass spectrometer of special design has provided successfully, over a large altitude range in Earth's thermosphere, quantitative measurements of atomic nitrogen and oxygen, as well as of the more familiar components, molecular nitrogen and oxygen, argon, and helium. Instruments having some of the same features are carried on the Viking spacecraft for study of the Martian atmosphere during the entry phase of the landings on Mars during the summer of 1976. A closely related instrument also will be carried on the 1978 Pioneer-Venus mission for studying the upper atmosphere of Venus as a probe descends to the planet.

Presented as Paper 2 at the 10th International Symposium on Rarefied Gas Dynamics, July 19-23, 1976, Aspen, Colo. The author is indebted to his colleagues D. C. Kayser, K. Mauersberger, and W. E. Potter for helpful discussions concerning the work. The research was supported by NASA Contract NAS-5-11438 between Goddard Space Flight Center and the University of Minnesota.

*Regents' Professor of Physics, School of Physics and Astronomy.

Introduction

The quantitative analysis of a complex gas mixture is, at best, a difficult problem. The in situ analysis of the upper atmosphere of a planet from a fast-moving spacecraft is even more difficult. In laboratory investigations, a variety of instruments have been employed. These include infrared and ultraviolet spectrometers, thermal conductivity gages, magnetic resonance spectrometers, gas chromatographs, and others. Some have unusual sensitivity or accuracy for measuring specific gases. As all-around instruments, capable of analyzing complex gas mixtures under the difficult conditions of space flight, none can match mass spectrometers in performance. A successful instrument, employed in upper atmosphere or planetary entry studies, must have rather unusual properties in order to meet some severe constraints. These include 1) make measurements over a large pressure range, e.g., from 10^{-9} to 10^{-4} Torr; instruments thus must be sensitive and also have a large dynamic range; 2) quantitatively sample the ambient atmosphere from hypersonic spacecraft; 3) make a complete analysis with a repetition rate of a few seconds to determine density profiles or structure in an atmosphere through which one is moving at high speed; 4) meet stringent weight, power, and reliability conditions; typically instruments should weigh less than 6 kg, consume less than 10 Watts of power, and be able to withstand severe vibration and temperature environments; 5) make isotope as well as gas analyses; 6) measure chemically reactive gases such as atomic oxygen; and 7) measure relatively rare constituents in the presence of much more abundant ones.

In practice, various types of neutral mass spectrometers have been employed in space investigations. Among the most successful have been instruments employing collimated electron beams to produce ions of the ambient gas, and magnetic deflection of the ions to accomplish mass discrimination. The use of a quasiopen source[1-6] has made possible a qualitative measurement of reactive gases such as atomic oxygen. Through the use of proper electric field conditions in the source, it also has been possible to distinguish quantitatively between true ambient particles and those that have struck the spacecraft surface and reacted, or those emitted from the spacecraft or instrument itself.[7-9] Philbrick et al.[10] earlier used a retarding potential analyzer without a mass analyzer and were able to obtain some limited results. Offermann and Grossmann[11] employed a liquid-helium-cooled ion source in an attempt to eliminate particles that strike ion source surfaces.

Because of their compactness, ability to measure readily several masses of ions simultaneously, and other desirable characteristics, instruments employing the Mattauch-Herzog[12] geometry were adopted for the open-source mass spectrometers (OSS) on the Earth satellites Atmosphere Explorer C, D, and E,[13] for the entry phase of the mission to Mars[14] (Viking project), and for the upper atmosphere measurements during entry into the Venusian atmosphere (Pioneer-Venus mission).[15]

Figure 1 is a schematic drawing of the Atmosphere Explorer instrument and shows essential features common to the Mars and Venus instruments as well. The high-speed gas beam enters the ion source from the left, and passes through several grids before reaching the region between the bar magnets M, shown on end in the figure. The solid dot between magnets represents the electron beam, perpendicular to the figure, which is collimated by the magnetic field produced by the bar magnets. Typically, the electron beam current is 100 μA and the energy of the electrons 75 eV. The energy of the electrons can be reduced to 25 eV to produce a different cracking pattern of ion fragments and hence help resolve ambiguities when several gases have fragments with the same molecular mass number.

In the flight instruments discussed here, a permanent magnet is employed to perform the mass analysis, and mass spectra can be swept by continuously varying the ion accelerating potential and the potential across the electric analyzer. More commonly, the potentials are switched automatically to a sequence of values corresponding to masses of

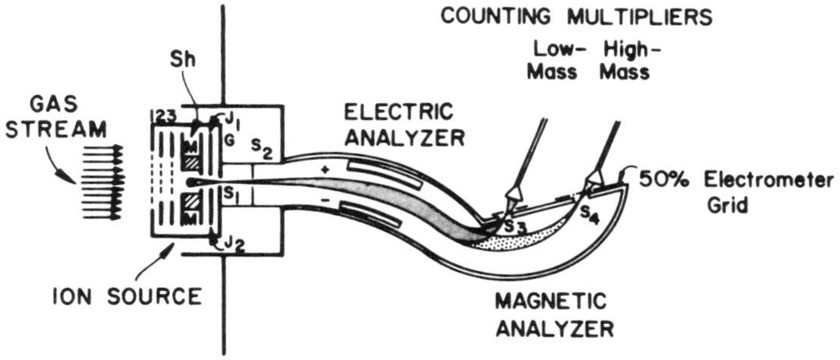

Fig. 1 Schematic drawing of the Atmosphere Explorer open source mass spectrometer (OSS). The spacecraft moves to the left, and so the ambient atmosphere appears as a high-speed molecular beam moving to the right.

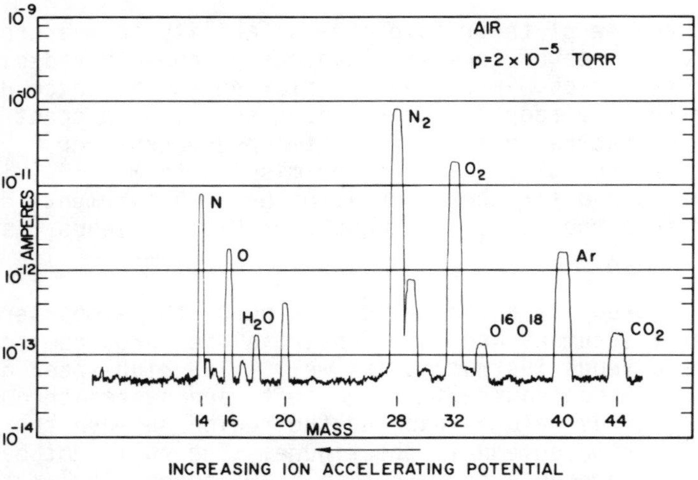

Fig. 2 Mass spectrum of air obtained in laboratory with open source instruments flown on Atmosphere Explorer C, D, and E satellites.

particular interest. This makes more efficient use of the time available in the event that one is interested in only certain masses.

In the Atmosphere Explorer instrument, the two collectors measure ions differing in mass by a factor of 8. The low-mass collector measures masses from 1-6 amu, the high-mass collector from 7-48. To increase the dynamic range of the high-mass collector, the ions, after passing through slit S_4, go through a grid having a transmission of approximately 50% before going on to the counting multiplier. This grid is attached to an electrometer amplifier and is used to measure ion currents beyond the counting range of the multiplier. This arrangement greatly increases the measurement range of the high-mass collector.

In the interest of simplicity and economy, but at the expense of sensitivity and dynamic range, only electrometer amplifiers were used in the case of the Viking instruments. The Venus instrument employs a collector system very similar to that described for the Atmosphere Explorer instruments.

Figure 2 is a mass spectrum of air obtained in the laboratory with one of the Atmosphere Explorer OSS instruments. Clearly seen are peaks at 28, 32, and 40, corresponding to N_2, O_2, and Ar, respectively, the principal constituents of the atmosphere. Also seen are peaks at 14 and 16, ions resulting from the dissociation of N_2 and O_2, respectively, by the ionizing electrons. Because of the isotopic nature of oxygen, nitrogen, and argon, ion peaks are seen at mass 34 ($^{16}O^{18}O$), at mass 29 ($^{14}N^{15}N$), and at mass 36 (^{36}Ar). The peak at mass 20 is due to Ar^{2+}, doubly charged ions appearing in the mass spectrum at half the mass of the parent particle. The 44 peak is due to a CO_2 impurity, whereas the 18 and 17 peaks are produced from water vapor. Other small peaks are associated mainly with impurities.

In a CO_2 atmosphere, the main peak seen is that at mass 44, CO_2^+. Other prominent peaks occur at masses 12, 16, 22, and 28, corresponding to C^+, O^+, CO_2^{2+}, and CO^+, ions produced when CO_2 is bombarded by electrons having sufficient energy to produce both ionization and dissociation. It is believed that the Martian atmosphere may contain an appreciable amount of argon. It will be easy to settle this question, as the mass 40 position does not have any peaks due to CO_2. On the other hand, N_2 detection will pose some problem, since the 28 position also is occupied by a CO_2 fragment. If the N_2 concentration is low, the change in the 28 peak due to its presence may be too small to detect. If so, the 14 peak may be a better indicator. Here, however, the picture may be confused by the presence of doubly charged ions from CO, a low-abundance but likely constituent of the Martian atmosphere. From the cracking patterns of the several molecules, one can determine the relative amounts of N_2 and CO present, but not to the same accuracy as would be possible were either one present alone.

Earth Satellite-Borne Mass Spectrometers

The quantitative measurement of nonreactive constituents with mass spectrometers carried on Earth satellites has been reduced to a routine. Prior to the Atmosphere Explorer satellites C, D, and E, successful measurements had been made from a number of other satellites.[16-19] In a typical instrument, the mass spectrometer ion source is mounted in a volume connected to the ambient atmosphere through a small orifice. When the orifice "looks" forward, the high speed of the satellite produces a ram effect, resulting in an increase in pressure in the ion source region. Conversely, for a spinning satellite the orifice periodically looks backward, and so there will be a drastic diminution of signal under these conditions. The overall modulation of the signal from the wake to the ram position will be by a factor that may run into the thousands. Even so, the phenomenon is well understood in terms of elementary kinetic theory, and ambient particle densities may be deduced from the modulated data to remarkable accuracy.[20]

It long has been known that atomic oxygen is a major constituent of the thermosphere. It first was identified mass spectroscopically with an "open" source mass spectrometer flown on a sounding rocket.[1] Because atomic oxygen is highly reactive, it was appreciated that its quantitative measurement in a mass spectrometer ion source was subject to considerable uncertainty. There followed a number of rocket flights[21-24] in which quasiopen and closed ion-source mass spectrometers were carried on the same rocket. For the nonreactive gases such as N_2, O_2, Ar, and He, the results were in remarkable agreement, showing that the quasiopen source employed (called open source from here on) acted very nearly like an ideally closed one insofar as having a calculable relation between ambient particle densities and those measured in the ionizing region of the ion source. The open source had the property that it could measure, at least qualitatively, reactive species such as atomic oxygen. This led to the adoption of the open-source mass spectrometer[13] (OSS), along with a conventional closed-source instrument,[25] as part of the Atmosphere Explorer (AE) C, D, and E payloads.

A laboratory study in which a high-speed molecular beam was incident on the AE-C open ion source, thus simulating satellite flight, showed that the geometry of the source was such that it had, within a few percent, the desirable stagnation characteristics of an ideally closed source.[7] Its "openness," on the other hand, made it possible, in a special mode of operation, to take advantage of the speed of the vehicle and thus distinguish between true ambient particles and those that had struck instrument surfaces and possibly changed chemical form as a result.

Whereas atomic oxygen measurements made with rocket-borne mass spectrometers (open or closed source) were regarded with suspicion because of the uncertainty of the fate of atomic oxygen striking the instrument surfaces, the situation was clarified unexpectedly at high altitudes in the case of closed-source instruments flown on the OGO-6 satellite. It was found that after a number of orbits the surfaces of the enclosure became covered with atomic oxygen, with the result that additional atomic oxygen entering the instrument was converted quantitatively to molecular oxygen and measured as such.[16] This phenomenon was verified on the San Marco 3[17] and ESRO 4[18] satellites, as well as on the more recent Atmosphere Explorer satellites.[13,25] The instruments hence measured total oxygen: atomic plus molecular. Above 200 km, where O_2 is but a minor constituent when compared with O, the major constituent, the instruments indeed measured atomic oxygen quantitatively. Below 200 km, where atomic and molecular oxygen have comparable abundances, the relative amounts of each is left in an ambiguous state.

Nier et al.[8] were able to distinguish between O and O_2 by taking advantage of the experience gained in molecular beam studies[7] and periodically operated their Atmosphere Explorer C open-source spectrometer in a mode that rejected particles that had hit instrument surfaces, and hence were slowed down. Under normal conditions, the grid 3 (Fig. 1) is at a slightly higher potential and the focusing plates J_1 and J_2 at a lower potential than the ionizing region Sh. As a result, ions produced in the electron beam obtain some acceleration as soon as they are formed. If the grid 3 and the plates J_1 and J_2 are tied to Sh or to points on the voltage divider having potentials close to that of Sh, the ions actually will experience a retarding potential because of the negative space charge of the electron beam.

At satellite speeds ($\sim$ 8.5 km/sec), incoming ambient particles have an energy of approximately 0.3 eV/amu. For atomic oxygen (16 amu), this is an energy of over 5 eV. This initial energy of ions formed in the electron beam is sufficient to overcome the small retarding field in the ionizing region and allows the ions to pass on into the accelerating region between plates J_1, J_2 and ground (G). On the other hand, ambient particles that strike the ion source plates are largely accommodated and lose most of their incoming energy. They fail to leave the ionizing region and hence are not measured.

The OSS mass spectrometers on the Atmosphere Explorer satellites are mounted to "look" radially outward from the spin axis. This latter is maintained perpendicular to the plane of the orbit. During each spin cycle, the normal to the opening of the ion source thus sweeps through the forward direction. The instruments have various modes of operation. Of particular interest in the present discussion is a mode in which the collection toggles between masses 16 and 32 (atomic and molecular oxygen) at 1/16-sec intervals. Figure 3 shows the results obtained at two different altitudes, 176 and 247 km, on orbit 1165 of satellite AE-C. We see a number of interesting features. First of all, we note that the count rates have rather sharp maxima as the instrument "looks" forward. In this particular instrument the backgrounds, at positions other than the forward one, are actually higher than they could have been had the potentials in the ion source been chosen slightly differently so as to reject more fully slow particles that had struck surfaces. Nevertheless, we can discern several interesting points. In going from 247 to 147 km, both the 16 and 32 peaks have increased in intensity,

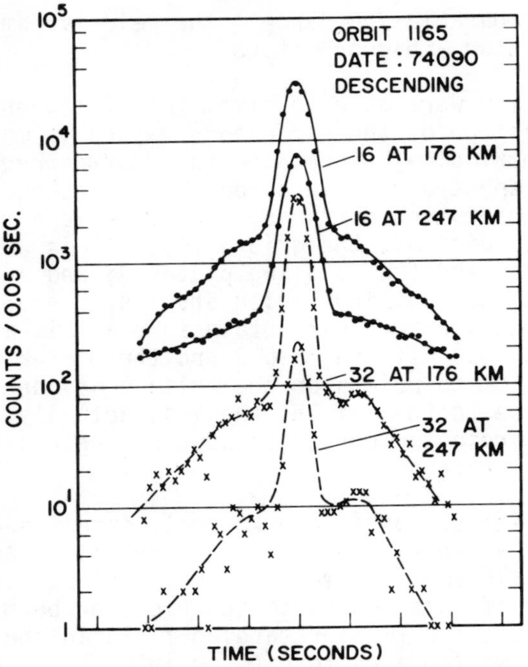

Fig. 3 Mass spectrometer electron multiplier output for masses 16 and 32 when mass spectrometer is in "fly-through" mode and satellite is in spinning mode. Spin period is 15 sec.

as one would expect. Because of the difference in scale heights of the two constituents, the 32 peak has increased more rapidly than the 16 peak. The ordinate scale in Fig. 3 is given in terms of number of ions reaching the high-mass collector of Fig. 1 per 0.05 sec, the collection interval. In practice, these numbers are converted to particle densities, making use of laboratory calibrations and comparisons with the normal mode of operation of the mass spectrometer.

An interesting point to note in Fig. 3 is the difference in width of the mass 16 and 32 peaks, the former being approximately $\sqrt{2}$ times wider than the latter at half height. Although part of the width is due to the acceptance angle of the instrument itself,[7] the largest part arises as a result of the finite temperature of the atmosphere. Particles reaching the instrument have a velocity distribution made up of a unidirectional component of about 8.5 km/sec, upon which is superimposed an omnidirectional Maxwellian distribution having a maximum in the neighborhood of 1 km/sec, the exact value depending upon the mass, the atmospheric constituent, and the local temperature of the atmosphere. Potter et al.[26] have studied systematically the shape and width of peaks such as shown in Fig. 3 for a large number of orbits of the Atmosphere Explorer D satellite and have verified that the peak width can be used to measure in situ atmospheric temperature on a routine basis.

Mass Spectrometers and Planetary Entry Probes

At the time of preparing the present paper, only one attempt had been made to obtain an atmospheric composition profile with a terrestrial entry probe. In this case, a high-speed rocket was fired into the Earth's atmosphere from a high altitude.[27] A closed-source mass spectrometer was employed, with measurements intended in a much higher pressure region of the atmosphere than the thermospheric determination described in the previous section. Gas was admitted to the instrument through a small constriction, the instrument being pumped by a small sputter pump. Although the nitrogen and argon measurements turned out as expected, the clean surfaces of the instrument acted as a high-speed pump for molecular oxygen, with the result that this constituent was lost and hence not measured except near the end of the flight. Since atomic oxygen is not a constituent of the lower atmosphere, considerations such as discussed in the previous section are not applicable.

In the case of early orbits of the Atmosphere Explorer C satellite, apogee occurred at 4000 km and perigee near 150 km. The descending trajectory hence approximated that of a planetary entry probe. After laboratory calibration, the OSS instrument had been sealed under ultrahigh vacuum conditions, not to be opened again until apogee of an orbit several days after launch of the satellite. By means of a pyrotechnic device, combined with a cutter mechanism,[28] a cap covering the ion source was removed, exposing the source to the ambient atmosphere, as shown in Fig. 1. Because the opening took place near 4000 km, the response expected as the satellite moved to lower altitudes was similar to that which one would find for a planetary probe entering the Earth's atmosphere. Data were obtained for the first orbit when the instrument was open, as well as for many of the succeeding early orbits. Figures 4-7 show some of the results obtained for both the descending and ascending phases of first and third orbits after removal of the cap covering the source.

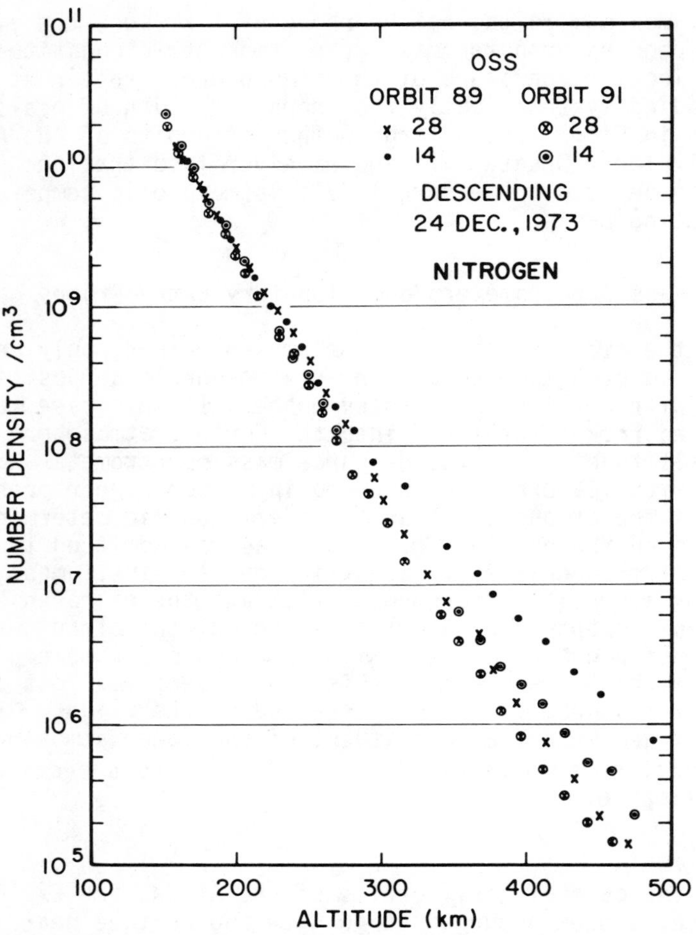

Fig. 4 N_2 densities computed from mass 28 and mass 14 peaks in descending legs of first and third orbits of AE-C after exposing ion source of open-source mass spectrometer to ambient atmosphere.

Figure 4 gives the N_2 density for the descending legs of the orbits as calculated from the mass 28 peak and as calculated from the 14 peak, assuming that all of the 14 peak arises from the dissociation of N_2 in the ionization process in the instrument. Figure 5 provides the same data for the ascending legs of the orbits. A comparison of the four profiles (two in Fig. 4 and two in Fig. 5) for N_2 derived from the 28 peaks shows that they are in very good agreement, and, had there been only one entry, the result obtained from it

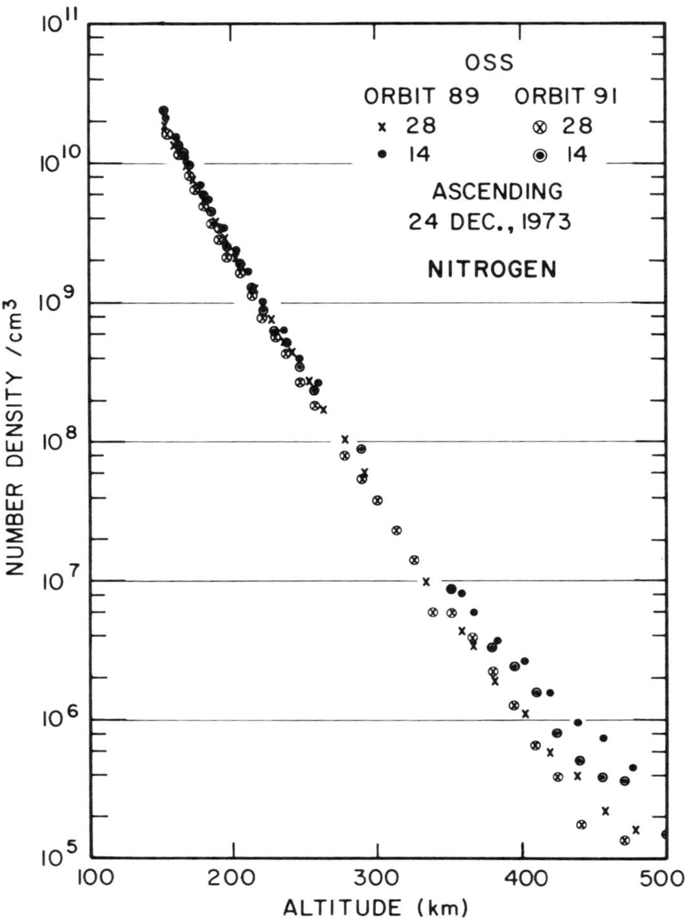

Fig. 5 N_2 densities computed from mass 28 and mass 14 peaks in ascending legs of first and third orbits of AE-C after exposing ion source of open-source mass spectrometer to ambient atmosphere.

would have been entirely satisfactory. Chemical effects on surfaces leading to the creation of CO (also having a mass of 28 amu) as discussed by Hedin et al.[16] and Moe and Moe[29] play no significant role with an open-source instrument in the ambient density range covered in the present experiment.

The N_2 densities computed from the 14 peaks show an inconsistency that can be explained in terms of the thermospheric atomic nitrogen determinations of Mauersberger et al.[30]

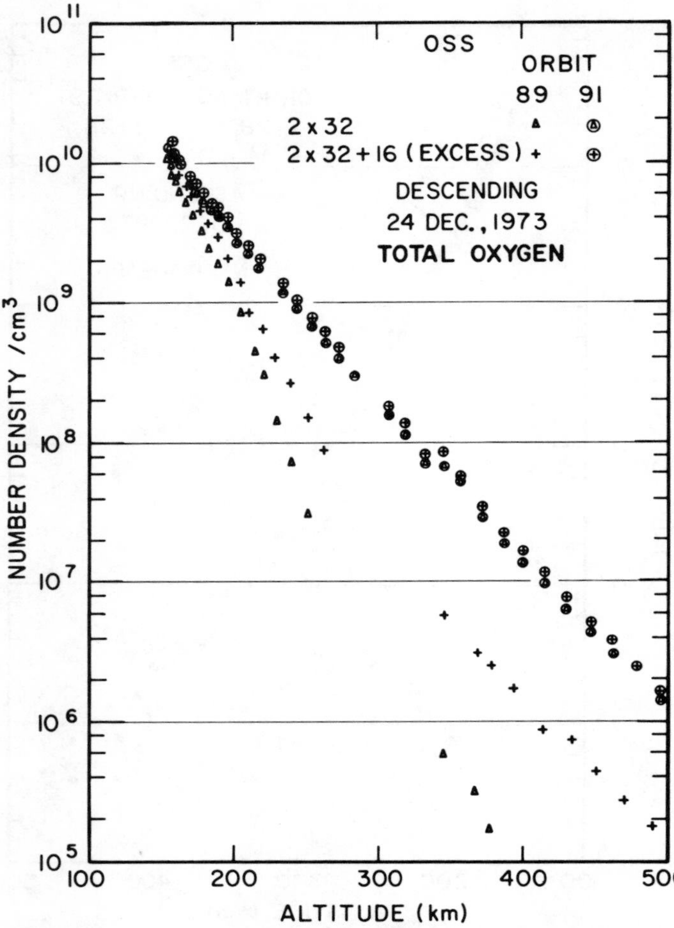

Fig. 6 Total oxygen densities computed from 32 and 16 peaks in descending legs of first and third orbits of AE-C after exposing ion source of open-source mass spectrometer to ambient atmosphere.

For example, at high altitudes all of the N_2 values predicted from the mass 14 peaks are higher than those predicted from the mass 28 peaks. This is what one would expect, since the atmosphere contains some atomic nitrogen, and so it is incorrect to attribute all of a 14 peak to a dissociation product of N_2 in the instrument. The first orbit (89) is especially interesting. At high altitude on the downleg (and to a lesser extent on the upleg), the N_2 predicted from the 14 peak is clearly too high. In the work of Mauersberger et al.,[30] it was shown that most of the atomic nitrogen enter-

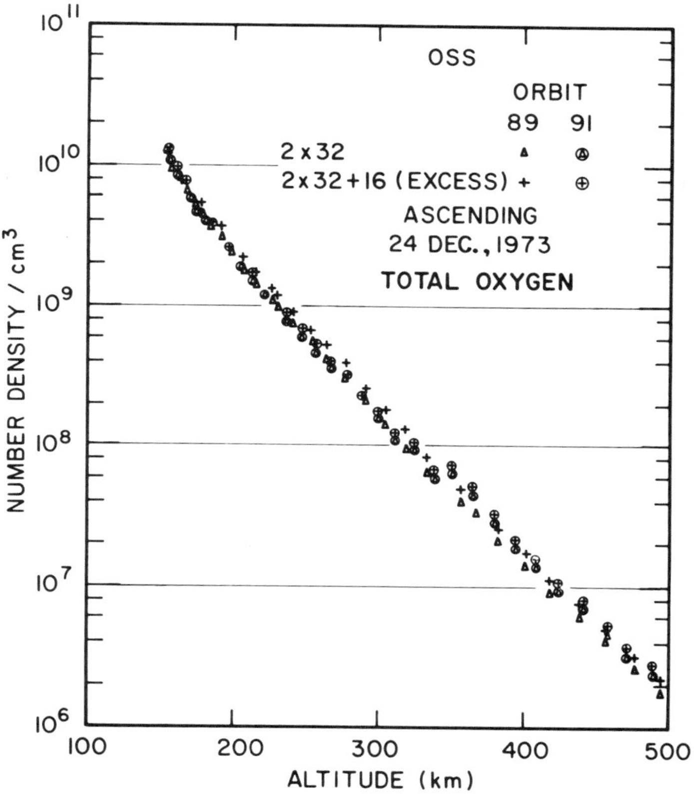

Fig. 7 Total oxygen densities computed from 32 and 16 peaks in ascending legs of first and third orbits of AE-C after exposing ion source of open-source mass spectrometer to ambient atmosphere.

ing the mass spectrometer reacted with atomic oxygen residing on the surfaces of the ion source, forming NO, which then was released and measured. In the first orbit, especially on the downleg, there has not been an opportunity to form an atomic oxygen layer, with the result that much of the atomic nitrogen is reflected as such and measured. Some of the effect remains in the upleg of the first orbit. By the third orbit, the effect largely has disappeared. It is unfortunate that the mass 30 peak was not monitored during these orbits. In the case of the OSS instrument on Atmosphere Explorer D, launched on Oct. 6, 1975, the mass 30 peak was monitored, and, as expected, in the downleg of the first orbit after exposing the instrument to the ambient atmosphere, the mass 30 peak

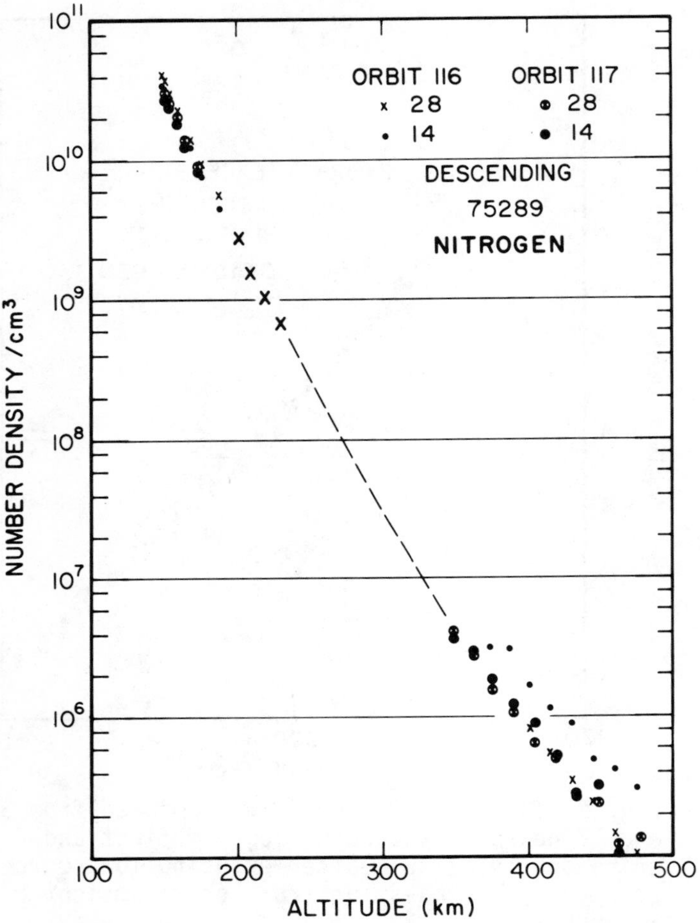

Fig. 8 N_2 densities computed from mass 28 and mass 14 peaks in descending legs of first and second orbits of AE-D after exposing ion source of open-source mass spectrometer to ambient atmosphere. Results obtained from fly-through mode in orbit 116 are shown as squares. Fly-through values for orbit 117 were not obtained.

was lower than in subsequent legs, showing that its abundance was indeed complementary to that of the abundance of the atomic nitrogen in the ion source.

Figures 6 and 7 provide the oxygen data that complement the nitrogen data shown in Figs. 4 and 5. The descending leg (Fig. 6) of the first orbit (89) clearly shows the transition

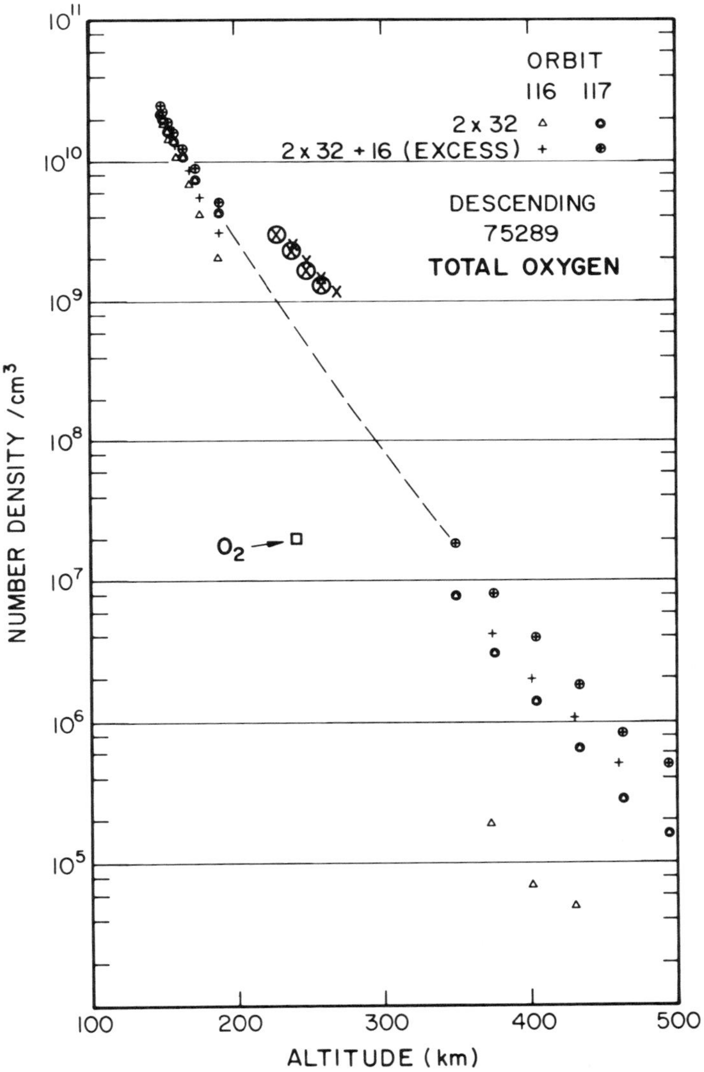

Fig. 9 Total oxygen densities computed from 32 and 16 peaks in descending legs of first and second orbits of AE-D after exposing ion source of open-source mass spectrometer to ambient atmosphere. Fly-through atomic oxygen values appear as large X's for orbit 116 and as large circled X's for orbit 117. The single O_2 point obtained in orbit 116 by the fly-through method appears as a square. It represents the number of ambient O_2 molecules rather than total number of atoms, as in the other plots.

of O to O_2 as time proceeds and atomic oxygen begins to cover the ion source surfaces. At 350 and 250 km, the O_2/N_2 abundance ratio has the approximate values 0.04 and 0.07 respectively. This is consistent with what one expects from ambient O_2/N_2 determinations at lower altitudes[31] and the assumption that the observed O_2 peak is due entirely to ambient molecular oxygen. This is also consistent with the laboratory observations of Lake and Mauersberger,[32] who showed that, when a reasonably clean surface was exposed to atomic oxygen, the initial effect was to form CO_2, and only after some time, when a layer of atomic oxygen had had an opportunity to form on the surface, did molecular oxygen appear.

When O_2 is bombarded with 75-eV electrons, as in the present ion source, atomic as well as molecular ions are formed, the O/O_2 ratio being 0.065. Some of the observed 16 peak thus is due to the presence of O_2. If this contribution is subtracted from the 16 peak, one can attribute the remainder (called excess 16 in this paper) to atomic oxygen. The excess can be added to the amount of oxygen represented by the 32 peak alone. The sum then represents total oxygen. This is shown as crosses in the graphs.

We note that below 250 km in the downleg of orbit 89 the curves for oxygen, measured as molecular oxygen, and for total oxygen rapidly converge, so that by perigee (150 km) they are virtually together. The agreement continues in the ascending mode of orbit 89 (Fig. 7) and for both the descending and ascending modes of later orbits, such as 91 and others.

It long has been known, when molecular oxygen is introduced into a mass spectrometer, the 44(CO_2) and 18(H_2O) peaks rise.[33] The effect apparently is caused by reactions on the surfaces. The same effect is observed in satellite-borne mass spectrometers as the satellite approaches perigee. In the case of satellites it is probably more pronounced, since atomic oxygen, the main constituent of the atmosphere at altitudes above approximately 175 km, is chemically more active than molecular oxygen. Even with the relatively open ion source of the Atmosphere Explorer OSS instruments, the CO_2 formation is considerable. At perigee of the initial orbit 89 discussed here, the CO_2 abundance in the ion source was approximately 10% of that of N_2; by the 1200th orbit, it has dropped to about 0.1% of the N_2. Although it gradually decreased with time, the effect was still present, but at an attenuated level, even after 12,000 orbits.

Figures 6 and 7 illustrate the difficulties encountered when one attempts to measure a reactive component such as atomic oxygen with a conventional mass spectrometer. In the case of the open-source mass spectrometer on AE-C, no immediate attempt was made to employ the fly-through mode. On AE-D it was used even in parts of the initial orbits, including orbit 116, the first one in which the ion source was exposed to the atmosphere. Figures 8 and 9 show some of the results obtained in the descent legs of the first two orbits. At altitudes above 350 and below 200 km, where the instrument was in its normal mode of operation, the curves are very similar to those shown in Figs. 4 and 6. In the intermediate altitude range, the instrument was in various fly-through modes, including ones in which the mass 16 (atomic oxygen), mass 32 (molecular oxygen), and mass 28 (molecular nitrogen) were observed in at least part of the interval between 250 and 350 km.

In Fig. 8, we note that the N_2 fly-through points fall on the same profile as the points taken while the instrument was in the normal mode of operation. In Fig. 9, one sees that around 400 km in orbit 116, where the instrument is in the normal mode of operation and there has been no previous exposure to atomic oxygen, the O_2/N_2 ratio lies near 0.04, as one would expect if the 32 peak is due only to ambient molecular oxygen. Of most interest, perhaps, is the atomic oxygen measurement by the fly-through method. As shown in Fig. 9, measurements from about 225 to 275 km were made for both orbits 116 and 117. The results obtained agree closely, as they should. The atomic oxygen values lie well above those shown in Fig. 6 for total oxygen in the case of the instrument on AE-C. It is clear that in this altitude range of an initial orbit the loss of atomic oxygen on instrument surfaces is considerable, but the surfaces are not yet so covered with atomic oxygen as to produce a quantitative conversion of incoming atomic oxygen to molecular oxygen, which then is released as measured. It apparently requires several satellite passes down to 150 km before the exposure is sufficient to produce the quantitative conversion.[31]

Conclusions

An investigation of the results obtained in the initial orbits after exposing a satellite-borne mass spectrometer to the ambient atmosphere confirms that chemical reactions on the surfaces of the instrument make measurement of reactive atmospheric species such as atomic oxygen very difficult. Only with an instrument that can distinguish between true ambient particles and ambient particles that have struck surfaces does it appear that quantitative measurements are possible. The interpretation of data obtained with mass spectrometers carried on planetary probes must, accordingly, be approached with caution.

References

[1]Schaefer, E. J. and Nichols, M. H., "Neutral Composition Obtained from a Rocket-borne Mass Spectrometer," Space Research, Vol. 4, 1964, pp. 205-234.

[2]Nier, A. O., Hoffman, J. H., Johnson, C. Y., and Holmes, J. C. "Neutral Composition of the Atmosphere in the 100- to 200-Kilometer Range," Journal of Geophysical Research, Vol. 69, March 1964, pp. 979-989.

[3]Mauersberger, K., Müller, D., Offermann, D., and von Zahn, U., "A Mass Spectrometric Determination of the Neutral Constituents in the Lower Thermosphere above Sardinia," Journal of Geophysical Research, Vol. 73, Feb. 1968, pp. 1071-1076.

[4]Philbrick, C. R. and McIsaac, J. P., "Measurements of Atmospheric Composition near 400 km," Space Research, Vol. 12, 1972, pp. 743-749.

[5]Niemann, H. B., Spencer, N. W., and Schmitt, G. A., "A Thermosphere Composition Measurement Using a Quadrupole Mass Spectrometer with a Side Energy Focusing Quasi-Open Source," Journal of Geophysical Research, Vol. 78, May 1973, pp. 2265-2277.

[6]Krankowsky, D., Lämmerzahl, P., Bonner, F., and Wieder, H., "The AEROS Neutral and Ion-Mass Spectrometer," Journal of Geophysics, Vol. 40, 1974, pp. 601-611.

[7]Hayden, J. L., Nier, A. O., French, J. B., Reid, N. M., and Duckett, R. J., "The Characteristics of an Open Source Mass Spectrometer under Conditions Simulating Upper Atmosphere Flight," International Journal of Mass Spectroscopy and Ion Physics, Vol. 15, Sept. 1974, pp. 37-47.

[8]Nier, A. O., Potter, W. E., Kayser, D. C., and Finstad, R. G., "The Measurement of Chemically Reactive Atmospheric Constituents by Mass Spectrometers carried on High-Speed Spacecraft," Geophysical Research Letters, Vol. 1, Sept. 1974, pp. 197-200.

[9]French, J. B., Reid, N. M., Nier, A. O., and Hayden, J. L., "Rarefied Gas Dynamic Effects on Mass Spectrometric Studies of Upper Planetary Atmospheres," AIAA Journal, Vol. 13, Dec. 1975, pp. 1641-1646.

[10]Philbrick, C. R., Narcisi, R. S., Baker, D. W., Trzcinski, E., and Gardner, M. E., "Satellite Measurements of Neutral Composition with a Velocity Mass Spectrometer," Space Research, Vol. 13, 1973, pp. 321-325.

[11]Offermann, D. and Grossmann, K. U., "Thermospheric Density and Composition as Determined by a Mass Spectrometer with CryoIon Source," Journal of Geophysical Research, Vol. 78, Dec. 1973, pp. 8296-8304.

[12]Mattauch, J. and Herzog, R., "Uber einen neuen Massenspektrographen," Zeitschrift Physik, Vol. 89, 1934, pp. 786-795.

[13]Nier, A. O., Potter, W. E., Hickman, D. R., and Mauersberger, K., "The Open-Source Neutral-Mass Spectrometer on Atmosphere Explorer-C, -D, and -E," Radio Science, Vol. 8, April 1973, pp. 271-276.

[14]Nier, A. O., Hanson, W. B., McElroy, M. B., Seiff, A., and Spencer, N. W., "Entry Science Experiments for Viking 1975," Icarus, Vol. 16, Feb. 1972, pp. 74-91.

[15]von Zahn, U., Krankowsky, D., Nier, A. O., Mauersberger, K., and Hunten, D., "A Gas Analyzer for the Investigation of the Venus Upper Atmosphere," Space Science Review (in press).

[16]Hedin, A. E., Hinton, B. B., and Schmitt, G. A., "Role of Gas-Surface Interactions in the Reduction of OGO 6 Neutral Particle Mass Spectrometer Data," Journal of Geophysical Research, Vol. 78, Aug. 1973, pp. 4651-4668.

[17]Newton, G. P., Kasprzak, W. T., and Pelz, D. T., "Equatorial Composition in the 137- to 225-km Region from the San Marco 3 Mass Spectrometer," Journal of Geophysical Research, Vol. 79, May 1974, pp. 1929-1941.

[18]Trinks, H. and von Zahn, U., "The ESRO 4 Gas Analyzer," Review of Scientific Instruments, Vol. 46, Feb. 1975, pp. 213-217.

[19]Philbrick, C. R., "Satellite Measurements of Neutral Atmospheric Composition in the Altitude Range 150 to 450 km," Space Research, Vol. 14, 1975, pp. 151-156.

[20]Hedin, A. E., Avery, C. P., and Tschetter, C. D., "An Analysis of Spin Modulation Effects on Data Obtained with a Rocket-Borne Mass Spectrometer," Journal of Geophysical Research, Vol. 69, Nov. 1964, pp. 4637-4648.

[21]Hedin, A. E. and Nier, A. O., "A Determination of the Neutral Composition, Number Density, and Temperature of the Upper Atmosphere from 120 to 200 kilometers with Rocket-Borne Mass Spectrometers," Journal of Geophysical Research, Vol. 71, Sept. 1966, pp. 4121-4131.

[22]Kasprzak, W. T., Krankowsky, D., and Nier, A. O., "A Study of Day-Night Variations in the Neutral Composition of the Lower Thermosphere," Journal of Geophysical Research, Vol. 73, Nov. 1968, pp. 6765-6782.

[23]Krankowsky, D., Kasprzak, W. T., and Nier, A. O., "Mass Spectrometer Studies of the Composition of the Lower Thermosphere during Summer 1967," Journal of Geophysical Research, Vol. 73, Dec. 1968, pp. 7291-7306.

[24]Müller, D. and Hartmann, G., "A Mass Spectrometric Investigation of the Lower Thermosphere above Fort Churchill with Special Emphasis on the Helium Content," Journal of Geophysical Research, Vol. 74, March 1969, pp. 1287-1293.

[25]Pelz, D. T., Reber, C. A., Hedin, A. E., and Carignan, G. R., "A Neutral Atmosphere Composition Experiment for the Atmosphere Explorer-C, -D, -E," Radio Science, Vol. 8, April 1973, pp. 277-285.

[26]Potter, W. E., Kayser, D. C., Knutson, J., and Nier, A. O., "Measurements of Molecular Oxygen on AE-D," Transactions of the American Geophysical Union, Vol. 57, April 1976, p. 295.

[27]Seiff, A., Reese, D. E., Sommer, S. C., Kirk, D. B., Whiting, E. E., and Niemann, H. B., "PAET, An Entry Probe Experiment in the Earth's Atmosphere," Icarus, Vol. 18, April 1973, pp. 525-563.

[28]Thorness, R. B. and Nier, A. O., "Device for Remote Opening of a Vacuum System," The Review of Scientific Instruments, Vol. 33, Sept. 1962, pp. 1005-1007.

[29]Moe, M. M. and Moe, K., "The Roles of Kinetic Theory and Gas-Surface Interactions in Measurements of Upper Atmosphere Density," Planetary and Space Science, Vol. 17, May 1969, pp. 917-922.

[30]Mauersberger, K., Engebretson, M. J., Potter, W. E., Kayser, D. C., and Nier, A. O., "Atomic Nitrogen Measurements in the Upper Atmosphere," Geophysical Research Letters, Vol. 2, Aug. 1975, pp. 337-340.

[31]Nier, A. O., Potter, W. E., and Kayser, D. C., "Atomic and Molecular Oxygen Densities in the Lower Thermosphere," Journal of Geophysical Research, Vol. 18, Jan. 1976, pp. 17-24.

[32]Lake, L. R. and Mauersberger, K., "Investigation of Atomic Oxygen in Mass Spectrometer Ion Sources," International Journal of Mass Spectroscopy and Ion Physics, Vol. 13, April 1974, pp. 425-436.

[33]Nier, A. O., Hoffman, J. H., Johnson, C. Y., and Holmes, J. C., "Neutral Constituents of the Upper Atmosphere: The Minor Peaks Observed in a Mass Spectrometer," Journal of Geophysical Research, Vol. 69, Nov. 1964, pp. 4629-4636.

MCLEOD PRESSURE GAGE AND MERCURY DRAG EFFECT

Lloyd B. Thomas*, C. Leon Krueger+, and Robert E. Harris‡

University of Missouri, Columbia, Mo.

Abstract

Neglect of the mercury vapor pumping effect, inherent with the widely used McLeod gage/cold-trap combination, has raised doubt concerning the accuracy of results in numerous scientific papers. From the measurements of the pumping error available it is difficult to obtain a feeling of confidence in the magnitude of the effect for application of corrections. We need such knowledge for the work of this laboratory and have undertaken to obtain it experimentally. We have isolated the effect under carefully controlled conditions of temperature, length, and diameter of uniform tubing, and are able to initiate or cease pumping readily and to measure the resulting pressure changes beyond the cold-trap with a very sensitive Pirani gage. Results are given for the five inert gases and nitrogen over the pressure range needed for complete description of the effect. Comparison with results of other investigators is attempted.

Introduction

The McLeod gage is in principle a beautifully simple "Apparatus for Measurement of Low Pressures of Gas." The quotation is the title of the original paper of Professor McLeod[1] of the Indian Civil-Engineering College, Coopers Hill, read before the Physical Society, June 13, 1874. The gage employs the principle of the constancy of the pressure-volume product at constant temperature. It amplifies the pressure of a gas to a range, readable by the difference in height of two mercury columns, by reducing the volume by ratios known by calibration. Since the final pressure of the compressed gas is seldom over a few centimeters there is little need, for most gases, to compensate for deviation from ideality. The gage has had extremely wide use. It has undergone numerous modifications to meet special needs and has served over the century as the basic standard for submillimeter pressure measurement throughout the breadth of scientific work. Many doubts have arisen concerning its validity, and irritating

---

Presented as Paper 79 at the 10th International Symposium on Rarefied Gas Dynamics, Aspen, Colo., July 19-23, 1976. We thank the U.S. Airforce Office of Scientific Research for support of this work.

* Professor of Chemistry, Physical Chemical Laboratory.

+ Research Associate, Physical Chemical Laboratory.

‡ Associate Professor of Chemistry, Physical Chemical Laboratory.

characteristics show up when its precision is pushed toward the limit, but the former have been largely imaginary and the latter surmountable by patient adaptation and optimization. The gage has the tremendous advantage of relative ease of construction and calibration so that it is readily available to any worker who is moderately skilled in ordinary laboratory techniques. In the opinion of the authors, the McLeod gage has furnished, and will continue to furnish, the most practical "absolute" standard, either for direct measurement or for calibration of the various other gages needed for continuous monitoring or for extension to lower pressure ranges than the McLeod can reach.

The presence of mercury vapor, which is inherent with the McLeod gage, and often also arises from other liquid mercury in the vacuum system, usually cannot be tolerated in the region of experimentation, and so must be removed. For work involving "permanent" gases this ordinarily is done by interposing a cold-trap between the mercury source and the region of experiment. This causes a diffusive flow of mercury vapor through the passage from the mercury liquid toward the cold-trap, which flow increases the permanent gas pressure in the working region beyond the cold-trap above the partial pressure of the pumped gas as measured by the McLeod. The effect was described by Gaede[2] in 1915, but little or no heed was paid to the error caused until this was encountered as discrepancies in experimental work by Podgurski and Davis[3], and independently by Ishii and Nakayama[4] in the 1960's. In the ensuing period scientists generally have become aware of the phenomenon and the pitfall that exists. About a dozen papers describing attempts at measurement of the magnitude of the pressure differential, and one paper since Gaede on theory of the effect have come to our attention. Attempts to bring together the published data, varying in the gases studied, pressure ranges, radii of pumping tubing, etc., for comparison indicate wide discrepancies of results and, thus, apparent difficulty in measurement of the effect. The magnitude of the pressure difference set up may be described roughly as follows. For usual tube diameters, and gas pressures above a few tenths of a millimeter, the pressure difference is the vapor pressure of mercury, i.e. ~0.002 mm at 25°C. For light gases, H_2 and He, as the pressure diminishes below 0.001 mm, the error for 25°C and a tube diameter of 12 mm, approaches the flat value of 2 or 3% of the McLeod reading. For the heavier gases with large collision cross sections (with Hg), the percentage errors (for 25°C and 12 mm tube diam.) rise as the pressure diminishes, and, well below 0.001 mm, approach constant values which may be as high as 50%–i.e., the actual pressure beyond the trap is 50% higher than the pressure at the McLeod. For Xe this value is about 40%. As the diameter of the connecting tubing increases, the values above indicated increase, and as the vapor pressure of Hg increases the values indicated increase nearly proportionately. The length of connecting tubing has little if any effect. One infers readily that all low-pressure work that aims at absolute accuracy and involves the determination of pressure (or concentration or number density) and which has not taken cognizance of this effect is open to suspicion if the McLeod/cold-trap combination was used either directly or to calibrate ionization gages or other types that may have been employed. We suspect that the number of such papers in the literature may run into the thousands. For example, all thermal accommodation coefficient papers before the mid-sixties involving the low-pressure Knudsen method are subject to this error. Knowledge of the pumping pressure correction is needed for our own work on tangential-momentum accommodation by the magnetic suspension method[5] and for our earlier work on photochemical and electron impact kinetics–e.g., Ref. 6. A number of ways have been suggested to avoid the pressure differential, but in practice the cure is often worse than the disease and one would rather tolerate the effect and then correct for it. The measurements described herein are needed in the work of this Laboratory, and we hope they may be of use to others, at least in helping to clarify the phenomenon and to furnish more extensive data of sufficient reliability to provide an adequate test of theory than now exists, and perhaps better values of diffusion coefficients, intermolecular collision cross sections, etc., which must enter into the theories. The McLeod gage/cold-trap combination has so many factors in its favor that, with no substitute of comparable versatility in view, it will no doubt be with us for at least many decades to come. These considerations provide the incentive for this work.

Experimental Design and Procedure

The opportunity to design an experiment "out of whole cloth" where one selects the disposition of apparatus from almost unlimited sets of possibilities is rare. Dereliction in the study of the literature of the area before settling on a design contributes to this sense of freedom, in that one is not prejudiced by that which has gone before. This design, selected without such restricting prejudice turns out to contain certain components similar to those used by others, but the method as a whole appears to be quite unique. It satisfies certain general requirements which we felt should be met: 1) applicability to all gases noncorrosive toward mercury and of sufficient volatility to survive temperatures needed to condense mercury vapor, 2) permission of clean vacuum conditions to an extent compatible with use of stopcocks with "Apiezon T" grease, 3) an adequately sensitive pressure sensor which should indicate the change in pressure beyond the cold trap by continuous monitoring (rather than by difference of two discontinuous measurements, as by a McLeod gage, for example), 4) capability of initiating or terminating the mercury pumping at will, with rapid response of the pressure sensor, and 5) feasibility of frequent calibration of the sensor in the circumstances peculiar to each run.

The third modification of the method, as used in most of the work here presented, is described with reference to Fig. 1. (Earlier modifications are described in Refs. 7 and 8.) The

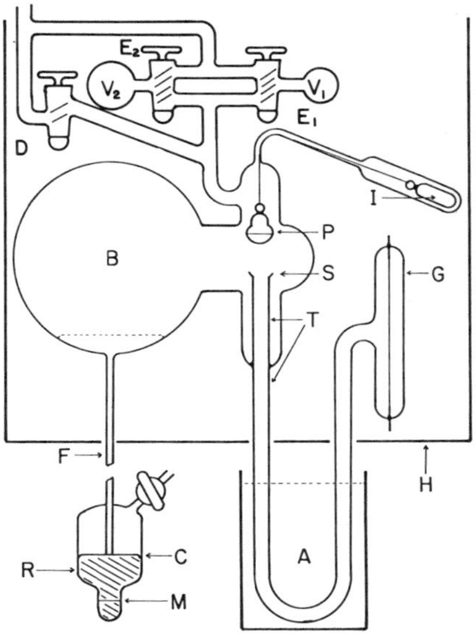

Fig. 1 The apparatus for mercury vapor pumping pressure measurement.

apparatus diagrammed is contained in a water-filled thermostat bath H, except for the parts shown that are extended through the bottom of the bath. The three stopcocks near the top are of the mercury-seal type with the cups extending above the water surface but with the barrels and connecting tubing below. The bath is held to ±0.01° of the selected temperature. Stopcock D connects through a cold trap to a conventional vacuum system with a McLeod gage and gas supply manifold. The apparatus centers around the 3-liter bulb B. The mercury pumping occurs in tube T, 1.169 cm i.d. and 25 cm in length, which leads through the bottom of the thermostat to the cold-trap immersed in a dewar filled with liquid N_2, liquid O_2 or dry ice-acetone. Beyond this cold-trap, and back up in the bath, is the Pirani gage G. This consists of 14 cm of 0.0007-in. tungsten filament in a pyrex tube about 1 cm i.d. This gage is connected as one of four equal arms of a Wheatstone bridge and is operated at 43 ohms, about 120° above the 25°C bath temperature. The off-balance of the bridge is read on a Dana digital volt-meter (DVM) on the range with the last digit reading 10^{-7} V. The sensitivity of the gage is 2 or 3 in the last digits per 10^{-7} Torr pressure change, i.e. 2 or 3 V/Torr, the narrow range for the various gases and pressure ranges resulting from several compensating factors. The DVM prints the voltage on paper tape upon command. The mercury pumping is initiated by letting the mercury up 80 cm through the 2-mm capillary tubing F from the reservoir F to spread over the bottom of B to form a pool with a fresh surface of 20-cm^2 area. The glass pendant P and seat S are portions of a spherical ground joint. P may be seated on S by sliding upward the iron-cored glass capsule I, attached to P by a fine tungsten wire, by use of a small magnet. E_1 and E_2 and 3-way stopcocks allowing evacuation of the calibrated volumes V_1 and V_2 or allowing expansion of the gas in B and connected volumes (about 3300 ml) by about 1.25% and 4.14%, depending to a slight extent on the temperature of the cold bath, A. "Obnoxious volumes" in uniform temperature regions (in the U-tube in A, the connecting tube to G, and in G itself) were minimized to keep small the necessary "corrections" in the calculation of $\Delta P/P$ from the observations.

The procedure in measuring the pumping pressure is as follows. With trap A cold, and P down on S, and the cold-trap beyond D (not shown) at room temperature so the mercury vapor pervades the whole system, the subject gas is admitted and adjusted within the desired range of pressure; D is closed; and the pressure is read carefully on the McLeod gage. Lifting P then quickly condenses the vapor from the system. Liquid N_2 is put on the external cold-trap; the system beyond D is evacuated; and E_1 and E_2 are set to evacuate V_1 and V_2. The bridge current is adjusted until the DVM reads near zero on the low-pressure side (to avoid crossing null in the run) and the timer is set to print the DVM reading at 1-min. intervals. When the drift in the DVM slows (due primarily to approach to steady-state adsorption conditions on the tungsten surface) to a satisfactory rate, calibration of the Pirani is begun. The system is expanded alternately into V_1, V_2, back into V_1 (after its evacuation), etc., at 3-min. intervals, which is ample time to allow the DVM to complete its excursion and to evacuate fully the unused V_1 or V_2. As many as 13 expansions are required (for the largest pumping pressures) to calibrate the Pirani over the decreasing pressure range, which is retraced (not quite fully or the DVM would cross null) when mercury pumping is initiated. The last calibration expansion is always into V_1, the volume of which is precisely that of the mercury in R which is let into the system when the level is set at the calibration mark M. Thus, upon initiating the pumping the compression just cancels the last expansion and the pumping pressure developed is indicated by the DVM excursion from the reading after the next-to-last expansion back to a steady value near the beginning of calibration. For He, only 4 expansions into the smaller V_1 are required, at the most. Large-scale calibration plots of the DVM readings vs the fraction of P (initial) remaining after each expansion, were made. The DVM reading, with pumping at its steady value, with due allowance for drift, then is used to locate the fraction of the P (initial) in the small dead-end from the cold-trap to and including the Pirani (about 50 cc). A small correction is applied for the diminished pressure in the large volume due to the pumping compression into the dead-end region. It will be noted that in this work we are not investigating a McLeod gage as such, but we have isolated the pumping phenomenon under much more precisely controlled, relevant parameter conditions than are possible with the gage itself. The pressure in B, of course, corresponds to the pressure at the McLeod and that at G corresponds to the pressure in the experimental region beyond the cold-trap (as discussed in the Introduction). Calling these, respectively, P and P′, we extract (P′–P)/P or ΔP/P as the pumping pressure fractional increment. Series of values of this fractional increment at P's from $\sim 5 \times 10^{-4}$ to $\sim 1 \times 10^{-1}$ Torr were measured for each of the six gases reported. With all components behaving perfectly, it is possible to take a set of eight points for one gas and to calculate the results in about two full days.

We thought possibly we could provide an alternate way of turning the pumping on or off by lifting or seating "P". The joint, when closed, must let the pumped gas pressure equalize in B and below S, but still cut off essentially all flow of Hg vapor. At the lowest gas pressures the seating of P on S cuts the pumping pressure about 95%, but definitely not completely, and at the higher pressures only about 70%. However this device was quite useful in testing for absence of droplets of Hg left in B, or clinging to the walls of F near its junction with B when the mercury was drawn down into R. We would include this feature if we were to build a revised apparatus for such studies as these.

Experimental Results and Discussion

Our experimental data points are plotted for He, Ne, and N_2 in Fig. 2 and for Ar, Kr, and Xe in Fig. 3. The ordinates are the pumping pressure increments expressed as a percentage of P, the permanent gas pressure in B, or at the McLeod gage or other liquid mercury source of vapor. P′ is the pressure at G, or, in any case, that in the mercury-free space beyond the cold-trap. The abscissas are the P values, defined above. The curves are plotted from empirical equations with constants obtained from the data points as explained below.

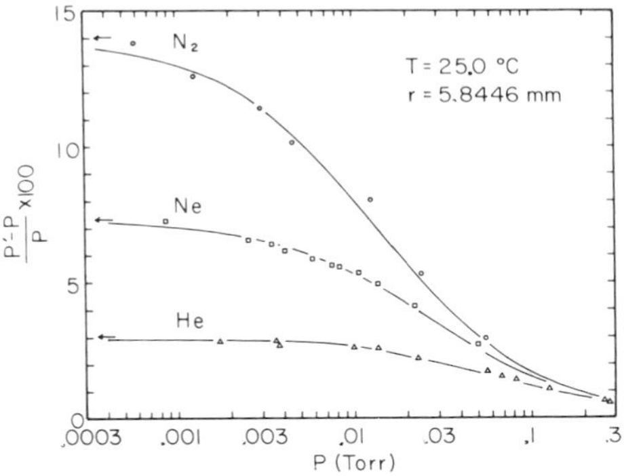

Fig. 2 The mercury streaming effect in He, Ne and N_2 gases. The pressure difference caused by the mercury flow is expressed as a percentage of the permanent gas pressure at the mercury source, P. The points represent experimental data and the smooth curves are obtained from least-squares fitting of Eq. (1). The arrows indicate the low-pressure limiting values according to the curves.

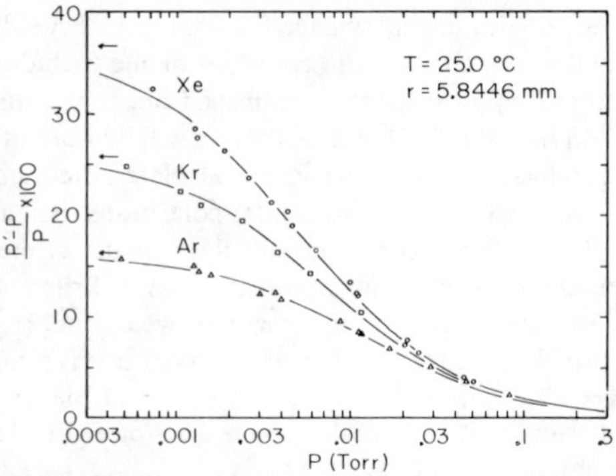

Fig. 3 The mercury streaming effect in Ar, Kr, and Xe gases (see caption of Fig. 2; also applicable to Fig. 3).

There are two papers available which propose theories of the mercury vapor pumping effect–those of Gaede[2] and Takaishi.[9] Resulting equations from both theories may be represented in the form

$$P'-P = P_{Hg}-(K/r)\,[\ln(P'/P)] \qquad (1)$$

In the Gaede Equation

$$K = (3D_{12}/8)\,(2\pi m_{Hg}/kT)^{1/2} \qquad (2)$$

In the Takaishi equation

$$K = (4kT/3\pi^2 a d^2)\,[(m_g + m_{Hg})/m_g]^{1/2} \qquad (3)$$

In Eqs. (1) to (3), P′ and P are as above, P_{Hg} is the equilibrium vapor pressure of mercury, r is the radius of the pumping tube, and m's are the molecular or atomic masses of the pumped gas and of mercury, D_{12} is the mutual diffusion coefficient at unit pressure for Hg and the gas, d is the collision diameter for the gas with Hg, k is the Boltzmann constant, and a is a constant between the limits 1 and 4/3. Both of the theoretical equations are derived under the assumptions which should result in more accurate representation of the pumping phenomenon toward the low-pressure limit. In applying Eq. (1) we have taken the Ernsberger and Pitman[10] expression [$\log_{10}P_{Hg}(\mathrm{mm}) = 8.0372-(3204/T)$] for the vapor pressure of Hg (0.001954 mm at 298.15°K) and our pumping tube radius (5.8446 mm) and obtained for each gas a least squares value of K, using for each gas all data points for which the pumping pressures were less than 0.5 P_{Hg}. The K values so obtained are shown for the six gases in Table 1.

The limiting values of (P′–P)/P as P → 0 are shown in the second row, and also indicated (as percent) by the arrows in Figs. 2 and 3. Rows 3 to 5 are sample values calculated using Eq. (1) for (P′–P)/P with P = 0.001, 0.01, and 0.1 Torr. The pumping pressures for any value of P for each of the gases, assuming our r and P_{Hg}, can be obtained by substitution in Eq. (1) using the appropriate K value from Table 1. The points (not shown) used to plot the curves of Figs. 1 and 2 were so obtained. As noted, the curves were plotted from Eq. (1) with the K's

Table 1 K values for Eq. (1) with P_{Hg} = .001954 Torr and r = 5.8446 mm and some fractional pumping increments

		He	Ne	Ar	Kr	Xe	N_2
	K	0.3853	0.1611	0.0755	0.0497	0.0366	0.0871
$(\Delta P/P)$	$P \to 0$	0.0301	0.0734	0.1633	0.2586	0.3665	0.1402
"	$P = 10^{-3}$	0.0296	0.0707	0.1499	0.2256	0.3021	0.1302
"	$P = 10^{-2}$	0.0260	0.0531	0.0873	0.1082	0.1229	0.0803
"	$P = 10^{-1}$	0.0118	0.0153	0.0173	0.0180	0.0184	0.0170

determined by least-squares fits to the points toward the left sides of the plots. The points on the right sides appear to fit the curves even better than those used to determine the curves. This is somewhat misleading, apparently implying that the "empirical" expression (1) does an excellent job of describing the pumping phenomenon over the whole transition regime. The apparent enhanced agreement to the right of Figs. 2 and 3 is due to the division of the pumping increment by ever-increasing values of P. (1000-fold in the range plotted). The same data points and the curves obtained in the same way are plotted as (P′–P) vs P for He, N_2, and Xe in Fig. 4. Plotted thus, the fits are much better to the left side and the data for the three gases show consistently higher pumping pressures at the higher abscissa pressures than indicated by the empirical expression. All the curves of Fig. 4 are headed toward the asymptotic limit as $P \to \infty$ of $P' - P = P_{Hg}$, as indicated by the arrow in the upper right corner. The P_{Hg} we used[10] is about 6% higher than that from a source[11] used by many others, and it is interesting to note that if the correct value were about 10% higher still than that from Ref. 10, the curves of Fig. 4 would fit the data points quite well over the whole range. We do not consider this at all likely, and since Eq. (1), as noted, has a basis in each of the theories, the consistent disagreement between points and curves, would seem to indicate shortcomings of the theories as expected, in the higher ranges of P.

We attempt a brief comparison of the results of this work with those of ten other papers available in the literature. Emphasis has been preponderantly at the lower pressure end of our range (and beyond) because of involvement of the effect with ion-gage calibration. Data cov-

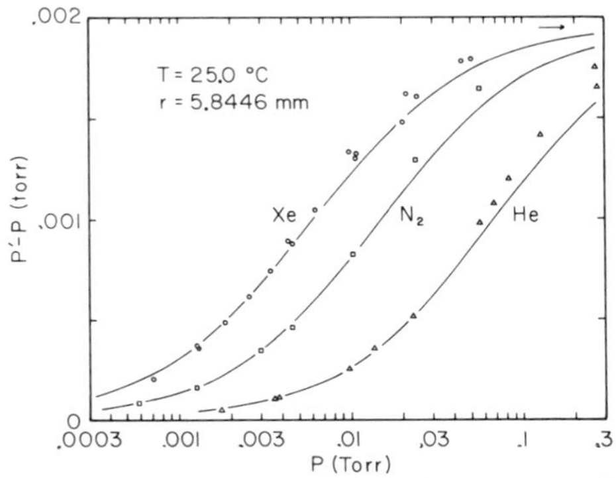

Fig. 4 Deviations of experimental data from data from fitted curves. Data and curves used in Figs. 2 and 3 are plotted as pressure difference vs pressure at the mercury source. Only data below $P' - P = \frac{1}{2} P_{Hg}$ were used in the fitting. The arrow indicates the P_{Hg} used in Eq. (1).

Table 2 Summary of available measurements of pumping pressure

	He	Ne	Ar	Kr	Xe	N_2
12	0.218[a] (0.0458)[b]	...	0.0746 (0.140)	...	0.0399 (0.277)	...
13	0.364 (0.0272)	0.201 (0.0498)	0.0766 (0.136)	0.0489 (0.221)	0.0328 (0.347)	0.0813 (0.128)
14	...	0.201 (0.0498)	0.0789 (0.132)	0.0598 (0.177)	0.0503 (0.214)	0.0964 (0.107)
15	0.435 (0.0227)	0.117 (0.0871)	0.0991 (0.104)	0.0431 (0.254)	0.0335 (0.399)	0.0553 (0.193)
16	...	0.169 (0.0595)	0.0837 (0.124)	0.0529 (0.203)	0.0360 (0.312)	...
17	...	...	0.0793 (0.131)	0.0489 (0.221)	0.0412 (0.268)	...
18	...	...	...	0.0333 (0.341)	0.0260 (0.456)	0.0580 (0.183)
19	...	...	...	...	0.0357 (0.315)	...
20	...	...	...	...	...	0.0700 (0.150)
4	...	...	...	...	...	0.0716[a] (0.146)[b]
Average of above	0.339 (0.0292)	0.172 (0.0584)	0.0820 (0.127)	0.0478 (0.227)	0.0369 (0.303)	0.0721 (0.145)
Present work	0.385 (0.0257)	0.161 (0.0626)	0.0755 (0.138)	0.0497 (0.217)	0.0366 (0.306)	0.0871 (0.119)

[a]The upper number of each pair is the K obtained from fit to Eq. (1). K is in Torr (pressure) x mm (length).

[b]The lower number (in parenthesis) is the limiting, low pressure, value of $\Delta P/P$ for P_{Hg} = 0.001954 Torr and r = 5 mm.

ering the principal range of change in $\Delta P/P$ down to ΔP itself approaching P_{Hg} are relatively scarce. We make use of Eq. (1) to bring the diverse data of the ten papers to a common basis for comparison. Data, as best we could extract it from the papers, were used with Eq. (1) to determine the 33 values of K given in Table 2. The lower numbers (in parentheses) are the low-pressure limiting values of $\Delta P/P$ calculated for the tabulated K's, the 25°C value of P_{Hg},[10] and a pumping tube radius of 5 mm. It is interesting to note the relation between the number of available measurements on each member of the "homologous series" and the magnitude of the pumping pressure (related to ease of measurement). For the individual gases, variation in limiting pumping pressure among investigations from low to high of 100% is typical. However, taking the shotgun approach and averaging all available values of either K or limiting $\Delta P/P$ gives average values reasonably close to those of this paper. The limiting $\Delta P/P$ values for the present work in Table 2 are lower than indicated by the arrows in Figs. 2 and 3 because of the 5.0-mm r used for Table 2. Nakayama[18] and Takaishi and Sensui[19] present results in terms of "B" which is our $r \cdot P_{Hg}/K$ and our entries in Table 2 are converted from their "B" values. Rambeau's results are given as our estimate of K's which best fit the data he presents. K values for the other references have been extracted similarly. We have used the Ref. 10 values for P_{Hg} in each case. One point of doubt in comparison of the various investigation results by Eq. (1), as we have done, is that the effect of the tube radius expressed in Eq. (1) clearly has not been substantiated experimentally. Evidence from Refs. 14, 13, and, especially, 19 suggest that P'/P increases with r more rapidly than stated in Eq. (1). We propose to test this with our apparatus soon. We hope that the original work presented here, and the summary of previous work, will help in narrowing the limits of uncertainty in applying corrections to McLeod gage readings, at least, and to help in providing a better experimental basis for test of the mercury pumping effect.

References

[1]McLeod, H., "Apparatus for Measurement of Low Pressures of Gas," Philosophical Magazine, Vol. 48, No. 11D, 1874, pp. 110-113.

[2]Gaede, W., "Die Diffusion der Gase durch Quecksilberdampf bei niederen Drucken und die Diffusionsluftpumpe," Annalen der Physik, Vol. 46, 1915, pp. 357-392.

[3]Podgurski, H.H., and Davis, F.N., "A Precision McLeod Gage for Volumetric Gas Measurement," Vacuum, Vol. 10, 1960, pp. 377-381.

[4]Ishii, H., and Nakayama, K., "A Serious Error Caused by Mercury Vapour Stream in the Measurement with a McLeod Gauge in the Cold-Trap System (Effect of the Diffusion of Nitrogen in the Mercury Vapour Stream)," Transactions of the Eighth National Vacuum Symposium, Pergamon Press, Great Britain, 1961, pp. 519-524.

[5]Thomas, L.B., and Lord, R.G., "Comparative Measurements of Tangential Momentum and Thermal Accomodations on Polished and on Roughened Steel Spheres," Eighth International Symposium on Rarefied Gas Dynamics, Academic Press, NewYork, 1972, pp. 405-412.

[6]Thomas, L.B., and Gwinn, W.D., "The Quantum Efficiency of the Mercury Sensitized Photochemical Decomposition of Hydrogen," Journal of the American Chemical Society, Vol. 70, 1948, pp. 2643-2648.

[7]Thomas, L.B., Harris, R.E., and Krueger, C.L., "Observations and Measurement of the Mercury Pumping Effect on Several Gases by Four New Methods," Japanese Journal of Applied Physics, Supplement 2, Part 1, 1974, pp. 151-154.

[8]Thomas, L.B., and Krueger, C.L., "Some Measurements of Hg Vapor Pumping Pressures in Pairs of Isotopic Species," Journal of Vacuum Science Technology, Vol. 13, No. 1, January/February, 1976, pp. 490-493.

[9]Takaishi, T., "Kinetic Theory of the Mercury Drag Effect in Vacuum Measurements with a McLeod Gauge," Transactions of the Faraday Society, Vol. 61, No. 509, Part 5, May, 1965, pp. 840-853.

[10]Ernsberger, F.M., and Pitman, H.W., "New Absolute Manometer for Vapor Pressures in the Micron Range," Review of Scientific Instruments, Vol. 26, No. 6, June, 1955, pp. 584-589.

[11]International Critical Tables of Numerical Data, Physics, Chemistry, and Technology, Vol. III, edited by E.W. Washburn, McGraw-Hill Book Company, Inc., New York, 1928, p. 209.

[12]Meinke, C., and Reich, G., "Influence of Diffusion on the Measurement of Low Pressure with the McLeod Vacuum Gauge. Based on a Paper by Gaede," Vacuum, Vol. 13, 1963, pp. 579-581.

[13]Dadson, R.S., Elliott, K.W.T., and Woodman, D.M., "The McLeod Gauge Vapour Stream Effect," Proceedings of the Fourth International Vacuum Congress, 1968, pp. 679-682.

[14]deVries, A.E., and Rol, P.K., "Theoretical and Experimental Determination of an Error in the Pressure Indication of a McLeod Manometer," Vacuum, Vol. 15, No. 3, 1965, pp. 135-139.

[15]Utterback, N.G., and Griffith, T., "Reliable Submicron Pressure Readings with Capacitance Manometer," The Review of Scientific Instruments, Vol. 37, No. 7, July 1966, pp. 866-870.

[16]Rambeau, G., "Étude de l'entrainement moléculaire par la vapeur de mercure dans l'utilisation de la jauge McLeod," LeVide, Vol. 23, No. 142, July, August, 1969, pp. 219-225.

[17]Tunnicliffe, R.J., and Reese, J.A., "An Investigation into the Effect of Mercury-Vapour Pumping Errors on McLeod Gauge Measurements, Vacuum, Vol. 17, No. 8, May 6, 1967, pp. 457-459.

[18]Nakayama, K., "An Accurate Measurement of the Mercury Vapour Drag Effect in the Pressure Region of Transition Flow," Shinku, Vol. 12, 1969, pp. 325-334.

[19]Takaishi, T., and Sensui, Y., "Study of the Mercury Vapour Drag Effect in Vacuum Measurement," Vacuum, Vol. 20, No. 11, July 20, 1970, pp. 495-499.

[20]Berman, A., "Graphical Evaluation of the Mercury Vapour Pumping Effect on McLeod Gauge Measurements," Vacuum, Vol. 24, No. 6, 1974, pp. 241-243.

[21]Sensui, Y., "Kinetic Analysis of Mercury Vapour Drag Effect in Vacuum Measurements," Vacuum, Vol. 20, No. 12, 1970, p. 539.

LINE WIDTH OF THE N_2^+ FIRST NEGATIVE BAND SYSTEM

Frank Robben*
University of California, Berkeley, Calif.

Robert Cattolica†
Sandia Laboratories, Livermore, Calif.

Donald Coe‡ and Lawrence Talbot§
University of California, Berkeley, Calif.

Abstract

The N_2^+ first negative band system was excited by a high energy electron beam in a rarefied, underexpanded free jet. The line width was measured using a Fabry-Perot interferometer with a grating monochromator filter to isolate single rotational lines. The original purpose of the measurements was to obtain the translational temperature of nitrogen from the width and shape of the Doppler broadened lines. However, at gas temperatures approaching 10^oK the temperature deduced from the line width approached about 75^oK; that is, a Gaussian distribution function at 75^oK, when convoluted with the measured interferometer instrument function, was in good agreement with the experimental measurement. We believe that the most likely explanation is the hyperfine structure of nitrogen which theoretically can split the odd rotational levels into 13 components. Whatever the cause, accurate measurements of translational temperatures in nitrogen, using the electron beam fluorescence technique, do not appear feasible for temperatures below about 100^oK.

Presented as Paper 122 at the 10th International Symposium on Rarefied Gas Dynamics, Aspen, Colo., July 19-23, 1976. This work was supported by the National Science Foundation, Grant ENG74-02058.

*Physicist, Energy and Environment Division, Lawrence Berkeley Laboratory; Research Engineer, Mechanical Engineering Department.

†Member of the Technical Staff, Gas Dynamics Department.

‡Graduate Research Assistant, Mechanical Engineering Department.

§Professor, Mechanical Engineering Department.

Introduction

Analysis of the fluorescence excited by a high-energy electron beam has become a standard technique for measurement of density and rotational temperature[1] of nitrogen, and translational temperature of helium[2] and argon[3] in rarefied gas dynamics. To obtain translational temperature the Doppler broadening of the fluorescence is determined by measuring the spectral line shape with a Fabry-Perot interferometer. The original purpose of this work was to study translational nonequilibrium in an underexpanded free jet of nitrogen using the electron beam fluorescence technique.

Measurements and Results

To apply this technique to nitrogen, a single rotational line must be selected from the band spectrum for analysis by the Fabry-Perot interferometer. This was accomplished by incorporating a diffraction grating as an integral part of the interferometer as shown in Fig. 1. Figure 2 shows the spectral line shape of the K' = 3, R branch, 0-0 vibrational transition of the N_2^+ first negative band system at 3908.3 Å. The line is a doublet due to spin doubling.[4] Also shown in Fig. 2 is a model fit to the experimental line shape consisting of the convolution of two Gaussian distributions separated by the doublet splitting with the instrument function. The best fit as shown in Fig. 2 was obtained with a temperature of 90°K for the two Gaussians. However, this measurement was

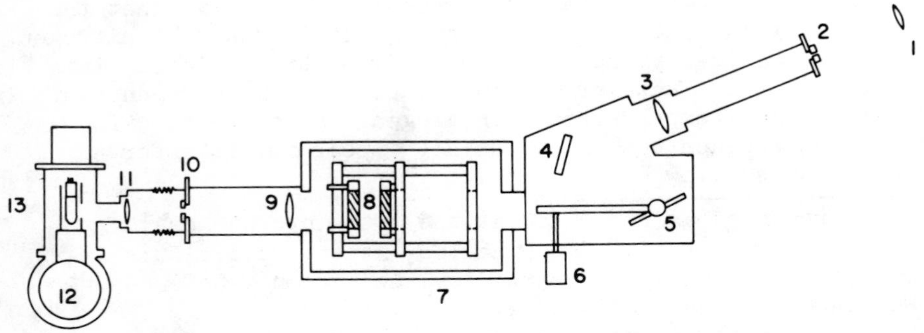

Fig. 1 Fabry-Perot interferometer and grating monochrometer. 1-lens; 2-rectangular aperture; 3-lens; 4-mirror; 5-diffraction grating 600 lines per mm and 48° blaze angle; 6-grating stepping motor; 7-constant temperature, interferometer chamber; 8-piezoelectrically scanned Fabry-Perot interferometer; 9-lens; 10-rectangular aperture; 11-lens; 12-liquid nitrogen reservoir; 13-photomultiplier used in photon counting mode.

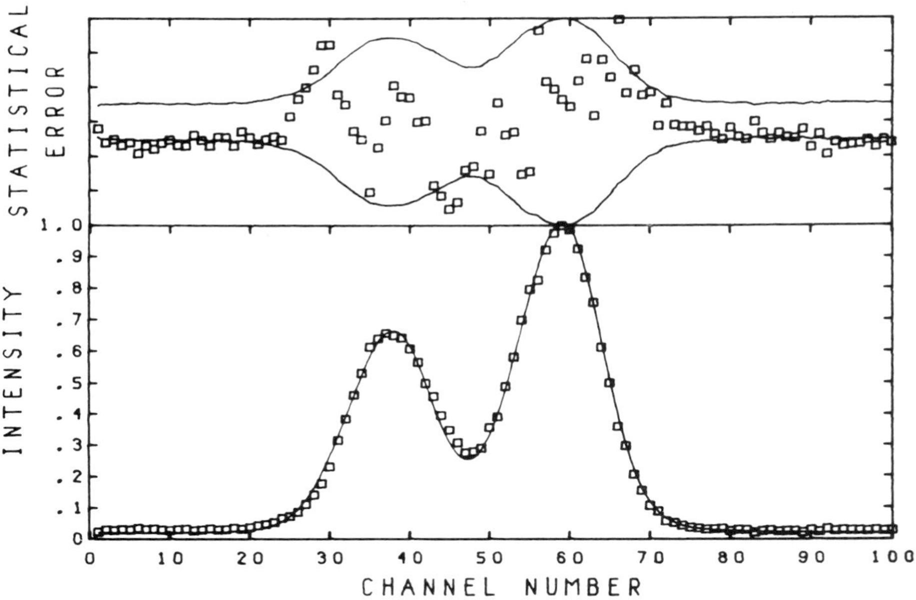

Fig. 2 Experimental and Gaussian modeled spectral profiles for the nitrogen K' = 3, 0-0 vibrational transition of the N_2^+ first negative band system. The lower figure shows the experimental data (squares) and the modeled fit (curve), and the upper figure shows the deviation between the experimental data and the modeled fit, with zero deviation on the center-line. The curves indicate $\pm$ 3 standard deviations of error (for a single point) based on Poisson photon counting statistics. The free jet conditions were P_oD = 158 Torr-mm, X/D = 8, and D = 3.18 mm. The best fit effective temperature is 90°K, whereas the isentropic temperature is 23°K. The doublet splitting used was 0.078 cm^{-1}. The Fabry-Perot mirror separation was 15.1 mm, the spectra has 100 channels, the free spectral range was 88 channels, and the working finesse of the interferometer was approximately 14.

made at X/D = 8 in a free jet under conditions in which an equilibrium (isentropic) translational temperature approximately equal to 25°K is expected (D is the orifice diameter and X is the distance downstream from the orifice on the jet centerline). In Fig. 3 the same experimental data is compared with a model fit with 25°K Gaussians, indicating the magnitude of the difference in the experimental measurement from a 25°K Doppler line shape. As supported by extensive additional measurements, there is a broadening of the rotational lines of this nitrogen band system with a width equivalent to about a

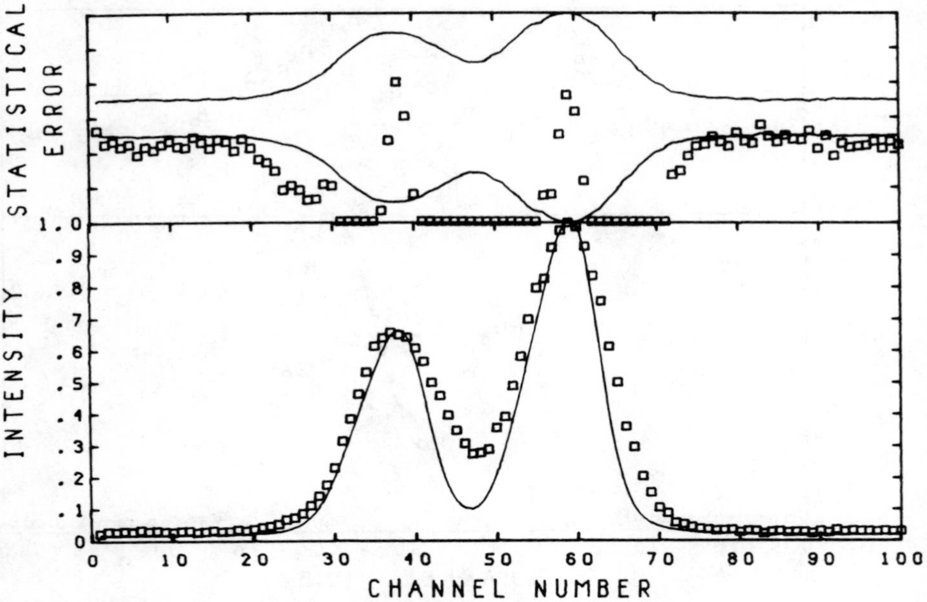

Fig. 3 Experimental and Gaussian modeled spectral profiles. All data are the same as for Fig. 2, except that a temperature of 25°K, approximately equal to the true gas temperature, was used for the modeled fit.

75°K translational temperature of nitrogen. In other words these lines, when excited by electron collision, have an inherent line width of about 0.03 cm^{-1} with a measured line shape (after convolution with the instrument function) which is nearly indistinguishable from a Gaussian distribution within experimental error.

Figure 4 shows the apparent translational and rotational temperatures as a function of X/D for P_oD = 158 Torr-mm. (P_o is the stagnation pressure.) Under these conditions the flow is expected to be nearly isentropic with translational temperature equal to the calculated isentropic temperature for a free jet expansion. Thus the high temperatures measured are not the true gas translational temperatures, and the lines are broadened by some effect other than the Doppler broadening. (The deviation between the measured rotational and isentropic temperatures shown in Fig. 4 is well known; under these conditions it probably is caused by momentum transfer from low-energy secondary electrons.)

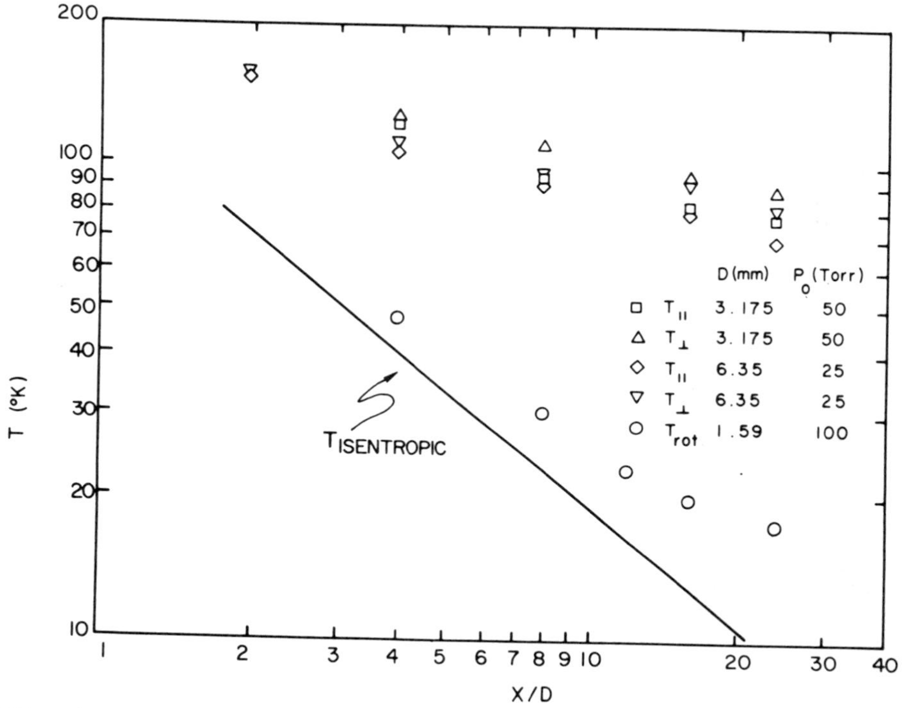

Fig. 4 Isentropic, rotational, and effective translational temperatures vs X/D for P_0D = 158 Torr-mm. Translational temperatures were obtained from the apparent Doppler broadening of the K' = 3, R branch, 0-0 vibrational transition of the N_2^+ first negative band system.

Discussion

We have conducted careful experiments to show that the measured line width is not an experimental artifact. Measurements of helium free jets agree with those obtained by Cattolica et al.[2] The rotational line broadening is apparently independent of gas density, electron beam current and voltage, position in the halo of the beam, and fluorescence direction with respect to both the electron beam and flow directions. Further, similar results have been obtained by Rocket and Brundin.[5] Thus the anomalous width does not appear to be the result of quenching collisions, plasma effects, or electron beam collisional effects but rather a real property of isolated, excited nitrogen ion molecules.

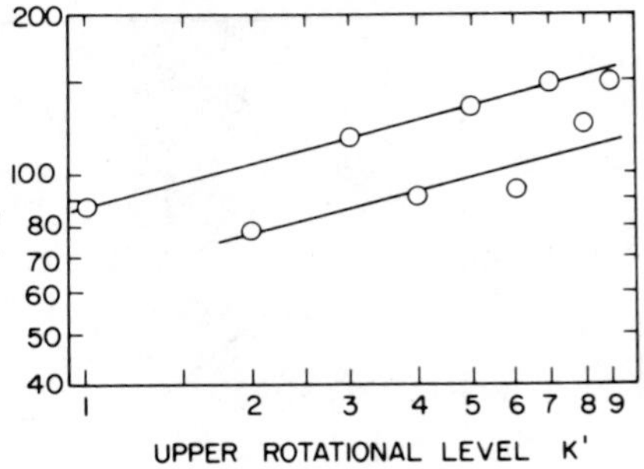

Fig. 5 Effective translational temperatures of lines originating from different rotational levels of the 0-0 vibrational transition of the N_2^+ first negative band system. The free jet flow conditions were P_0D = 158 Torr-mm, X/D = 4, and D = 3.175 mm.

We did find that the line width was a function of rotational level K' and, as shown in Fig. 5, is broader for odd upper K' values. The levels with odd upper K' are those with higher nuclear spin multiplicity (nitrogen has nuclear spin 1) which indicates the possibility that the hyperfine structure may be the source of the excessive line width. In general, hyperfine structure in molecules is of the order of atomic hyperfine structure[4] and in this case would be small compared to the Doppler broadening. However, $^2\Sigma$ - $^2\Sigma$ transitions belong to Hund's case b and may have hyperfine structure of the order of the fine structure.[6] In this nitrogen band the fine structure doublet splitting is of the order of 0.08 cm^{-1}, as shown in Fig. 2.

From consideration of the rules governing hyperfine structure[4], the strong levels (odd upper K') are expected to split into 13 components and the weak levels into 6 components. The calculation of the detailed hyperfine spectrum in diatomic molecules is difficult and is further complicated by the possibility that both magnetic and quadrupole effects may be important.[7] We might anticipate that some of these components may have equal energies and others negligible intensity, but apparently the hyperfine structure of a given rotational line will be composed of a number of components. Also we would expect the strong levels (odd upper K') to have more components (and therefore be broader) than the weak levels. This effect was observed and is shown in Fig. 5.

If the rotational lines are broadened by the presence of several hyperfine components, it might seem that the resulting line profile would deviate from a Gaussian distribution. Indeed, close examination of Figs. 2 and 3 indicates some asymmetry in the weaker of the two components which may be due to the presence of several hyperfine components. By modeling the hyperfine structure with several components (seven were used with various spacings and intensities) and convoluting these Doppler broadened lines with an instrument function, it was found that the resulting profile generally could be fit by a single Gaussian within the present experimental error.

Conclusions

Without a rigorous quantum mechanical calculation of the detailed hyperfine structure we cannot be certain that it is the source of the excessive broadening of the rotational lines; however, it appears to be the most likely, especially in the absence of other possibilities. Whatever the source, it results in a broadening of the rotational lines of the N_2^+ first negative band system equivalent to about a 75°K Doppler width. Thus, to determine the translational temperature from rotational line width this additional broadening must be convoluted with the instrument function and the Doppler broadening. Given the accuracy of the measurements, the use of the electron beam fluorescence technique for translational temperature measurements in nitrogen does not appear feasible for gas temperatures below about 100°K.

References

[1] Marrone, P. V., "Temperature and Density Measurements in Free Jets and Shock Waves," The Physics of Fluids, Vol. 10, March 1967, pp. 521-538.

[2] Cattolica, R., Robben, F., Talbot, L., and Willis, D. R., "Translational Nonequilibrium in Free Jet Expansions," The Physics of Fluids, Vol. 17, October 1974, pp. 1793-1807.

[3] Muntz, E. P., "Static Temperature Measurement in a Flowing Gas," The Physics of Fluids, Vol. 5, January 1962, pp. 80-90.

[4] Herzberg, G., Molecular Spectra and Molecular Structure I, D. Van Nostrand Co., Inc., Princeton, N.J., 1950.

[5] Rockett, P., and Brundin, C. L., "A Study of Some Thermal States of Nitrogen by the Method of Electron Beam Induced Fluorescence," in Rarefied Gas Dynamics, Vol. II, edited by M. Becker and M. Fiebig, DFVLR Press, Porz-Wahn, 1974, pp. C.2-1.

[6] Frosch, R. A., and Foley, H. M., "Magnetic Hyperfine Structure in Diatomic Molecules," The Physical Review, Vol. 88, December 1952, pp. 1337-1349.

[7] Townes, C. H. and Schawlow, A. L., Microwave Spectroscopy, McGraw-Hill, New York, N.Y., 1955.

DIRECT MEASUREMENT OF DENSITY AND ROTATIONAL TEMPERATURE IN A CO_2 JET BEAM BY RAMAN SCATTERING

Isaac F.Silvera,* F.Tommasini,+ and R.J.Wijngaarden≠

Natuurkundig Laboratorium der Universiteit van Amsterdam, Amsterdam, The Netherlands

Abstract

A sensitive intracavity crossed jet-beam laser-beam system for the study of expanding jets by means of light scattering has been built. Rotational and vibrational Raman scattering have been used to study the axial dependence of the rotational temperature and density of a CO_2 jet for values of x/D up to $\sim$ 10, where x is the axial distance from the nozzle of diameter D. Both temperature and density follow the theoretical curves for an isentropic expansion for small x/D; however, the temperature deviates as x/D increases, falling much more slowly than the isentrope predicts. Deviations are more pronounced for higher flow conditions and correlate with conditions favoring condensation in the beam. Results are presented for two sonic nozzles and a number of stagnation conditions to show the systematic behavior. This technique provides a powerful experimental method for analyzing jet beams without disturbing the distribution.

Introduction

We have developed a sensitive light scattering system to study the properties of expanding atomic and molecular jet beams.[1] In this paper we describe some of the systematic

---

Presented as paper 103 at the 10th International Symposium on Rarefield Gas Dynamics, July 19-23, 1976, Aspen, Colo. We thank Dr. A.J.Berlinsky for interesting discussions.

*Professor of Experimental Physics.

+Visiting faculty member, Permanent address: Instituto di Scienze Fisische, Universita di Genova, Italy.

≠Masters Candidate.

behavior that has been observed in an expanding CO_2 jet by measurement of the rotational and vibrational Raman spectra of isolated CO_2 molecules. In particular, the rotational temperature and the absolute density have been determined as a function of x/D, where x is the axial distance from the nozzle of diameter D. We present results of measurements for two sonic nozzles of D = 0.21 and 0.10 mm, for stagnation temperatures from 217° to 423° K, and stagnation pressures of 0.3 to 2.0 atm.

Experimental Apparatus

Previously, diagnostic beam work of this type has been carried out by means of electron beam fluorescence. More recently, Lewis et al.[2] have studied condensation and the rotational temperature in a CO_2 beam by means of Rayleigh and Raman scattering. We have built an intracavity system in which the molecular beam and laser beam cross at right angles within the laser cavity, as shown in Fig.1. This has the advantage of providing substantially higher useable laser power, and thus proportionally higher scattered light signals. A maximum laser cavity power of 25 W has been obtained, as compared to the nominal 1.5 W output of the Spectra Physics 165 Ar^+ laser at 5145 Å; higher cavity power is expected, but we are evidently limited at this time by an unidentified loss. Since the crossed beams interact very weakly, measured properties are characteristic of the undisturbed distribution of the jet beam. The scattered laser light is collected from a focal region of width $\sim$ 10 μ to provide local information. The collection optics, not shown in the figure, also consist of a mirror and lens pair, similar to the laser focussing optics, but at right angles, the mirror doubling the collected light signal. Lens L has a 15 mm focal length and the diameter of the optical tail of the stainless steel cryostat is 8 cm. Cryogenic pumping is used,resulting in a very compact beam enclosure, 62 cm in length; a small diffusion pump is used to preevacuate the system. The scattered light is analyzed with a Spex double-grating monochromator and detected with an RCA C31034 cooled photomultiplier tube.

In order to study axial variations of the jet beam properties, the nozzle, which has provision for centering within the focal region of the laser beam, could be translated vertically over a distance of 4 mm. In this way the optical system remains stable and allows for accurate absolute and relative measurements. The nozzle temperature is regulated by means of a resistance heater and a thermal link to the cryogenic pump; temperature is measured with a thermocouple, which also acts as the sensor for an electronic temperature controller.

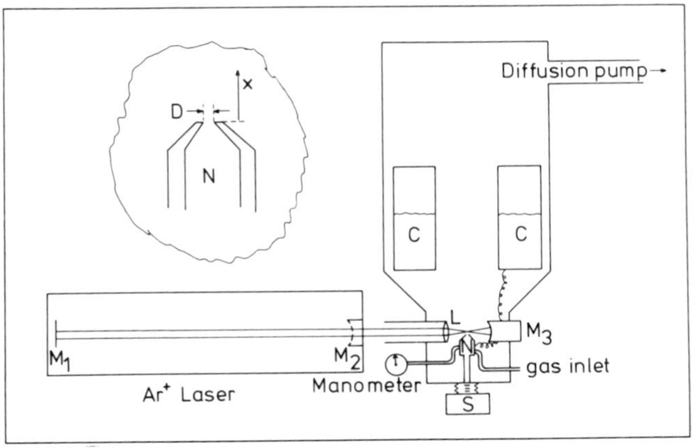

Fig.1 Schematic diagram of the experimental apparatus. The light collection optics, normal to the plane, are omitted in the drawing. The cryogenic pump actually consists of two reservoirs so that a liquid helium cooled pumping surface can be thermally shielded by a liquid nitrogen cooled surface. Provision has also been made for a mass analyzer to be mounted on top, in line with the jet axis.

The entire system has worked remarkably well and trouble free and no major problems were encountered. Care was taken to design a rugged, stable apparatus, including the adjustable lens and mirror mounts. Optical alignment was achieved by using the laser beam with mirror M2 installed. In fact, when aligned in this fashion, the system lases in a dual cavity mode with sufficiently high power in the scattering region and consequently it usually was not advantageous to remove mirror M_2. Once aligned the entire laser-nozzle beam system was stable for days.

Analyses

The Raman scattering efficiency is given in the polarizability approximation by [3]

$$I_{\rho\sigma}(i \rightarrow f) = r'(\nu\nu'^3/c^4)\rho_i|\langle f|\alpha_{\rho\sigma}|i\rangle|^2 \tag{1}$$

where ν is the frequency of the incident light, ν' the scattered light, i and f the initial and final molecular states, ρ_i the density of scatterers in state i, and $\alpha_{\rho\sigma}$ is the $\rho\sigma$ component of the polarizability tensor; r' is a numerical factor. For rotational Raman scattering from CO_2 gas in thermodynamic equilibrium in which the molecules are excited (Stokes transitions) from rotational quantum states J to J' = J+2

$$I(J\rightarrow J') = \frac{r\nu\nu'^3}{c^4}(\alpha_{\parallel}-\alpha_{\perp})^2 \frac{\rho_1 S_J e^{-E_J/kT_R}}{\sum_{\text{even } J}(2J+1)e^{-E_J/kT_R}} \quad (2)$$

with
$$S_J=[3(J+1)(J+2)]/[2(2J+3)] \quad (3)$$

and $E_J = B\,J(J+1)$. For CO_2, B = 0.39027 cm^{-1} (Ref.4), and the anisotropy in the polarizability $(\alpha_{\parallel} - \alpha_{\perp})=2.1 \times 10^{-24} cm^3$.[5] The constant r includes polarization factors that are not indexed in Eq.2. We have shown previously[1] that CO_2 gas in an expanding beam can be described by Eq.(2) locally, and thus characterized by an equilibrium rotational temperature T_R, and that information from anti-Stokes transitions is redundant to that from the Stokes transitions. Equation (2) can be linearized so that a plot of $\ln[I(J\rightarrow J')/S_J]$ vs $-BJ(J+1)$ provides a straight line for which the slope is proportional to T_R^{-1} and the intercept is proportional to ρ_1, the density of CO_2 monomers. Absolute density is obtained by measuring the rotational spectrum from a stagnant (nonflowing) gas of known pressure, to calibrate the system.

We compare our experimental results to theoretical curves for an isentropic expansion with stagnation temperature T_o and density ρ_o [6]

$$T/T_o = \{1+m^2[(\gamma-1)/2]\}^{-1} \quad (4)$$

$$\rho_1/\rho_o = \{1+m^2[(\gamma-1)/2]\}^{-1/(1-\gamma)} \quad (5)$$

Here m is the Mach number and γ is the specific heat ratio, C_p/C_v = 7/5, for CO_2. Equations (4) and (5) can be determined as a function of the reduced axial distance x/D by means of the graph of Owen and Thornhill[7] for m vs x/D. These theoretical curves are given by the solid lines in Figs. 2-4.

Experimental Results for Rotational Spectra

A few hundred runs were made over a period of several months for two nozzle diameters and various stagnation conditions of temperature between 205° and 423° K and pressures up to 2.5 atm. Each run consisted of measurement of up to 30 rotational lines, whose initial rotational state was identified by the frequency shift. The rotational quantum number and peak heights of each transition were the input data to a computer

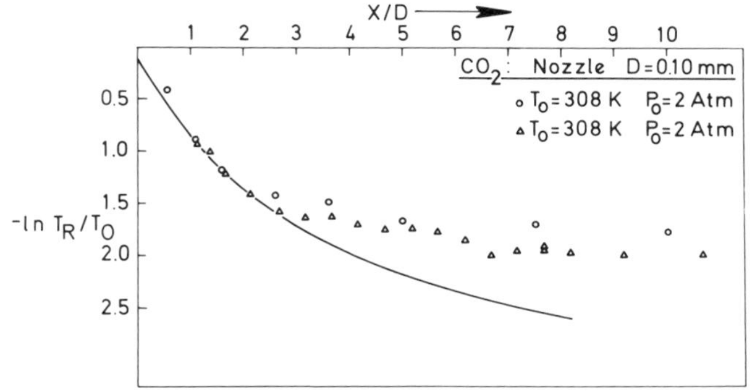

Fig.2 A plot of the natural logarithm of T_R/T_o vs x/D to demonstrate reproducibility of measurements.

program in which each line was given a statistical weight based on the Poisson statistics for Raman intensities; least-squares fits to the linearized version of Eq.(2), described in the previous section, were found, directly providing the temperature and relative density. A misidentification of rotational levels by $\Delta J=2$ was identified easily by the curvature in the locus of data points. Computer outputs included the straight line fits, and plots of rotational temperature and absolute density as a function of x/D. Computer simulations were performed to analyze situations in which the Raman signal could originate from both the jet beam and a penetrating background gas of CO_2 at a different temperature and density. Briefly, a significant influence on the determined temperature occurs when the penetrating gas has a density of roughly 10^{-2} that of the jet or greater. In this case there are very clear systematic deviations from the straight-line behavior for a single "component" gas, and so such effects could be detected if they had a significant influence.

Error bars shown in the graphs correspond to a standard deviation of the least-squares fits of the spectra, and are indicated only when they are larger than the symbol marking the data point. Possible fluctuations or errors in the stagnation conditions are not included in the error bars. Relative accuracy of the x measurement was $\sim 10\ \mu$, whereas absolute determination of the nozzle position, i.e., the position x=0, was about 25 μ.

In order to demonstrate the reproducibility of the system, we show, in Fig.2, the results of the determination of T_R for two runs, under the same stagnation conditions and measured several months apart. Between these runs the pressure measuring system had been changed, as well as the temperature regulation; these data represent one of the worst cases of reproducibility. The failure of the data from the two runs to re-

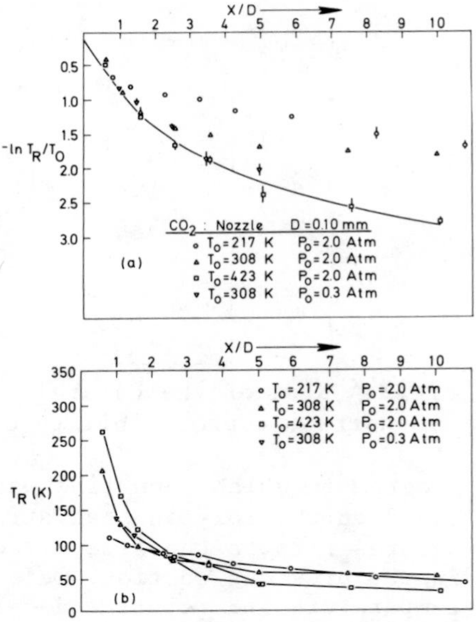

Fig.3 The axial dependence of the rotational temperature for the nozzle diameter given in the figure. a) The solid line is the theoretical isentrope. b) The lines here are a guide for the eye.

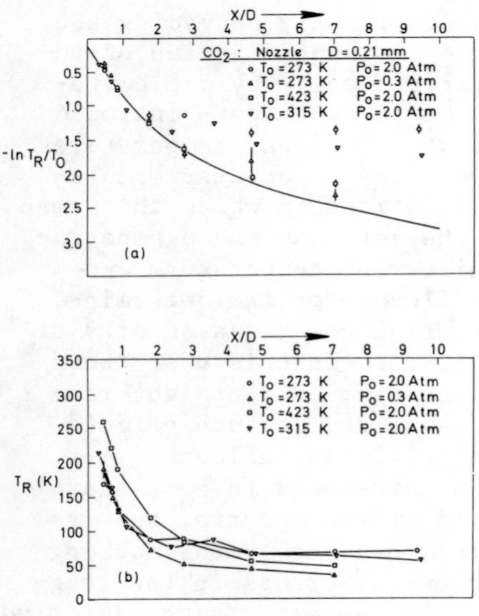

Fig.4 Axial dependences as described in Fig.3

produce still was puzzling. The only important difference that we noted in the conditions was that the background pressure for the data marked by circular symbols was 80 times lower than that for the triangles (1×10^{-5} Torr, compared to 8×10^{-4} Torr).

In Figs. 3a and 4a we show the logarithm of T_R/T_O as a function of x/D for both nozzles. We see that T_R follows the theoretical isentrope for small x/D and then deviates, decreasing much more slowly than the isentrope. Contrary to expectations, the deviations begin for smaller x/D the higher the flow or the lower the stagnation temperature: for (T_o, P_o) = (217 K, 2 atm) deviations occur for $x/D < 1$, whereas for (423° K, 2 atm) the isentrope was followed to at least $x/D = 10$ for the 0.1 mm nozzle. If the deviations from isentropic behavior arose from freezing of the distribution due to a decreasing number of collision as the beam expands, then one would expect to find just the opposite behavior. In Figs. 3b and 4b we plot the same data, but here T_R vs x/D. This is to demonstrate that for each expansion T_R approaches some low value in the region of ∿ 25 to 60° K, but there appears to be no characteristic temperature. For most expansions, the temperature was still decreasing slowly at $x/D=10$.

The possibility that the deviations arose from interaction with the (CO_2) background gas was checked by allowing the background pressure P_B to rise from 10^{-5} to ∿ 20 Torr by pumping with a throttled diffusion pump instead of with the cryogenic surfaces shown in Fig.1. The result shown in Fig.5 was, that, at a given x/D, the temperature and density of the CO_2 in the beam remained constant as the pressure rose until the mach disc was pushed through the measuring region, whereupon T_R and ρ_1 rose sharply. The pressure for which this occurs is in excellent agreement with the empirical relation $x/D=.67\ (P_o/P_B)^{1/2}$ for the position of the Mach disc.[6] These results exclude interference by the background gas as the source of the deviations of T_R from isentropic behavior as the calculated position of the Mach disc was $x/D>1000$ in all cases.

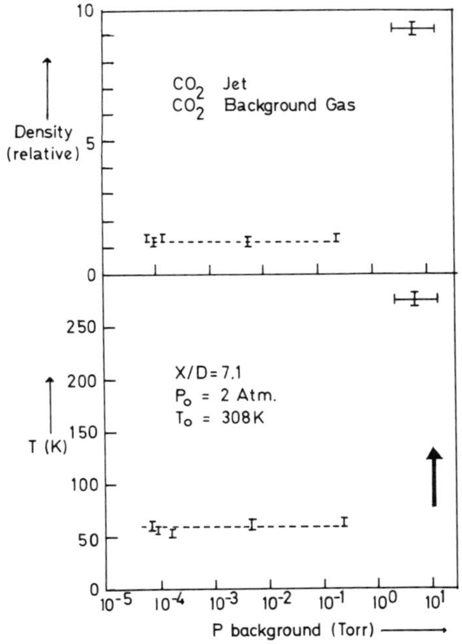

Fig.5 Plot of the measured rotational temperature and density of CO_2 in the expanding beam at a fixed distance x as a function of the background gas pressure. The arrow indicates the calculated pressure at which the Mach Disk comes into the measuring region.

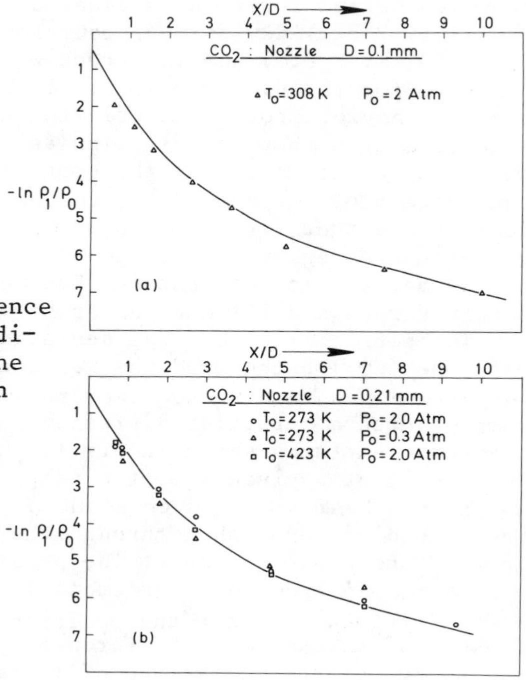

Fig.6 Axial density dependence for several stagnation conditions. The solid line is the theoretical behavior for an isentropic expansion.

The most likely reason for the deviations of T_R is condensation or clustering in the beam[8], a phenomenom which has been detected at small values of x/D by Lewis, et al.[2] by means of Rayleigh scattering. The deviations would then be attributed to the large heat of condensation of CO_2 which becomes available to the beam particles as clustering commences and must be accounted for in the energy balance for an adiabatic expansion.[9] The formation of clusters is also enhanced by higher flow (lower stagnation temperature) conditions in agreement with the experimental observations.

In Fig.6 we show plots of the absolute density of CO_2 monomers as a function of x/D; these data correspond to the runs of Figs.3 and 4 and appear to follow the isentrope quite well. Raman scattering measurements should measure only the monomer density[1], and thus, if condensation takes place at the expense of monomers as indicated by the data of Van Deursen, et.al[10,11], one would expect to observe a deviation of ρ_1 below the isentrope. However only a small percentage of the beam need condense in clusters to give a large deviation of the temperature from the isentrope[12], whereas the detection of small deviation in the density would require more accurate measurements.

Vibrational Transitions

In addition to rotational transitions, weaker Raman active molecular vibrational transitions also can be observed.[13] In Fig.7 we show the more important vibrational transitions. The notation here is (n_1, n_2, n_3), where n_i's are the vibrational occupation quantum numbers; n_2 corresponds to infrared active bending mode. The lowest excited vibrational state is the (01^00) state, which lies at $h\nu/k \simeq 960^o$ K above the ground state. As a consequence, at room temperature, almost all molecules are in the ground vibrational state. The two Raman active Fermi resonance lines have been observed as a function of x/D, their intensity being directly proportional to the beam density (in the approximation that there is no thermal excitation). These results are in good agreement with density determined from the rotational transitions, and are not shown. It also is possible to determine vibrational temperatures by observing Raman transitions from the thermally populated (01^00) state to the (03^10) and (11^10) states. In order to have sufficient intensity, one must expand from a higher stagnation temperature, so that the (01^00) state is sufficiently populated. A recent attempt to measure the vibrational temperature in this way was frustrated by an outgassing from the surface of the heated nozzle, which contaminated the coated optics of the intracavity system, reducing the signal below measurable levels.

Future work is planned to measure the vibrational and translational temperatures of such an expanding beam directly, as well as spectra from clusters, which is one of the principle objectives of this research program.

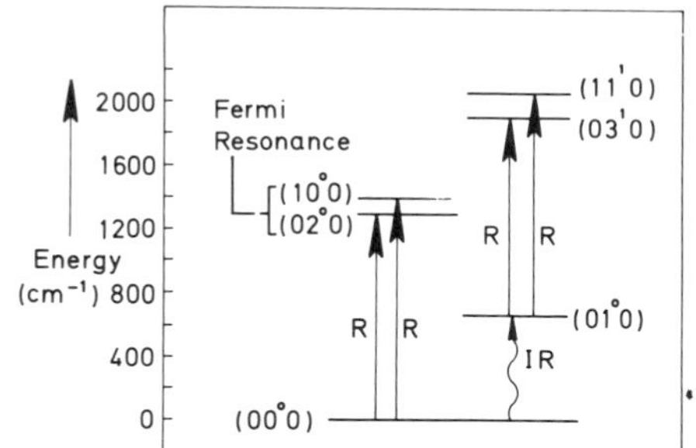

Fig.7 Some important vibrational energy levels for CO_2. R refers to Raman active transitions and IR to infrared active transitions. The notation (n_1, n_2, n_3) refers to the occupation numbers of the three vibrational modes.

References

[1]Silvera, I.F. and Tommasini, F., "Intracavity Raman Scattering from Molecular Beams: Direct Determination of Local Properties in an Expanding Jet Beam", Physical Review Letters, Vol.37, July 1976, p.136.

[2]Lewis, J.W.L., Williams, W.D., and Powell, H.M., "Laser Diagnostics of a Condensing Binary Mixture Expansion Flow Field", 9th Rarefied Gas Dynamics, edited by M.Becker and M.Fiebig, DFVLR - Press, Porz. Wahn, 1974, Vol.2, p.F7-1.

[3]Herzberg, G., Spectra of Diatomic Molecules, 2nd ed., Van Nostrand Reinhold, New York, 1950, p.127.

[4]Barrett, J.J. and Weber, A., "Pure Rotational Raman Scattering in a CO_2 Electric Discharge", Journal of the Optical Society of America, Vol.60, January 1970, p.70.

[5]Hirschfelder, J.O., Curtis, C.F., and Bird, R.B., Molecular Theory of Gases and Liquids, Wiley, New York, 1964, p.950.

[6]Anderson, J.B., "Molecular Beams from Nozzle Sources", Molecular Beams and Low Density Gas Dynamics, edited by P.P. Wegener, Dekker, New York, 1974, Chap. I, and references therein.

[7]Owen, P.L. and Thornhill, C.K., "Report and Memorandum No.2616", Aeronautical Research Council (U.K.) 1948.

[8]Hagena, O.F., "Cluster Beams from Nozzle Sources", Molecular Beams and Low Density Gas Dynamics, edited by P.P. Wegener, Dekker, New York, 1974, Chap.2; also Hagena, O.F. and Obert, W., "Cluster Formation in Expanding Supersonic Jets: Effect of Pressure, Temperature, Nozzle Size, and Test Gas", Journal of Chemical Physics, Vol.56, March 1972, p.1793; also Yealland, R.M., Deckers, J.M., Scott, I.D., and Tuori, C.T., "Dimer Concentrations in Supersonic Free Jets", Canadian Journal of Physics, Vol.50, October 1972, p.2464.

[9]Gallagher, R.J. and Fenn, J.B., "Relaxation Rates from Time of Flight Analysis of Molecular Beams", Journal of Chemical Physics, Vol.60, April 1974, p.3487.

[10]Van Deursen, A., Van Lumig, A., and Reuss, J., "Intensities and Cross-sections of Ar Clusters in a Molecular Beam", International Journal of Mass Spectrometry and Ion Physics, Vol.18, October 1975, p.129.

[11]Van Deursen, A., private communication, 1975.

[12]Fenn, J.B., private communication, 1976.

[13]Herzberg, G., Infrared and Raman Spectra of Polyatomic Molecules, Van Nostrand, New York, 1960, p.272.

INFLUENCE OF ELECTRON BEAM-BODY INTERACTIONS ON VIBRATIONAL TEMPERATURE MEASUREMENTS

Knut Erhardt* and Gustav Schweiger*

D F V L R, Institute fuer Angewandte Gasdynamik, Koeln, Federal Republic of Germany

Abstract

The effect of an electron beam (acceleration voltage 20 kV) on the population distribution function of the vibrational levels of N_2 is investigated under static conditions. A metallic plate was used as slow electron source by directing the electrons from the gun against this plate. The concentration of slow secondary electrons was altered by changing plate material, by changing the angle between electron beam and plate, and by analyzing the fluorescence light at different distances from the plate. It was found that the population number is relatively strongly affected by the slow electrons. Neglecting this effect and determining the vibrational temperature T_v from the measured intensity ratios would result in an incorrect measurement of T_v by as much as 300^0 K or more. Attempts have been made to correlate the observed departure from a thermal distribution in the vibrational levels with secondary electron concentration.

Presented as Paper 82 at the 10th International Symposium on Rarefied Gas Dynamics, Aspen,Colo., July 19-23, 1976. The authors gratefully are indebted to W. Hoyer for his accurate and responsible work during construction and test of the experimental setup. The project was sponsored by the DFG under Schw.184/1.

* Scientist

Introduction

In the past 15 years the electron beam technique has been used extensively for local gas temperature and density measurements in low density test facilities. In spite of numerous results concerning rotational temperatures very few measurements of vibrational temperature have been made. Recently the measurement of population numbers in the vibrational levels of molecular nitrogen for example seems to earn increasing interest for the investigation of mixing processes in nonequilibrium flow for example. In most of the published results the vibrational temperature measured by the electron beam technique deviates from aerodynamic calculations.[1-3]

At present it is not clear whether or not this effect is due to the electron/gas-particle interaction. Electron-beam-induced effects on the distribution function in the rotational structure have been observed in the past. Whereas the effect of the electron beam on the rotational distribution function was investigated by several authors, no similar work seems to have been done for the vibrational distribution.

It is well known that slow secondary electrons influence the population distribution in the rotational levels. Therefore, it is reasonable to expect similar effects in the vibrational levels, especially with respect to the resonance-like excitation cross sections for the excitation of vibrational levels by slow electrons[4]. In this investigation an attempt was made to study the effect of slow electrons on the measurements of N_2 vibrational population numbers by simulating different secondary electron concentrations; particular attention was given to the neighborhood of aerodynamic models. The interest was focused not only on the examination of the effect of secondary electrons but also on finding correlations that make possible a prediction or correction of the secondary electron influence.

Experimental Setup

The experimental device is shown in Fig. 1. An electron beam of 20-keV energy is directed onto the surface of a flat plate mounted in a vacuum chamber.

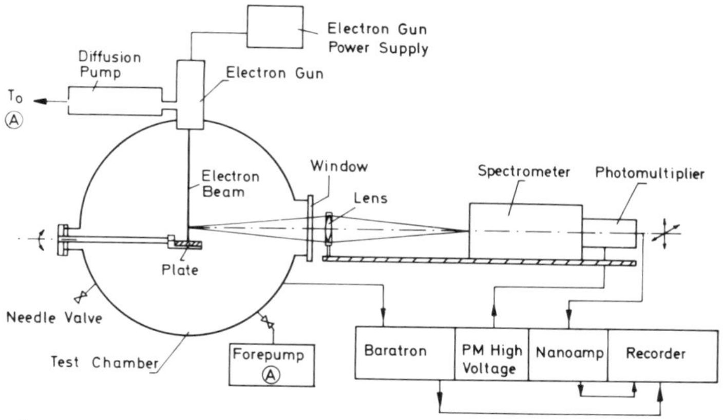

Fig. 1 Test arrangement for static investigation of electron beam-model interaction.

Plate materials were brass and aluminium. The model could be tilted to change the angle between electron beam and model. The electron-beam-induced fluorescence near the model surface is focused 1:1 to the entrance slit of the monochromator. Entrance and exit slits were set to 0.25 mm, resulting in a bandwith of about 10 Å; the slit height was 18 mm for intensity reasons. The fluorescence spectrum then is measured by photomultiplier, nanoamperemeter, and line recorder. The optical system (lens, monochromator, multiplier) can be moved parallel and perpendicular to the electron beam axis.

By this system, a spectrally resolved investigation of the electron-beam-induced fluorescence was possible as a function of plate material, beam incidence angle, gas density, and position of the test volume relative to beam and model. Air was used as gas medium in the chamber.

Results

The electron-beam-excited nitrogen bands of the 1N and 2P systems have been investigated. The excitation cross section of the 1N system is about 10^2 times as large as that of the 2P system for excitation by fast electrons[5,6]. It can be assumed, therefore, that the 2P band intensities are proportional

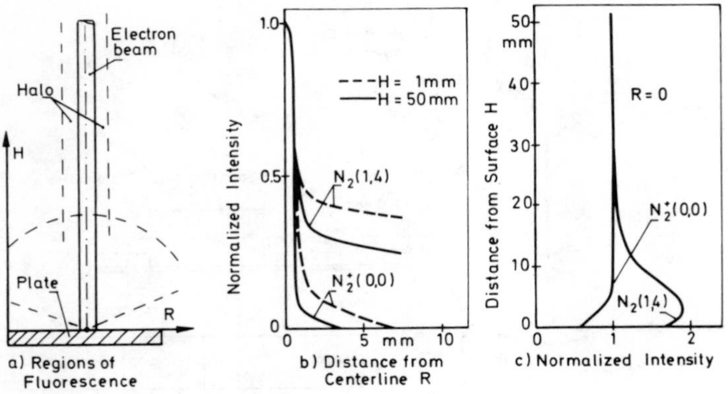

Fig. 2 Schematic diagramm of secondary electron effects: a) geometric positions for electron beam; b) representative intensity profiles in radial direction showing the difference between the 1N and 2P systems; and c) intensity profiles on the same bands in centerline direction.

to the secondary electron concentration. Fig. 2a shows schematically the fluorescence intensity distribution when the electron beam impinges perpendicularly on a flat plate. Fig. 2b shows profiles of the intensity of the (0,0) band in the 1N system and (1,4) band of the 2P system across the electron beam. The (0,0) band intensity can be interpreted as a first approximation to be proportional to the concentration of primary electrons and, correspondingly, the (1,4) band to be proportional to the concentration of secondary electrons. Fig. 2c gives results of fluorescence intensity measurements on the centerline of the beam as a function of distance from the plate. No corrections for slit height and optical shadowing by the plate have been applied to the data shown in Fig. 2c.

Because of the strong increase in intensity of the (1,4) band in the 2P system, it is a reasonable assumption to use it in a first approximation as a measure for secondary electron concentration. The N_2 (1,4) band intensity related to the corresponding N_2^+ (0,0) band intensity is plotted as a function of distance from surface in Fig. 3 for various test conditions. For the determination of vibrational temperatures, the intensity ratio (0,1)/(1,2) in

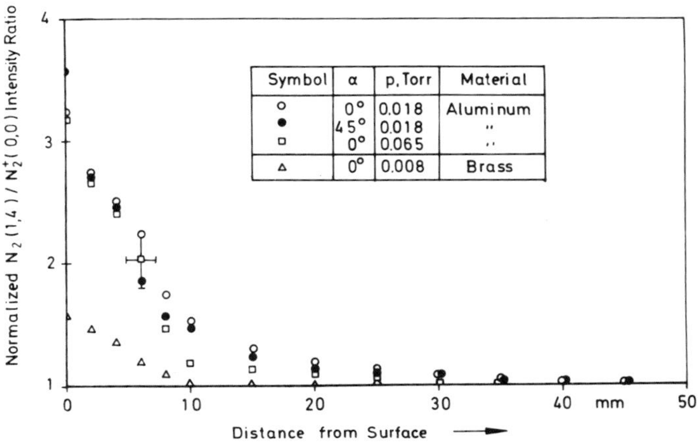

Fig. 3 Variation of N_2 (1,4)/N_2^+ (0,0) ratio, used as a measure for relative secondary electron concentration with distance from the model surface.

the 1N system usually is measured. Applying the excitation-emission model of Lewis[7] to the evaluation of the vibrational temperature, the measurements would yield a vibrational temperature that is dependent on the distance from the surface of the model, as shown in Fig. 4. The actual gas temperature was in all cases about 300° K.

Two conclusions can be drawn immediately from these results:
first, the population numbers in the vibrational levels are affected by the secondary electrons at

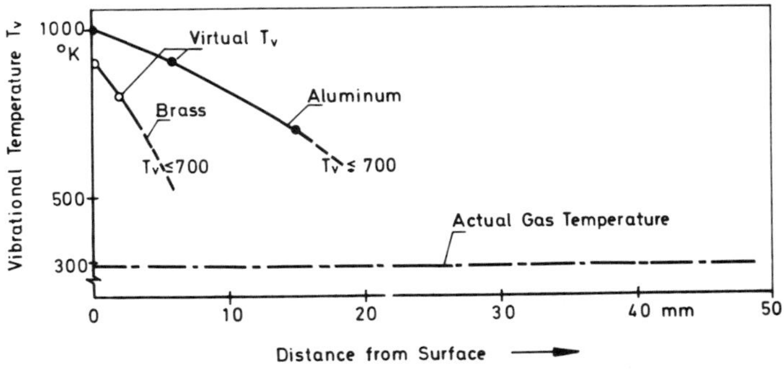

Fig. 4 Deviation of virtual vibrational temperature, as derived from measured intensity ratios, from actual gas temperature.

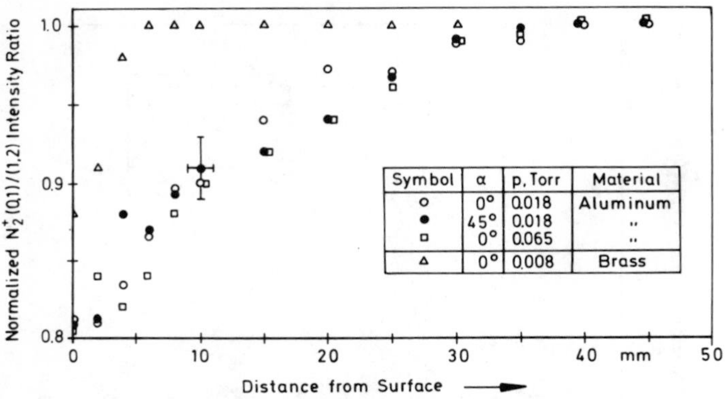

Fig. 5 Variation of (0,1)/(1,2) intensity ratio as used for vibrational temperature determination with distance of test volume from body surface.

least in the ionized molecule N_2^+, because the intensity ratio I (0,1)/I (1,2) of the 1N system is directly proportional to the population numbers of the ground and first vibrational levels in the upper electronic state of 1N system;
second, the departure from thermal distribution (indicated by the variation of the (0,1)/(1,2) intensity ratio in spite of the constant gas temperature) is dependent on the number of slow electrons. Measurements of the 1N (0,1)/(1,2) ratio have been made for different plate materials and different angles of incidence. Fig. 5 shows some of these results.

By analyzing these and similar profiles, it turned out that the observed deviation from Boltzmann distribution in the vibrational levels does not seem to depend on the absolute concentration of neither slow nor fast electrons but on the ratio of primary to secondary electrons, as shown by Fig.6. This figure shows the relative intensity of the (0,1) to the (1,2) band, which is related to the population numbers in the ground and first vibrational level as a function of the ratio of the (1,4) band of the 2P system to the (0,0) band of the 1N system. It is an experimental fact, that all data, which have been taken for widely differing absolute concentrations of fast primary and slow secondary electrons, could be correlated within the experimental accuracy using the N_2 (1,4)/N_2^+ (0,0) ratio.

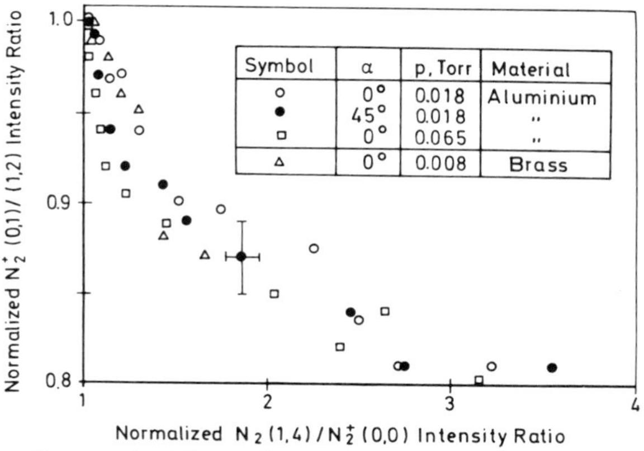

Fig. 6 Correlation between the intensity ratio representing a measure for vibrational temperature (ordinate) and the intensity ratio assumed to be proportional to the relative secondary electron concentration (abscissa). Diagramm obtained by combination of Figs. 3 and 5.

Neclecting double excitation processes, the observed 1N intensity I (1N) is assumed to be composed of two parts excited by primary respectively secondary electrons

$$I\ (1N) = I\ (1N)_p + I\ (1N)_s$$

Accepting the assumption that under the present experimental conditions the 2P system is proportional to the secondary electron density n_s it follows

$$I\ (1N)_s \quad n_s \quad \text{and} \quad I\ (2P) \quad n_s$$

Therefore

$$I\ (2P) = C\ I\ (1N)_s$$

and $$I\ (1N)/I(2P) = I\ (1N)_p/I(1N)_s + 1/C$$

The ratio I (1N)/I(2P) is therefore a measure of 1N excitation ratio by high energetic primary to slow secondary electrons. It is known[8], that electrons with energies below 300 eV will preferably populate the first vibrational level. Excitation of 1N system by slow electrons results in a changed (0,1)/(1,2) intensity ratio; but since only the sum of excitation processes by fast and slow electrons can be measured, the resulting variation of the intensity ratio will be a function of relative secondary electron concentration.

The basic requirement for validity of this consideration is that no double excitation takes place. In principle it is possible that a molecule is first vibrationally excited in its ground state by secondary electrons and then ionized and excited by a fast primary electron. However, a rough estimation shows only a very small probability for the descript double excitation process to occure in the test volume. Even if the vibrational levels in the N_2 electronic ground state are influenced by slow electrons, this would be therefore without effect on the fluorescence intensity distribution. At present, attempts are made to formulate an excitation-emission model to explain the experimental findings more exactly. It is hoped that this results in a quantitative description of the disturbance of vibrational distribution function by the interaction of the electron beam and the gas molecules.

References

[1]Dankert, C. and Bütefisch, K.A., "Influence of Nozzle Geometry on Vibrational Relaxation in an Expanding Flow," Proceedings of the 9th International Symposium on Rarefied Gasdynamics, edited by Becker, M. and Fiebig, M., Vol. I, DFVLR-Press, Porz-Wahn, 1974, pp. B 20-1 - B 20-10.

[2]Lewis, J.H., "An Experimentally Determined Model for the Noncontinuum Flow at the Leading Edge in a High Speed Ratio Stream," Ph. D. Thesis, Feb. 1971, Pricetown University

[3]Harbour, P.J., "Absolute Determination of Flow Parameters in a Low Density Hypersonic-Tunnel," Proceedings of the 6th International Symposium on Rarefied Gasdynamics, edited by Trilling and H.Y. Wachmann, Academic Press, New York, 1969, pp. 1713-1722.

[4]Schulz, G.J., "Vibrational Excitation of Nitrogen by Electron Impact," The Physical Review, Vol. 125, Jan. 1962, pp.229-232.

[5]Borst, W.L. and Zipf, E.C., "Cross Section for Electron-Impact Excitation of the (0,0) First Negative Band of N_2^+ from Threshold to 3 keV," Physical Review A, Vol. 1, No. 3, March 1970, pp. 834-840.

[6]Burns, D.J., Simpson, F.R., and McConkey, J.W., "Absolute cross sections for electron excitation of the second positive bands of nitrogen," Journal of Physics B, ser. 2, vol. 2, 1969, pp. 52-64.

[7]Lewis, J.W.L. and Williams, W.D., "Vibrational Temperature Measurements Using the Electron Beam," AIAA-Journal, Vol. 7, June 1964, pp. 1202-1203.

[8]Polyakova, G.N., Fogel, Ya. M., and Zats, A.V., "Rotational and Vibrational Energy Level Distributions of N_2^+ Ions produced in collisions between Electrons and Nitrogen Molecules," Soviet Physics JETP, Vol. 25, No. 6, December 1967, pp. 993-997.

[illegible], N. [illegible] and [illegible], "[illegible] Cross Section for [illegible] by Electron Impact [illegible] First [illegible]," Physical Review, [illegible], pp. 87-98.

[illegible], and McConkey, J. W., "Absolute cross sections for electron excitation of the second positive bands of nitrogen," Journal of Chemical Physics, [illegible], pp. [illegible].

Lewis, J. [illegible] and Williams, W. D., "Vibrational Temperature Measurements Using the Electron Beam," AIAA Journal, Vol. [illegible], June 1964, pp. 1202-[illegible].

[illegible], "[illegible] Vibrational Energy Distribution [illegible] produced in collisions between electrons [illegible] Molecules," Soviet Physics Technical Physics, Vol. [illegible], December 1967, pp. 973-[illegible].

THE EFFECT OF ROTOR CLEARANCE ON PERFORMANCE OF AXIAL FLOW MOLECULAR PUMP

A.E. Dabiri*

Arya-Mehr University of Technology,
Teheran, Iran

Summary

The axial or radial flow molecular pump has been developed as a new vacuum pump which provides high and clean vacuum. The problem of backstreaming, which occurs in diffusion pumps, is eliminated and, therefore, the ultimate vacuum is limited only by leakage, outgassing and diffusion through the walls.

The problem of free molecular flow in axial flow molecular pumps has been investigated experimentally and theoretically by Kruger and Shapiro.[1,2] The study was mostly concerned with parallel flat blades, assuming infinite blade height, and calculations were made upon single-row and multi-row blades using analytical and Monte-Carlo methods.

The purpose of the present investigation is to study the performance of a single-row and multi-row flat blades with finite height for different parameters and blade speeds. The study also includes the effect of rotor clearance, which exists between a rotor tip and rotor housing on the pressure ratio and the pumping speed of a single blade, and that of a multistage pump.

To find the performance of a single-row blade with finite height, a three-dimensional analysis, considering the collisions of upstream and downstream molecules with a rotor housing wall was performed. The clearance between the rotor tip and the rotor housing was considered to be like a blade with a blade angle of 90 degrees and zero blade speed. To

Summary of Paper 142 presented at 10th International Symposium on Rarefied Gas Dynamics, Aspen, Colo., July 19-23, 1976.

*Associate Professor.

calculate the effect of clearance, only molecules which are transmitted directly through the clearance from one region to another region and molecules which are transmitted by a single collision with rotor housing were considered. The transmission of molecules by collision with the rotor was neglected.

A multistage pump, having alternate moving rotor rows and stationary stator rows, was considered. Kruger and Shapiro[1,2] used an approximate method to find the transmission coefficients based upon the assumption of infinite blade height. They assumed that the velocity distribution of molecules between blade rows is Maxwellian. Here we use their method to find both the performance of a multistage pump with finite height blades and the effect of clearance in a multistage pump.

The results show that maximum pumping speed and maximum pressure ratio increase when the heights of the rotor blades are considered to be finite. On the other hand, maximum pressure ratio decreases with clearance and this drop increases as the blade angle or spacing chord ratio decreases.

In conclusion, the results indicate that the values of the blade angle and spacing chord ratio are enough for the prediction of the performance of the pump by assuming infinite blade height for large blade angles, large spacing chord ratios and small clearances. But the exact prediction of the performance for the small values of blade angles, small chord ratios and large clearances must be made by considering the effect of rotor clearance.

References

[1] Kruger, C.H. and Shapiro, A.H., "The Axial-Flow Compressor in the Free Molecular Range", Rarefied Gas Dynamics, edited by L. Talbot, Academic Press, New York, 1961.

[2] Kruger, C.H., "The Axial-Flow Compressor in the Free-Molecular Range", Ph.D. Thesis, 1960, Massachusettes Institute of Technology, Cambridge, Mass.

Index to Contributors to Volume 51: Parts I and II